# Guide to Wireless Ad Hoc Networks

The **Computer Communications and Networks** series is a range of textbooks, monographs and handbooks. It sets out to provide students, researchers and non-specialists alike with a sure grounding in current knowledge, together with comprehensible access to the latest developments in computer communications and networking.

Emphasis is placed on clear and explanatory styles that support a tutorial approach, so that even the most complex of topics is presented in a lucid and intelligible manner.

For other titles published in this series, go to
http://www.springer.com/series/4198

Sudip Misra · Isaac Woungang ·
Subhas Chandra Misra
Editors

# Guide to Wireless
# Ad Hoc Networks

 Springer

*Editors*

Sudip Misra, PhD
School of Information Technology
Indian Institute of Technology
Kharagpur
India

Isaac Woungang, PhD
Department of Computer Science
Ryerson University
350 Victoria St.
Toronto ON
Canada

Subhas Chandra Misra, PhD
Department of Industrial
  & Management Engineering
Indian Institute of Technology
Kanpur
India

*Series Editor:*
Professor A.J. Sammes, BSc, MPhil, PhD, FBCS, CEng
Centre for Forensic Computing
Cranfield University
DCMT, Shrivenham
Swindon SN6 8LA
UK

CCN Series ISSN 1617-7975
ISBN 978-1-84996-785-3        e-ISBN 978-1-84800-328-6
DOI 10.10007/978-1-84800-328-6

British Library Cataloguing in Publication Data
A catalogue record for this book is available from the British Library

Printed on acid-free paper

Springer Science+Business Media
springer.com

*Dedicated to*
*the newborns: Tultuli (Subhas's daughter)*
*and Babai (Sudip's son)*

*and*

*Isaac's grand ma: Maria Happi*

# Preface

## Overview and Goals

Wireless communication technologies are undergoing rapid advancements. The past few years have experienced a steep growth in research in the area of wireless ad hoc networks. The attractiveness of ad hoc networks, in general, is attributed to their characteristics/features such as ability for infrastructure-less setup, minimal or no reliance on network planning and the ability of the nodes to self-organize and self-configure without the involvement of a centralized network manager, router, access point or a switch. These features help to set up a network fast in situations where there is no existing network setup or in times when setting up a fixed infrastructure network is considered infeasible, for example, in times of emergency or during relief operations.

Even though ad hoc networks have emerged to be attractive and they hold great promises for our future, there are several challenges that need to be addressed. Some of the well-known challenges are attributed to issues relating to scalability, quality-of-service, energy efficiency and security.

This handbook attempts to provide a comprehensive guide on fundamental concepts, new ideas and results in the areas of mobile ad hoc and other ad hoc networking systems. This book has been prepared keeping in mind that it needs to prove itself to be a valuable resource dealing with both the important core and the specialized issues in this area. We have attempted to offer a wide coverage of topics. We hope that it will be a valuable reference for students, instructors, researchers and industry practitioners. We believe this is particularly an attractive feature of this book, as the very limited selection of books available on ad hoc networks we are aware of are written primarily for academicians/researchers. We have attempted to make this book useful for both the academicians and the practitioners alike.

## Organization and Features

Altogether there are 21 chapters in the book. Chapter 1 discusses some of the properties of general multihop networks in theory and in practice. The authors discuss some of the important issues relating to simulation, testbed and theoretical studies in these types of networks. Chapters 2 and 3 discuss some of the characteristic features of ad hoc networks that make them distinct from other types of networks and why/how the issue of cooperation plays an important role in their successful functioning. Chapters 4–8 relate to the important issue of routing (including multicasting and broadcasting) in ad hoc networks. A chapter on formal verification of routing protocols, Chapter 8, written by Câmara et al., will definitely be a chapter of particular attraction to many readers. Chapters 9–11 deal with mobility issues in these networks. Chapters 12 and 13 discuss the quality of service issues, which have been long considered to be challenging in this type of network. In particular, Chapter 13, discussing about delay management is a topic of great interest currently amongst researchers and practitioners alike. Chapter 14 discusses the address allocation mechanisms in ad hoc networks and Chapter 15 discuss congestion control. Chapters 16–19 are dedicated to security and trust issues in ad hoc networks, which again, have drawn considerable concern amongst researchers. Chapters 20 and 21 discuss some trendier issues – vehicular ad hoc networks and the integration of mobile ad hoc networks into IP-based access networks.

We list below some of the important features of this book, which, we believe, would make this book a valuable resource for our readers:

- Most of the chapters of the book are authored by prominent academicians/researchers/practitioners in wireless ad hoc networks, who have been working with these topics for several years and are supposed to have thorough understanding of the concepts.
- The authors of this book are distributed in a large number of countries and most of them are affiliated with institutions of worldwide repute. This gives this book an international flavor. The readers of this book can get absorbed by perspectives, suggestions, experiences and issues projected forward by authors from different countries.
- Almost all the chapters in this book have a distinct section providing *directions for future research*, which particularly targets researchers working in these areas. We believe, this section in each chapter should provide insight to the researchers about some of the current research issues.
- The authors of each chapter have also attempted, to the extent possible, to provide a comprehensive bibliography, which should greatly help the researchers and readers interested further to dig into the topic.
- Almost all chapters of this book have a separate section outlining *thoughts for practitioners*. We believe, this section in every chapter will be particularly useful for industry practitioners working directly with the practical aspects behind, enabling these technologies in the field.
- Most of the chapters provide a list of important terminologies and their brief definitions.

- Most of the chapters also provide a set of questions at the end, which can help in assessing the understanding of the readers.
- In order to make the book useful for pedagogical purposes, almost all chapters of the book also have a corresponding set of presentation slides. The slides can be obtained as a supplementary resource by contacting the publisher, Springer.

We have made attempts, in all possible way we could, to make the different chapters of the book look as much coherent and synchronized as possible. However, it cannot be denied that, as the chapters were written by different authors, it was not possible to fully achieve this goal. We believe that this is a limitation of most edited books of this sort.

## Target Audience

The book is written by primarily targeting the student community. This includes the students of all levels – those getting introduced to these areas, those having an intermediate level of knowledge of the topics and those who are already knowledgeable about many of the topics. In order to keep up with this goal, we have attempted to design the overall structure and content of the book in such a manner that makes it useful at all learning levels. To aid in the learning process, almost all chapters have a set of questions at the end of the chapter. Also, in order that teachers can use this book for classroom teaching, the book also comes with presentation slides and sample solutions to exercise questions, which are available as supplementary resources.

The secondary audience for this book is the research community, whether they are working in the academia or in the industry. To meet the specific needs to this audience group, most chapters of the book also have a section in which attempts have been made to provide directions for future research.

Finally, we have also taken into consideration the needs to those readers, typically from the industries, who have the quest for getting insight into the practical significance of the topics, i.e., how the spectrum of knowledge and the ideas are relevant for real-life working of wireless ad hoc networks.

## Supplementary Resources

As mentioned earlier, the book comes with the following supplementary resources:

- Solution manual, having sample solutions to most questions provided at the end of the chapters.
- Presentation slides, which can be used for classroom instruction by teachers.

Teachers can contact the publisher, Springer, in order to get access to these resources.

## Acknowledgments

We are extremely thankful to the 49 authors of the 21 chapters of this book, who have worked very hard to bring this unique resource forward for help of the student, researcher and practitioner communities. The authors were very much interactive at all stages of preparation of the book from initial development of concept to finalization. We feel it is contextual to mention that, as the individual chapters of this book are written by different authors, the responsibility of the contents of each of the chapters lies with the concerned authors.

We are also very much thankful to our colleagues in the Springer publishing and marketing teams, in particular, Mr. Wayne Wheeler and Ms. Catherine Brett, who tirelessly worked with us and guided us in the publication process. Special thanks also go to them for taking special interest in publishing this book, considering the current worldwide market needs for such a book.

Finally, we would like to thank our parents, Prof. J. C. Misra, Mrs. Shorasi Misra, Mr. John Sime, Mrs. Christine Seupa, our wives Satamita, Sulagna and Clarisse, and our children, Babai, Tultuli, Clyde, Lenny, and Kylian, for the continuous support and encouragement they offered during this project.

June 10, 2008

Dr. Sudip Misra
Dr. Isaac Woungang
Dr. Subhas C. Misra

# Contents

# Contributors

**Nael Abu-Ghazaleh** Department of Computer Science, Thomas J. Watson School of Engineering and Applied Science, State University of New York at Binghamton, Binghamton, NY, USA, nael@cs.binghamton.edu

**Venkat Balakrishnan** Information Networked System Security (INSS) Research Group, Department of Computing, Division of ICS, Macquarie University, North Ryde, Sydney, NSW, Australia, venkat@ics.mq.edu.au

**Doina Bein** Department of Computer Science, University of Texas at Dallas, Richardson, TX, USA, siona@utdallas.edu

**Mike Burmester** Computer Science Department, Florida State University, Tallahassee, FL, USA, burmester@cs.fsu.edu

**Daniel Câmara** Computer Science Department, Federal University of Minas Gerais, Belo Horizonte, Brazil, danielc@dcc.ufmg.br

**Dazhi Chen** Department of EECS, Syracuse University, Syracuse, NY, USA, dchen02@syr.edu

**Xiaowen Chu** Department of Computer Science, Hong Kong Baptist University, Kowloon Tong, Kowloon, Hong Kong, chxw@comp.hkbu.edu.hk

**John A. Clark** Department of Computer Science, University of York, Heslington, York, UK, jac@cs.york.ac.uk

**Amitabha Das** Division of Computer Communications, School of Computer Science and Engineering, Nanyang Technological University, Singapore, asadas@ntu.edu.sg

**Fethi Filali** Institut Eurécom, Sophia-Antipolis France, fethi.filali@eurecom.fr

**Mesut Güneş** Institute of Computer Science, Computer Systems and Telematics (CST), Distributed, embedded Systems (DeS), Freie Universität Berlin, Berlin, Germany, guenes@inf.fu-berlin.de

**Wenbo He**  Department of Computer Science, University of Illinois at Urbana-Champaign, Siebel Center, Urbana, IL, USA, wenbohe@uiuc.edu

**Choong Seon Hong**  Department of Computer Engineering, Kyung Hee University, Korea, cshong@khu.ac.kr

**Jiangyi Hu**  Department of Computer Science, Florida State University, Tallahassee, FL 32306, USA, jiangyhu@cs.fsu.edu

**Ying Huang**  Department of Computer Science, University of Illinois at Urbana-Champaign, Siebel Center, Urbana, IL, USA, huang23@cs.uiuc.edu

**John Felix Charles Joseph**  Center for Multimedia and Networks, Division of Computer Communications, School of Computer Science and Engineering, Nanyang Technological University, Singapore, john0007@ntu.edu.sg

**Vinay Kolar**  Department of Wireless Networks, RWTH Aachen University, Aachen, Germany, vko@mobnets.rwth-aachen.de

**Anis Laouiti**  Telecom SudParis, 9 rue Charles Fourier, 91011 Evry Cedex, France, anis.laouiti@it-sudparis.eu

**Bu-Sung Lee**  Division of Computer Communications, School of Computer Science and Engineering, Nanyang Technological University, Singapore, ebslee@ntu.edu.sg

**Feng Li**  Department of Computer Science and Engineering, Florida Atlantic University, Boca Raton, FL, USA, fli4@fau.edu; lifengg2008@gmail.com

**Fan Li**  Department of Computer Science, University of North Carolina at Charlotte, Charlotte, NC, USA, fli@uncc.edu

**Justin Lipman**  Intel Asia Pacific, Research and Development Ltd., Shanghai, PRC, justin.lipman@gmail.com

**Hai Liu**  School of Information Technology & Engineering, University of Ottawa, Ottawa, Ontario, Canada, hailiu@site.uottawa.ca

**Jiangchuan Liu**  School of Computing Science, Simon Fraser University, BC, Canada, jcliu@cs.sfu.ca

**Antonio A.F. Loureiro**  Computer Science Department, Federal University of Minas Gerais, Belo Horizonte, Brazil, loureiro@dcc.ufmg.br

**Miroslaw Malek**  Institute for Informatics, Humboldt University, Berlin, Germany, malek@informatik.hu-berlin.de

**Bratislav Milic**  Institute for Informatics, Humboldt University, Berlin, Germany, milic@informatik.hu-berlin.de

**Pascale Minet**  INRIA, Rocquencourt, 78153 Le Chesnay cedex, France, pascale.minet@inria.fr

**Klara Nahrstedt** Department of Computer Science, University of Illinois at Urbana-Champaign, Siebel Center, Urbana, IL, USA, klara@cs.uiuc.edu

**Al-Sakib Khan Pathan** Networking Lab, Department of Computer Engineering, Kyung Hee University, Korea, spathan@networking.khu.ac.kr; pathan_sakib@yahoo.com

**Musfiq Rahman** Department of Computer Science, American International University Bangladesh, Bangladesh, musfiq.rahman@gmail.com

**Ashfaqur Rahman** Department of Computer Science, American International University Bangladesh, Bangladesh, ashfaqur.rahman.omi@gmail.com

**Boon-Chong Seet** Department of Electrical and Electronic Engineering, Auckland University of Technology, New Zealand, boon-chong.seet@aut.ac.nz

**Sevil Şen** Department of Computer Science, University of York, Heslington, York, UK, ssen@cs.york.ac.uk

**Mihail L. Sichitiu** Department of Electrical and Computer Engineering, North Carolina State University, Raleigh, NC, USA, mlsichit@ncsu.edu

**Ivan Stojmenović** EECE, The University of Birmingham, Birmingham, UK; School of Information Technology & Engineering, University of Ottawa, Ottawa, Ontario, Canada, stojmenovic@storm.ca

**Yi Sun** School of Computing Science, Simon Fraser University, BC, Canada, sunyi@cs.sfu.ca

**Sameer Tilak** University of California, San Diego Supercomputer Center, La Jolla, CA, USA, sameer@sdsc.edu

**Alicia Triviño-Cabrera** University of Málaga, Spain, atrica@gmail.com

**Uday Tupakula** Information Networked System Security (INSS) Research Group, Department of Computing, Division of ICS, Macquarie University, North Ryde, Sydney, NSW, Australia, uday@ics.mq.edu.au

**Vijay Varadharajan** Information Networked System Security (INSS) Research Group, Department of Computing, Division of ICS, Macquarie University, North Ryde, Sydney, NSW, Australia, vijay@ics.mq.edu.au

**Pramod K. Varshney** Department of EECS, Syracuse University, Syracuse, NY, USA, varshney@ecs.syr.edu

**Yu Wang** Department of Computer Science, University of North Carolina at Charlotte, Charlotte, NC, USA, yu.wang@uncc.edu

**Martin Wenig** Department of Comuter Science, Informatik 4, RWTH Aachen University, Aachen, Germany, wenig@cs.rwth-aachen.de

**Jie Wu** Department of Computer Science and Engineering, Florida Atlantic University, Boca Raton, FL, USA, jie@cse.fau.edu

**Yuan Xue** Vanderbilt University, Nashville, TN, USA, yuan.xue@vanderbilt.edu

**Yinying Yang** Department of Computer Science and Engineering, Florida Atlantic University, Boca Raton, FL, USA, yyang4@fau.edu

# Chapter 1
# Properties of Wireless Multihop Networks in Theory and Practice

Bratislav Milic and Miroslaw Malek

**Abstract** Simulation and testbeds are frequently used for the validation of wireless networking protocols, but several assumptions regarding node place-ment, wireless signal propagation, and traffic type must be made. We compare common models with the measurements made in Berlin's and Leipzig's free multihop wireless networks. It is shown that the properties observed in reality are different than in commonly used models: network is connected but with low average node density; it has large number of bridges and articulation points that can compromise its connectivity; and the traffic distribution over nodes is highly asymmetrical. As an illustration of the discrepancy between reality and synthetic models, we present issues of reactive route discovery process that cannot be observed in simulation that use common placement and propagation models. This chapter focuses on the understanding of limitations of simulation methodologies. It also provides general guidelines on ways of reducing the gap between simulation theory and practice of wireless multihop networks.

## 1.1 Introduction

Wireless multihop networks (WMNs) include static and dynamic networks such as stationary mesh, mobile ad hoc, and sensor networks. They can be complemented with the stationary (wired) infrastructure, creating hybrid networks. They are envisioned for various usages like everyday Internet access, disaster management, and military applications. Each of these use cases has different assumptions and protocol requirements. It is difficult to build a general-purpose testbed that can fulfill all these assumptions. Simulation is

B. Milic (✉)
Institut für Informatik, Humboldt-Universität zu Berlin, Berlin, Germany
e-mail: milic@informatik.hu-berlin.de

M. Malek (✉)
Institut für Informatik, Humboldt-Universität zu Berlin, Berlin, Germany
e-mail: malek@informatik.hu-berlin.de

S. Misra et al. (eds.), *Guide to Wireless Ad Hoc Networks*,
Computer Communications and Networks, DOI 10.1007/978-1-84800-328-6_1,
© Springer-Verlag London Limited 2009

cheaper and easier to use than a testbed, so majority of protocols are verified in simulators. In order to create high-quality simulators, the models used in them should correspond as close as possible to reality.

The goal of the chapter is to help the reader in quality improvement of simulation studies. We compare the topological and link reliability properties in open multihop wireless networks in Berlin [1] and Leipzig [2] with the common theoretical and simulation models such as uniform, grid, and the random way-point model (RWM). The properties of real, user-initiated, and maintained networks are significantly different than in common simulation models. The synthetic node distributions and topologies induced by them are appropriate for certain use cases, but they should not be used exclusively: our work shows that these models are not appropriate for urban scenarios and networks that are built for everyday use and Internet access. As an illustration of the importance of field data on protocol quality, at the end of the chapter we present problems of reactive route discovery protocols that are caused by real topologies and that have not been observed in simulation studies due to the properties of synthetic topologies.

Simulation and theoretical studies of WMNs require a model as a starting point. Due to the complexity of WMNs, the general WMN model is composed of six sub-models:

- **Node model** describes node properties such as: number of network interfaces, energy source, memory capacity, processing capabilities, whether node knows its geographic location (GPS module), duty cycling, existence of nodes with extended capabilities (super-nodes), etc.
- **Node deployment and node mobility models.** Deployment (placement) models describe the placement area and the number of nodes deployed in the area. They provide node position in case of static networks or initial node positions for mobile networks. Some of the most popular deployment models are uniform and grid models. Mobility models describe movement patterns of nodes. Node movement creates dynamic network topologies – because of mobility, links between nodes are created and broken.
- **Radio model** defines the characteristics of the radio used by the node, such as its operating frequency, bandwidth, output power, reception thresholds, error correction, MAC (Medium Access Control) layer functionality, energy consumption for packet reception and transmission, etc.
- **Wireless signal propagation model** describes the signal propagation and influence of environment on its quality. In simulators this model is used to calculate the signal-to-noise and interference ratio (SNIR) at the receiver. It is then assumed that if the SNIR is higher than some prescribed threshold (defined in the wireless radio model), the packet is successfully received. Frequently used models for the signal propagation in WMNs are path loss, two-ray ground, shadowing, and Rayleigh fading models.
- **Packet loss model.** The losses may be caused only by wireless channel properties, and packet collisions on channel or additional packets can be dropped,

for instance, in accordance with uniform or Markov error models [3]. These models were mostly used to offset the effects of low packet losses created by the unrealistic path-loss-only models. With realistic wireless radio and signal propagation models, the need for them is substantially reduced.

- **Traffic models** define which nodes send (sources) and which receive (destinations) traffic in the network, as well as the properties of traffic flows. For wireless mesh and mobile ad hoc networks, these properties are defined in layers three (transport) and four (application) of the TCP/IP stack. The application layer is usually modeled as a constant bit rate (CBR) flow or a file transport protocol (FTP) transfer. The application layer traffic determines the transport layer to be used: CBR flows are associated with UDP (User Datagram Protocol) and FTP flows with TCP (Transmission Control Protocol). The duration of CBR flow or size of the file to transfer is defined as the traffic parameter.

The joint model is built by through composition of individual sub-models: for example, static wireless ad hoc network is composed of 100 nodes placed uniformly on a square kilometer area; in addition to path loss, there exists Rayleigh fading on the wireless channel; nodes are equipped with 802.11b network cards; nodes are unaware of their geographical locations; no additional packets at nodes are lost; there are 10 FTP flows transferring 10 MB file between randomly selected source-destination pairs.

Some of the sub-models are built from real data measurements (e.g., wireless signal propagation in telecommunication research, see [4] for a detailed survey). We address the sub-models that are WMN-specific: topological, link reliability (packet loss), and traffic properties in real, user-initiated, and maintained networks.

The importance of real datasets for WMN research has been already noticed and the CRAWDAD site [5] is the WMN community archive of various measurements. However, in all CRAWDAD data sets observed network either has a single wireless hop or it was collected from a testbed, leaving the problem of realistic, user-initiated topologies open. So the lack of real topological data forces researchers to use synthetic node placement distributions in simulation, emulation, and testbeds. Our work is of particular importance from topological perspective, since node placement and behavior is not pre-determined like in simulation and testbeds, but user-initiated and controlled. We hope that it is a step forward toward realistic topology and placement models.

The chapter is organized as follows: Section 1.2 provides detailed motivation and related work. Section 1.3 defines the metrics, topological properties, and placement models, which are referred in the text. Section 1.4 gives our measurement and simulation methodologies. In Section 1.5 we present the measured and the simulation data and discuss the differences between them. Guidelines for practitioners and for future work are contained in Sections 1.6 and 1.7. Finally, we conclude the chapter by recapitulating the most important facts in Section 1.8. All examples in text are numbered and end with ◆ sign.

## 1.2 Background

In protocol development process, there are two main stages. First is theoretical, providing the basis of a protocol: for instance, if a proactive routing protocol is developed, we must first define the topology dissemination algorithm (e.g., plain flooding) and route calculation algorithm (e.g., Dijkstra's shortest path first).

The second phase consists of protocol implementation, testing, and tuning of parameters. It can be performed in simulators, emulators, and testbeds.

Network simulators, such as ns2 [3], OMNET++ [6], SWANS [7], are favored for their fast prototyping and cost efficiency, so they are preferred to emulators and testbeds in the early stages of protocol implementation. Simulators require careful modeling of the end system if we want to obtain realistic results. The worst that can happen is that we make a conclusion that a protocol is good enough to be implemented and used in real systems, just to find out that we have omitted important aspects within our simulation study and that protocol is actually unusable.

*Example 1.1* The discrepancy between model and reality can be very subtle as noticed in an IBM study of Mote nodes [8]. They have implemented the IEEE 802.15.4 on Motes and planned to build a WMN for logistics management. The issue they faced is that the Mote nodes do not have sufficient processing power to timely respond to all interrupts created by MAC layer. As soon as the traffic in network grows above certain limit (which was very low – 1 packet/node/s), the nodes cannot operate MAC layer and the network crashes. Such behavior is untraceable in most of the simulators, which do not account for processing speed of individual nodes. ◆

Emulators are at the borderline between simulators and reality. Nodes in emulator are real; they can run real software instead of simulation code; protocols built for them are deployable to real networks without or with minor modifications. However, the packets are not sent but packet delivery is calculated, like in simulation. Emulators can be large and flexible – for instance, the ORBIT testbed [9] consists of an indoor radio grid emulator consisting of 400 nodes in a 20×20 grid. The grid can be dynamically configured to form differently shaped topologies.

Network testbeds are the best choice for protocol testing because there are no hidden assumptions in them like in simulators and emulators. They are widely used for MAC, networking/routing, or transport protocols verification. However, building and maintaining dedicated testbeds is a time- and resource-consuming task. They are limited spatially and in node count and they cannot scale to hundreds of nodes like real networks: MIT Roofnet has between 30 and 40 active nodes [10]; wireless mesh testbed at the University of California in Santa Barbara [11] has 25 nodes in one building distributed over five floors; Kotz et al. [12] created a 33 node mobile testbed at the University of Darmouth.

Up to now, real data collection for WMNs' analysis has been mainly performed in testbeds. Data sampling in testbeds is easy because of the complete control over nodes that allows detailed logging of events of interest and their forwarding to a central collection server. However, if considered for the analysis of topological properties, testbeds have several serious drawbacks:

- Network topology in testbeds does not originate from user behavior, but from designers assumptions such as initial node placement or movement patterns. These may or may not follow the real user behavior.
- Due to the limited size of a testbed it may be restricted to a single type of propagation environment (e.g., an office building in [11]) that in turn creates topologies specific for that type of propagation environment.
- In order to reduce the maintenance time and costs, it is common to use the same equipment for the whole testbed. In real networks, with the growth of a network grows the diversity of propagation environment and of the equipment deployed in it.

Other measurements in testbeds can be influenced, but to a smaller degree. For instance, traffic patterns and testbed's topology are related with interference patterns in the testbed, providing partially synthetic link reliability data.

## 1.3 Terminology and Models

### 1.3.1 Definitions

This section defines some of the important notions used in the chapter. As it is common in the literature, connectivity graphs (weighted digraphs) are used to model WMNs. For simulation, connectivity graph is calculated based on the placement and communication models. For the real networks, it is reconstructed from the data sampled in them. In the text, terms "connectivity graph"' and "network", "vertex" and "node", "edge," and "link" are used interchangeably so the definitions in this section apply both to graphs and to networks.

In communication graph, nodes are represented as vertices. If node $p$ is able to communicate with node $q$, there exists (directed) arc $pq$ in the communication graph. There are no self-loops in the connectivity graph and there can exist only one arc between a pair of nodes. Arc's weight $w_{pq}$ represents the quality of the link $pq$ in direction from node $p$ to node $q$ (probability that a packet traverses it successfully). Link quality is asymmetrical and we use arcs to distinguish the qualities in two directions.

We consider bidirectional communication links only: if there exists an arc from $p$ to $q$, there exists an arc in reverse direction as well. However, $w_{pq}$ need not be equal as $w_{qp}$ (links are asymmetrical). This assumption is in accordance with most of MAC the protocols, which cannot function properly without direct bidirectional links (e.g., if MAC layer implements retry mechanism, upon packet reception receiver informs the sender that the packet was received, otherwise sender retries the packet transmission).

A pair of arcs $pq$ and $qp$ creates an edge in the graph. The edge represents bidirectional, asymmetrical communication link in the network. So, although the connectivity graph is directed, definitions and theorems [13] that are valid for undirected graphs also apply.

**Definition 1.1** Two vertices $a$ and $b$ are adjacent if there exists an edge between them ($ab \in E(G)$). If vertex $a$ belongs to an edge $e$, $e$ and $a$ are incident.

**Definition 1.2** A walk of length $k$ is a sequence $v_0, e_1, e_2, \ldots, e_k, v_k$ of vertices and edges such that $e_i = v_{i-1}v_i$ for all $i$'s. A trail is a walk with no repeated edge. A path is a walk with no repeated vertex.

**Definition 1.3** The degree of a vertex $v$ in a graph $G$, written $d_{G(v)}$ or $d(v)$ is the number of edges incident on $v$. A pendant vertex is a vertex of degree 1.

**Definition 1.4** Components of a graph (partitions of a network) $G$ are its maximally connected subgraphs. A bridge in a graph (cut-edge) is an edge whose deletion increases a number of components. An articulation point in a graph (cut-vertex) is a vertex whose deletion increases the number of components in the graph.

Consequence of the Definition 1.4 is that a vertex incident to a bridge is an articulation point unless it is a pendant vertex. If a pendant vertex is removed from a graph (a node decides to leave the network or it fails), its removal does not affect other nodes in the network and the number of connected components in the network does not change.

**Definition 1.5** Link quality $w_{pq}$ is the probability that a packet sent by node $p$ is successfully received at node $q$, within one communication cycle (no retries).

Probability that node $q$ receives a packet sent by node $p$ is $w_{pq}$. In commonly used MAC implementations, node $q$ responds with a reply to $p$, indicating successful reception of the packet. Node $p$ receives reply with probability $w_{qp}$. Probability that the communication cycle is successfully completed is $w_{pq} \times w_{qp}$.

**Definition 1.6** The estimated number of packet retransmissions (ETX) for wireless link between nodes $p$ and $q$ is calculated as $ETX = 1/(w_{pq} \times w_{qp})$.

**Definition 1.7** A stationary process is a stochastic process whose probability distribution at a fixed time or position is the same for all times or positions. Such distribution is called stationary distribution.


### 1.3.1.1 Node Placement Models

In order to build a connectivity graph for simulation, it is needed to place the nodes in an area to determine the existence of a link between each pair of nodes and its quality. There exists a number of different placement models that shape the connectivity graphs. Three sample topologies created by these models are shown in Fig. 1.1. It is obvious that they differ substantially from one another.

In uniform placement model, a placement area (usually rectangular, on rare occasions circular) of size $|A|$ is chosen and $n$ nodes are placed over it with uniform probability $p_{\text{uniform}} = n / |A|$.

Grid placement model places nodes at intersections of a rectangular grid. Usually, the grid has quadratic-shaped cells with cell edge length that is close to

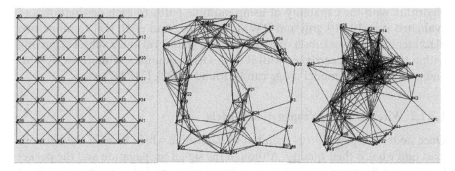

**Fig. 1.1** Topologies created by grid (left), uniform (center), and RWM (right) models

the communication radius of a node: e.g., distance between nodes is 220 m and the free-space communication radius is 250 m. It creates networks that are regular in shape and provides excellent connectivity (there are neither bridges nor articulation points in this model), good resilience to node and link failures (in particular in the central parts of the grid), and large set of disjoined paths between node pairs.

Movement models change the connectivity graph over time: because of node movement their distance varies, which in turn breaks and establishes communication links, and changes their quality. They can also alter the initial distribution of nodes.

One of the most frequently used movement models is RWM. In this model we define minimum ($v_{min}$) and maximum ($v_{max}$) allowed speeds of nodes and pause time between two movements. A node chooses uniformly random a point in the placement area and heads toward it with a speed selected uniformly between the minimum and the maximum allowed speeds. Once the destination is reached, node waits for pause time and then repeats the process.

When this model was proposed, it was believed that RWM preserves uniform distribution and that the average speed of nodes is the arithmetical mean of minimum and maximum speed $v_{avg} = (v_{min} + v_{max})/2$. Both of these assumptions were false. The average speed is substantially lower than the arithmetic mean [14]. RWM also increases node density in the central parts of the placement area and decreases it in the vicinity of area borders. This happens because nodes always choose the shortest path to the selected destination and this shortest path tends to cross the central section of the placement area. The precise stationary node density probability distribution function can be found in [15]. These properties of RWM do not make it "good" or "bad," but they must be kept in mind during simulation so that the intended use case and the actual simulation fit together.

These models are, for example, used in the following studies: Wei and Zakhor [16] evaluate their multipath selection algorithm on 7×7 grid; Souryal and Moayeri [17] simulate their routing algorithm, which adapts itself to link fading in 8×8 grid scenario and in a combination of uniform and RWM scenarios; Jansen et al. [18] have developed a proactive distance-vector routing algorithm with the capability of using multiple paths to each destination and evaluated it in $10 \times 10$ grid and RWM scenarios. Aad et al. [19] combine two placement models – a subset of nodes forms a static grid and a subset is moving within the grid in accordance with RWM. Other examples of uniform, RWM, and grid placement model usage can be found in [20], [21], [22], etc.

### 1.3.1.2 Wireless Signal Propagation Models

Once nodes are placed, it is required to determine which links in the network exist and what is their quality. Although the signal propagation and the packet loss are not the same, simulators usually calculate packet losses based only on the wireless signal propagation models. Other effects that affect successful packet reception are ignored, unless the link layer is simulated (this is extremely rare in WMN research). For instance, the popular simulator ns2 [3], if used with the shadowing propagation model, calculates SNIR ratio at the receiver at the start of packet reception and that decides whether the packet is successfully received or not, ignoring the packet coding and bit errors probabilities. This is done in order to reduce computation complexity and simulation run-time. In real systems even if SNIR was high at the start, the packet can be dropped due to SNIR decrease later on, or due to redundancy in packet coding scheme a packet can be successfully received even if some of its bits were garbled by noise.

If a sender transmits the signal of strength $p_t$ and if received signal has strength $p_r$, the attenuation of the wireless channel is $a = p_t/p_r$. We assume that the packet is successfully received if the channel attenuation is less than some threshold attenuation $a_t$. The channel attenuation has different components and can be expressed as $a = a_{PL} + a_{SH} + a_{FA}$ [dB] where $a_{PL}$ represents attenuation due to the path loss, $a_{SH}$ due to the shadowing, and $a_{FA}$ is the attenuation due to the fading. Path loss is deterministic while shadowing and fading are stochastic processes.

In the commonly used path loss model, shadowing and fading factors are ignored. It is assumed that signal strength decreases with inter-node distance $d$ in proportion to $d^{-\alpha}$ where $\alpha$ stands for path loss factor. Factor $\alpha$ is equal to two for vacuum or higher if there are obstacles between sender and receiver. At certain distance, the signal strength falls below the reception threshold and the communication link ceases to function. This breaking point is called the communication radius $R$.

This means that a link exists with probability one if the distance is less than or equal to the communication radius $R$, and with probability zero if the distance is larger than the communication radius. Link quality, assuming rather low traffic in the network, is close to one even for nodes on inter-node distances close to the $R$. The path loss model is a rough approximation of reality as it was illustrated on a 33 node testbed by Kotz et al. in [12].

Shadowing and fading factors improve quality of the propagation model. In [23], measurement shows that for the same transmitter–receiver pair, fixed distance, frequency, and transmission power, the mean value of received signal power is not deterministic but varies due to the objects in and around the signal path. The shadowing model abstracts different phenomena affecting the wireless signal propagation that can increase or decrease signal strength: diffraction, reflection, self-interference, scattering, absorption. The shadowing variations (expressed in *dB*) are given by the normal distribution with zero mean, path loss $\alpha$, and shadowing variance $\sigma^2$:

$$a_{SH}(\alpha, \sigma) = \frac{1}{\sigma\sqrt{2\pi}}e^{-\frac{\alpha^2}{2\sigma^2}}. \tag{1.1}$$

Detailed study of shadowing is out of scope for this chapter and can be found in [4] and [23]. For us, it is important to understand that it does not have the Heaviside-function behavior of path-loss-only models – even if two nodes are close, they may experience problems with packet reception; and even if they are far away, they may communicate (for instance, there exists line-of-sight between nodes).

The shadowing model has its limitations as well. One of the most important is that it does not address the correlated shadowing. The correlated shadowing model that partially solves this issue has been proposed in the literature, but it is intended for single sender–multiple receivers scenarios, and it is not applicable for WMNs.

*Example 1.2* Example of correlated shadowing is a high and thick concrete building that impacts all communication links that are traversing it, substantially shortening the communication range of nodes, increasing packet loss or even disabling all communications. At the same time, in proximity of the building there exists large open space, enabling excellent communication over it over long distances.

This phenomenon can be seen in Fig. 1.2.[1] The figure shows a small portion of the network in Berlin. Between nodes are buildings, streets, parks, even a river, creating a very complex propagation environment. Nodes are placed in buildings of different heights, some of them are on roofs and some are inside apartments. Some nodes use standard and some directional, high-gain antennas. This influences the signal propagation as it can be seen in the figure where some of the links are very short-ranged, while some reach over long distances. ◆

## 1.3.2 Data Sampling and Simulation Methodology

For data sampling, we were limited by the existing, built-in capabilities of Berlin and Leipzig networks. We had no control over the networks, so we did not know what is happening in the network at the sampling moment: some users might be experimenting with new equipment, or new and possibly incorrect

---

[1] The Figure does not show all the nodes and links in the area since coordinates of approximately 1/3 of nodes in Berlin's network are unknown and cannot be shown on the map. Due to it, it seems that network is substantially sparser than it really is.

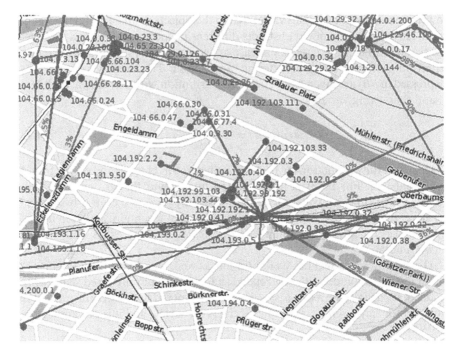

**Fig. 1.2** Part of Berlin's Network

protocol versions. The implication of this uncertainty is that a single sample cannot be trusted, but that large series of measurements is needed to reason about topological properties. The same, stochastic methodology is applied to simulation to eliminate the effects of individual atypical results within a simulation scenario.

Table 1.1 shows important parameters of measurements and simulation. Number of samples in all cases was approximately 1500. In real networks number of participating nodes varies over time as shown in column with minimal and maximal number of nodes encountered over all samples. All simulation scenarios had 400 nodes. Number of edges varies with the number of nodes in real networks or with simulation parameters.

**Table 1.1** Network and simulation parameters

| Scenario | Samples | Avg.# Nodes | Min–Max Nodes | Avg. # Edges | Min–Max Edges | Area | $R$ | $\alpha$ | $\sigma$ |
|---|---|---|---|---|---|---|---|---|---|
| Berlin | 1465 | 315.29 | 199–419 | 633.79 | 291–951 | – | – | – | – |
| Leipzig | 1589 | 586.66 | 452–615 | 1277.91 | 1006–1396 | – | – | – | – |
| Uniform | 1500 | 400 | 400–400 | 1061 | 945–1188 | $1000^2$ | 67 | 4 | 0 |
| Uniform(S) | 1500 | 400 | 400–400 | 1063.33 | 958–1190 | $1000^2$ | 40 | 4 | 7 |
| RWM | 1500 | 400 | 400–400 | 1524.48 | 1337–1799 | $1000^2$ | 40 | 4 | 7 |
| Grid | – | 400 | 400–400 | 2560 | 2560–2560 | $1000^2$ | 40 | 4 | 7 |

### 1.3.3 Data Sampling Methodology

Sampling in real networks does not allow high intrusion level that would be ideal for sampling purposes: it is not possible to modify the software running on nodes so that it collects and sends the captured data to a central repository. The data had to be extracted from running protocols, as they were. Both networks used extended Optimized Link State Routing (OLSR) routing protocol [24]. OLSR is a proactive protocol, and each node has the global topology knowledge. Due to several issues of OLSR, discussed on networks' websites [1, 2], the protocol used in networks does not fully comply with the standard defined in RFC 3626:

- The topology control (TC) packages disseminate the link quality data, not just the topology information.
- In order to utilize the networks resources better, ETX metric (Definition 1.6) is used for packet routing [25] – packets are not routed by minimizing the hop count, but by minimizing the ETX.
- Multipoint relays are not used – instead of them each node is disseminating the topology on its own, using plain network flooding. Multipoint relays in OLSR are nodes that disseminate topology information for their neighbors, reducing contention on the wireless channel.
- In order to reduce the overhead produced by dissemination of TC packets, fisheye algorithm is applied for information dissemination: each TC packet has time to live (TTL) field, which specifies how many hops a packet should be forwarded. Nodes send TC packets every 0.5 s, setting the following values in TTL field: 255, 3, 2, 1, 2, 1, 1, 3, 2, 1, 2, 1, 1.

TC packets are also used for link quality estimation. A node knows that it should receive a TC packet from a neighbor every 0.5 s, with some small jitter. There is a sliding window mechanism that counts how many packets should have been sent by node $q$ and how many actually reached node $p$. Based on these numbers, node $p$ estimates the link quality $w_{qp}$ as #received/#planed. Neighbor $q$ estimates the link quality $w_{pq}$ using the same approach. Since TC packets include this information, both nodes are aware of the link qualities in both directions. Although this method has some drawbacks (e.g., it does not capture burstiness of losses, or causes of packet drops), it provides a good estimate of link quality and improves network throughput [25].

The differences between the protocol in free networks and the OLSR standard grew, so the protocol has been officially differentiated from the OLSR and it is known as B.A.T.M.A.N. [26].

Frequent topology updates and static nodes allow us to take samples from a single node. The samples are taken every 10 min in Berlin and every 15 min in Leipzig network. The extracted samples include topology and ETX data. Successive samples differ from one another since nodes are joining and leaving network, and network traffic is changing, which in turn generates different interference patterns and different link quality data. Additional changes in link quality are created by the interference with network-unrelated wireless access points and WLAN cards.

In Berlin, we have installed a node in the network, running the extended OLSR protocol, Version 0.4.10. Taking of topology samples from a node running the OLSR daemon is rather simple, since the daemon can output the topology table to a textual file. In Leipzig, we could not install a node that directly participates in the network, so we have used the data available from their website. The network in Leipzig is not connected like in Berlin, but it has several components. Since it is used for Internet access, each of the components can forward the topology data to the central server that builds the joint topology. The topology data available at the server is in .dot format [27], and each link is weighted with the ETX value. The topology file includes the "virtual" links as well. They do not have ETX'es associated with them so they are easily recognized and ignored in analysis.

For easier manipulation of topological and link quality data, it was parsed and placed in a relational database [28]. The data processing was done in custom-written Java application. Statistical processing and most of the figures have been made in R-Environment [29].

### 1.3.3.1 Simulation Methodology

For simulation we use our, custom-built simulator specialized for topology creation and analysis, which we have already used in [30] and [31]. We have simulated the following scenarios:

- Uniform: nodes are placed uniformly in the area. There is no shadowing on the links; the path-loss model is used.
- Uniform (S): nodes are placed uniformly, but there exists shadowing on the channel.
- RWM: nodes are initially uniformly placed and then they move in accordance with the RWM. Minimum speed is $0.5 \, \text{m/s}$, maximum is $10 \, \text{m/s}$, pause time is zero in order to speed up the convergence to the steady-state distribution. Topology snapshot is taken after 2000 s of movement, which is enough to reach the steady state distribution [15].
- Grid: nodes are placed in a square grid, 20 by 20 nodes. Inter-node distance is slightly smaller (36 m) than the nominal communication radius (40 m). Because of the shadowing and the large inter-node distance, usable links exist only with direct neighbors in the grid. Due to its simplicity, we do not simulate the grid placement but calculate the properties of interest in Section 1.5.

Except in first scenario, link qualities are calculated based on the shadowing model (Section 1.3). For the calculation of shadowing probabilities, we have refactored the source code from ns2 simulator so that it can be used in our simulator. Path-loss exponent for shadowing model is 4 and standard deviation is 7, which is in accordance with the measurements in urban areas for low-height antennas [4].

For the sake of simplified presentation, instead of defining the threshold $a_t$ and the sending power $p_t$ we use the nominal communication radius of a node

$R$ – communication radius that would exist in the presence of path-loss attenuation only ($\sigma = 0$). In our simulation we impose the log-normal variations to this nominal communication radius in order to determine the existence of a link between two nodes and its quality.

The nominal communication radius of a node is calculated in the following manner: It is known that the attenuation of the wireless signal depends on the distance which the signal travels $d$ and the path-loss coefficient $\alpha$: $a_{PL} \sim d^{-\alpha}$. The threshold attenuation value $a_t$ is then $a_t = 10 \log (p_r/p_t)$ [dB] $= 10 \log [p_r/(R^{\alpha}p_r)]$ [dB]. Finally, we get (note that the attenuation $a_t$ in dB is negative so that the $R$ is positive): $R = 10^{-\frac{a_t}{10\alpha}}$

The stochastic nature of shadowing always provides some probability that a packet will be successfully received, no matter the inter-node distance, although this probability can be arbitrarily low. If it is applied directly to our topology simulator, it results in a complete graph with $n(n-1)/2$ edges, where majority of them are unusable for communication, having link quality very close to zero. We avoid this paradoxical situation by defining the quality threshold that decides which links remain in the graph and which are removed: if a link has ETX value higher than 100, it does not belong to the connectivity graph. We apply the same rule to network samples, in order to eliminate the links that are not functional, but the routing protocol has not realized it yet (a node on a link has left the network, but the link still remains in the topology table).

In uniform placement model, in order to create a connected network the communication radius of nodes has to be rather large, increasing the average node degree throughout the graph. The results of [32] show that in order to have a connected network with high probability, the average degree of nodes should be more than 10. This increases the number of disjoint paths and decreases the number of bridges and articulation points in the network. We have selected the simulation parameters for both uniform scenarios such that the average node degree is approximately 5.5. This does not guarantee the connectivity and there exist isolated nodes or smaller groups of nodes. The number of nodes that do not belong to the main partition is small and its impact on our results is minor. By trial and error, it can be shown that further reduction of the average node degree increases bridge count but disconnects the network to partitions consisting of 5 to 10 nodes. For RWM scenario we use the same parameters as for the uniform scenario with shadowing. The analysis in Section 1.5 shows that same set of parameters used with different node distribution creates completely different topological properties of a network.

### 1.3.4 Topological Structure of the Network and Link Reliability Analysis

The goal of this section is to provide guidelines for researchers which parameters and details are of importance for conducting simulation that are closer to reality. We analyze the data collected from real networks and compare

**Table 1.2** Mean values for measured and simulated data

|              | Freifunk Berlin | Freifunk Leipzig | Uniform (S) | Uniform  | RWM      |
| ------------ | --------------- | ---------------- | ----------- | -------- | -------- |
| Node degree  | 4.0204          | 4.3565           | 5.3167      | 5.3049   | 7.6222   |
| Bridge share | 0.1506          | 0.07943          | 0.0212      | 0.0212   | 0.01249  |
| Art. Points  | 75.9338         | 93.3285          | 32.7673     | 32.8905  | 21.3662  |
| Bridge ETX   | 5.8824          | 4.0795           | 28.4579     | $\sim$1  | 27.3775  |
| No-bridge ETX| 17.5832         | 18.2664          | 20.4782     | $\sim$1  | 20.8674  |
| $P_{bt}$     | 0.743           | 0.468            | 0.352       | 0.35     | 0.127    |

it with uniform, uniform shadowing, RWM, and grid models. The summary of the measurements can be found in Table 1.2.

### 1.3.4.1 Node Degree Distributions

The cumulative node degree distribution function is shown in Fig. 1.3. It can be seen that synthetic placement models have substantially fewer low-degree nodes than real networks, which reduces the issues created by low-connectivity nodes (see Section 1.6). The tail of the real node degree distributions is more pronounced (nodes with large number of neighbors are not uncommon) than in uniform and in particular than in grid scenarios. The RWM distribution has a very heavy tail, thus the protocols simulated with the use of this model are more prone to suffer from contention issues than the protocols simulated with other models or deployed in reality.

*Example 1.3* Let us assume that the node and wireless propagation parameters are set so that the topology of the grid reassembles the one shown in Fig. 1.1. Let the nodes be organized in 20 by 20 grid. The grid is square-shaped with $n = 400$ nodes and $a = \sqrt{n} = 20$ nodes form the edge of the square. The grid has four corners, so the share of nodes with degree 3 is $P_3 = 4/n = 1\%$. Nodes on the outside edge but not in the corners have degree 5 and their share is $P_5 = (4\sqrt{n} - 8)/n = 18\%$. Remaining nodes have degree 8 and their share is $P_8 = (n - 4\sqrt{n} + 4)/n = 81\%$. The average node degree is: $8 + (4 - 12\sqrt{n})/n = 7.46$. ◆

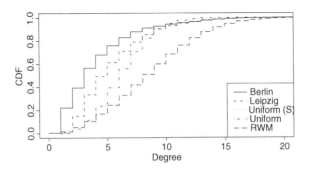

**Fig. 1.3** Comparison of cumulative node degree distributions

### 1.3.4.2 Bridges and Articulation Points Analysis

Figure 1.4 shows the distribution of articulation points' count. Each articulation point is critical for connectivity, if node is turned off, the node's software fails or its energy source is depleted, the network gets partitioned. Network in Leipzig has more articulation points than Berlin's network and simulated scenarios. Although network in Leipzig has more nodes than simulation scenarios, the increase in the number of articulation points is disproportional to the increase in total node count. If we observe it relatively to the network size, Berlin's network has highest share of articulation nodes – 23.8% on the average.

Bridge share in the network is also shown in Fig. 1.4. Bridge share is the ratio of bridge count and total edge count. Again, real networks exhibit poorer connectivity and their bridge share is considerably larger than for synthetic scenarios. Berlin's network has the largest bridge share: most of the samples from Berlin have between 12% and 20% of bridges. Although the share of bridges in Leipzig is half the share in Berlin, it is still 3.8 times larger than for uniform and 6.5 times than for RWM scenarios. Grid placement model has neither bridges nor articulation points.

The degree analysis shows that real topologies have high share of pendant nodes – resulting in reduced connectivity on borders of the network. However, the bridges are also present in central parts of the network where their impact on network functionality is even more emphasized.

*Example 1.4* To illustrate the importance of bridge position in a network, let us assume that there exists a single bridge in a network, forming two subgraphs connected over it. Each of the subgraphs is either at least two-connected or consists of a single node. If one of the subgraphs consists of $k$ nodes, the other has $n–k$ nodes.

In such a network, number of node pairs that, while communicating, avoid the bridge traversal is $k(k-1)/2 + (n-k)(n-k-1)/2$. Number of node pairs that have to use the bridge while communicating is $k(n-k)$. Therefore, the probability that two nodes selected randomly from the network will have the bridge

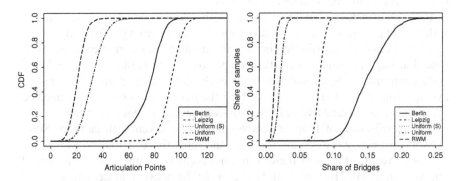

**Fig. 1.4** Cumulative distribution of articulation points count and bridge share

as a part of their communication path is (probability of bridge traversal $P_{bt}$):
$P^2_{tb}(k) = \frac{2k(n-k)}{n(n-1)}$.

If the bridge is placed on the border of the network, connecting a single node with the rest of the network, its impact will be negligible $P^2_{tb}(1) = 2/n$. However, if a bridge divides the network in two equal parts, it creates the maximum number of paths traversing the bridge, resulting in the probability of bridge traversal: $P^2_{tb}(n/2) = n/[2(n-1)] \sim 1/2$. This can already seriously impact the global functionality of the network – even with moderate traffic, it can be expected that the bridge will be heavily congested. Not only that this congestion reduces the available throughput per flow but also that it can compromise other properties: for instance, due to the long waiting times of packets to traverse a bridge, delivery deadlines for the real-time traffic may not be met. Obviously, this ratio is increasing if we add more bridges – for instance, in a graph with two bridges, which create three equally sized subgraphs, the probability of bridge traversal is $P^3_{tb}(\frac{n}{3}) = \frac{2n}{3(n-1)} \sim \frac{2}{3}$. ◆

For Berlin's network, this probability is particularly high. The probability in Leipzig is smaller than in Berlin, but larger than in synthetic scenarios. It clearly indicates the existence of bridges in the central parts of real networks.

In order to study the size and relation of these network components connected over bridges, we artificially divide the connectivity graph by removing the bridges in it, thus obtaining disconnected graph components. Each of the components obtained in such a way is either at least two-connected or consists of a single node.

The vast majority of components has size of one, and the distribution directly obtained from component count could not tell us much. To offset this effect and get clearer picture, we weight each component by its size, relatively to the network size: $C_{rel} = C_{count}|C|/n$. For instance, if a network has 100 nodes and 10 of three-node components, weighted impact of three-node components is: $3 \cdot 10/100 = 0.3$.

The weighted distribution of these components can be seen in Fig. 1.5. Because of its property of grouping nodes in the central part of the placement area, RWM distribution is skewed to the right: not only that it has few bridges, but they are all placed on network borders.

Both uniform placement scenarios have rather equally distributed components, with a slight preference for very small (below 20) and large (300–350) components.

Berlin network's distribution is bimodal: majority of its large components is placed around 110 and rest of the distribution weight is placed on small, 1–5 node components. Since samples from Berlin have more than 300 nodes on average, this means that for most of time the network has two well-connected components connected over a bridge (or even several bridges in series).

The distribution of Leipzig samples has to be taken with caution. Since network in Leipzig is disconnected from the start, this distribution cannot tell us much except for sizes of these components (three components with sizes of approximately 80, 210, and 270). However, as the network in Leipzig grows, it is to expect that it will get connected and that the component structure will

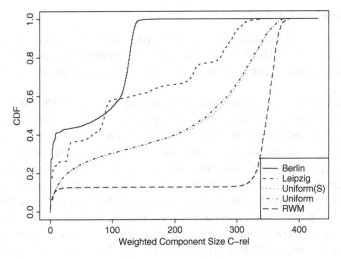

**Fig. 1.5** Distribution of weighted network components obtained by bridge removal

not change substantially: lots of pendant nodes and internally well-connected components that can communicate with each other only over bridges, just like in Berlin. Similar properties can be observed in free wireless network in Hanover. We did not make measurements in it, but the topology visualization on the website of the network clearly shows clustering of nodes as well as the weak connectivity among three clusters [33].

From this analysis we can conclude that the user-initiated networks have similar topological properties. We attribute these properties to sociological reasons:

- New participants in the network are more likely to join the network in city areas where the connectivity is already good. Improved connectivity (several links to the network) increases resilience to individual node failures and provides higher total connection uptime for the user. Also, information about the network existence is spread through word of mouth, attracting new participants.
- A participant in the network expects to have at least a single communication link to the remainder of the network, creating large number of pendant nodes.
- Pendant node joins the network if the bridge link connecting it with network is of sufficient quality. User subjectively decides what is sufficient, but everything with ETX value over 10 is likely to be rejected.
- A pendant node might become a seed for a new, larger, and well-connected subnetwork.

Since these are not technological but sociological reasons, we believe that these types of topologies are to be frequently encountered in other open multihop networks.

### 1.3.4.3 Link Quality Analysis

Figure 1.6 shows the cumulative distribution of ETX values for bridges and ordinary links. The ETX distributions for ordinary links in Berlin and Leipzig are almost identical, while bridge ETX distributions are slightly different.

The distributions from real networks incline to lower ETX values (which means higher link quality) than the synthetic placement models. Also, in real networks, bridges have considerably better quality than ordinary links. In synthetic scenarios the situation is opposite – ordinary links have better quality on the average. This goes in hand with typical theoretical assumption that bridges are always long, stretched communication links and due to their length, their quality suffers. Mathematically, it is more likely that a bridge is also a long edge. But in reality, mathematical rules are not the only rules that apply. If a user is bound to use a bridge to access the network (e.g., in the simplest, pendant node scenario), unless the link provides certain quality of service (e.g., throughput, delay), the user will not participate in the network at all. This way, users prune bridge links with low quality.

Important property to notice is that some links have high ETX values, larger than 40. The probability of successful packet transmission over such links is very low. To make the situation worse, these links could be bridges, and if they are it is impossible to avoid their usage. This property particularly influences optimistic routing and broadcasting schemes, which rely on the presence of multiple nodes in vicinity, so even if some packets are lost, the message will be delivered using another path. In this case, we have only one link connecting two subnetworks, and if the message is not delivered over it, it is lost. In Section 1.6 we study this scenario in detail.

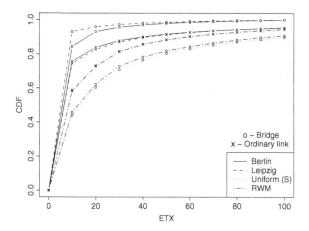

**Fig. 1.6** ETX cumulative distributions

#### 1.3.4.4 Traffic Analysis

Networks in Berlin and Leipzig are mainly used for Internet access – some of the nodes are sharing their Internet connectivity with other network participants. In [34] we have provided the distribution of incoming and outgoing traffic on the major Internet gateway in Berlin's network. The majority of nodes used network moderately, generating less than 2 GB of traffic per month, but there were some selfish nodes as well – one node generated astonishing 55 GB of traffic in a single month.

In this section we devote more attention on comparison of traffic volumes in real networks and simulation. Table 1.3 shows the traffic statistics for this gateway. Only a subset of Berlin's nodes uses this gateway to access the Internet, but the generated traffic is still impressive.

The common approach in simulation is that traffic source and destination are selected randomly and each pair exchanges the same (or similar) amount of data. Table 1.4 shows traffic flow examples from the literature. In order to be able to calculate average flow sizes, we have chosen only the scenarios with CBR flows. In case that authors used more than one set of flow parameters, we have chosen the flow that creates largest load in the network.

Besides the number of flows and nodes in the simulation, Table 1.4 shows the simulation and flow duration. They are important because an intensive and short flow puts less stress on the network on the long run than several, less-intensive, long-duration, parallel flows.

*Example 1.5* Let us assume that there exists a small, static, 20-node WMN, each node equipped with a 1 Mbit/s radio. Nodes create CBR traffic and each packet has 100 bytes payload. A flow is created between two randomly selected nodes. In this sample network we observe two traffic scenarios lasting 100 s each:

**Table 1.3** Incoming and outgoing traffic measured at the major gateway in Berlin's network

| Month | Nodes | Total traffic | Average traffic per node |
|---|---|---|---|
| November 2006 | 193 | 252.3 GB | 1338.7 MB |
| December 2006 | 213 | 278.4 GB | 1338.8 MB |
| February 2007 | 188 | 231.4 GB | 1260 MB |

**Table 1.4** Typical traffic scenarios in simulation

| Research study | Simulation duration (SD) | Nodes (n) | Flow duration (FD) | Flows (FL) | Packet size (PS) | Packet frequency (f) | Monthly traffic (max) | Average traffic per node |
|---|---|---|---|---|---|---|---|---|
| Souryal et al. [17] | 900 s | 200 | 93 s | 100 | 64 B | 4/s | 632.8 MB | 65.4 MB |
| Aad et al. [19] | 400 s | 200 | 400 s | 50 | 100 B | 1/s | 247.2 MB | 247.2 MB |
| Srinath et al. [35] | 300 s | 80 | 30 s | 40 | 512 B | 4/s | 5062.4 MB | 506.24 MB |
| Das et al. [36] | 500 s | 100 | 500 s | 40 | 512 B | 2/s | 2531.2 MB | 2531.2 MB |

- **Intensive:** Total four flows are created and each flow sends 100 packets/s. A flow lasts 10 s and then stops. Flow start times are uniformly distributed in interval [0, 90]s to ensure that all flows end within 100 s.
- **Balanced**: Four flows are created in the network and each sends 35 packets/s. They last for the whole observation period.

The first impression is that the *intensive* scenario creates more load: it has same number of flows as the second and each flow sends three times more packets than in the *balanced* scenario.

Let us first calculate the total number of packets that scenario injects in the network: for the *intensive* it is $4 \times 100$ pac/s $\times 10$ s $= 4000$ packets. For *balanced* it is $4 \times 35$ pac/s $\times 100$ s $= 14{,}000$ packets. So, the balanced scenario puts more load on the network in the long run.

That was to expect, but our intuition still believes that the intensive scenario creates larger peak loads in the network. The *intensive* can create load of up to 400 new packets/s and the *balanced* has constant peak of "only" 140 new packets per second. However, the intensive flows are started at random times and their overlap is probabilistic. Let us calculate the probabilities of overlapping flows in *intensive* scenario:

We start by observing the placement process of $f$ flows determined by their starting point in time, each having duration $d$ on time interval of total length $l$. The probability that an event (another flow starts) happens during the time interval $d$ within the interval $l$ is $P(d) = d/l$. The probability $P^k(d)$ that exactly $k$ out of $n$ events occur in $d$ is:

$$P^k(d) = ( c\, nk ) P(d)^k [1 - P(d)]^{n-k} .$$

If we take a flow and observe whether it overlaps with other flows, it can overlap with up to $n = f - 1 = 3$ other flows. Also, $P(d) = 10/90 = 1/9$. If we substitute it in derived equation, we get the following probabilities: $P^0(10) = 0.702$, $P^1(10) = 0.264$, $P^2(10) = 0.033$, $P^3(10) = 0.001$.

This is just a simplified analysis and it does not account for the duration of overlap, but it is sufficient to show that in more than 70% of cases *intensive* scenario creates smaller peak load than the *balanced* scenario.

Additionally, it is to expect that for the *intensive* scenario there will be long periods without activity in the network, enabling the network to deliver all scheduled packets, exchange routing information, and prepare for the next peak. Finally, if we calculate the expected values of peak intensities, we get 140 packets/s for *balanced* and 133.3 for *intensive* scenario.♦

For comparison with the measurements from Berlin's network, two metrics are introduced: maximal monthly traffic in bytes generated by a node for a given scenario PS $\times f \times$ Month, assuming that traffic flows are active during the whole month and the average traffic load that is put on the network per node $f(\frac{FD}{SD})$Month$(\frac{FL}{n})$PS. The second metric accounts that only a subset of nodes is generating traffic (FL/$n$) and that flows do not exist for the entire period (FD/SD). We assume that a month has 30 days and "Month" is duration of it in seconds.

Table 1.4 shows that only one scenario creates significant load while others are well below the average node load observed in real network. That does not mean that the simulation methodology in these papers was inadequate: after all, almost 75% of nodes in Berlin's network also generated less than 1 GB of traffic for a month, but increasing stress on the network would improve results and give clearer picture to the reader what are the limitations of proposed protocols.

The conclusion is that the researchers should use more diverse traffic types and uneven traffic distributions – most of the load in Berlin's network came from minority of nodes. The goal of community is to develop robust protocols and the researchers should shy away of putting high stress and load on their protocols.

### 1.3.5 Thoughts for Practitioners

The research community should increase the cooperation with open communities such as those in Berlin or in Leipzig and help them overcome real issues that they face on daily basis. A protocol that identifies and solves a real problem will have great impact on the future of WMNs. Furthermore, such protocol will be deployed in the network and tested under real conditions, giving the final proof of its validity.

Even a simple protocol that was thoroughly studied in the literature can display unexpected behavior in practice because of the discrepancy between the simulation scenarios and the real world. That should remind us that our simulation is as good as our simulation model, and that we must always check how close the model corresponds to the reality.

*Example 1.6* Let us observe the distributed Breadth First Search (dBFS or flooding). dBFS is an important part of routing protocols. It is used as the route discovery mechanism in reactive routing protocols (e.g., [37, 38, 39]) and for topology data dissemination in proactive protocols [26, 40, 41]. dBFS is used in routing protocols because it is robust, it is simple to implement and test, and it does not require additional information like geographical location of a node.

In reactive routing protocols dBFS is used for route discovery: the search is initiated from source and it should reach all nodes in the network. Once the search reaches the destination, it replies back to the source revealing the shortest path to itself. Reactive routing protocols have a threshold defining how many times dBFS route search is performed before declaring that destination is not reachable. For instance, in AODV(Ad-hoc On Demand Distance Vector Routing) [39], this threshold is set to four: one initial search followed by three retries.

dBFS as a route discovery mechanism is criticized for its high redundancy that is notable in well-connected parts of the network. The issues arise when dBFS faces a bridge in the network – it is the only path connecting two, possibly large network components. Such situation exists in Berlin: high share of bridges and low quality of some bridges (5.3% of bridges had link quality less than 0.1 and 22.6% of them less than 0.5). Because of it, less than 30% of nodes on the average are reached in the first route discovery – in literature, this ratio was approximately 60% with highest node mobility and highest traffic load and well over 80% for other scenarios with lower load and/or mobility [42].

In [30] we have proposed distributed bridge detection algorithm and unicasting of route request messages over the bridges in the network. This limits retries only to network sections where it is crucial to unicast messages, and in other parts, with redundant paths between nodes, it is not performed in order to limit the contention. The route discovery success ratio is improved considerably. Without unicasting, probability of finding a valid route is only 0.469 after four successive attempts. The unicasting scheme increases the probability of successful route discovery to over 0.9. ◆

Unfortunately, at the moment there exists slight mistrust between purely practical work in these open communities and researchers. It was disappointing for the open community that was working on the precise implementation of different protocols, once they discovered that in practice they do not perform as good as in simulation. Improved communication between communities should remove the mistrust and advance the research efforts as well as the individual practice.

So even the seemingly correct simulation parameters, independently chosen by different researchers, can lead to undesired effects. Therefore, we propose that researchers should create a small checklist prior to starting the simulation:

- Inherent property of wireless links is loss of packets. Your protocol should be able to deal with packet losses. If it is not loss-resilient, you must improve it.
- Ask yourself: Is the scenario I want to simulate the same as the scenario I have put in the simulator? Is the scenario realistic?
- Carefully check the wireless propagation model. Are the parameters in it in accordance with the simulation scenario? Notice should be taken that path loss and shadowing coefficients are dependant on antenna height.
- Make several tests and visually (or statistically in the manner that we have shown) check the resulting topologies. Attention should be put on edge density, network diameter, number of partitions, bridge, and articulation point existence.
- Test your protocols under various traffic loads and traffic distributions. Use the example from Section 1.5 as a starting point for the calculation of peak and average loads in the network. If calculation is too complex for the traffic model you are using, sample it from simulation and analyze it afterwards.
- If any of the parameters does not fit with your simulation scenario, change it until you find it satisfactory, provided it corresponds to reality.

### 1.3.6 Directions for Future Research

One of the major tasks of research community is the further improvement of WMNs' models. At the moment, focus of the community is placed mostly on the improvement of wireless signal propagation models and their implementation in simulators. The results of this effort are already noticeable as the shadowing propagation model is now supported by all major simulators. However, its penetration in simulation studies is not sufficient – quite a few of recently published papers are still based on simple path-loss models. Small-scale (Rayleigh) fading is also implemented in some of the simulators, and combination of the path-loss, shadowing, and fading will create high-quality propagation model.

Situation with node placement models is not so clear and it will require additional research. The number of proposed placement models is high, but they are not verified in reality. In this chapter, we have shown that the usual models are not appropriate for urban scenarios where nodes are participating in network in order to have Internet access.

Problem that remains open is to find appropriate mathematical model that captures the topological properties of real networks. It is to expect that a military WMN has different topological structure than the open networks observed in big cities, so we cannot expect that a single model captures all topologies of interest. So for each use case the process of data sampling, model fitting, and verification will have to be repeated. That will be a tedious job, but a job that has to be done.

## 1.4 Conclusions

We have analyzed Berlin's and Leipzig's open WMNs: their topological, link quality, and traffic properties. They are among the largest known WMNs with over 300 and close to 600 participating nodes, respectively. Our analysis reveals interesting properties that are not captured by the frequently used grid, uniform, and RWMs: node degree distribution and average node degree differ significantly; bridges and articulation points in real networks are common, which can negatively impact multipath routing or optimistic routing protocols. Generated network traffic is not distributed equally over all participating nodes, and the amount of generated traffic also exceeds the typical simulation scenarios.

These insights should encourage building of new, precise models that reflect the reality better. Obviously, the methodologies that are used in simulation should be improved and we have provided general guidelines that should set the reader on the right track.

## Keywords

*Cut-edge (bridge)* in network is a link whose removal increases the number of connected network components.

*Cut-vertex (articulation point)* is the node in network whose removal increases the number of connected components in the network.

*Quality of a link* is the probability that a packet will be successfully received at the end of the link.

*Simulation scenario* defines the input for a simulator, covering six WMN sub-models.

*Simulator* is a dedicated software that behaves or operates like a given system.

*Testbed* is a dedicated network, built for purpose of component and protocol testing, as well as for measurements of natural phenomena.

*Topology* of a network is defined by its participating nodes and communication links established between them.

*Traffic flow* in a network is defined by its source, destination, traffic protocol, and (optionally) intensity.

*Vertex (node) degree* is the number of edges incident on the vertex.

*Wireless multihop network model* consists of six sub-models: node, node deployment, radio, wireless signal propagation, packet loss, and traffic model.

## Problems

1. What does the Wireless Multihop Network model consist of?
2. Discuss issues of RWM that have not been mentioned in the text? Which of these issues apply to other movement models? Is there a "perfect" movement model?
3. Select a respectable journal or conference proceedings. Analyze simulation setups over several consecutive years. Are there differences in assumptions and used simulation models?
4. Search the literature for other placement models besides the models mentioned in the chapter. Discuss their benefits and drawbacks.
5. What are the issues of the shadowing wireless signal propagation model?
6. What are the benefits and drawbacks of simulators, emulators, and testbeds?
7. Search for information on other open wireless networks (besides the Freifunk communities in Berlin, Hanover, and Leipzig). What can you tell about their structure? What sort of routing protocol do they use? If possible, search for user discussions and locate several issues emphasized by the users. Are these topics addressed in research literature?
8. How do you define a traffic flow?
9. Let there be two networks, each made of $n$ nodes. One is organized in a ring structure, and the other forms a complete graph. What are the benefits and drawbacks of these topologies?
10. Let us observe a grid with $n$ nodes and the same connectivity structure as in Fig. 1.1. What is the probability that random removal of exactly three vertices will disconnect the graph?

# References

1. Berliner freifunk-community, olsrexperiment.de/.
2. Leipziger freifunk-community, leipzig.freifunk.net/.
3. K. Fall and K. Varadhan, Error model, in The ns Manual, 2007, pp. 126–130, www.isi. edu/nsnam/ns/ns-documentation.html.
4. A. Aguiar and J. Gross, Wireless channel models, Technical Report TKN-03-007, 2003.
5. Crawdad – a community resource for archiving wireless data at Dartmouth, crawdad.cs. dartmouth.edu/.
6. Omnet++ simulator, www.omnetpp.org/.
7. Jist/SWANS simulator, jist.ece.cornell.edu/.
8. J. Thomsen and D. Husemann, Evaluating the use of motes and TinyOS for a mobile sensor platform, in Proceedings of Parallel and Distributed Computing and Networks, Insbruck, Austria, 2004.
9. Orbit: Open-access research testbed for next-generation wireless networks, www.orbit-lab.org/.
10. MIT roofnet, pdos.csail.mit.edu/roofnet.
11. UCSB meshnet, moment.cs.ucsb.edu/meshnet/.
12. D. Kotz, C. Newport, R. S. Gray, J. Liu, Y. Yuan, and C. Elliott, Experimental evaluation of wireless simulation assumptions, in Proceedings of the ACM/IEEE International Symposium on Modeling, Analysis and Simulation of Wireless and Mobile Systems (MSWiM), October 2004, pp. 78–82.
13. D. B. West, Introduction to Graph Theory. Prentice Hall, 1996.
14. J. Yoon, M. Liu, and B. Noble, Random waypoint considered harmful, in Proceedings of IEEE INFOCOM, San Francisco, US, 2003.
15. C. Bettstetter, G. Resta, and P. Santi, The node distribution of the random waypoint mobility model for wireless ad hoc networks, IEEE Transactions on Mobile Communications, 2003, pp. 257–269.
16. W. Wei and A. Zakhor, Path selection for multi-path streaming in wireless ad hoc networks, in Proceedings of International Conference on Image Processing, Atlanta, USA, September 2006.
17. M. Souryal and N. Moayeri, Channel-adaptive relaying in mobile ad hoc networks with fading, in Proceedings of The First IEEE Conference on Sensor and Ad Hoc Communications and Networks, SECON2004, Santa Clara, USA, October 2004.
18. R. Jansen, S. Hanemann, and B. Freisleben, Proactive distance-vector multipath routing for wireless ad hoc networks, in Proceedings of Communication Systems and Networks, CSN2003, Benalmadena, Spain, 2003.
19. I. Aad, J.-P. Hubaux, and E. W. Knightly, Denial of service resilience in ad hoc networks, in Proceedings of the 10th Annual International Conference on Mobile Computing and Networking, MOBICOM 2004, Philadelphia, USA, September 2004.
20. N. Aboudagga, M. T. Refaei, M. Eltoweissy, L. DaSilva, and J.-J. Quisquater, Authentication protocols for ad hoc networks: Taxonomy and research issues, in Proceedings of the 8th International Symposium on Modeling Analysis and Simulation of Wireless and Mobile Systems, MSWiM 2005, Montreal, Canada, October 2005.
21. L.-J. Chen, T. Sun, G. Yang, M. Sanadidi, and M. Gerla, Adhoc probe: Path capacity probing in wireless ad hoc networks, in Proceedings of the First International Conference on Wireless Internet, WICOM, Budapest, Hungary, July 2005.
22. Y. Zhang and Q. Huang, Adaptive tree: A learning-based meta-routing strategy for sensor networks, in Proceedings of IEEE Consumer Communications and Networking Conference, CCNC2006, Las Vegas, USA, January 2006.
23. J. B. Andersen, T. S. Rappaport, and S. Yoshida, Propagation measurements and models for wireless communication channels, IEEE Communications Magazine, 1995.
24. OLSR implementations, www.olsr.org/.

25. D. DeCouto, D. Aguayo, J. Bicket, and R. Morris, A high-throughput path metric for multihop wireless routing, in Proceedings of the 9th annual international conference on Mobile computing and networking, MOBICOM 2003, September 2003, pp. 134–146.

26. B.A.T.M.A.N. (Better Approach To Mobile Ad-hoc Networking) routing protocol, https://www.open-mesh.net/batman.

27. E. Gansner and S. North, An open graph visualization system and its applications to software engineering, Software Practice and Experience, vol. 30, no. 11, 2000, pp. 1203–1233.

28. Mysql, www.mysql.com.

29. R Development Core Team, R: A language and environment for statistical computing, R Foundation for Statistical Computing, Vienna, Austria, 2005, ISBN 3-900051-07-0. Available: www.R-project.org

30. B. Milic and M. Malek, Adaptation of breadth _rst search algorithm for cut-edge detection in wireless multihop networks, in Proceedings of 10th ACM-IEEE International Symposium on Modeling, Analysis and Simulation of Wireless and Mobile Systems (MSWIM 2007), Chania, Greece, 2007.

31. B. Milic and M. Malek, Dropped edges and faces' size in gabriel and relative neighborhood graphs, in Proceedings of The Third IEEE International Conference on Mobile Ad-hoc and Sensor Systems (MASS 2006), Vancouver, Canada, 2006.

32. X. Li, P. Wan, Y. Wang, and C. Yi, Fault tolerant deployment and topology control in wireless networks, in Proceedings of the 4th ACM international symposium on Mobile ad hoc networking and computing, Maryland, USA, 2003.

33. Hannover free network map, map.freifunk-hannover.de/map.php.

34. B. Milic and M. Malek, .Analyzing large scale real-world wireless multihop network, IEEE Communication Letters, vol. 11, no. 7, July 2007.

35. P. Srinath, P. Abhilash, and I. Sridhar, Router handoff: a preemptive route repair strategy for AODV, in Proceedigns of IEEE International Conference on Personal Wireless Communications, New Delhi, India, 2002.

36. S. Das, C. Perkins, and E. Royer, Performance comparison of two on-demand routing protocols for ad hoc networks, in Proceedings of the IEEE Conference on Computer Communications (INFOCOM), Tel Aviv, Israel, 2000.

37. I. Chakeres and C. Perkins, Dynamic manet on-demand (DYMO) routing (IETF Draft), March 2007.

38. D. Johnson, D. Maltz, and Y.-C. Hu, The dynamic source routing protocol for mobile ad hoc networks (RFC 4728), February 2007.

39. C. Perkins, E. Belding-Royer, and S. Das, .Ad hoc on-demand distance vector (AODV) routing (RFC 3561), July 2003.

40. T. Clausen and P. Jacquet, .The optimized link state routing protocol (RFC 3626), www.ietf.org/rfc/rfc3626.txt, October 2003.

41. C. E. Perkins and P. Bhagwat, Highly dynamic Destination-Sequenced Distance-Vector routing (DSDV) for mobile computers, in Proceedings of the Conference on Communications architectures, protocols and applications. London, UK: ACM Press New York, NY, USA, 1994, pp. 234–244.

42. B. Williams and T. Camp, Comparison of broadcasting techniques for mobile ad hoc networks, in Proceedings of the ACM International Symposium on Mobile Ad Hoc Networking and Computing (MOBIHOC), Lausanne, Switzerland, 2002, pp. 194–205.

# Chapter 2
# Self-Configuring, Self-Organizing, and Self-Healing Schemes in Mobile Ad Hoc Networks

**Doina Bein**

**Abstract** The evolution of technology, the expansion of the Internet, and the tendency of systems to become more software-dependent make computing environments and networks more complicated and less humanly controlled. In this chapter, we consider the problem of organizing a set of mobile nodes, with unique IDs, that communicate through a wireless medium, into a connected network, in order to obtain a *self-configuring* or *self-organizing* network. Additionally, we address the issue of how a reliable structure, once acquired by self-configuring, can be maintained when topological changes occur, due to node failure, node motion, or link failure, in order to obtain a *self-healing* network. We discuss these concepts and present a brief history of self-configuring or self-healing algorithms, respectively, for wireless mobile networks. We detail a number of representative algorithms used in practice. We then go on to address the current theoretical results on self-configuring networks for which we propose directions for future research.

## 2.1 Introduction

Beginning as a military application, Mobile Ad hoc Networks (MANETs) had become largely used for personal use: e.g., personal area network (PAN), for short-range communication of user devices, wireless local area network (WLAN), and in-house digital network (IHDN), for video and audio data exchange.

A MANET is a peer-to-peer, multihop connected network, and is composed usually of tens to hundreds of mobile nodes. The nodes have transmission ranges of up to hundreds of meters and each individual node must be able to act both as a host, which generates user and application traffic, and as a router,

---

D. Bein (✉)

Department of Computer Science, University of Texas at Dallas, 800 W. Campbell Road; MS EC31, Richardson, TX 75080, USA

e-mail: siona@utdallas.edu

S. Misra et al. (eds.), *Guide to Wireless Ad Hoc Networks*,
Computer Communications and Networks, DOI 10.1007/978-1-84800-328-6_2,
© Springer-Verlag London Limited 2009

which carries out network control and routing protocols. The mobility of the nodes makes self-configuring of the network much harder. Although the nodes are battery-powered, the energy consumption is of second importance (a difference to wireless sensor networks), since each device could have its battery recharged or replaced when needed. Providing quality of service (QoS), which is the routing of packets, is the most important issue and has to be scalable in the context of a changing topology, limited bandwidth, and limited transmission power. In contrast, a cellular network is a large network consisting of stationary and mobile nodes, where the mobile nodes largely outnumber the stationary ones. The stationary nodes are called *base stations*; they have unlimited power supply, and are placed to cover a large area with little overlap. For cellular networks, the self-configuring aspect is needed for handling mobile-to-mobile node communication when the mobile nodes move within the area.

In a MANET, the nodes must share the wireless communication medium efficiently and be able to perform routing for the transmitted data. Channel access poses a difficult problem, namely the so-called *hidden terminal problem* (Tobagi and Kleinrock [55]). In this problem, a node is considered to be a terminal: transmissions from different nodes that use the same communication channel at the same time may interfere by colliding with one another, and as a result either a truncated message or corrupted data are received.

Inspired by biological systems, IBM introduced the term *autonomic system* for a system that is less dependent on human intervention and able to cope by itself with the complexity and heterogeneity of its life cycle (Horn [25]) – in other words is *self-manageable*. As defined by IBM, an autonomic system has four major and four minor characteristics [27]. There are four major characteristics, referred to as "self-CHOP":

- *Self-configure* is the property to implement specific strategies to change the relations among the components to guarantee either survivability in changing environments or a higher performance.
- *Self-heal* is the property to detect (or predict) faults and automatically correct faults (events that cause the entire system or parts of it to malfunction).
- *Self-optimize* is the property of monitoring its components and fine-tune the resources automatically to optimize the performance.
- *Self-protect* is the property of anticipating, detecting, identifying, and protecting itself from attacks in order to maintain overall integrity.

Additionally, the four minor characteristics are:

- *Self-aware* means knowing itself (its components, resources, the relations among them, and the limits) in a detailed manner.
- *Self-adapt* means automatically identifying the environment, generating strategies on how to interact with neighboring systems, and adapt its behavior to a changing environment.
- *Self-evolve* means generating new strategies and implementing open standards.

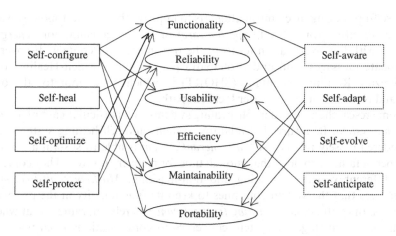

**Fig. 2.1** A self-managing system and the relationship between the basic factors of software evaluation and the characteristics of an autonomous system

- *Self-anticipate* means anticipating the requests for resources from the users without involving them in the complexity of its functionality.

The international standard ISO 9126 evaluates the software. It is divided into four parts, which address the quality model, external metrics, internal metrics, and quality in use metrics. The quality model established in the first part of the standard, ISO 9126-1, classifies software quality in a structured set of six characteristics (see e.g., the wikipedia page [58]). The relations between these characteristics and the quality factor of a system are depicted in Fig. 2.1.

## 2.2 Background

In this context we cite Robertazzi and Sarachik [48]: "A self-organizing network based on radio communications will create its own connections, topology, transmission schedules, and routing patterns in a distributed manner. It may establish local hubs, a backbone network, gateways, and relays."

The hidden terminal problem, presented in Section 2.2.1, creates a common pitfall for any self-configuring, self-healing, or self-optimizing MANET.

A self-configuring and self-organizing wireless network has two mechanisms implemented: *discovery* of routes between pair of nodes and *update* the current topology, by first detecting the node or link failures and secondly by optimizing the routes obtained through discovery. The discovery mechanism can be done proactively, when routes between any pairs of nodes are sought, periodically, or on-demand, when only certain routes are required. On updating the current topology, either single or multiple routes are maintained between a pair of nodes (more details are given in Section 2.2.2).

A self-optimizing mechanism attempts to improve the current topology with respect to either route length (*path-aware*) or energy consumption (*energy-aware*), with low overhead in terms of the transmission of control, discovery, or update messages. Gui and Mohapatra [23] presented a Self-Healing and Optimizing Routing Technique (SHORT) in which nodes monitor the on-demand routes to obtain a better local subpath.

Some researchers consider self-healing systems as a particular case of a *fault tolerant system*. Be aware that adaptive systems and self-healing systems are considered closely related. A *fault tolerant system* is able to sustain a certain number of faults, provided that enough time for recovery is given. They come in two different varieties: masking and non-masking. *Masking fault-tolerance* guarantees that the system continues to keep its functionality in the presence of faults. In contrast, *non-masking fault-tolerance* merely guarantees that when faults stop occurring, the system converges to configurations from where it continues to function. (More details are given in Section 2.2.3.)

### 2.2.1 Hidden Terminal Problem

For medium access control (MAC), the hidden terminal problem can be formulated as follows. Given four nodes, A, B, C, and D, such that A and C are not in the communication range of each other, A sends a packet to B on the same channel and at the same time when C sends a packet to D. Neither A nor C is able to determine, by itself, that a collision has occurred (see Fig. 2.2).

Numerous solutions [2, 7, 14, 19, 21, 29, 42, 47, 50, 53, 54] have been proposed for this problem. We present two representative solutions, namely the RTS/CTS approach and the timeslot approach.

The RTS/CTS (Request-to-Send/Clear-To-Send) approach [2, 7, 29], used in IEEE 801.11 wireless LAN standard, is similar to the Ethernet model. When node A wants to send a packet to node B, it first sends a RTS message to B, and

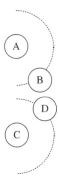

**Fig. 2.2** Hidden terminal problem

waits for B's reply as a CTS message. If B is busy or the channel is dedicated for other communication, A will not receive a CTS message, and will keep retrying until some timeout limit is reached or a CTS message is received from B. On receiving B's CTS message, A sends the data and waits for the acknowledgment.

The timeslot approach (TDMA) [21, 47, 50] provides a transmission schedule for each node in the network, which needs to be recalculated each time a node joins or leaves the network. Similar approaches concern the frequency bands and the spread spectrum code (CDMA).

The RTS/CTS approach works better for networks with unpredictable load transmission, while the timeslot approach works better for networks with more uniform load transmission per node. The timeslot approach guarantees quality of service, at the cost of high computational time for computing the schedule (when a group of nodes joins, the schedule for the whole network needs to be recomputed) and large time difference between two transmissions of the same node for densely distributed networks.

## 2.2.2 Self-Configuring and Self-Organizing MANETs

A self-configurable system must be able to extract the necessary information supporting its software intelligence from the data it collects. The need for automatic changes (self-configuring) ought to take the following design issues for a self-configuring network in consideration:

- *Ad hoc deployment*: the nodes may not be positioned in a regular pattern (grid, honeycomb, 3D grid, 3D honeycomb, etc.).
- *Error-prone wireless medium*: the wireless medium is more error-prone than the wired medium, and collisions could occur more frequently.
- *Limited resources and energy constraint*: a unit has limited resources (battery, memory, computational power). The number of actions a node executes and the time consumed by an action must be minimized, in order to prolong its battery lifetime.

Adapting to the changing environment can be done periodically or gradually. The first proposed self-configuring protocols for a MANET are protocol LCA of Baker and Ephremides [4, 5, 12], protocol DEA by Post et al. [45, 46], and protocol Layer Net, proposed by Bhatnagar and Robertazzi [8]. They periodically discard the network topology information and rebuild the network from scratch. Later protocols consider a gradual approach to self-configure a MANET, for example, the protocol SWAN by Scott and Bambos [49].

The first self-configuring (self-organizing) wireless network was proposed as a two-tier hierarchical model by Baker and Ephremides [4, 5, 12]. The nodes are classified as ordinary, cluster heads, and gateways, with the restriction that a node belongs to a single cluster and it is one hop away from its cluster head.

Since selecting the minimum number of such cluster heads is NP hard, they proposed a link cluster algorithm (LCA) for categorizing the nodes and a link activation algorithm (LAA) to schedule (activate) the links between nodes.

The LCA algorithm uses a dominating set partitioning of the network based on node ID and works as follows: the node with the highest identity number among a group of nodes without a cluster head within one hop declares itself as a cluster head. The other nodes become either gateways (if there are connected to two or more cluster heads) or ordinary nodes. Variations of the LCA algorithm are to consider either the lowest ID or the highest connected node [19, 39] instead of the highest ID node.

The distributed evolutionary algorithm (DEA) proposed by Post et al. [45, 46] is based on a clique partitioning of the network and is uniform (the same for each node in the network). It works as follows: a *starter* node activates all its neighbors, which are part of some clique as itself (so-called clique neighbors), to begin communication based on a schedule decided by itself. Then these nodes become the starter nodes for the rest of the network.

A tree layer model proposed by Bhatnagar and Robertazzi [8] builds a spanning tree starting at some starter node. Each node has a layer number equal to its distance from the root (measured in number of hops). The schedules for transmission are made one layer at a time in a kind of a breadth-first search.

In protocol SWAN (Scott and Bambos [49]), new connections are sought during random access periods; and after a timeout, connections that do not respond to a control call are declared unusable.

Konstantinou et al. [30] propose a four-layer architecture for a self-configuring network, which could be applied to a MANET. From top to down, the *Application Layer* is the one where the request for data is initiated and the collected data are processed. The *Self-configuration Layer* generates the rules (methods) to be applied in case of disconnection. The *Configuration Model Layer* is responsible for the discovery and maintenance of the network topology, and monitoring the network elements. The mobile nodes play the role of *Network Elements*.

Sehrabi and Pottie [52] present a self-configuration scheme, similar to a TDMA schedule. Li and Rus [37] present a scheme where mobile nodes modify their trajectory to transmit messages in case of disconnected ad hoc networks. Gao et al. [15, 17] propose a randomized algorithm for constructing and maintaining a CDS with low overhead, by splitting the area into grids and selecting one active node per grid. Another connected topology proposed by Gao et al. [16] is the restricted Delaunay graph (RDG) where only Delaunay edges with a limited fix transmission range are included. Alzoubi et al. [3] describe a distributed construction of a minimum connected dominating set (MCDS) for the unit-disk graphs with a constant approximation ratio. Krishnamachari et al. [33] examine some self-configuration problems, specifically a partition of the network into coordinating cliques, Hamiltonian cycle formation, and conflict-free channel allocation, related to the formation of specialized structures on the network connectivity graph.

### 2.2.3 Self-Healing

The concept of self-healing was inspired by the study of immune systems in biology (Forrest et al. [13]). *Self-healing* is the property of a system to detect that it is not operating correctly and, with or without user intervention, makes the necessary adjustments to restore itself to normal. Healing systems that require external intervention are called *assisted-healing systems*. Self-healing systems (Ghosh et al. [20]) are generally sought for software agents, grid and middleware computing. The first type of a self-healing network was proposed by Grover [22] for a digital cross-connect system (DCS). The network is a dynamic restoration-control, employing autonomous restoration algorithms. The paper of Zhang and Arora [59, 60] focuses specifically on the self-healing algorithms for cellular networks.

The states through which a system transits during its lifetime are depicted in Fig. 2.3. When the system functions properly we say that it is in an acceptable state. When a fault occurs, generally considered to be a disconnection of the network due to node movement or crashes, then the system enters a degraded state, where some parts of the network continue to function properly but some parts do not. If the functionality of the network depends on the failed portion, then the system enters a failed state. A recovery mechanism is able to bring the network into an acceptable state, for example, by moving nodes to the affected part.

A recent survey of self-healing systems of Ghosh *et al.* [20] identifies three critical issues for self-healing systems: *(a)* maintenance of the system health, *(b)* detection of system failure mechanism, and *(c)* recovery to an acceptable ("healthy") state from a degradable or failed state.

a) Maintenance of the system health can be done by:
   - *Maintaining redundancy*: Replicating components.
   - *Probing*: A special component of the system collects updated information about the other components. Michiels et al. [41] propose a DMonA architecture in which two types of sensors gather information from the functional layer of a system: *State sensors (SS)* collect data related to the internal state, and *analysis sensors (AS)* collect data related to the messages flowing through the system.

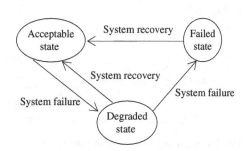

**Fig. 2.3** Transitions from the functional point of view of a MANET

- *Performance log analysis*: Self-evaluate the performance and self-fine-tune (Kaiser et al. [28]), monitor periodical for typical symptoms (Hong et al. [24])

b) Detection of system failure can be divided into:
  - *Something is missing* (i.e., a missing component, a missing response): Nagpal et al. [43] propose a strategy in which the agents can produce replications when they sense the disappearance of their neighbors. System failure can be assessed when messages (George et al. [18]) or responses to a query are not received (Aldrich et al. [1]).
  - *Some monitored value is out of range*: Merideth et al. [40] propose a model in which proactively probing a system and collecting data when a fault is detected could improve the survivability of it.
  - *A foreign element is detected*: A proactive containment strategy notifies to the system the presence of a malicious replica (Merideth et al. [40]).

c) Recovery mechanism could employ:
  - *Redundancy techniques as replicating components*: Nagpal et al. [43] propose a strategy in which the agents can produce replications when they sense the disappearance of their neighbors. During the healing process, a biological system produces more cells than necessary to combat the intrusion, so that some may survive the attack (George et al. [18]).
  - *Repair strategies*: In some cases, the faulty components are isolated and the system is reconfigured (de Lemos et al. [36]).
  - *Byzantine agreement*: For some systems, where the function of the system is to produce output results and the non-faulty components always produce the same value for the output, voting could take place and majority is applied to the output results of all processes. In that way, faulty processes can be detected and tolerated (Merideth et al. [40]). Another strategy is to delegate several agents to do the same task, and either select directly or vote on the agent that achieves the best complexity (on time, space, etc.) (Huhns et al. [26]).

Periodically, the system may check for malfunctions and, based on the existing threshold values of certain parameters, may decide that a failure occurs or a malicious agent is present. The criteria of what is considered an acceptable state may vary in time (Shaw [51]), and the key properties of a system such as its performance and accuracy may have dynamic thresholds. The steps are given in Fig. 2.4.

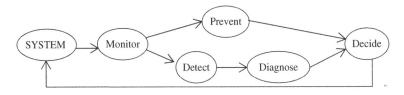

**Fig. 2.4** Steps taken by a self-healing system when a fault occurs

*Monitor*: observe the behavior of a system and the values of specific parameters (called *indicators*).

*Detect*: automatically establish when the observed behavior and indicators deviate from an established, acceptable range.

*Diagnose*: determine whether a detected deviation of a set of detected deviations could be characterized as a *fault*.

*Decide*: automatically change or intervene to repair a diagnosed fault.

*Prevent*: anticipate possible faults based on monitored indicators.

The failure detection mechanism, besides detecting the fault, can accomplish some or all of the following: gauge the degree of malfunction, assert whether the recovery mechanism is needed, assert whether the recovery mechanism is able to correct the state, and bring the system to an acceptable state.

## 2.3 Thoughts for Practitioners

To provide high-quality Internet access to mobile users using portable devices such as cell phones, laptops, PDAs, when the user is not in the proximity of a stationary access point (such as a base station), it is desirable to reach the access point via the portable devices of other users in a multihop fashion. In this way, extended coverage and better throughput for the same bandwidth are provided; consider, for example, a small area where a large number of users try to use their cell phones at the same time. The route to that access point, and the way the network of portable devices adapts when a particular access point is removed from the network, is just an example of how the current theoretical research on MANET can help design self-configuring MANETs.

In an ad hoc environment, the set of neighbors can change at arbitrary moments of time. Thus in order to increase the throughput on routing the packets, nodes should use fewer routes; these routes form the *communication backbone* of the network. At the same time, since in an ad hoc network a node can fail or move somewhere else with high frequency, the nodes that are part of the backbone must be provided with enough redundancy for communication. This is done by letting them act as alternative routers, and alternative routes have to be available before crashes affect the communication backbone. Unlike in the wired environment, bidirectional communication is not guaranteed between any pairs of nodes, since their communication range is not fixed and can also vary based on node power.

There is a tradeoff between selecting a smaller set of nodes and the power consumption of the selected nodes. Maintaining the routing infrastructure to only a subset of nodes reduces the routing overhead (the schedule of the MAC) and excessive broadcast redundancy. At the same time, to save energy, unused non-backbone nodes can go into a sleeping mode and wake up only when they have to forward data or are selected due to the failure or movement of nodes. Selecting fewer nodes to act as routers has a power consumption disadvantage:

The routers will deplete their power faster than the non-router nodes. Thus once the power level of these nodes falls below a certain threshold, those nodes can be excluded as routers, and some other nodes have to replace them.

Selecting the nodes in the communication backbone can be modeled as a linear program where for each node $i$ in the network a binary variable $y_i \in \{0,1\}$ is associated, such that $y_i = 1$ if and only if node $i$ is selected to be part of the backbone. The objective is to minimize the sum of all $y$-variables such that some properties are achieved for every node $i$ (e.g., for the connected dominating set the sum of $y_i$ and all the $y$-variables of $i$'s neighbors is greater than 1; and if two $y$-values are 1, then there is a connection between the corresponding nodes).

For all the above reasons, hierarchical structures such as link-cluster architecture and dominating sets (with varieties as connected dominating set, weakly connected dominating set [11], $k$-fold dominating set [31, 32, 34, 38, 56], $k$-dominating set [35, 44, 57], and the $k$-connected $k$-dominating set [9, 10]) are not able to provide sufficiently fast redundancy or fault-tolerance for high-speed or real-time networks, where the latency in every node should be very short. The decision time at every node needed to decide how to route messages should be comparable to the time of the propagation delay between neighboring nodes.

The $k$-fold cover $t$-set [6] provides a favorable alternative to hierarchical structures by selecting not only nodes from the immediate neighborhood but also nodes located at two or more hops. Thus the network can tolerate up to $k$ node failures without losing the routing infrastructure. Unfortunately, the $k$-fold cover set model can fit better for wireless networks with low traffic or without power constraints.

Since most of the dominating set-related problems are NP-complete, heuristics or approximation algorithms are given for some, but no implementation in an actual architectural model. It is for practitioners to implement them and compare them for the practical point of view by simulations.

## 2.4 Directions for Future Research

The topological connectivity of a MANET is variable due to the mobility of the nodes and the limited battery power at each node. This variation motivates the necessity of generating rules or actions to be executed when a change occurs in the network, and preference is given to the distributed algorithms that are able to cope with the change and adapt to it, in shorter time, with the minimum cost. It is preferred that the time to correct the failure should be proportional to the number of nodes involved in the failures.

Algorithms that require a large amount of time or number of messages for a small number of faults, thus an excessive amount of energy and bandwidth, do not scale well. At the same time, algorithms that have a low cost, but require reactivation each time a failure occurs, may disrupt the functionality of the network more frequently. Thus an amount of redundancy is required in selecting in selecting the nodes that are part of the backbone communication of the network and the routing procedure at each node, at a time when the self-healing actions are not executed, unless the network is already in a degraded state.

## 2.5  Conclusions

In this chapter, we have presented the problem of organizing a set of mobile, radio-equipped nodes, nodes that communicate through a wireless medium, into a connected network, in order to obtain a *self-configuring* or *self-organizing* network. We address the issue of how a reliable structure, once acquired by self-configuring, can be maintained when topological changes occur due to node failure, node motion, or link failure, in order to obtain a so-called *self-healing* network. In Section 2.2, we have detailed some protocols; simulations that illustrate different cases are available on the papers that have proposed them. A number of theoretical results presented in Section 2.3 have not been implemented in real networks; no protocols have been proposed; thus there is work left for practitioners to implement, and compare, from the practical point of view by simulations.

## Terminologies

*Assisted-healing system*: a healing system that requires external intervention.

*Autonomous system*: a system that is less dependent on human intervention and able to cope by itself with the complexity and heterogeneity of its life cycle.

*Energy-aware optimizing mechanism*: mechanism that attempts to improve the current topology with respect to energy consumption.

*Hidden terminal problem*: transmission from different nodes that use the same communication channel at the same time may interfere by colliding with each other, and as a result either a truncated message or some garbage is received.

*On-demand discovery mechanism*: only certain routes, required by the application layer, are sought by the discovery mechanism.

*Path-aware optimizing mechanism*: mechanism that attempts to improve the current topology with respect to route length.

*Proactive discovery mechanism*: the routes between any pairs of nodes are sought by the discovery mechanism, periodically.

*Self-configuring system*: system that changes the relations among the components to guarantee either survivability in changing environments or higher performance.

*Self-healing system*: system that detects or predicts faults, and automatically corrects faults (events that cause the entire system or parts of it to malfunction).

*Self-optimizing mechanism*: mechanism that improves the current topology with respect to either route length (*path-aware*) or energy consumption (*energy-aware*), with a low overhead in terms of the transmission of control messages; that is, messages used by discovery or updating mechanism.

## Questions

1. What are the self-CHOP characteristics of an autonomous system?
2. Exemplify one difference between a self-configuring and a self-healing system.
3. Can the hidden terminal problem occur when the message sent by node A reaches node C as well? What if the nodes B and C are the same?
4. Exemplify one difference between discovery mechanism and updating mechanism for self-configuring networks.
5. For the discovery mechanism, which is more energy-consuming, proactive or on-demand discovery?
6. A self-optimizing mechanism, which is part of an updating mechanism for self-configuring networks, attempts to improve the current topology with respect to what?
7. A scalable self-configuring mechanism must have a low overhead in terms of transmission of control messages. Why?
8. LCA is a heuristic approach to the dominating set problem, and the dominating set problem is NP-complete. DEA is a heuristic approach to the clique partition problem, and the clique partition problem is NP-complete as well. Which of the two, LCA or DEA, is better for a good approximation?
9. Exemplify two critical issues for the self-healing systems, as identified by Ghosh et al. [20].
10. What are the steps a self-healing mechanism takes when a fault occurs?

## References

1. Aldrich J, Sazawal V, Chambers C, Notkin D (2002) Architecture-centric programming for adaptive systems. Proc 1st Workshop on Self-healing Syst, 93–95
2. Alwan A, Bagrodia R, Bambos N, Gerla M, Kleinrock L, Short J, Villasenor J (1996) Adaptive mobile multimedia networks. IEEE Pers Comm 3(2):34–51
3. Alzoubi KM, Wan P-J, Frieder O (2002) Message-optimal connected-dominating-set construction for routing in mobile ad hoc networks. Proc 3rd ACM Intl Symp on Mobile Ad Hoc Networking and Computing, 157–164

4. Baker DJ, Ephremides A (1981) A distributed algorithm for organizing mobile radio telecommunication networks. Proc 2nd Intl Conf Distrib Computing Syst
5. Baker DJ, Ephremides A (1981) The architectural organization of a mobile radio network via a distributed algorithm. IEEE Trans Commun **29**:11
6. Bein D (2008) Fault-tolerant k-fold pivot routing in wireless sensor networks. Hawaii Intl Conf. Syst. Sci (HICSS), 245
7. Bharghavan V, Demers A, Shenker S, Zhang L (1994) MACAW: a media access protocol for wireless LAN's. Proc Conf Comm Architectures, Protocols, and Applications, 212–225
8. Bhatnagar A, Robertazzi TG (1990) Layer Net: a new self-organizing network protocol. Mil Comm Conf (Milcom) **2**:845–849
9. Dai F, Wu J (2005) On constructing k-connected k-dominating set in wireless networks. Proc IEEE Intl Parallel and Distributed Processing Symp (IPDPS)
10. Dai F, Wu J (2006) On constructing k-connected k-dominating set in wireless ad hoc and sensor networks. J Parallel Distrib Computing **66**(7):947–958
11. Dunbar JE, Grossman JW, Hattingh JH, Hedetniemi ST, McRae AA (1997) On weakly-connected domination in graphs. Discrete Math **167/168**:261–269
12. Ephremides A, Baker DJ (1981) An alternative algorithm for the distributed organization of mobile users into connected networks. Conf Inf Sci Syst
13. Forrest S, Hofmeyr SA, Somyaji A (1997) Computer immunology. Comm ACM **40**(10):88–96
14. Gallager R (1985) A perspective on multiaccess channels. IEEE Trans Inf Theory **31**(2):124–142
15. Gao J, Guibas LJ, Hershberger J, Zhang L, Zhu A (2001) Discrete Mobile Centers. Proc 7th Annual Symp on Computational Geom (SCG), 188–196
16. Gao J, Guibas LJ, Hershberger J, Zhang L, Zhu A (2001) Geometric spanner for routing in mobile networks. Proc 2nd ACM Intl Symp on Mobile Ad Hoc Networking and Computing (MobiHoc), 45–55
17. Gao J, Guibas LJ, Hershberger J, Zhang L, Zhu A (2003) Discrete mobile centers. Discrete Comput Geom **30**(1):45–65
18. George S, Evans D, Marchette S (2003) A biological programming model for self-healing. 1st ACM Workshop on Survivable and Self-regenerative Syst, 72–81
19. Gerla M, Tsai JT-C (1995) Multicluster, mobile, multimedia radio network. Wireless Networks **1**(3):255–265
20. Ghosh D, Sharman R, Rao HR, Upadhyaya S (2007) Self-healing systems – survey and synthesis. Decis Support Syst **42**:2164–2185
21. Goodman DJ, Valenzuela RA, Gayliard KT, Ramamurthi B (1989) Packet reservation multiple access for local wireless communications. IEEE Trans Comm **37**(8):885–890
22. Grover WD (1987) The selfhealing network: a fast distributed restoration technique for networks using digital cross-connect machines. IEEE Globecom
23. Gui C, Mohapatra P (2003) SHORT: self-healing and optimizing routing techniques for mobile ad hoc networks. Proc 4th ACM Intl Symp on Mobile Ad Hoc Networking and Computing (MobiHoc), 279–290
24. Hong Y, Chen D, Li L, Trivedi K S (2002) Closed loop design for software rejuvenation. Workshop on Self-healing, Adaptive and Self-managed Syst (SHAMAN), 159–170
25. Horn P (2001) Autonomic computing: IBM's perspective on the state of information technology.    http://www-1.ibm.com/industries/government/doc/content/bin/auto.pdf. Accessed 20 February 2008
26. Huhns MN, Holderfield VT, Gutierrez RLZ (2003) Robust software via agent-based redundancy. Proc 2nd Intl Joint Conf on Autonomous Agents and Multiagent Syst (AAMAS), 1018–1019
27. IBM (2008) Autonomic computing: The 8 Elements. http://researchweb.watson.ibm.com/autonomic/overview/elements.html. Accessed 20 February 2008

28. Kaiser G, Gross P, Kc G, Parekh J, Valeto G (2002) An approach to autonomizing legacy systems. Workshop on Self-healing, Adaptive and Self-managed Syst (SHAMAN)

29. Karn P (1990) MACA-a new channel access method for packet radio. 9th Computer Networking Conf- ARRL/CRRL Amateur Radio

30. Konstantinou AV, Florissi D, Yemini Y (2002) Towards self-configuring networks. Proc DARPA Active Networks Conf and Exposition (DANCE), 143

31. Kratochvil J (1995) Problems discussed at the Workshop on Cycles and Colourings, http://univ.science.upsj.sk/c&c/rhistory/cc95prob.htm, Accessed 20 February 2008

32. Kratochvil J, Manuel P, Miller M, Proskurowski A (1998) Disjoint and fold domination in graphs. Australas J Combinatorics 18:277–292

33. Krishnamachari B, Wicker S, Bejar R, Fernandez C (2003) On the complexity of distributed self-configuration in wireless networks. Telecomm Syst 22(1–4): 33–49

34. Kuhn F, Moscibroda T, Wattendorf R (2006) Fault-tolerant clustering in ad hoc and sensor networks. Proc 26th IEEE Intl Conf on Distributed Computing Syst (ICDCS), 68

35. Kutten S, Peleg D (1998) Fast distributed construction of small k-dominating sets and applications. J Algorithms, 40–66

36. de Lemos R, Fiadeiro JL (2002) An architectural support for self-adaptive software for treating faults. Proc 1st Workshop on Self-healing Syst, 39–42

37. Li Q, Rus D (2000) Sending messages to disconnected users in disconnected ad hoc mobile networks. Proc 6th Annual ACM/IEEE Intl Symp on Mobile Computing and Networking (MobiCom), 44–55

38. Liao CS, Chang GJ (2003) k-tuple domination in graphs. Inf Processing Letters 87:45–50

39. Lin CR, Gerla M (1997) Adaptive clustering for mobile wireless networks. IEEE J Selected Areas of Comm 15:1265–1275

40. Merideth MG, Narasimhan P (2003) Proactive containment of malice in survivable distributed systems. Proc Intl Conf on Security and Management, 3–9

41. Michiels S, Desmet L, Janssens N, Mahieu T, Verbaeten P (2002) Self-adapting concurrency: the DMonA architecture. Proc 1st Workshop on Self-healing Systems, 43–48

42. Morrow Jr. RK, Lehnert JS (1992) Packet throughput in slotted ALOHA DS/SSMA radio systems with random signature sequences. IEEE Trans Comm 40(7):1223–1230

43. Nagpal R, Kodancs A, Chang C (2003) Programming methodology for biologically-inspired self-assembling systems. AAAI Spring Symp on Computational Synthesis

44. Penso LD, Barbosa VC (2004) A distributed algorithm to find k-dominating sets. Discrete Applied Math 141(1–3):243–253

45. Post MJ, Kershenbaum AS, Sarachik PE (1985) A distributed evolutionary algorithm for reorganizing network communications. Mil Comm Conf (Milcom)

46. Post MJ, Kershenbaum AS, Sarachik PE (1985) A biased greedy algorithm for scheduling multi-hop radio networks. Proc Conf Inf Sci Syst, 564–572

47. Rajendran V, Obraczka K, Garcia-Luna-Aceves JJ (2006) Energy-efficient, collision-free medium access control for wireless sensor networks. Wireless Networks 12(1):63–78

48. Robertazzi TG, Sarachik PE (1986) Self-organizing communication networks. IEEE Comm Magazine 24(1):28–33

49. Scott K, Bambos N (1997) Formation and maintenance of self-organizing wireless networks. 31st Asilomar Conf on Signals, Syst and Computers 1:31–35

50. Shakkottai S, Rappaport TS, Karlsson PC (2003) Cross-layer design for wireless networks. IEEE Comm Magazine 41(10):74–80

51. Shaw M (2002) Self-healing: softening precision to avoid brittleness. Workshop on Self-Healing Syst (WOSS), 111–114

52. Sohrabi K, Pottie G (1999) Performance of a novel self-organizing protocol for wireless ad hoc sensor networks. Proc IEEE Vehicular Technology Conf, 1222–1226

53. Sousa ES, Silvester JA (1988) Spreading code protocols for distributed spread-spectrum packet radio networks. IEEE Trans Comm 36(3):272–281

54. Storey JS, Tobagi FA (1989) Throughput performance of an unslotted direct-sequence SSMA packet radio network. IEEE Trans Comm **37**(8):814–823
55. Tobagi FA, Kleinrock L (1975) Packet switching in radio channels: Part II – The hidden terminal problem in carrier sense multiple access and the busy tone solution. IEEE Trans Comm **23**:1417–1433
56. Vazirani V (2001) Approximation Algorithms. Morgan Kaufmann Publishers, Springer Verlag
57. Wang FH, Chang JM, Wang YL, Huang SJ (2003) Distributed algorithms for finding the unique minimum distance dominating set in split-stars. J Parallel Distrib Comput **63**:481–487
58. Wikipedia (2008) ISO 9126. http://en.wikipedia.org/wiki/ISO_9126. Accessed 20 February
59. Zhang H, Arora A. (2002) GS$^3$: scalable self-configuration and self-healing in wireless sensor networks. 21st ACM Symp on Principles of Distributed Computing (PODC)
60. Zhang H, Arora A (2003) GS$^3$: scalable self-configuration and self-healing in wireless sensor networks. Computer Networks **43**(4):459–480

# Chapter 3
# Cooperation in Mobile Ad Hoc Networks

**Jiangyi Hu and Mike Burmester**

**Abstract** Mobile ad hoc networks (MANETs) are collections of self-organizing mobile nodes with dynamic topologies and no fixed infrastructure [1, 2]. Cooperation among nodes is fundamental to the function of a MANET. However, nodes in a MANET are autonomous and independent wireless devices. Due to the lack of infrastructure, the constraints of resources at each node, and the ad hoc nature of nodes, we cannot assume that every node behaves as the protocol requires. This chapter presents a detailed study on the recent advances in stimulating cooperation in MANETs. Virtual currency systems and reputation systems are described, followed by a discussion of the directions for future research.

## 3.1 Introduction

As wireless devices get smaller, cheaper, and more sophisticated, they become more ubiquitous and organizations are looking for inexpensive ways to connect them. Mobile ad hoc networks (MANETs) are paradigms for wireless communication in which mobile nodes are dynamically and arbitrarily located. Such networks are self-forming and self-organizing as shown in Fig. 3.1.

In a MANET, nodes are free to move randomly and organize themselves arbitrarily; thus, the network's wireless topology may change rapidly and unpredictably. In such networks, communication is achieved by forwarding packets via intermediate nodes on routes that link the source and the destination. Routes are typically determined by using on-demand routing protocols, such as the Dynamic Source Routing (DSR) [3] or the Ad hoc On-Demand Distance Vector Routing (AODV) [4], that generate routing information only when a source node initiates a transmission.

J. Hu (✉)

Computer Science Department, Florida State University, Tallahassee, FL 32306, USA

e-mail: jiangyhu@cs.fsu.edu

S. Misra et al. (eds.), *Guide to Wireless Ad Hoc Networks*,
Computer Communications and Networks, DOI 10.1007/978-1-84800-328-6_3,
© Springer-Verlag London Limited 2009

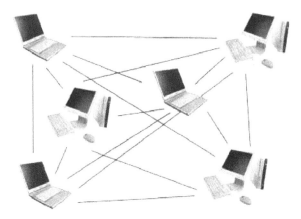

**Fig. 3.1** Mobile Ad hoc Network

Nodes in a MANET do not have a priori knowledge of the network topology. They have to discover it. A node will find its local topology by broadcasting its presence, and listening to broadcast announcements from its neighbors. As time goes on, each node gets to know about all other nodes and finds one or more ways to reach them. End-to-end communication in a MANET does not rely on any underlying static network infrastructure but requires routing via several intermediate nodes.

MANETs are suited for use in situations where a network infrastructure is unavailable. For example, in some business environments, the need for collaborative computing might be more important outside the office environment than inside. A MANET can also be used to provide crisis management services applications, such as in disaster recovery, where the entire communication infrastructure is destroyed and re-establishing communication quickly is crucial. MANETs also meet the requirements for military applications, such as rapid network formation, extended operating range, and survivability. Supporting flexible and adaptive applications with no fixed infrastructure, MANETs are expected to play an important role in the future.

Although all networking functions of a MANET, such as routing and packet forwarding, are realized through node cooperation, this cannot be taken for granted. The nodes in MANETs are usually battery-powered mobile devices with constrained resources, and energy is a precious resource for them. In addition, nodes (users) are self-interested and rational; there is no guarantee that they are willing to consume resources even though they may not wish to violate the protocol instructions [5]. A node spends energy in transmitting, receiving, and processing traffic: providing service to other nodes consumes energy and does not have any direct advantages. Thus, service provision is not in the interest of the nodes. Nodes may behave selfishly to maximize their own welfare. Since the functionality of MANETs is highly dependent on the cooperation of all available nodes, selfish behavior may affect the performance of the network severely. Indeed, selfishness is the dominant type of uncooperative behavior in a civilian MANET.

Selfish nodes in MANETs may gain distinct advantages [5], such as:

- Exploit the service of cooperating nodes;
- Exploit incentive measures to gain monetary benefits;
- Preserve resource and power;
- Prevent other nodes from getting proper service.

## 3.2 Background

Schemes proposed in the literature that stimulate cooperation and mitigate the detrimental effect of uncooperative behavior in a MANET can be classified as (1) *virtual currency systems* and (2) *reputation systems.*

### 3.2.1 Virtual Currency Systems

*Virtual currency systems* [6–11] use some forms of incentive to compensate for the service of a node. A node receives a virtual payment for serving the network and uses the incentives to gain service from the network. Two examples of such systems are *Nuglets* [6–10] and *Sprite* [11].

#### 3.2.1.1 Nuglets

Buttyan and Hubaux introduced a virtual currency, called *nuglets*, and presented a mechanism of charging/rewarding service usage/provision to stimulate cooperation in MANET.

They assumed that each node belongs to a different authority and tries to maximize the benefit it gets from the network. A node that uses a service must pay for it (in nuglets) to the nodes that provide it. Every node is interested in increasing its number of nuglets, and the only way to achieve this is to provide service to others. Two models were presented for using nuglets: a *packet purse model*, in which the source node of the packet is charged, and a *packet trade model*, in which the destination node is charged [6, 7].

In the *packet purse model*, when sending the packet, the source loads it with a number of nuglets sufficient to reach the destination. Each intermediate node takes some nuglets for the forwarding service depending on the amount of energy used, the current battery status of the forwarding node, and its current number of nuglets. If a packet does not have enough nuglets to be forwarded, then it is discarded. The main advantages of this model are: (1) it stimulates cooperation; (2) it deters nodes from sending useless data and overloading the network. The major disadvantage is that it is difficult for the source node to estimate the number of nuglets that are required to reach a given destination.

In the *packet trade model*, packets are traded for nuglets by intermediate nodes. Each intermediary node "buys" the packet from the previous node for some nuglets and "sells" it to the next node for more nuglets. In this way, every intermediate node gains nuglets for forwarding, and the total cost of forwarding the packet is paid by the destination node. The advantage of this model is the source node does not have to know in advance the number of nuglets required to deliver a packet. Instead, the destination node, which gets the packet, will pay for the service. A serious disadvantage is that this method cannot directly prevent the nodes from overloading the network. Malicious nodes can overload the network by sending the useless data without being charged.

To control the number of nuglets that are charged for packet forwarding, the authors proposed two methods: *fixed-per-hop charge* and *auctions*. We will discuss them using the *packet purse model* as an example.

With the fixed-per-hop charge, each forwarding node acquires exactly the same number of nuglets for the forwarding operation. The reward does not depend on the amount of energy used, the state of the battery, or other factors. This approach is simple to implement and can easily be added to any existing routing algorithm. But it is not flexible.

With auctions, each forwarding node runs a sealed bid price auction to determine the next hop. The bidders, which are the potential next hop neighbors to the destination node of the packet, determine a price for which they are willing to forward the packet, and send it to the forwarding node in a sealed form. When the forwarding node receives all the bids, it determines the winner of the auction, which offers the lowest bid. The advantage of this approach is that it tries to minimize the number of nuglets spent during the delivery of packets, and the lifetime of the network can be lengthened by routing the traffic in such a way that the energy consumption is balanced among the nodes. But this approach is complex and causes considerable overhead in terms of both bandwidth and latency. In addition, it can only be incorporated with multipath routing algorithms in which the nodes have multiple paths for a same destination.

### 3.2.1.2 Sprite

Zhong et al. proposed Sprite [11], a simple, cheat-proof, credit-based system for MANETs. In Sprite, every node is economically rational and its objective is to maximize the benefit of its actions minus the cost of its actions. Sprite uses credit to provide incentives for mobile nodes to cooperate and report actions honestly.

A *Credit Clearance Service* (CCS) is introduced to determine the charge and credit to each node involved in the transmission of a message. When a node receives a message, the node keeps a receipt of the message and later reports it to the CCS when the node has a fast connection with the CCS. Payments and charges are determined from a game theory perspective. Figure 3.2 shows the architecture of Sprite.

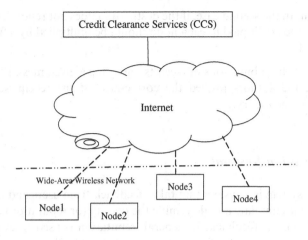

**Fig. 3.2** The Architecture of Sprite

In order to prevent a denial-of-service attack to the destination by sending it a large amount of traffic, the sender of the packet is charged instead of the destination. A node that has tried to forward a message is compensated, but the credit that a node receives depends on whether or not its forwarding action is successful. Forwarding is considered successful if and only if the next node on the path reports a valid receipt to the CCS.

Three selfish actions and the corresponding countermeasures are discussed in Zhong et al. [11]:

- After receiving a message, a selfish node may save a receipt but does not forward the message. To prevent this, the CCS should give more credit to a node that forwards a message than to a node that does not forward a message. If the destination does not submit a receipt, the CCS first determines the last node on the path that has ever received the message. Then the CCS pays this last node less than it pays each of the predecessors of the last node.
- After receiving a message a node may not report the receipt. This is possible if the sender colludes with the intermediate nodes, so that the sender can pay the node a behind-the-scene compensation, which is a little bit more than the CCS will pay, and the sender still get a net gain. In order to prevent this cheating action, the CCS charges the sender an extra amount of credit if the destination does not report the receipt so that colluding groups get no benefit.
- Since reporting a receipt to the CCS is sufficient for getting credit, a group of colluding nodes may forward only the receipt of a message, instead of forwarding the whole message, to its successor. Two cases are considered: (1) the destination colludes with the intermediate nodes; (2) the destination does not collude with the intermediate nodes. In the first case, if the destination really submits the receipt, the intermediate nodes and the destination should be paid as if no cheating had happened since the message is for the

destination. In the second case, if the destination does not report a receipt of a message, the credit paid to each node should be multiplied by a fraction, $r$, where $r < 1$.

By modeling the submissions of receipts regarding a given message as a one-round game, the authors proved the correctness of the receipt-submission system using game theory.

## *3.2.2 Reputation Systems*

A reputation system is a system that takes feedback from users and provides a mechanism to accumulate and determine the quality (or reputation) of a given source based on this feedback. In general, reputation is used to evaluate the trust of an entity. The goals of a reputation system are [12]:

- To provide information to distinguish a trustworthy principal from an untrustworthy principal.
- To encourage principals to act in a trustworthy manner.
- To discourage untrustworthy principals from participating in the service that the reputation mechanism protects.

Reputation mechanisms that are applied to MANETs to address threats arising from uncooperative nodes rely on neighbor monitoring to dynamically assess the trustworthiness of neighbor nodes and exclude untrustworthy nodes.

Several reputation systems have been proposed to mitigate selfishness and stimulate cooperation in MANET, including CONFIDANT [12–15], CORE [16] and OCEAN [17].

### 3.2.2.1 CONFIDANT

CONFIDANT [12–15] stands for Cooperation Of Nodes: Fairness In Dynamic Ad-hoc Network. This scheme works as an extension to on-demand routing protocols. It aims at detecting and isolating uncooperative nodes, thus making it unattractive to deny cooperation. Reputation is used to evaluate routing and forwarding behavior according to the network protocol.

With CONFIDANT, each node has four components: a monitor, a trust manager, a reputation system, and a path manager. These components interact to provide and process protocol information as shown in Fig. 3.3.

- The monitor is the equivalent of a "neighbor watch", where nodes locally monitor deviating behavior.
- The trust manager makes decisions about providing or accepting route information, accepting a node as part of a route, or taking part in a route originated by another node. It consists of the following components:

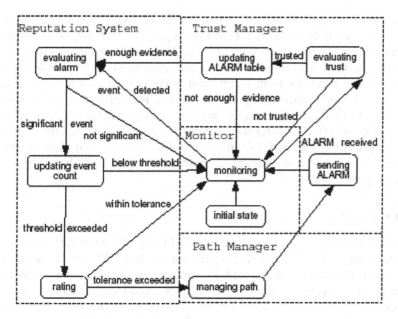

**Fig. 3.3** The interaction of the components in CONFIDANT

   o  An alarm table containing information about received alarms.

   o  A trust table managing the trust levels for nodes to determine the trustworthiness of an alarm.

   o  A friends list containing all the "friends" that the node may send
alarms to.

• The reputation system in CONFIDANT manages a table consisting of
entries for nodes and their rating. The rating is changed only when there
is sufficient evidence of uncooperative node behavior that has occurred
sufficiently many times (exceeding a threshold) to rule out coincidences.

• The path manager performs the following functions:

   o  Re-ranks path according to the reputation of the nodes in the path.

   o  Deletes paths containing uncooperative nodes.

   o  Takes actions on receiving a request for a route from an uncooperative
node (e.g., ignore, do not send any reply).

   o  Takes actions on receiving a request for a route containing an uncooperative node in the source route (e.g., ignore, alter the source).

ALARM messages are sent by the trust manager of a node to warn others of
uncooperative nodes. Outgoing ALARM messages are generated by the node
itself after having experienced, observed, or received a report of uncooperative
behavior. The recipients of these ALARM messages are so-called friends, and
are administered in a friends list.

As shown in Fig. 3.3, each node monitors the behavior of its neighbors. If a suspicious event is detected, the information is given to the reputation system. If the event is significant for the node, it is checked whether it has occurred more often than a predefined threshold that is high enough to distinguish deliberate uncooperative behavior from simple coincidences such as collisions. What constitutes a significance rating can be defined for different types of nodes according to their security requirements. If a certain threshold is exceeded, the reputation system updates the rating of the node that caused the event. If the rating turns out to be intolerable, the information is relayed to the path manager, which proceeds to delete all routes containing the misbehaving node from the path cache.

### 3.2.2.2 CORE

Michiardi et al. proposed a mechanism called CORE (COllaborative REputation mechanism) to enforce node cooperation in a MANET [16]. It is a generic mechanism that can be integrated with any network function like packet forwarding, route discovery, network management, and location management.

CORE stimulates node cooperation by using a collaborative monitoring technique and a reputation mechanism. In this mechanism, reputation is a measure of someone's contribution to network operations. Members that have a good reputation can use the resources while members with a bad reputation, because they refused to cooperate, are gradually excluded from the community.

CORE defines three types of reputation [16, 17]:

1. Subjective reputation, that is a reputation value which is locally calculated based on direct observation.
2. Indirect reputation, that is second-hand reputation information which is established by other nodes. To eliminate an attack where a malicious node disseminates false-negative reputation information, only positive reputation information is distributed in CORE.
3. Functional reputation, that is related to a certain function, where each function is given a weight as to its importance. For example, data packet forwarding may be deemed to be more important than forwarding packets with route information, so data packet forwarding will be given greater weight in the reputation calculations.

Each node computes a reputation value for every neighbor using a sophisticated reputation mechanism that differentiates between subjective reputation, indirect reputation, and functional reputation. In contrast to CONFIDANT, which uses only negative values to represent reputation, CORE uses a negative to positive range to assess reputation. Each calculation is normalized so that reputation ranges from $-1$ (bad) to $+1$ (good). 0 represents a neutral view and is used when there are not enough observations to make an accurate assessment of a node's reputation. Having a positive to negative range allows good behavior to be rewarded and bad behavior to be punished. By placing more weight on past behavior, the CORE scheme is tolerant of sporadically bad behavior. However, such an approach is vulnerable to an attack where a node can build up a good reputation before behaving uncooperatively for a period. The better the reputation the uncooperative node can build up, the more time the node will have in which it can fail to meet its obligations while its reputation is still positive.

CORE consists of two basic components: a watchdog mechanism and a reputation table.

The watchdog mechanism [18] is used to detect misbehavior nodes. When a node forwards a packet, the node's watchdog listens promiscuously to monitor if the next node in the path also forwards the packet. If the next node does not forward the packet, then it is considered as misbehaving.

The reputation table is a data structure stored in each node. Each row of the table consists of four entries: the unique identifier of the entity, a collection of recent subjective observations made on that entity's behavior, a list of the recent indirect reputation values provided by other entities, and the value of the reputation evaluated for a predefined function.

A reputation table is updated in the following ways:

- Local calculation of a node's subjective reputation. If a node refuses to cooperate, then the CORE scheme will decrease the reputation of the node, leading to its exclusion if the non-cooperative behavior persists.
- Global distribution of reputation takes place within a reputation dissemination phase. This phase involves sending messages containing a list of nodes that have successfully cooperated in providing a function, i.e., a list of nodes with positive reputations.
- Reputations are gradually decreased to a null value if there is no interaction with the observed node.

### 3.2.2.3 OCEAN

S. Bansal et al. proposed an Observation-based Cooperation Enforcement in Ad hoc Networks (OCEAN) [17]. In contrast to CONFIDANT and CORE, OCEAN avoids indirect (second-hand) reputation information and uses only direct first-hand observations of other nodes' behavior. A node makes routing decisions based solely on the direct observations of its neighboring nodes' interaction.

In OCEAN, the rating of each node is initialized to Neutral (0), with every positive action resulting in an increment ($+1$) of the rating, and every negative action resulting in a decrement ($-2$) of the rating. Once the rating of a node falls below a certain faulty threshold ($-40$), the node is added to a faulty list. The faulty list represents a list of misbehaving nodes.

OCEAN has five components residing in each node to detect and mitigate misbehavior.

- NeighborWatch observes the behavior of the neighbors of a node. It works the same way as watchdog [18]. Whenever misbehavior is detected, NeighborWatch reports to the RouteRanker, which maintains ratings of the neighbor nodes.
- RouteRanker maintains a rating for each of its neighboring nodes. The rating is initialized to Neutral and is incremented and decremented based on the observed events from the NeighborWatch component.
- Rank-Based Routing uses the information from NeighborWatch to make the decision of selection of routes. An additional field, called the avoid-list, is added to the DSR Route-Request Packet (RREQ) to avoid routes containing nodes in the faulty list.
- Malicious Traffic Rejection rejects traffic from nodes that are considered misbehaving. All traffic from a misbehaving node is rejected so that a node is not able to relay its own traffic under the guise of forwarding it on.
- Second Chance Mechanism allows nodes previously considered misbehaving to become useful again. A timeout approach is used where a misbehaving node is removed from the faulty list after a fixed period of inactivity. Even though the node is removed from the faulty list, its rating is not increased so that it can quickly be added back to the faulty list if it continues the misbehavior.

OCEAN focuses on the robustness of packet forwarding: maintaining the overall packet throughput of a MANET with the existence of misbehaving nodes at the routing layer. OCEAN's approach is to disallow any second-hand reputation exchanges.

Routing decisions are made based solely on the direct observations of neighboring nodes' behavior. This eliminates most trust management complexity.

## 3.3 Thoughts for Practitioners

Currently, virtual currency systems and reputation systems are the main solutions to prohibit selfish behavior and stimulate node cooperation in MANETs. Virtual currency systems use incentives to encourage node providing service for others. Reputation systems deter node selfishness by evaluating and updating the reputation of every node. However, there are some issues with these systems.

For virtual currency systems, how to represent and distribute virtual currency is an inevitable problem. In Nuglets, to implement either the *packet purse model* or the *packet trade model*, tamper-proof hardware is required at each node to prevent the node from illegitimately increasing its own nuglets and to ensure that the correct amount of nuglets is deducted or credited at each node. Sprite uses a Credit Clearance Server to determine the charge and credit to each node. However, a centralized server is not only inappropriate for self-organized MANETs but also may become a target for single point of failure attack.

Another issue with virtual currency systems is the location privilege problem [19]. Nodes in the center of the network are privileged compared to the nodes in the periphery of the network, which have fewer packets to forward, even if they are willing to do so. Nodes in the periphery of the network will have less chance to earn virtual currency and will run out of their credit soon.

Reputation systems use reputation as a prediction for future quality of service. Since reputation is not a tangible property, there are some issues that should be considered when using it in MANETs. First, in most reputation systems, reputation information is shared and dissimulated among the nodes. Without a priori "trust" relationship or a trusted management mechanism, these systems may be destabilized by false rating, either false accusation or false praise. Second, a node in a reputation system has to keep, update, and share reputation information; this causes more overhead both for individual nodes and for the network.

## 3.4 Directions for Future Research

Since MANETs are uniquely different from the networks with fixed infrastructure, they are more vulnerable to security attacks [20, 21]. There are some other factors that could be taken into account when designing systems aimed at stimulating cooperation:

*Identification.* The distinctness of entities is a fundamental requirement for a MANET, and identity is used to represent entities. However, MANETs are subject to the Sybil attack [22] in which a node presents multiple identities for malicious intent. The Sybil attack undermines the assumed mapping between identities and entities. The vulnerability of a MANET to the Sybil attacks depends on how cheaply identities can be generated, the degree to which the MANET accepts inputs from entities that do not have a chain of trust linking them to a trusted entity, and whether the MANET treats all entities identically. The lack of fixed infrastructure and the dynamic network topology require that the identity validation for a MANET does not rely on any centralized authority and works efficiently even with frequent node movement.

*Trust management.* Trust is an essential requirement to enable security in open network environment. In a MANET where there is no fixed infrastructure and the topology changes dynamically, communication can only be achieved via trusted routes. Traditional security mechanisms typically focus on protecting resources from malicious users and restricting access to authorized users only. However, in MANETs, network entities may act deceitfully by providing false or misleading information. Traditional security mechanisms are unable to protect against this type of threat. In addition, due to the increased possibilities of exploiting existing vulnerabilities and creating new threats, it poses new problems on MANETs in terms of possible attack scenarios, threats, menaces, and damages. In order to achieve security and survivability in MANETs, there is a demand for a reliable establishment and maintenance of trust relationships among the users. Trust management is very important for authentication, authorization, privacy, and other security applications in MANETs.

*Complexity.* Systems that stimulate node cooperation in MANETs inevitably cause overhead to the network. As some or all of the nodes in a MANET may rely on batteries or other exhaustible means for their energy, the most important system design criteria for optimization may be energy conservation. This also applies to nodes that do not have any internal power and rely on induced power or scavenging. A successful approach strikes the right balance between sufficient stimulation among the node to achieve better performance, without unnecessarily wasting too much resource.

*Special cases.* Although node selfishness has a detrimental effect on network performance, it can be justified in certain circumstances. A node may behave selfishly to save power for critical applications or to extend its lifetime. This is especially true if the role or position of the node is critical. For example, if the node is the only one connecting two groups, then it may put high priority on critical communications by sacrificing other traffic.

## 3.5  Conclusions

Selfish node behavior is common in MANETs because the resource of a mobile node is limited. It degrades the network performance and may even disable the whole network system. Currently, virtual currency systems and reputation systems are the main solutions proposed in the literature that stimulate cooperation and mitigate the detrimental effect of uncooperative behavior in MANETs.

Virtual currency systems use some form of incentive to enforce nodes' cooperation. Nodes get the incentives upon serving the network and use them to gain service from the network. Reputation systems, on the other hand, use the nodes' reputation to mitigate selfish behavior. Nodes maintain the reputation of other nodes based on direct observation or the exchange of reputation messages with other nodes. Future research on node cooperation should take other factors into account, such as identification, trust management, complexity, etc.

## Terminologies

*Selfish behavior:* In MANETs, nodes are often battery-powered and energy is a precious resource for them. In addition, nodes (users) are self-interested and rational; there is no guarantee that they are willing to consume resource in providing service to other nodes. Selfish behavior refers to non-cooperation in certain network operations.

*Nuglets:* Nuglets is a virtual currency system that stimulates node cooperation in a MANET. Nodes are rewarded or charged (in muglets) for providing or getting services. There are two models that use nuglets: packet purse model and packet trade model.

*Location privilege problem:* In virtual currency systems, nodes in the center of the network are privileged compared to the nodes in the periphery of the network, which have fewer packets to forward, even if they are willing to do so. Nodes in the periphery of the network will have less chance to earn virtual currency and will run out of their credit soon.

*Virtual currency systems:* Virtual currency systems use credit or micro payments to compensate for the service of a node, thus stimulating node cooperation in MANETs. A node receives a virtual payment for forwarding messages of another node, and this payment is charged from the sender. Examples of virtual currency systems are Nuglets and Sprite.

*Reputation systems:* A reputation system is a system that takes feedback from users and provides a mechanism to accumulate and determine the quality (or reputation) of a given source based on this feedback. Reputation systems used in MANETs to deter node selfishness include: CONFIDANT, CORE, OCEAN, etc.

*Watchdog:* The communication between wireless devices is broadcast. That is, when a node sends a message, all nodes within broadcast range distance can receive it. The watchdog mechanism takes advantage of this feature to detect misbehavior nodes. Usually, when a node forwards a packet, the node's watchdog listens promiscuously to monitor if the next node in the path also forwards the packet.

*CONFIDANT:* CONFIDANT stands for Cooperation Of Nodes: Fairness In Dynamic Ad-hoc Network. It works as an extension to on-demand routing protocols. It aims at detecting and isolating uncooperative nodes, thus making it unattractive to deny cooperation. Reputation is used to evaluate routing and forwarding behavior according to the network protocol. With CONFIDANT, each node has four components: a monitor, a trust manager, a reputation system, and a path manager. The components work interactively to detect and isolate selfish nodes.

*CORE:* CORE stands for COllaborative REputation mechanism. It stimulates node cooperation by using a collaborative monitoring technique and a reputation mechanism. It is a generic mechanism that can be integrated with any network function like packet forwarding, route discovery, network management, and location management.

*Sprite:* Sprite is a simple, cheat-proof, credit-based system for MANETs. In Sprite, every node is economically rational and its objective is to maximize the benefit of its actions minus the cost of its actions. Sprite uses credit to provide incentives for mobile nodes to cooperate and report actions honestly.

*The Sybil attack:* In MANETs, a node may present multiple identities for malicious intent. The Sybil attack undermines the assumed mapping between identities and entities.

## Questions

1. Why do nodes in a MANET behave selfishly?
2. What advantages do selfish nodes gain?
3. Why is node cooperation important in a MANET?
4. How does virtual currency systems stimulate cooperation in a MANET?
5. What are the models for nuglets?
6. How does packet purse model and packet trade model work?
7. What is the purpose of the CCS in Sprite?
8. How do reputation systems detect selfish nodes?
9. What are the advantages and disadvantages of using indirect reputation information in a reputation system?
10. In what circumstances can selfish behavior be justified?

## References

1. E. M. Belding-Royer and C. K. Toh. A review of current routing protocols for ad-hoc MANETs. IEEE Personal Communications Magazine, pp. 46–55, 1999.
2. C. E. Perkins. Ad hoc networking. Addison-Wesley, 2001.
3. D. B. Johnson and D. A. Maltz. Dynamic source routing in ad hoc wireless networks. Mobile Computing. pp. 153–181, 1996.
4. C. E. Perkins and E. M. Royer. Ad hoc on-demand distance vector routing. Proceedings of the 2nd IEEE Workshop on Mobile Computing Systems and Applications, 1999.
5. M. Conti, E. Gregori, and G. Maselli. Cooperation issues in mobile ad hoc networks. 24th International Conference on Distributed Computing Systems Workshops, pp. 803–808, 2004.
6. L. Buttyan and J.-P. Hubaux. Stimulating cooperation in self-organizing mobile ad hoc networks. Technical Report DSC/2001/046, Swiss Federal Institute of Technology, 2001.
7. L. Buttyan and J.-P. Hubaux. Enforce srvice availability in mobile ad-hoc networks. Proceedings of MOBIHOC 2000, 2000.
8. J.-P. Hubaux and L. Buttyan. Toward mobile ad-hoc wans: Terminodes. Technical Report DSC/2000/006, Swiss Federal Institute of Technology, 2000.
9. J.-P. Hubaux and L. Buttyn. Nuglets: A virtual currency to stimulate cooperation in self organized mobile ad hoc networks. Technical Report DSC/2001/001, Swiss Federal Institute of Technology, 2001.
10. J.-P. Hubaux, M. Jakobsson, N.-B. Salem, and L. Buttyan. A charging and rewarding scheme for packet forwarding in multi-hop cellular networks. Proceedings of MOBIHOC 2003, pp. 13–24, 2003.
11. S. Zhong, J. Chen, and Y. R. Yang. Sprite: A simple, cheat-proof, credit-based system for mobile ad hoc networks. Proceedings of IEEE Infocom '03, 2003
12. S. Buchegger and J.-Y. Le Boudec. Performance analysis of the CONFIDANT protocol: Cooperation of nodes fairness in dynamic ad-hoc networks. Proceedings of IEEE/ACM Symposium on Mobile Ad Hoc Networking and Computing, 2002.
13. S. Buchegger and J. Y. Le Boudec. The effect of rumor spreading in reputation systems for mobile ad-hoc networks. Proceedings of WiOpt'03: Modeling and Optimization in Mobile, Ad Hoc and Wireless Networks, 2003.
14. S. Buchegger and J.-Y. Le Boudec. Coping with false accusations in misbehavior reputation systems for mobile ad-hoc network. Technical Report IC/2003/31, EPFL, 2003.

15. S. Buchegger and J.-Y. Le Boudec. A robust reputation system for p2p and mobile ad-hoc networks. Proceedings of the Second Workshop on the Economics of Peer-to-Peer Systems, 2004.
16. R. Molva and P. Michiardi. Core: A collaborative reputation mechanism to enforce node cooperation in mobile ad hoc networks. IFIP-Communicatin and Multimedia Securtiy Conference, 2002.
17. S. Bansal and M. Baker. Observation-based cooperation enforcement in ad hoc networks. Technical report 072003, Stanford University, 2007.
18. S. Marti, T. J. Giuli, K. Lai, and M. Baker. Mitigating routing misbehavior in mobile ad hoc networks. Proceedings of MOBICOM 2000, pp. 255–265, 2000
19. Y. Wang, V. C. Giruka, and M. Singhal. A fair distributed solution for selfish nodes problem in wireless ad hoc networks. Proceedings of Ad-Hoc, Mobile, and Wireless Networks: Third International Conference, ADHOC-NOW, pp. 211–224, 2004
20. F. Y. Loo. Ad hoc network: Prospects and challenges, Technical Report D1, AML, University of Tokyo, Japan, 2004.
21. P. Michiardi and R. Molva. Ad hoc networks security. ST Journal of System Rsearch, Volume 4, 2003.
22. J. R. Douceur. The Sybil attack. Proceedings of the 1st International Workshop on Peer-to-Peer System, 2002.

# Chapter 4
# Routing in Mobile Ad Hoc Networks

**Al-Sakib Khan Pathan and Choong Seon Hong**

**Abstract** A Mobile Ad Hoc Network (MANET) is built on the fly where a number of wireless mobile nodes work in cooperation without the engagement of any centralized access point or any fixed infrastructure. Two nodes in such a network can communicate in a bidirectional manner if and only if the distance between them is at most the minimum of their transmission ranges. When a node wants to communicate with a node outside its transmission range, a multi-hop routing strategy is used which involves some intermediate nodes. Because of the movements of nodes, there is a constant possibility of topology change in MANET. Considering this unique aspect of MANET, a number of routing protocols have been proposed so far. This chapter gives an overview of the past, current, and future research areas for routing in MANET. In this chapter we will learn about the following things:

- The preliminaries of mobile ad hoc network
- The challenges for routing in MANET
- Expected properties of a MANET routing protocol
- Categories of routing protocols for MANET
- Major routing protocols for MANET
- Criteria for performance comparison of the routing protocols for MANET
- Achievements and future research directions
- Expectations and reality

## 4.1 Introduction

With the staggering growth of wireless handheld devices and plummeting costs of mobile telecommunications, mobile ad hoc network has emerged as a major area of research for both the academic and the industrial sectors. A mobile ad

C.S. Hong (✉)
Department of Computer Engineering, Kyung Hee University, 1 Seocheon, Giheung, Yongin, Gyeonggi 449701, Korea
e-mail: cshong@khu.ac.kr, sakib.pathan@gmail.com

S. Misra et al. (eds.), *Guide to Wireless Ad Hoc Networks*,
Computer Communications and Networks, DOI 10.1007/978-1-84800-328-6_4,
© Springer-Verlag London Limited 2009

hoc network (MANET) is built on the fly where a number of mobile nodes work in cooperation without the engagement of any centralized access point or any fixed infrastructure. MANETs are self-organizing, self-configuring, and dynamic topology networks, which form a particular class of multi-hop networks. Minimal configuration, absence of infrastructure, and quick deployment make them convenient for combat, medical, and other emergency situations. All nodes in a MANET are capable of movement and can be interconnected in an arbitrary manner.

The issue of routing in MANET is somewhat challenging and non-trivial. Due to the mobility of the nodes, connectivity between any two nodes in the network is considered intermittent and often it is very difficult, if not impossible to use traditional wired network's routing mechanisms. Basically, the major challenges for routing in MANET are imposed by the resource constraints and mobility of the nodes participating in the network. As there is no fixed infrastructure in such a network, we consider each node as a host and a router at the same time. Hence, during routing of data packets within the network, at each hop, each host also has to perform the tasks of a router. In fact, these special aspects of mobile ad hoc networks have attracted many researchers to work on solving the routing issues in MANET. A sample model of mobile ad hoc network is presented here in Fig. 4.1, which consists of some mobile devices with wireless communication facilities.

So far, a significant number of proposals for routing in MANET have seen the daylight. However, it is apparent that there could not be a single solution for routing in MANETs. Different deployment scenarios and application-dependent requirements need the employment of different types of routing mechanisms. In this chapter, we will learn about the routing protocols for MANET, their features, advantages, drawbacks, and future expectations.

Let us start this chapter with a brief background of MANET. We will know about how the practitioners, researchers, scientists, and industrialists have tried

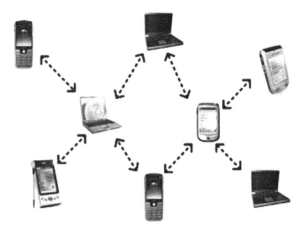

**Fig. 4.1** An Example of Mobile Ad hoc Network (MANET)

to solve this challenging issue for MANET. We will know various types of routing schemes those are already proposed or those could be applied for these types of networks. Considering the practical scenarios, we will also discuss how the reality might betray the expectations.

## 4.2 Background

From the advent of packet radio network up to today's MANET, the whole life cycle of ad hoc networks can be categorized mainly into three parts: first generation, second generation, and third generation. Today's ad hoc networks are considered as the third-generation networks.

The first generation goes back to 1972. At that time, they were called PRNET (Packet Radio Networks). In 1973, the Defense Advanced Research Projects Agency (DARPA) initiated research on the feasibility of using packet-switched radio communications to provide reliable computer communications [1, 2]. This development was motivated by the need to provide computer network access to mobile hosts and terminals, and to provide computer communications in a mobile environment.

The second generation of ad hoc networks emerged in 1980s, when the ad hoc network systems were further enhanced and implemented as a part of the SURAN (Survivable Adaptive Radio Networks) program [3]. This provided a packet-switched network to the mobile battlefield in an environment without infrastructure. This program proved to be beneficial in improving the radio performance by making them smaller, cheaper, and resilient to electronic attacks.

In the 1990s, the concept of commercial ad hoc networks arrived with handheld computers and other small portable communication equipments. At the same time, the idea of a collection of mobile nodes was proposed at several research conferences. From then up to today, research works have been going on for solving various issues of mobile ad hoc networks.

We mentioned the formal definition of a mobile ad hoc network earlier. Let us investigate how the unique characteristics of MANET make the task of routing complicated. So far we have learnt that the major features of this type of network are each node is considered both as a host and as a router; the nodes in the network are allowed to move while participating in the network; for their connectivity they use wireless communications; there is no centralized entity in the network; and the nodes are mainly battery-powered. Now, let us consider the following network structure for starting our discussion on routing in MANET.

*Example 4.1* In Fig. 4.2, a sample model of MANET is presented where there are three nodes; A, B, and C. The radio transmission ranges of the nodes are shown as circles.

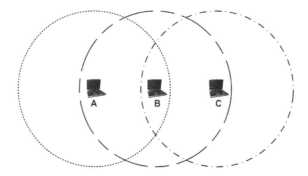

**Fig. 4.2** A MANET with three nodes

In the figure, node A and node B are within the transmission ranges of each other. We call any of these nodes as a neighbor of the other. Likewise, B and C are neighbors. But, A and C are not neighbors as none of their transmission ranges covers other node. In this setting, the neighbors can communicate directly and no routing is required. But, if node A and C want to communicate with each other, they must seek help from node B, who can help them by forwarding their data packets. Here, we can reach this decision that it is quite natural. Yes, it could be done as node A knows about B and C knows about B, so both A and C can use B as an intermediate node for their communications! Simple neighbor information could be used in such a case.

*Example 4.2* Now, the task of routing data packets becomes more complicated if we consider a model like that presented in Fig. 4.3.

With the addition of node D, we have several options to exchange data between A and C. For example, a packet from A can take the path, A-B-C or A-D-C or A-D-B-C or A-B-D-C. This is where we need to employ efficient mechanism or logic for routing the packet in the best possible way. The whole

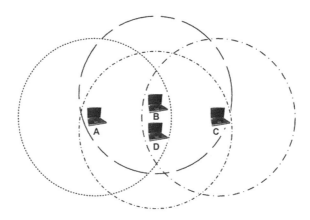

**Fig. 4.3** A MANET with four nodes

scenario gets even more complicated with the increase of the number of nodes in the network. If two nodes are far from each other and if they must have to communicate using a path involving multiple intermediate nodes, in that case, neighbor information might not be enough to solve the problem. Even if neighbor information is used, it is not possible or inefficient for a MANET to provide the full topological information to each node in the network. Because of the mobility of nodes within the network, the scenario becomes more and more complex. Hence, to allow a MANET to operate successfully maintaining all the properties of ad hoc networks, different routing protocols were developed by the practitioners. Sometimes, choosing a single routing protocol does not provide the complete solution, rather the system and environment settings require different approaches of routing. As we have seen in Figs. 4.2 and 4.3, based on the situation we can apply different routing mechanisms. While only neighbor information is enough for solving the routing problem in Fig. 4.2, some extra mechanism is necessary for efficient routing in case of Fig. 4.3.

## 4.3  Routing Protocols

From the very beginning of the concept of mobile ad hoc network, the researchers took the issue of routing as a major challenge. With the course of time, many routing protocols have been proposed. In this section, we will learn about various routing protocols for MANET, their major aspects, and their relative pros and cons.

### 4.3.1  Expected Properties of MANET Routing Protocols

Considering the special properties of MANET, when thinking about any routing protocol, we generally expect the following properties, though all of these might not be possible to incorporate in a single solution:

- A routing protocol for MANET should be distributed in manner in order to increase its reliability. Where all nodes are mobile, it is unacceptable to have a routing protocol that requires a centralized entity. Each node should be intelligent enough to make routing decisions using other collaborating nodes. A distributed but virtually centralized protocol might be a good idea.
- The routing protocol should assume routes as unidirectional links. Wireless medium may cause a wireless link to be opened in unidirection only due to physical factors. It may not be possible to communicate bidirectionally. Thus a routing protocol must be designed considering unidirectional links.
- The routing protocol should be power-efficient. It should consider every possible measure to save power, as power is very important for small battery-powered devices. To save power, the routing-related loads could be distributed among the participating nodes.

- The routing protocol should consider its security. MANET routing protocols in many cases lack proper security. Generally, a wireless medium is
  highly vulnerable and susceptible to various sorts of threats and attacks.
  Because of the use of wireless technology in MANETs, the methods of
  attacks against such networks are larger in scale than those of their wired
  counterparts [4, 5]. At physical layer, denial of service attacks may be
  avoided using coded or frequency hopping spread spectrum; however,
  at routing level, we need authentication for communicating nodes, non-
  repudiation, and encryption for private networking to shun hostile entities.
- Hybrid protocols, which combine the benefits of different routing protocols
  can be preferred in most of the cases. A protocol should be much more
  reactive (which reacts on demand) than proactive (which uses periodic
  refreshment of information) to avoid protocol overhead.
- A routing protocol should be aware of Quality of Service (QoS). It should
  know about the delay and throughput for the route of a source–destination
  pair, and must be able to verify its longevity so that a real-time application
  may rely on it.

## 4.3.2  Categorizing the Routing Protocols for MANET

One of the most interesting aspects for routing in MANET, which many
research works have tried to solve is, whether or not the nodes in the network
should keep track of routes to all possible destinations, or instead keep track of
only those destinations of immediate interest. Generally, a node in MANET
does not need a route to a destination until the node is necessarily be the
recipient of packets, either as the final destination or as an intermediate node
along the path from the source to the destination. As this is still a controversial
issue, we can assume that the mechanism should not be fixed for all types of
settings, instead based on the situation and application at hand, any of the
methods could be chosen.

Though there is no common consensus about the method of keeping the
information about routes in the network, many routing protocols have been
proposed by this time on the basis of all the available methods. The routing
protocols for MANET could be broadly classified into two major categories:

- Proactive Routing Protocols
- Reactive Routing Protocols

### 4.3.2.1  Proactive Routing Protocols

Proactive protocols continuously learn the topology of the network by exchanging topological information among the network nodes. Thus, when there is a
need for a route to a destination, such route information is available immediately. The main concern regarding using a proactive routing protocol is: if the

network topology changes too frequently, the cost of maintaining the network might be very high. Moreover, if the network activity is low, the information about the actual topology might even not be used and, in such a case, the investment with such limited transmission ranges and energies is lost, which might result in a shorter lifetime of the network than that is expected. Proactive protocols are sometimes called as table-driven routing protocols.

### 4.3.2.2 Reactive Routing Protocols

The reactive routing protocols, on the other hand, are based on some sort of *query-reply* dialog. Reactive protocols proceed for establishing route(s) to the destination only when the need arises or on demand basis. They do not need periodic transmission of topological information of the network; hence, they primarily seem to be resource-conserving protocols. Reactive protocols are also known as on-demand routing protocols.

### 4.3.2.3 Hybrid Routing Protocols

Often reactive or proactive feature of a particular routing protocol might not be enough; instead a mixture might yield better solution. Hence, in the recent days, several hybrid protocols are also proposed. The hybrid protocols include some of the characteristics of proactive protocols and some of the characteristics of reactive protocols.

Based on the method of delivery of data packets from the source to destination, classification of the MANET routing protocols could be done as follows:

- *Unicast Routing Protocols*: The routing protocols that consider sending information packets to a single destination from a single source.
- *Multicast Routing Protocols*: Multicast is the delivery of information to a group of destinations simultaneously, using the most efficient strategy to deliver the messages over each link of the network only once, creating copies only when the links to the destinations split. Multicast routing protocols for MANET use both multicast and unicast for data transmission.

Multicast routing protocols for MANET can be classified again into two categories:

- Tree-based multicast protocol
- Mesh-based multicast protocol

Mesh-based routing protocols use several routes to reach a destination while the tree-based protocols maintain only one path. Tree-based protocols ensure less end-to-end delay in comparison with the mesh-based protocols. Besides all of these categories, recently some geocast [6] routing protocols are also proposed, which aim to send messages to some or all of the wireless nodes within a particular geographic region. Often the nodes know their exact physical positions in a network, and these protocols use that information for transmitting packets from the source to the destination(s).

### 4.3.3 Proposed Routing Protocols: Major Features

In this section, we will investigate the major routing protocols for MANET. We will explore their distinctive features with easily understandable examples wherever necessary.

#### 4.3.3.1 Proactive Routing Protocols

Dynamic Destination-Sequenced Distance-Vector Routing Protocol

Dynamic Destination-Sequenced Distance-Vector Routing Protocol (DSDV) [7] is developed on the basis of Bellman–Ford routing [8] algorithm with some modifications. In this routing protocol, each mobile node in the network keeps a routing table. Each of the routing table contains the list of all available destinations and the number of hops to each. Each table entry is tagged with a sequence number, which is originated by the destination node. Periodic transmissions of updates of the routing tables help maintaining the topology information of the network. If there is any new significant change for the routing information, the updates are transmitted immediately. So, the routing information updates might either be periodic or event-driven. DSDV protocol requires each mobile node in the network to advertise its own routing table to its current neighbors. The advertisement is done either by broadcasting or by multicasting. By the advertisements, the neighboring nodes can know about any change that has occurred in the network due to the movements of nodes.

The routing updates could be sent in two ways: one is called a "*full dump*" and another is "*incremental.*" In case of *full dump*, the entire routing table is sent to the neighbors, whereas in case of *incremental* update, only the entries that require changes are sent. Full dump is transmitted relatively infrequently when no movement of nodes occur. The incremental updates could be more appropriate when the network is relatively stable so that extra traffic could be avoided. But, when the movements of nodes become frequent, the sizes of the incremental updates become large and approach the network protocol data unit (NPDU). Hence, in such a case, full dump could be used. Each of the route update packets also has a sequence number assigned by the transmitter. For updating the routing information in a node, the update packet with the highest sequence number is used, as the highest number means the most recent update packet. Each node waits up to certain time interval to transmit the advertisement message to its neighbors so that the latest information with better route to a destination could be informed to the neighbors. Let us explain DSDV routing protocol with an example.

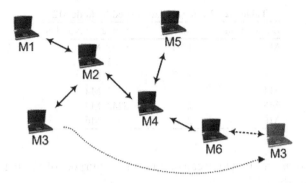

**Fig. 4.4** A sample MANET using DSDV

*Example 4.3*  Figure 4.4 shows a sample network consisting of six mobile nodes. Table 1.1 shows a sample structure of the forwarding table maintained in node M2. The *Install* time field helps determine when to delete a stale route. As in DSDV, any change in the routing path is immediately propagated throughout the network, it is very rare that deletion of stale routes occur. The *Stable_data* field contains the pointers that are needed to be stored when there is a competition with other possible routes to any particular destination. Table 1.2 shows a sample advertisement table of node M2 using DSDV.

Now, in Fig. 4.4, if a node, say M3, moves close to M6, only the entry for M3 needs to be changed. After some time M2 will get the information of M3 from M4, as M4 will get the information about M3 from M6, and accordingly M2 can adjust the entry for M3 in its own routing (forwarding) table. If M3 quits the network after some time interval, its entry will be deleted from M2's routing table.

## Wireless Routing Protocol

Wireless Routing Protocol (WRP) [9] belongs to the general class of path-finding algorithms [8, 10, 11], defined as the set of distributed shortest-path algorithms that calculate the paths using information regarding the length and second-to-last hop of the shortest path to each destination. WRP reduces the number of cases in

**Table 1.1**  Structure of node M2's forwarding table

| Destination | Next Hop | Metric | Sequence Number | Install | Flags | Stable_data |
|---|---|---|---|---|---|---|
| M1 | M1 | 1 | S593_M1 | T001_M2 | – | Ptrl_M1 |
| M2 | M2 | 0 | S983_M2 | T001_M2 | – | Ptrl_M2 |
| M3 | M3 | 1 | S193_M3 | T002_M2 | – | Ptrl_M3 |
| M4 | M4 | 1 | S233_M4 | T001_M2 | – | Ptrl_M4 |
| M5 | M4 | 2 | S243_M5 | T001_M2 | – | Ptrl_M5 |
| M6 | M4 | 2 | S053_M6 | T002_M2 | – | Ptrl_M6 |

**Table 1.2**  Route table advertised by node M2

| Destination | Metric | Sequence Number |
|---|---|---|
| M1 | 1 | S593_M1 |
| M2 | 0 | S983_M2 |
| M3 | 1 | S193_M3 |
| M4 | 1 | S233_M4 |
| M5 | 2 | S243_M5 |
| M6 | 2 | S053_M6 |

which a temporary routing loop can occur. For the purpose of routing, each node maintains four things:

1. A distance table
2. A routing table
3. A link-cost table
4. A message retransmission list (MRL)

The distance table of node $x$ contains the distance of each destination node $y$ via each neighbor $z$ of $x$ and the predecessor node reported by $z$. The routing table of node $x$ is a vector with an entry for each known destination $y$, which specifies:

- The identifier of the destination $y$
- The distance to the destination $y$
- The predecessor of the chosen shortest path to $y$
- The successor of the chosen shortest path to $y$
- A tag to identify whether the entry is a simple path, a loop, or invalid
- Storing predecessor and successor in the table is beneficial to detect loops and to avoid count-to-infinity problems.

The link-cost table of node $x$ lists the cost of relaying information through each neighbor $z$, and the number of periodic update periods that have elapsed since node $x$ received any error-free message from $z$. The message retransmission list (MRL) contains information to let a node know which of its neighbors has not acknowledged its update message and to retransmit the update message to that neighbor.

WRP uses periodic update message transmissions to the neighbors of a node. The nodes in the response list of update message (which is formed using MRL) should send acknowledgments. If there is no change from the last update, the nodes in the response list should send an *idle Hello* message to ensure connectivity. A node can decide whether to update its routing table after receiving an update message from a neighbor and always it looks for a better path using the new information. If a node gets a better path, it relays back that information to the original nodes so that they can update their tables. After receiving the acknowledgment, the original node updates its MRL. Thus, each time the consistency of the routing information is checked by each node in this protocol, which helps to eliminate routing loops and always tries to find out the best solution for routing in the network.

Cluster Gateway Switch Routing Protocol

Cluster Gateway Switch Routing Protocol (CGSR) [12] considers a clustered mobile wireless network instead of a "*flat*" network. For structuring the network into separate but interrelated groups, cluster heads are elected using a cluster head selection algorithm. By forming several clusters, this protocol achieves a distributed processing mechanism in the network. However, one drawback of this protocol is that, frequent change or selection of cluster heads might be resource hungry and it might affect the routing performance. CGSR uses DSDV protocol as the underlying routing scheme and, hence, it has the same overhead as DSDV. However, it modifies DSDV by using a hierarchical cluster-head-to-gateway routing approach to route traffic from source to destination. Gateway nodes are nodes that are within the communication ranges of two or more cluster heads. A packet sent by a node is first sent to its cluster head, and then the packet is sent from the cluster head to a gateway to another cluster head, and so on until the cluster head of the destination node is reached. The packet is then transmitted to the destination from its own cluster head.

*Example 4.4* Figure 4.5 shows two clusters C1 and C2 each of which has a cluster head. A gateway is the common node between two clusters. Any source node passes the packet first to its own cluster head, which in turn passes that to the gateway.

The gateway relays the packet to another cluster head and this process continues until the destination is reached. In this method, each node must keep a "*cluster member table*" where it stores the destination cluster head for each mobile node in the network. These cluster member tables are broadcasted by each node periodically using the DSDV algorithm. Nodes update their

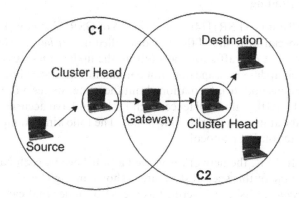

**Fig. 4.5** Clustered MANET

cluster member tables on reception of such a table from a neighbor. Also each node maintains a routing table that is used to determine the next hop to reach the destination.

## Global State Routing

In Global State Routing (GSR) protocol [13], nodes exchange vectors of link states among their neighbors during routing information exchange. Based on the link state vectors, nodes maintain a global knowledge of the network topology and optimize their routing decisions locally. Functionally, this protocol is similar to DSDV, but it improves DSDV in the sense that it avoids flooding of routing messages. In this protocol, each node maintains one list and three tables. They are:

- A neighbor list
- A topology table
- A next hop table
- A distance table

Neighbor list contains the set of neighboring nodes of a particular node $x$. Each destination $y$ has an entry in the topology table of $x$. Each entry in this topology table has two parts, one is the link state information reported by destination $y$ and the other is the timestamp indicating the time node $y$ has generated this link state information. Next hop contains the identity of the next hop node, to which a packet is to be forwarded to reach a particular destination. The distance table contains the shortest distance between $x$ and $y$.

Though the operational structure of GSR is similar to DSDV, it does not flood the link state packets. Instead, in this protocol nodes maintain link state table based on the up-to-date information received from neighboring nodes, and periodically exchange it with their local neighbors only. Information disseminated as the link state with larger sequence number replaces the one with smaller sequence number.

## Fisheye State Routing

Fisheye State Routing (FSR) [14] is built on top of GSR. The novelty of FSR is that it uses a special structure of the network called the "*fisheye*." This protocol reduces the amount of traffic for transmitting the update messages. The basic idea is that each update message does not contain information about all nodes. Instead, it contains update information about the nearer nodes more frequently than that of the farther nodes. Hence, each node can have accurate and exact information about its own neighboring nodes. The following example explains the fisheye state routing protocol.

*Example 4.5* In FSR, the network is viewed as a fisheye by each participating node. An example of this special structure is shown in Fig. 4.6.

Here, the *scope* of fisheye is defined as the set of nodes that can be reached within a given number of hops from a particular center node. In the figure, we

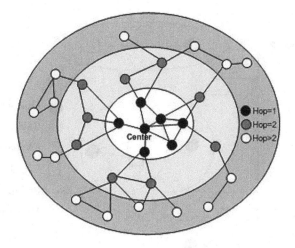

**Fig. 4.6** Fisheye structure

have shown three scopes with one, two, and three hops. The center node has the most accurate information about all nodes in the white circle and so on. Each circle contains the nodes of a particular hop from a center node. The advantage of FSR is that, even if a node does not have accurate information about a destination, as the packet moves closer to the destination, more correct information about the route to the destination becomes available.

Hierarchical State Routing

Hierarchical State Routing (HSR) [14] combines dynamic, distributed multilevel hierarchical clustering technique with an efficient location management scheme. This protocol partitions the network into several clusters where each elected cluster head at the lower level in the hierarchy becomes member of the next higher level. The basic idea of HSR is that each cluster head summarizes its own cluster information and passes it to the neighboring cluster heads using gateways. After running the algorithm at any level, any node can flood the obtained information to its lower level nodes. The hierarchical structure used in this protocol is efficient enough to deliver data successfully to any part of the network.

*Example 4.6*  Figure 4.7 shows the clustering and hierarchy used in HSR. Here, each node has a hierarchical address by which it could be reached. A gateway can be reached from the root via more than one path; hence it can have more than one hierarchical address.

Zone-Based Hierarchical Link State Routing Protocol

In Zone-Based Hierarchical Link State Routing (ZHLS) protocol [15], the network is divided into non-overlapping zones as in cellular networks. Each node knows the node connectivity within its own zone and the zone connectivity

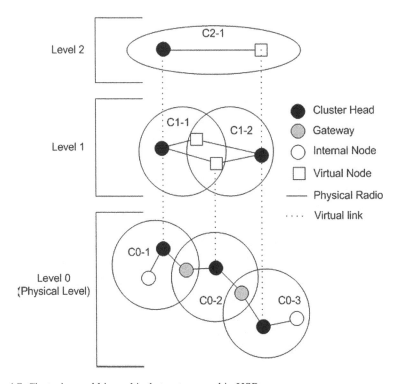

**Fig. 4.7** Clustering and hierarchical structure used in HSR

information of the entire network. The link state routing is performed by employing two levels: node level and global zone level. ZHLS does not have any cluster head in the network like other hierarchical routing protocols. The zone level topological information is distributed to all nodes. Since only zone ID and node ID of a destination are needed for routing, the route from a source to a destination is adaptable to changing topology. The zone ID of the destination is found by sending one *location request* to every zone.

Landmark Ad Hoc Routing

Landmark Ad hoc Routing (LANMAR) [16] combines the features of Fisheye State Routing (FSR) and Landmark Routing [17]. It uses the concept of *landmark* from Landmark Routing, which was originally developed for fixed wide area networks. A *landmark* is defined as a router whose neighbor routers within a certain number of hops contain routing entries for that router. Using this concept for the nodes in the MANET, LANMAR divides the network into several pre-defined logical subnets, each with a pre-selected *landmark*. All nodes in a subnet are assumed to move as a group, and they remain connected to each other via Fisheye State Routing (FSR). The routes to the landmarks, and hence the corresponding subnets, are proactively maintained by all nodes in the network through the exchange of distance-vectors. LANMAR could be regarded as an extension of FSR, which exploits group mobility by *summarizing* the routes to the group members with a single route to a *landmark*.

### Optimized Link State Routing

Optimized Link State Routing (OLSR) [18] protocol inherits the stability of link state algorithm. Usually, in a pure link state protocol, all the links with neighbor nodes are declared and are flooded in the entire network. But, OLSR is an optimized version of a pure link state protocol designed for MANET. This protocol performs hop-by-hop routing; that is, each node in the network uses its most recent information to route a packet. Hence, even when a node is moving, its packets can be successfully delivered to it, if its speed is such that its movements could at least be followed in its neighborhood. The optimization in the routing is done mainly in two ways. Firstly, OLSR reduces the size of the control packets for a particular node by declaring only a subset of links with the node's neighbors who are its *multipoint relay selectors*, instead of all links in the network. Secondly, it minimizes flooding of the control traffic by using only the selected nodes, called *multipoint relays* to disseminate information in the network. As only multipoint relays of a node can retransmit its broadcast messages, this protocol significantly reduces the number of retransmissions in a flooding or broadcast procedure.

*Example 4.7* Figure 4.8 shows a sample network structure used in OLSR. OLSR protocol relies on the selection of multipoint relay (MPR) nodes.

Each node calculates the routes to all known destinations through these nodes. These MPRs are selected among the one hop neighborhood of a node using the bidirectional links, and they are used to minimize the amount of broadcast traffic in the network.

### 4.3.3.2  Reactive Routing Protocols

All of the protocols mentioned in the previous section use periodic transmissions of routing information. In this section, we will investigate the working principles of some reactive routing protocols for mobile ad hoc networks. As stated earlier, unlike proactive protocols, reactive protocols proceed for finding a route to a destination only when a source node needs to transmit data to another node in the network.

### Associativity-Based Routing

Associativity-Based Routing (ABR) [19] protocol defines a new type of routing metric for mobile ad hoc networks. This routing metric is termed as *degree of association stability*. In this routing protocol, a route is selected based on the

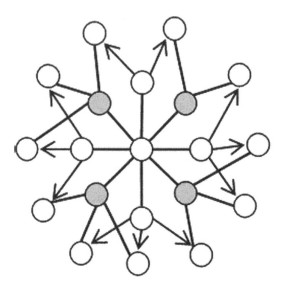

**Fig. 4.8** Multipoint Relays (MPRs) are in gray color. The transmitting node is shown at the center of the sample structure

degree of association stability of mobile nodes. Each node periodically generates *beacon* to announce its existence. Upon receiving the beacon message, a neighbor node updates its own associativity table. For each beacon received, the associativity tick of the receiving node with the beaconing node is increased. A high value of associativity tick for any particular beaconing node means that the node is relatively static. Associativity tick is reset when any neighboring node moves out of the neighborhood of any other node. ABR protocol has three phases for the routing operations:

- Route discovery
- Route reconstruction
- Route deletion

The route discovery phase is done by a broadcast query and await-reply (BQ-REPLY) cycle. When a source node wants to send message to a destination, it sends the query. All other nodes receiving the query append their addresses and their associativity ticks with their neighbors along with QoS information to the query packet. A downstream node erases its immediate upstream node's associativity tick entries and retains only the entry concerned with itself and its upstream node. This process continues and eventually the packet reaches the destination. On receiving the packet with the associativity information, the destination chooses the best route and sends the REPLY packet using that path. If there are multiple paths with same overall degree of association stability, the route with the minimum number of hops is selected. Route reconstruction is needed when any path becomes invalid or broken for the mobility or failure of any intermediate node. If a source or upstream node moves, a route

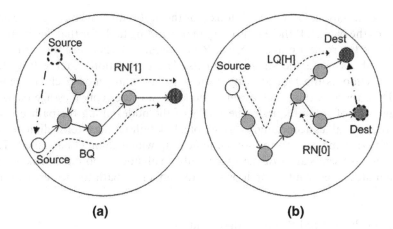

**Fig. 4.9** Route maintenance in ABR for two different scenarios

notification (RN) message is used to erase the route entries associated with downstream nodes. When the destination node moves, the destination's immediate upstream node erases its route. A localized query (LQ[H]) process, where H refers to the hop count from the upstream node to the destination, is initiated to determine whether the node is still reachable or not. Route deletion broadcast is done if any discovered route is no longer needed. Figure 4.9 shows the working principle of ABR protocol.

*Example 4.8* Figure 4.9 shows two different scenarios for route maintenance where ABR is used. In Fig. 4.9(a), the source moves to another place, as a result of which a new BQ request is used to find out the route to the destination. The RN [1] message is used to erase the route entries associated with the downstream nodes. In Fig. 4.9(b), the destination changed its position. Hence, immediate upstream node erases its route and determines if the node is still reachable by a localized query (LQ[H]) process.

Signal Stability–Based Adaptive Routing Protocol

Signal Stability–Based Adaptive Routing (SSA) [20] protocol focuses on obtaining the most stable routes through an ad hoc network. The protocol performs on-demand route discovery based on signal strength and location stability. Based on the signal strength, SSA detects weak and strong channels in the network. SSA can be divided into two cooperative protocols: the Dynamic Routing Protocol (DRP) and the Static Routing Protocol (SRP). DRP uses two tables: Signal Stability Table (SST) and Routing Table (RT). SST stores the signal strengths of the neighboring nodes obtained by periodic beacons from the link layer of each neighboring node. These signal strengths are recorded as weak or strong. DRP receives all the transmissions and, after processing, it passes those to the SRP. SRP passes the packet to the node's upper layer stack if it is the destination. Otherwise, it looks for the destination in routing table and forwards the packet. If there is no entry in the routing table for that destination, it initiates the route-finding process. Route-request packets are forwarded to the neighbors using the strong channels. The destination, after getting the request, chooses the first arriving request packet and sends back the reply. The DRP reverses the selected route and sends a route-reply message back to the initiator of route-request. The DRPs of the nodes along the path update their routing tables accordingly. In case of a link failure, the intermediate nodes send an error message to the source indicating which channel has failed. The source in turn sends an *erase* message to inform all nodes about the broken link and initiates a new route-search process to find a new path to the destination.

## Temporarily Ordered Routing Algorithm

Temporally Ordered Routing Algorithm (TORA) [21] is a reactive routing protocol with some proactive enhancements where a link between nodes is established creating a Directed Acyclic Graph (DAG) of the route from the source node to the destination. This protocol uses a "*link reversal*" model in route discovery. A route discovery query is broadcasted and propagated throughout the network until it reaches the destination or a node that has information about how to reach the destination. TORA defines a parameter, termed *height*. *Height* is a measure of the distance of the responding node's distance up to the required destination node. In the route discovery phase, this parameter is returned to the querying node. As the query response propagates back, each intermediate node updates its TORA table with the route and *height* to the destination node. The source node then uses the *height* to select the best route toward the destination. This protocol has an interesting property that it frequently chooses the most convenient route, rather than the shortest route. For all these attempts, TORA tries to minimize the routing management traffic overhead.

## Cluster-Based Routing Protocol

Cluster-Based Routing Protocol (CBRP) [22] is an on-demand routing proto-col, where the nodes are divided into clusters. For cluster formation, the following algorithm is employed. When a node comes up in the network, it has the *undecided* state. The first task of this node is to start a timer and to broadcast a HELLO message. When a cluster-head receives this HELLO message, it replies immediately with a triggered HELLO message. After that, when the node receives this answer, it changes its state into the *member* state. But when the node gets no message from any cluster-head, it makes itself as a cluster-head, but only when it has bidirectional link to one or more neighbor nodes. Otherwise, when it has no link to any other node, it stays in the *undecided* state and repeats the procedure with sending a HELLO message again.

Each node has a neighbor table. For each neighbor, the node keeps the status of the link and state of the neighbor in the neighbor table. A cluster head keeps information about all of its members in the same cluster. It also has a cluster adjacency table, which provides information about the neighboring clusters.

*Example 4.9* The network structure shown in Fig. 4.5 could be used to explain the clustering used in CBRP. However, while CGSR is a proactive routing protocol, CBRP is a reactive or on-demand routing protocol. Though the basic clustering mechanisms are same, the difference lies in the method of routing in the network. In case of CBRP, for sending data packets a source node floods route-request packet to the neighboring cluster heads. On receiving the request, a cluster head checks whether the destination node is its own cluster or not. If it is within that cluster, it sends the request to the node, and if not, it again sends the request to the neighboring cluster head. This process continues and the destination eventually gets the route request. The reply from the destination is sent using the reverse path of the route. In case of a route failure, a local repair mechanism is used. When a node finds the next hop is unreachable, it checks whether the next hop can be reached through any of its neighbors or whether the hop after the next hop can be reached via any other neighbor. If any of these works, the packet can be routed using the repaired path.

## Dynamic Source Routing

Dynamic Source Routing (DSR) [23] allows nodes in the MANET to dynamically discover a source route across multiple network hops to any destination. In this protocol, the mobile nodes are required to maintain route caches or the known routes. The route cache is updated when any new route is known for a particular entry in the route cache.

Routing in DSR is done using two phases: route discovery and route maintenance. When a source node wants to send a packet to a destination, it first consults its *route cache* to determine whether it already knows about any route to the destination or not. If already there is an entry for that destination, the source uses that to send the packet. If not, it initiates a route request broadcast. This request includes the destination address, source address, and a unique identification number. Each intermediate node checks whether it knows about the destination or not. If the intermediate node does not know about the destination, it again forwards the packet and eventually this reaches the destination. A node processes the route request packet only if it has not previously processed the packet and its address is not present in the route record of the packet. A route reply is generated by the destination or by any of the intermediate nodes when it knows about how to reach the destination. Figure 4.10 shows the operational method of the dynamic source routing protocol.

*Example 4.10* In Fig. 4.10, the route discovery procedure is shown where S1 is the source node and S7 is the destination node.

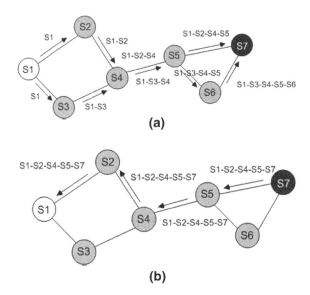

**Fig. 4.10 (a)** Route Discovery **(b)** Using route record to send the route reply

In this example, the destination gets the request through two paths. It chooses one path based on the route records in the incoming request packet and accordingly sends a reply using the reverse path to the source node. At each hop, the best route with minimum hop is stored. In this example, we have shown the route record status at each hop to reach the destination from the source node. Here, the chosen route is S1-S2-S4-S5-S7.

## Ad Hoc On-Demand Distance Vector Routing

Ad Hoc On-Demand Distance Vector Routing (AODV) [24] is basically an improvement of DSDV. But, AODV is a reactive routing protocol instead of proactive. It minimizes the number of broadcasts by creating routes based on demand, which is not the case for DSDV. When any source node wants to send a packet to a destination, it broadcasts a route request (RREQ) packet. The neighboring nodes in turn broadcast the packet to their neighbors and the process continues until the packet reaches the destination. During the process of forwarding the route request, intermediate nodes record the address of the neighbor from which the first copy of the broadcast packet is received. This record is stored in their route tables, which helps for establishing a reverse path. If additional copies of the same RREQ are later received, these packets are discarded. The reply is sent using the reverse path.

For route maintenance, when a source node moves, it can re-initiate a route discovery process. If any intermediate node moves within a particular route, the neighbor of the drifted node can detect the link failure and sends a link failure notification to its upstream neighbor. This process continues until the failure

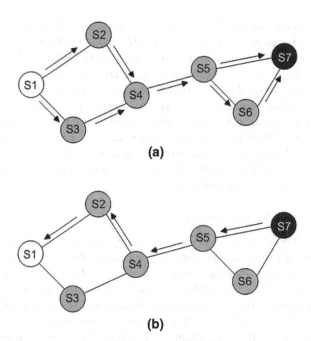

**Fig. 4.11** AODV protocol **(a)** Source node broadcasting the route request packet. **(b)** Route reply is sent by the destination using the reverse path

notification reaches the source node. Based on the received information, the source might decide to re-initiate the route discovery phase. Figure 4.11 shows an example of AODV protocol's operational mechanism.

*Example 4.11* In Fig. 4.11, S1 is the source node and S7 is the destination node. The source initiates the route request and the route is created based on demand. Route reply is sent using the reverse path from the destination.

#### 4.3.3.3 Hybrid Routing Protocols

Dual-Hybrid Adaptive Routing

Dual-Hybrid Adaptive Routing (DHAR) [25] uses the Distributed Dynamic Cluster Algorithm (DDCA) presented in [26]. The idea of DDCA is to dynamically partition the network into some non-overlapping clusters of nodes consisting of one parent and zero or more children. Routing is done in DHAR utilizing a dynamic two-level hierarchical strategy, consisting of optimal and least-overhead table-driven algorithms operating at each level.

DHAR implements a proactive least-overhead level-2 routing protocol in combination with a dynamic binding protocol to achieve its hybrid characteristics. The level-2 protocol in DHAR requires that one node generates an update on behalf of its cluster. When a level-2 update is generated, it must be flooded to all the nodes in each neighboring cluster. Level-2 updates are not transmitted beyond the neighboring clusters. The node with the lowest node ID in each cluster is designated to generate level-2 updates. The binding process is similar to a reactive route discovery process; however, a priori knowledge of clustered topology makes it significantly more efficient and simpler to accomplish the routing. To send packets to the desired destination, a source node uses the dynamic binding protocol to discover the current cluster ID associated with the destination. Once determined, this information is maintained in the dynamic cluster binding cache at the source node. The dynamic binding protocol utilizes the knowledge of the level-2 topology to efficiently broadcast a binding request to all the clusters. This is achieved using reverse path forwarding with respect to the source cluster.

## Adaptive Distance Vector Routing

Adaptive Distance Vector (ADV) [27] routing protocol is a distance-vector routing algorithm that exhibits some on-demand features by varying the frequency and the size of routing updates in response to the network load and mobility patterns. This protocol has the benefits of both proactive and reactive routing protocols. ADV uses an adaptive mechanism to mitigate the effect of periodic transmissions of the routing updates, which basically relies on the network load and mobility conditions. To reduce the size of routing updates, ADV advertises and maintains routes for the active receivers only. A node is considered active if it is the receiver of any currently active connection. There is a *receiver flag* in the routing entry, which keeps the information about the status of a receiver whether it is active or inactive. To send data, a source node broadcasts network-wide an *init-connection* control packet. All the other nodes turn on the corresponding *receiver flag* in their own routing tables and start advertising the routes to the receiver in future updates. When the destination node gets the *init-connection* packet, it responds to it by broadcasting a *receiver-alert* packet and becomes active. To close a connection, the source node broadcasts network-wide an *end-connection* control packet, indicating that the connection is to be closed. If the destination node has no additional active connection, it broadcasts a *non-receiver-alert* message. If the *init-connection* and *receiver-alert* messages are lost, the source advertises the receiver's entry with its *receiver flag* set in all future updates. ADV also defines some other parameters like trigger meter, trigger threshold, and buffer threshold. These are used for limiting the network traffic based on the network's mobility pattern and network speed.

## Zone Routing Protocol

Zone Routing Protocol (ZRP) [28] is suitable for wide variety of MANETs, especially for the networks with large span and diverse mobility patterns. In this protocol, each node proactively maintains routes within a local region, which is termed as routing zone. Route creation is done using a query-reply mechanism. For creating different zones in the network, a node first has to know who its neighbors are. A neighbor is defined as a node with whom direct communication can be established, and that is, within one hop transmission range of a node. Neighbor discovery information is used as a basis for Intra-zone Routing Protocol (IARP), which is described in detail in [29]. Rather than blind broadcasting, ZRP uses a query control mechanism to reduce route query traffic by directing query messages outward from the query source and away from covered routing zones. A covered node is a node which belongs to the routing zone of a node that has received a route query. During the forwarding of the query packet, a node identifies whether it is coming from its neighbor or not. If yes, then it marks all of its known neighboring nodes in its same zone as covered. The query is thus relayed till it reaches the destination. The destination in turn sends back a reply message via the reverse path and creates the route.

Sharp Hybrid Adaptive Routing Protocol

Sharp Hybrid Adaptive Routing Protocol (SHARP) [30] combines the features of both proactive and reactive routing mechanisms. SHARP adapts between reactive and proactive routing by dynamically varying the amount of routing information shared proactively. This protocol defines the proactive zones around some nodes. The number of nodes in a particular proactive zone is determined by the node-specific zone radius. All nodes within the zone radius of a particular node become the member of that particular proactive zone for that node. If for a given destination a node is not present within a particular proactive zone, reactive routing mechanism (query-reply) is used to establish the route to that node. Proactive routing mechanism is used within the proactive zone. Nodes within the proactive zone maintain routes proactively only with respect to the central node. In this protocol, proactive zones are created automatically if some destinations are frequently addressed or sought within the network. The proactive zones act as collectors of packets, which forward the packets efficiently to the destination, once the packets reach any node at the zone vicinity.

*Example 4.12* In Fig. 4.12, some proactive zones are shown in a sample MANET. Here, we have four destination nodes, A, B, C, and D. As destination D is not used heavily, no proactive zone is created within its surroundings.

But for the other three destinations, A, B, and C, proactive zones of different sizes are created. As node A has the highest number of calls within the network as a destination, its proactive zone is the largest among all the destinations. Any routing within the proactive zone is done using proactive routing mechanisms. But, outside of the proactive zones, reactive routings are employed. The zone radius acts as a virtual knob to control the mix of proactive and reactive routing for each destination in SHARP. For example, in case of destination D in the figure, reactive mechanism is used.

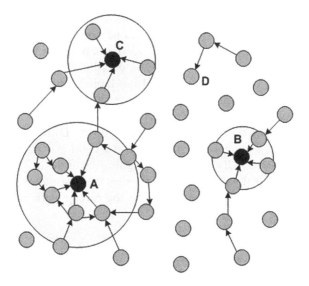

**Fig. 4.12** Proactive zones around the hot destinations in SHARP

Neighbor-Aware Multicast Routing Protocol

Neighbor-Aware Multicast Routing Protocol (NAMP) [31] is a tree-based hybrid routing protocol, which utilizes neighborhood information. The routes in the network are built and maintained using the traditional request and reply messages or on-demand basis. This hybrid protocol uses neighbor information of two-hops away for transmitting the packets to the receiver. If the receiver is not within this range, it searches the receiver using dominant pruning flooding method [32] and forms a multicast tree using the replies along the reverse path. Although the mesh structure is known to be more robust against topological changes, the tree structure is better in terms of packet transmission. As NAMP targets to achieve less end-to-end delay of packets, it uses the tree structure.

There are mainly three operations addressed in NAMP:

- Multicast tree creation
- Multicast tree maintenance
- Joining and leaving of nodes from the multicast group

All the nodes in the network keep neighborhood information of up to two-hop away nodes. This neighborhood information is maintained using a proactive mechanism. Periodic *hello* packet is used for this. To create the multicast tree, the source node sends a *flood request* packet to the destination with data payload attached. This packet is flooded in the network using dominant pruning method, which actually minimizes the number of transmissions in the network for a particular *flood request* packet. During the forwarding process of the packet, each node selects a forwarder and creates a secondary forwarder list (*SFL*). The secondary forwarder list (*SFL*) contains the information about the nodes that were primarily considered as possible forwarders but finally were not selected for that purpose. Each intermediate node uses the chosen forwarder to forward the packet, but keeps the knowledge about other possible forwarders in *SFL*. Secondary forwarder list is used for repairing any broken route in the network. In fact, link failure recovery is one of the greatest advantages of NAMP. The next example shows some figures to explain NAMP's operations in brief.

*Example 4.13* Figure 4.13 shows a sample network where NAMP has created the multicast tree consisting of the source, destination, and intermediate nodes (forwarders). Here, S1 is the source, S12 is the destination. Nodes S3, S6, S9, and S11 are the forwarding nodes. For each forwarding hop, each forwarder maintains the information of the neighboring nodes in the secondary forwarder list. In case of a link failure as shown in Fig. 4.13(b), S3 immediately finds an alternate path to repair the existing route for the S1-S12 source–destination pair. Figure 4.13(c) shows that S3 repairs the path to use the existing route to

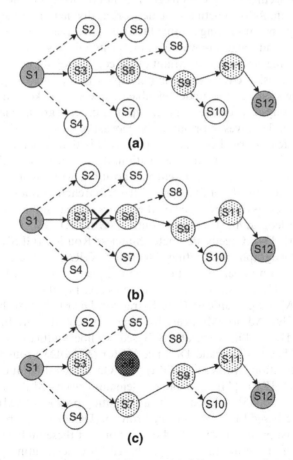

**Fig. 4.13 (a)** Network sample **(b)** Link failure **(c)** Link failure recovery in NAMP

reach the destination using the alternate node S7. Link failure recovery is done locally in NAMP, which is one of its greatest advantages.

#### 4.3.3.4 Other Routing Protocols

In addition to the mentioned routing protocols for MANET, there are some other routing protocols that do not rely on any traditional routing mechanisms, instead rely on the location awareness of the participating nodes in the network. Generally, in traditional MANETs, the nodes are addressed only with their IP addresses. But, in case of location-aware routing mechanisms, the nodes are often aware of their exact physical locations in the three-dimensional world. This capability might be introduced in the nodes using Global Positioning System (GPS) or with any other geometric methods. GPS is a worldwide, satellite-based radio navigation system that consists of 24 satellites in six orbital planes. By connecting to the GPS receiver, a mobile node can know its current physical location. Also sometimes the network is divided into several zones or geographic regions for making routing little bit easier. Based on these concepts, several geocast and location-aware routing protocols have already been proposed. Geocasting is basically a variant of the conventional multicasting where the nodes are considered under certain groups within particular geographical regions. In geocasting, the nodes eligible to receive packets are implicitly specified by a physical region; membership in a geocast group changes whenever a mobile node moves in or out of the geocast region.

The major feature of these routing protocols is that, when a node knows about the location of a particular destination, it can direct the packets toward that particular direction from its current position, without using any route discovery mechanism. Recently, some of the researchers proposed some location-aware protocols that are based on these sorts of idea. Some of the examples of them are Geographic Distance Routing (GEDIR) [33], Location-Aided Routing (LAR) [34], Greedy Perimeter Stateless Routing (GPSR) [35], Geo-GRID [36], Geographical Routing Algorithm (GRA) [37], etc. Other than these, there are a number of multicast routing protocols for MANET. Some of the mentionable multicast routing protocols are: Location-Based Multicast Protocol (LBM) [38], Multicast Core Extraction Distributed Ad hoc Routing (MCEDAR) [39], Ad hoc Multicast Routing protocol utilizing Increasing id-numberS (AMRIS) [40], Associativity-Based Ad hoc Multicast (ABAM) [41], Multicast Ad hoc On-Demand Distance-Vector (MAODV) routing [42], Differential Destination Multicast (DDM) [43], On-Demand Multicast Routing Protocol (ODMRP) [44], Adaptive Demand-driven Multicast Routing (ADMR) protocol [45], Ad hoc Multicast Routing protocol (AMRoute) [46], Dynamic Core-based Multicast routing Protocol (DCMP) [47], Preferred Link-Based Multicast protocol (PLBM) [48], etc. Some of these multicast protocols use location information and some are based on other routing protocols or developed just as the extension of another unicast routing protocol. For

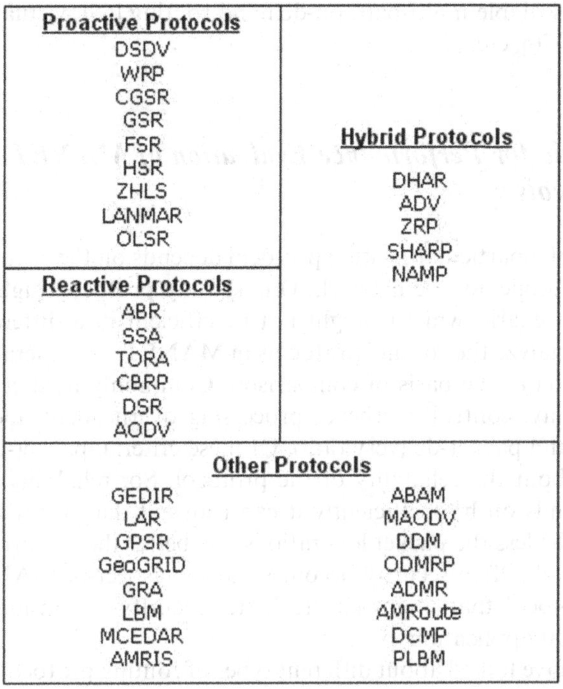

| | | |
|---|---|---|
| **Proactive Protocols** <br> DSDV <br> WRP <br> CGSR <br> GSR <br> FSR <br> HSR <br> ZHLS <br> LANMAR <br> OLSR | | **Hybrid Protocols** <br> DHAR <br> ADV <br> ZRP <br> SHARP <br> NAMP |
| **Reactive Protocols** <br> ABR <br> SSA <br> TORA <br> CBRP <br> DSR <br> AODV | | |

| **Other Protocols** | |
|---|---|
| GEDIR | ABAM |
| LAR | MAODV |
| GPSR | DDM |
| GeoGRID | ODMRP |
| GRA | ADMR |
| LBM | AMRoute |
| MCEDAR | DCMP |
| AMRIS | PLBM |

**Fig. 4.14** Major Routing Protocols for MANET at a glance

example, MAODV is the multicast-supporting version of AODV. Figure 4.14 shows the major routing protocols for MANET at a glance.

### 4.3.3.5 Other Recent Works on MANET Routing for Reference

In this section, we mention a list of references of the recent works on routing in MANET so that it could be used as a reference by the practitioners. Some of these works have taken the major routing protocols as their bases and some of them have enhanced various performances of the previous routing protocols. Mentionable recent works are: node-density-based routing [49], load-balanced routing [50], optimized priority-based energy-efficient routing [51], reliable on-demand routing with mobility prediction [52], QoS routing [53], secure distributed anonymous routing protocol [54], robust position-based routing [55], routing with group motion support [56], dense cluster gateway based routing protocol [57], dynamic backup routes routing protocol [58], gathering-based routing protocol [59], QoS-aware multicast routing protocol [60], recycled path routing [61], QoS multicast routing protocol for clustering in MANET [62], secure anonymous routing protocol with authenticated key exchange [63], self-healing on-demand geographic path routing protocol [64], stable weight-based on-demand routing protocol [65], fisheye zone routing protocol [66], on-demand utility-based power control routing [67], secure position-based routing protocol [68], scalable multi-path on-demand routing [69], virtual coordinate-based routing [70], etc.

## 4.3.4 Criteria for Performance Evaluation of MANET Routing Protocols

Performance of a particular routing protocol depends on the requirements and settings of a mobile ad hoc network. One routing protocol might seem to be efficient in a scenario while it might not be efficient in a different scenario. However, to analyze the routing protocols in MANET, we generally take some common criteria as the basis of comparison. Commonly used criteria are the end-to-end delay, control overhead, processing overhead of nodes, memory requirement, and packet-delivery ratio. Of these criteria, packet-delivery ratio mainly tells about the reliability of the protocol. So, reliability of a routing protocol depends on how efficiently it can transmit data from source to the destination. The less the packet loss ratio is, the better the performance of that routing protocol. Often security becomes the key aspect of MANET. In such cases, the protocol that might ensure better security is considered as more efficient for that application.

So far, we have talked about different types of routing protocols. We mainly categorized them into reactive, proactive, and hybrid protocols. Generally speaking, reactive protocols require less amount of memory, processing power, and energy than that of the proactive protocols. Having the knowledge of the MANET routing protocols and their comparison criteria, let us now investigate the key influencing factors for routing performance in different settings of MANETs.

### 4.3.4.1 Mobility Factors

- *Velocity of nodes*: The velocity of the mobile nodes within a MANET is not fixed. As there is no speed limitation of the wireless devices, high speed of nodes might affect the performance of many protocols. A protocol is considered good for MANET if it can perform well both in relatively static and in fully dynamic network state, though it is true that routing in a highly mobile MANET is a tough task.
- *Direction of mobility*: The direction of a node's mobility is not known in advance. It is a very common incident that a node travels to a direction where the number of neighbor nodes is less or there is no neighbor node. This is called drifting away of a node from a MANET. A hard-state approach or a soft-state approach could be used to handle such incidents. In hard-state approach, the node explicitly informs all the other nodes in the MANET about its departure or movement from a position, while in a soft-state approach a time out value is used to detect the departure.

- *Group or individual mobility*: MANETs are often categorized as Pure MANET and Military MANET. In a pure MANET, it is not obvious that the nodes should move in groups, but in case of military MANET, group mobility is the main concern. A military MANET can maintain a well-defined chain of commands, which is absent in case of a pure MANET. So the routing strategies could vary depending upon this factor. Two MANET protocols considered as good for supporting group mobility are: LANMAR [16], developed by University of California at Los Angeles, and OLSR [18], which is developed by the French National Institute for Research in Computer Science and Control (INRIA).
- *Frequency of changing of mobility model*: Routing strategy could also vary depending on the mobility model of the MANET. The topology of an ad hoc network could definitely change over time. But, the key factor here is the change of overall mobility model in a fast or relatively slow fashion. If the nodes change their relative positions too frequently, the maintenance cost of the overall network gets higher. For example, a MANET formed with war planes, tanks, helicopters, and ships is highly dynamic, while an ad hoc network formed with some laptops and palmtops carried by the participants in a conference is relatively less dynamic.

#### 4.3.4.2 Wireless Communication Factors

- *Consumption of power*: Power is a valuable resource in wireless networking. Especially for routing, power is highly needed. According to an experiment by Kravets and Krishnan (1998), power consumption caused by networking-related activities is approximately 10% of the overall power consumption of a laptop computer. This figure rises up to 50% in handheld devices [71]. In ad hoc network, every node has to contribute for maintaining the network connections. Hence, routing protocol should consider everything to save power of the participating battery-powered devices.
- *Bandwidth*: For any type of wireless communications, bandwidth available for the network is a major concern. An efficient routing protocol should try to minimize the number of packet-transmissions or control overhead for the maintenance of the network.
- *Error rate*: Wireless communication is always susceptible to high error rate. Packet loss is a common incident. So, the routing strategies should be intelligent enough to minimize the error rate for smooth communications among the nodes.
- *Unidirectional link*: Sometimes it is convenient for a routing protocol to assume routes as unidirectional links.

#### 4.3.4.3 Security Issues

- *Unauthorized access*: Security has recently become a major issue for ad hoc network routing. Most of the ad hoc network routing protocols that are currently proposed lack security. A wireless network is more vulnerable than a wired network. So, based on the requirement, sometimes preventing unauthorized access to the network becomes the major concern.

- *Accidental association with other networks*: Accidental associations between a node in one wireless network and a neighboring wireless network are just now being recognized as a security concern, as enterprises confront the issue of overlapping networks. At the routing level it should be ensured that the nodes can recognize their own network.

#### 4.3.4.4 Other Factors

- *Reliability of the network*: Reliability is sometimes defines as how efficiently a routing protocol can dispatch packets to the appropriate destinations. A routing protocol must be efficient enough to handle successful packet delivery so that an application may rely on it.
- *Size of the network*: The overall network size could be a crucial factor. A routing protocol might be good for a small network, but might not be fit for use in a large ad hoc network or vice versa.
- *Quality of service*: In the real-time applications, QoS becomes a key factor for evaluating the performance of a routing protocol.
- *Timing*: Regardless of the method of communication used, access time and tuning time must be considered. Tuning time is the measure of the amount of time each node spends in active mode. In the active mode a node consumes maximum power. So, minimizing the tuning time is one of the critical factors to conserve power.

### 4.4 Thoughts for Practitioners

It is still a matter of debate whether the routing protocols for mobile ad hoc networks should be predicted based on the network overhead or the optimization of the network path. In this chapter, we have learnt about a number of routing protocols for MANET, which are broadly categorized as proactive and reactive. Proactive routing protocols tend to provide lower latency than that of the on-demand protocols, because they try to maintain routes to all the nodes in the network all the time. But the drawback for such protocols is the excessive routing overhead transmitted, which is periodic in nature without much consideration for the network mobility or load. On the other hand, though reactive protocols discover routes only when they are needed, they may still generate a huge amount of traffic when the network changes frequently.

Depending on the amount of network traffic and number of flows, the routing protocols could be chosen. When there is congestion in the network due to heavy traffic, in general case, a reactive protocol is preferable. Sometimes the size of the network might be a major considerable point. For example,

AODV, DSR, OLSR are some of the protocols suitable for relatively smaller networks, while the routing protocols like TORA, LANMAR, ZRP are suitable for larger networks. Network mobility is another factor that can degrade the performance of certain protocols. When the network is relatively static, proactive routing protocols can be used, as storing the topology information in such case is more efficient. On the other hand, as the mobility of nodes in the network increases, reactive protocols perform better.

Overall, the answer to the debating point might be that the mobility and traffic pattern of the network must play the key role for choosing an appropriate routing strategy for a particular network. It is quite natural that one particular solution cannot be applied for all sorts of situations and, even if applied, might not be optimal in all cases. Often it is more appropriate to apply a hybrid protocol rather than a strictly proactive or reactive protocol as hybrid protocols often possess the advantages of both types of protocols.

## 4.5 Directions for Future Research

The structure of the Internet that is used today is based mainly on wired communications. The emerging technologies like fiber optics–based high-speed wired networks would flourish in the near future. With this existing network of networks, semi-infrastructure and infrastructure-less wireless networks will also be used in abundance. Figure 4.15 shows a conceptual view of the future global Internet structure. MANETs would definitely play an important role in the future Internet structure, especially for the mobile Internet. Hence, in some cases, it might be necessary that the routing protocols of MANET work in perfect harmony with their wired counterparts. Considering different approaches of routing, a hybrid approach might be more appropriate for such scenarios.

More and more efficient routing protocols for MANET might come in front in the coming future, which might take security and QoS (Quality of Service) as the major concerns. So far, the routing protocols mainly focused on the

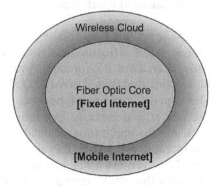

**Fig. 4.15** Future Global Internet Structure

methods of routing, but in future a secured but QoS-aware routing protocol could be worked on. We should keep this in mind that ensuring both of these parameters at the same time might be difficult. A very secure routing protocol surely incurs more overhead for routing, which might degrade the QoS level. So an optimal trade-off between these two parameters could be searched.

We saw that in the recent years some multicast routing protocols have been proposed. The reason for the growing importance of multicast is that this strategy could be used as a means to reduce bandwidth utilization for mass distribution of data. As there is a pressing need to conserve scarce bandwidth over wireless media, it is natural that multicast routing should receive some attention for ad hoc networks. So it is, in most of the cases, advantageous to use multicast rather than multiple unicast, especially in ad hoc environment where bandwidth comes at a premium. Another advantage of multicasting is that it provides group communication facility. A group of nodes can be addressed at the same time using only a group identifier. So it is an efficient communication tool for using in multipoint applications.

Ad hoc wireless networks find applications in civilian operations (collaborative and distributed computing) emergency search-and-rescue, law enforcement, and warfare situations, where setting up and maintaining a communication infrastructure is very difficult. In all these applications, communication and coordination among a given set of nodes are necessary. Considering all these, in future the routing protocols might especially emphasize the support for multicasting in the network.

## 4.6 Conclusions

In this chapter, we have talked about MANET, the challenges for routing in MANET, major routing protocols, the major features of MANET routing protocols, key aspects for routing in MANET, and future research issues for routing in MANET. We categorized the proposed routing protocols based on their working principles and discussed which type of protocol might be used in which situation.

The proliferation of mobile ad hoc networks is looming on the horizon. Exploitation of these types of infrastructure-less networks are expected to flourish in future, not only for civil but also for military reconnaissance scenarios. It is quite reasonable to think that the security and QoS (Quality of Service) requirements might differ largely for different types of civil and military applications. Based on these two critical aspects, appropriate routing protocols should have to be chosen for the application at hand. Some of the routing protocols proposed in the recent days for MANETs are considered as *promising* for use in real workplaces. However, *One cannot satisfy all*. This might also be true for any routing protocol that could emerge in the near future. So the ultimate solution is the use of different routing protocols for different situations. In that case, the cooperation among dissimilar routing protocols would be the major issue to address in future. Though the collaboration of different routing strategies is more or less well defined in case of wired networks, for mobile ad hoc networks there still remains a lot of scope of research on this issue.

## Terminologies

*MANET (Mobile Ad hoc Network)* – A Mobile Ad hoc Network (MANET) is a kind of wireless network that could be formed on the fly where a number of wireless mobile nodes work in cooperation, without the engagement of any centralized access point or any fixed infrastructure.

*QoS (Quality of Service)* – The ability of a network (including applications, hosts, and infrastructure devices) to deliver traffic with minimum delay and maximum availability.

*NPDU (Network Protocol Data Unit)* – A frame of data transmitted over the physical layer of a network.

*MRL (Message Retransmission List)* – In case of Wireless Routing Protocol (WRP), each node maintains a Message Retransmission List (MRL). MRL is used for confirming the reception of update messages by neighboring nodes.

*MPR (MultiPoint Relay)* – OLSR protocol relies on the selection of multipoint relay (MPR) nodes. MPRs are selected among the one-hop neighborhood of a node using the bidirectional links, and they are used to minimize the amount of broadcast traffic in the network.

*DRP (Dynamic Routing Protocol)* – Signal Stability–Based Adaptive Routing Protocol (SSA) uses DRP.

*SRP (Static Routing Protocol)* – Signal Stability–Based Adaptive Routing Protocol (SSA) uses SRP.

*DDCA* – Distributed Dynamic Cluster Algorithm

*IARP* – Intra-Zone Routing Protocol

*SFL* – Secondary Forwarder List

*DSDV* – Dynamic Destination-Sequenced Distance-Vector

*WRP* – Wireless Routing Protocol

*CGSR* – Cluster Gateway Switch Routing

*GSR* – Global State Routing

*FSR* – Fisheye State Routing

*HSR* – Hierarchical State Routing

*ZHLS* – Zone-Based Hierarchical Link State

*LANMAR* – Landmark Ad hoc Routing

*OLSR* – Optimized Link State Routing

*ABR* – Associativity-Based Routing

*SSA* – Signal Stability–based Adaptive

*TORA* – Temporarily Ordered Routing Algorithm
*CBRP* – Cluster Based Routing Protocol
*DSR* – Dynamic Source Routing
*AODV* – Ad hoc On-Demand Distance Vector
*DHAR* – Dual-Hybrid Adaptive Routing
*ADV* – Adaptive Distance Vector
*ZRP* – Zone Routing Protocol
*SHARP* – Sharp Hybrid Adaptive Routing Protocol
*NAMP* – Neighbor-Aware Multicast routing Protocol
*GEDIR* – GEographic DIstance Routing
*LAR* – Location-Aided Routing
*GPSR* – Greedy Perimeter Stateless Routing
*GeoGRID* – Geographical GRID
*GRA* – Geographical Routing Algorithm
*LBM* – Location-Based Multicast
*MCEDAR* – Multicast Core Extraction Distributed Ad hoc Routing
*AMRIS* – Ad hoc Multicast Routing protocol utilizing Increasing id-numberS
*ABAM* – Associativity-Based Ad hoc Multicast
*MAODV* – Multicast Ad hoc On-Demand Distance Vector
*DDM* – Differential Destination Multicast
*ODMRP* – On-Demand Multicast Routing Protocol
*ADMR* – Adaptive Demand-driven Multicast Routing
*AMRoute* – Ad hoc Multicast Routing
*DCMP* – Dynamic Core-based Multicast routing Protocol
*PLBM* – Preferred Link-Based Multicast

## Questions

1. What are the major challenges for routing in MANET?
2. Why do not we use the routing protocols for wired networks for MANETs?
3. Suppose that we have a MANET where the nodes are frequently moving from one place to another. If we use DSDV as the routing protocol for this network, which method of updates would be better? Why?
4. What is a gateway in cluster-based routing protocols for MANET?
5. What is a *scope* in Fisheye State Routing?
6. How is the fisheye concept beneficial for routing?
7. What is a *landmark* in LANMAR?
8. How does OLSR reduce traffic in case of a broadcast procedure?
9. What does "Height" mean in TORA?
10. What is a Hybrid routing protocol?

11. Look at the figure below. Construct the route table advertised by node N4 if
    DSDV is used as the routing protocol (three columns: *Destination*, *Metric*,
    and *Sequence Number*).

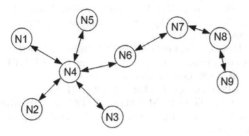

12. Which criteria could affect the performance of the routing protocols for
    MANET?
13. Which protocol is the best among all the proposed routing protocols for
    MANET? Why? Justify your answer.
14. In the figure below, which path will be chosen to reach the destination N4
    from the source N1, if Dynamic Source Routing is used? Why? Justify your
    answer.

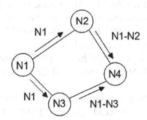

# References

1. Kahn RE (1977) The organization of computer resources into a packet radio network.
   IEEE Transactions on Communications, Volume COM-25, Issue 1:169–178
2. Jubin J, Tornow JD (1987) The DARPA Packet Radio Network Protocols. Proceedings of
   the IEEE, Volume 75, Issue 1:21–32
3. Freebersyser J, Leiner B (2001) A DoD Perspective on Mobile Ad Hoc Networks. In:
   Perkins CE (ed) Ad Hoc Networking, Addison-Wesley:29–51
4. Yang H, Luo H, Ye F, Lu S, Zhang, L (2004) Security in Mobile Ad Hoc Networks:
   Challenges and Solutions. IEEE Wireless Communications, Volume 11, Issue 1:38–47
5. Deng H, Li W, Agrawal DP (2002) Routing Security in Wireless Ad Hoc Networks. IEEE
   Communications Magazine, Volume 40, Issue 10:70–75
6. Maihöfer C (2004) A Survey of Geocast Routing Protocols. IEEE Communications
   Surveys & Tutorials, Volume 6, Issue 2:Q2:32–42
7. Perkins CE, Bhagwat P (1994) Highly Dynamic Destination-Sequenced Distance-Vector
   Routing (DSDV) for Mobile Computers. Proceedings of ACM SIGCOMM 1994:234–244
8. Cheng C, Riley R, Kumar SPR, Garcia-Luna-Aceves JJ (1989) A Loop-Free Extended
   Bellman-Ford Routing Protocol Without Bouncing Effect. ACM SIGCOMM Computer
   Communications Review, Volume 19, Issue 4:224–236

9. Murthy S, Garcia-Luna-Aceves JJ (1996) An Efficient Routing Protocol for Wireless Networks. Mobile Networks and Applications, Volume 1, Issue 2:183–197

10. Humblet PA (1991) Another Adaptive Distributed Shortest-Path Algorithm. IEEE Transactions on Communications, Volume 39, Issue 6:995–1003

11. Rajagopalan B, Faiman M (1991) A Responsive Distributed Shortest-Path Routing Algorithm Within Autonomous Systems. Journal of Internetworking Research and Experiment, Volume 2, Issue 1:51–69

12. Chiang C-C, Wu H-K, Liu W, Gerla M (1997) Routing in Clustered Multihop, Mobile Wireless Networks with Fading Channel. Proceedings of IEEE SICON:197–211

13. Chen T-W, Gerla M (1998) Global State Routing: A New Routing Scheme for Ad-hoc Wireless Networks. Proceedings of IEEE ICC 1998:171–175

14. Iwata A, Chiang C-C, Pei G, Gerla M, Chen T-W (1999) Scalable Routing Strategies for Ad Hoc Wireless Networks. IEEE Journal on Selected Areas in Communications, Volume 17, Issue 8:1369–1379

15. Jao-Ng M, Lu I-T (1999) A Peer-to-Peer Zone-Based Two-Level Link State Routing for Mobile Ad Hoc Networks. IEEE Journal on Selected Areas in Communications, Volume 17, Issue 8:1415–1425

16. Pei G, Gerla M, Hong X (2000) LANMAR: Landmark Routing for Large Scale Wireless Ad Hoc Network with Group Mobility. First Annual Workshop on Mobile and Ad Hoc Networking and Computing 2000 (MobiHoc 2000):11–18

17. Tsuchiya PF (1988) The Landmark Hierarchy: A New Hierarchy for Routing in Very Large Networks. Computer Communication Review, Volume 18, Issue 4:35–42

18. Jacquet P, Mühlethaler P, Clausen T, Laouiti A, Qayyum A, Viennot L (2001) Optimized Link State Routing Protocol for Ad Hoc Networks. IEEE INMIC 2001:62–68

19. Toh C-K (1996) A Novel Distributed Routing Protocol to Support Ad-Hoc Mobile Computing. Proceedings of the 1996 IEEE 15th Annual International Phoenix Conference on Computers and Communications:480–486

20. Dube R, Rais CD, Wang K-Y, Tripathi SK (1997) Signal Stability-Based Adaptive Routing (SSA) for Ad Hoc Mobile Networks. IEEE Personal Communications, Volume 4, Issue 1:36–45

21. Park VD, Corson MS (1997) A highly adaptive distributed routing algorithm for mobile wireless networks. Proceedings of IEEE INFOCOM 1997, Volume 3:1405–1413

22. Jiang M, Li J, Tay YC (1999) Cluster Based Routing Protocol (CBRP). IETF Draft, August 1999, available at http://tools.ietf.org/html/draft-ietf-manet-cbrp-spec-01. Accessed 21 February 2008

23. Broch J, Johnson DB, Maltz DA (1999) The Dynamic Source Routing Protocol for Mobile Ad Hoc Networks. IETF Draft, October, 1999, available at http://tools.ietf.org/id/draft-ietf-manet-dsr-03.txt. Accessed 21 February 2008

24. Perkins CE, Royer EM, Chakeres ID (2003) Ad hoc On-Demand Distance Vector (AODV) Routing. IETF Draft, October, 2003, available at http://tools.ietf.org/html/draft-perkins-manet-aodvbis-00. Accessed 21 February 2008

25. McDonald AB, Znati T (2000) A Dual-Hybrid Adaptive Routing Strategy for Wireless Ad-Hoc Networks. Proceedings of IEEE WCNC 2000, Volume 3:1125–1130

26. McDonald AB, Znati T (1999) A Mobility Based Framework for Adaptive Clustering in Wireless Ad-Hoc Networks. IEEE Journal on Selected Areas in Communications, Special Issue on Ad-Hoc Networks, Volume 17, Issue 8:1466–1487

27. Boppana RV, Konduru SP (2001) An Adaptive Distance Vector Routing Algorithm for Mobile, Ad Hoc Networks. Proceedings of IEEE INFOCOM 2001:1753–1762

28. Haas ZJ, Pearlman MR, Samar P (2002) The Zone Routing Protocol (ZRP) for Ad Hoc Networks. IETF draft, July 2002, available at http://tools.ietf.org/id/draft-ietf-manet-zone-zrp-04.txt. Accessed 21 February 2008

29. Haas ZJ, Pearlman MR, Samar P (2002) Intrazone Routing Protocol (IARP). IETF Internet Draft, July 2002, available at http://tools.ietf.org/wg/manet/draft-ietf-manet-zone-ierp/draft-ietf-manet-zone-ierp-02-from-01.diff.txt. Accessed 21 February 2008
30. Ramasubramanian V, Haas ZJ, Sirer, EG (2003) SHARP: A Hybrid Adaptive Routing Protocol for Mobile Ad Hoc Networks. Proceedings of ACM MobiHoc 2003:303–314
31. Pathan A-SK, Alam MM, Monowar MM, Rabbi MF (2004) An Efficient Routing Protocol for Mobile Ad Hoc Networks with Neighbor Awareness and Multicasting. Proceedings of IEEE E-Tech, July, 2004:97–100
32. Lim H, Kim C (2000) Multicast Tree Construction and Flooding in Wireless Ad Hoc Networks. Proceedings of the 3rd ACM International Workshop on Modeling, Analysis and Simulation of Wireless and Mobile Systems:61–68
33. Lin X, Stojmenovic I (1999) GEDIR: Loop-Free Location Based Routing in Wireless Networks. Proceedings of the IASTED International Conference on Parallel and Distributed Computing and Systems:1025–1028
34. Ko Y-B, Vaidya NH (2000) Location-Aided Routing (LAR) in Mobile Ad Hoc Networks. Wireless Networks, Volume 6:307–321
35. Karp B, Kung HT (2000) GPSR: Greedy Perimeter Stateless Routing for Wireless Networks. ACM MOBICOM 2000:243–254
36. Liao W-H, Tseng Y-C, Lo K-L, Sheu J-P (2000) GeoGRID: A Geocasting Protocol for Mobile Ad Hoc Networks based on GRID. Journal of Internet Technology, Volume 1, Issue 2:23–32
37. Jain R, Puri A, Sengupta R (2001) Geographical Routing Using Partial Information for Wireless Ad Hoc Networks. IEEE Personal Communications, Volume 8, Issue 1:48–57
38. Ko Y-B, Vaidya NH (1998) Location-based multicast in mobile ad hoc networks. Technical Report TR98-018, Texas A&M University
39. Sinha P, Sivakumar R, Bharghavan V (1999) MCEDAR: Multicast Core-Extraction Distributed Ad Hoc Routing. Proceedings of IEEE WCNC, Volume 3:1313–1317
40. Wu CW, Tay TC (1999) AMRIS: A Multicast Protocol for Ad Hoc Wireless Networks. IEEE MILCOM 1999, Volume 1:25–29
41. Toh C-K, Guichal G, Bunchua S (2000) ABAM: On-Demand Associativity-Based Multicast Routing for Ad Hoc Mobile Networks. Proceedings of IEEE VTS-Fall VTC 2000, Volume 3:987–993
42. Royer EM, Perkins CE (2000) Multicast Ad Hoc On-Demand Distance Vector (MAODV) Routing. IETF Draft, draft-ietf-manet-maodv-00, 15 July, 2000, available at http://tools.ietf.org/html/draft-ietf-manet-maodv-00. Accessed 21 February 2008
43. Ji L, Corson MS (2001) Differential Destination Multicast-A MANET Multicast Routing Protocol for Small Groups. Proceedings of IEEE INFOCOM 2001, Volume 2:1192–1201
44. Lee S, Su W, Gerla M (2002) On-Demand Multicast Routing Protocol in Multihop Wireless Mobile Networks. ACM/Kluwer Mobile Networks and Applications (MONET), volume 7, Issue 6:441–453
45. Jetcheva JG, Johnson DB (2001) Adaptive Demand-Driven Multicast Routing in Multi-Hop Wireless Ad Hoc Networks. Proceedings of ACM MobiHoc 2001:33–44
46. Xie J, Talpade RR, Mcauley A, Liu M (2002) AMRoute: Ad Hoc Multicast Routing Protocol. Mobile Networks and Applications, Volume 7, Issue 6:429–439
47. Das SK, Manoj BS, Murthy CSR (2002) A Dynamic Core Based Multicast Routing Protocol for Ad Hoc Wireless Networks. Proceedings of ACM MobiHoc 2002:24–35
48. Sisodia RS, Karthigeyan I, Manoj BS, Murthy CSR (2003) A Preferred Link Based Multicast Protocol for Wireless Mobile Ad Hoc Networks. Proceedings of IEEE ICC 2003, Volume 3:2213–2217
49. Quintero A, Pierre S, Macabéo B (2004) A routing protocol based on node density for ad hoc networks. Ad Hoc Networks, Volume 2, Issue 3:335–349
50. Saigal V, Nayak AK, Pradhan SK, Mall R (2004) Load balanced routing in mobile ad hoc networks. Computer Communications, Volume 27, Issue 3:295–305

51. Wei X, Chen G, Wan Y, Mtenzi F (2004) Optimized priority based energy efficient routing algorithm for mobile ad hoc networks. Ad Hoc Networks, Volume 2, Issue 3:231–239

52. Wang N-C, Chang S-W (2005) A reliable on-demand routing protocol for mobile ad hoc networks with mobility prediction. Computer Communications, Volume 29, Issue 1:123–135

53. Bür K, Ersoy C (2005) Ad hoc quality of service multicast routing. Computer Communications, Volume 29, Issue 1:136–148

54. Boukerche A, El-Khatib K, Xu L, Korba L (2005) An efficient secure distributed anonymous routing protocol for mobile and wireless ad hoc networks. Computer Communications, Volume 28, Issue 10:1193–1203

55. Moaveninejad K, Song W-Z, Li X-Y (2005) Robust position-based routing for wireless ad hoc networks. Ad Hoc Networks, Volume 3, Issue 5:546–559

56. Rango FD, Gerla M, Marano S (2006) A scalable routing scheme with group motion support in large and dense wireless ad hoc networks. Computers & Electrical Engineering, Volume 32, Issues 1–3:224–240

57. Ghosh RK, Garg V, Meitei MS, Raman S, Kumar A, Tewari N (2006) Dense cluster gateway based routing protocol for multi-hop mobile ad hoc networks. Ad Hoc Networks, Volume 4, Issue 2:168–185

58. Wang Y-H, Chao C-F (2006) Dynamic backup routes routing protocol for mobile ad hoc networks. Information Sciences, Volume 176, Issue 2:161–185

59. Ahn CW (2006) Gathering-based routing protocol in mobile ad hoc networks. Computer Communications, Volume 30, Issue 1:202–206

60. Sun B, Li L (2006) QoS-aware multicast routing protocol for Ad hoc networks. Journal of Systems Engineering and Electronics, Volume 17, Issue 2:417–422

61. Eisbrener J, Murphy G, Eade D, Pinnow CK, Begum K, Park S, Yoo S-M, Youn J-H (2006) Recycled path routing in mobile ad hoc networks. Computer Communications, Volume 29, Issue 9:1552–1560

62. Layuan L, Chunlin L (2007) A QoS multicast routing protocol for clustering mobile ad hoc networks. Computer Communications, Volume 30, Issue 7:1641–1654

63. Lu R, Cao Z, Wang L, Sun C (2007) A secure anonymous routing protocol with authenticated key exchange for ad hoc networks. Computer Standards & Interfaces, Volume 29, Issue 5:521–527

64. Giruka VC, Singhal M (2007) A self-healing On-demand Geographic Path Routing Protocol for mobile ad-hoc networks. Ad Hoc Networks, Volume 5, Issue 7:1113–1128

65. Wang N-C, Huang Y-F, Chen J-C (2007) A stable weight-based on-demand routing protocol for mobile ad hoc networks. Information Sciences: an International Journal, Volume 177, Issue 24:5522–5537

66. Yang C-C, Tseng L-P (2007) Fisheye zone routing protocol: A multi-level zone routing protocol for mobile ad hoc networks. Computer Communications, Volume 30, Issue 2:261–268

67. Min C-H, Kim S (2007) On-demand utility-based power control routing for energy-aware optimization in mobile ad hoc networks. Journal of Network and Computer Applications, Volume 30, Issue 2:706–727

68. Song J-H, Wong VWS, Leung VCM (2007) Secure position-based routing protocol for mobile ad hoc networks. Ad Hoc Networks, Volume 5, Issue 1:76–86

69. Reddy LR, Raghavan SV (2007) SMORT: Scalable multipath on-demand routing for mobile ad hoc networks. Ad Hoc Networks, Volume 5, Issue 2:162–188

70. Zhao Y, Chen Y, Li B, Zhang Q (2007) Hop ID: A Virtual Coordinate-Based Routing for Sparse Mobile Ad Hoc Networks. IEEE Transactions on Mobile Computing, Volume 6, Issue 9:1075–1089

71. Kravets R, Krishnan P (1998) Power Management Techniques for Mobile Communication. Proceedings of ACM MOBICOM 1998:157–168

# Chapter 5
# Multicasting in Mobile Ad Hoc Networks

Pascale Minet and Anis Laouiti

**Abstract** The success of wireless ad hoc networks and the increasing interest in multimedia applications explain the need of multicast protocols adapted to the wireless environment. In this chapter, we first introduce the design considerations for a multicast protocol in a wireless ad hoc network. We then present a classification of multicast protocols and another for reliable ones. Each family is illustrated by a representative example. Finally, we give some directions for the future research related to the multicast protocols: QoS (Quality of Service) support and network coding.

## 5.1 Introduction

In a wired or wireless network, there are three methods for transmitting a message:

- *Unicast*: when the message is sent to a single destination node. For instance, web client–server interactions use unicast transmissions.
- *Broadcast*: when the message is sent to all network nodes. For instance, a message advertising the foreseen network unavailability is broadcast to all nodes.
- *Multicast*: when the message is sent to a subset of the network nodes. For instance, users connect to a network to access the Internet services (e.g., video streaming, videoconference). Different groups of users sharing the same interests can coexist. Football fans receive the live video streaming of a match of the Super Bowl on their computer screens. Others watch the last episode of a well-known series, or attend a training seminar through the web using videoconference. The size of the group can be very high or on the contrary, very limited.

P. Minet (✉)
INRIA, Rocquencourt, 78153 Le Chesnay cedex, France
e-mail: pascale.minet@inria.fr

S. Misra et al. (eds.), *Guide to Wireless Ad Hoc Networks*,
Computer Communications and Networks, DOI 10.1007/978-1-84800-328-6_5,
© Springer-Verlag London Limited 2009

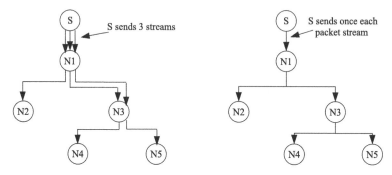

**Fig. 5.1** Unicast versus Multicast

To illustrate the differences between unicast and multicast transmissions, let us consider a source S wanting to send a stream to three receivers N2, N4, and N5. As illustrated in Fig. 5.1, with a unicast transmission, the source sends a separate stream to each receiver: three streams in this example to reach the three receivers N2, N4, and N5. With a multicast transmission, the source sends only one stream of data and the nodes are in charge of forwarding it to each interested receiver. In other words, with multicast, packets are only sent to where they are expected. In the example, N1 forwards a packet to N2 and N3. At its turn, N3 forwards to nodes N4 and N5.

As seen in the example, multicast reduces the number of streams transmitted by the sender, but do we really need multicast if simple broadcast is available? Using broadcast to transmit information interesting only a small number of users would lead to a big waste of network resources. Indeed, all nodes in the network receive the information, process it, and drop it if they are uninterested. Hence, the broadcast is greedy in:

- network bandwidth
- node processing power
- node energy
- node memory.

Furthermore, existing multimedia applications, such as Near Video On Demand, interactive games, collaborative work, are big consumers of multicast transmissions (high frequency of multicasts and big size of information to multicast), and more are coming. That is why an efficient multicast is required.

The goal of multicast protocols is to allow the network deliver the multicast information to interested users. The multicast protocol builds and maintains a structure that will provide routes to all nodes in the multicast group; hence, they will receive the information multicast in their group. This structure is updated according to the changes occurring in the multicast group membership:

- a node that wants to receive a particular stream of multicast information declares so to the network, a *join* message is generally used for that purpose; it allows the node to belong to the multicast group.
- in the same way, a node that is no longer interested in a multicast group declares it by sending a *leave* message; it is then removed from the multicast group.

Multicast protocols have been first designed for wired networks as the well-known multicast protocols: PIM (see [1] for the dense mode and [2] for the sparse mode), DVMRP [3] based on the distance vector, CBT [4] using core trees and MOSPF [5]. However, these protocols do not tackle the specificities of wireless networks such as MANETs (Mobile Ad hoc Networks):

- *Decentralized and self-organizing networks*: MANETs need neither an infrastructure nor a centralized entity to operate.
- *Semi-broadcast networks*: in a classical wired network, a node reaches in a single transmission all the nodes in its network. It is not the same anymore with MANETs, where a node reaches only the network nodes within its radio transmission range. That is why MANETs are sometimes called semi-broadcast networks. To reach all nodes in the network, a message must be forwarded.
- *Radio interferences*: they are inherent to radio transmission. Because of them, when a sender S transmits a message, no other node in the interference area of S can transmit another message. This interference area is generally limited to two transmission ranges from the sender. Notice that a one-hop transmission consumes bandwidth not only on the sender node and the destination nodes but also on all nodes located in the interference area of the sender.
- *Limited network resources*: the bandwidth in wireless networks is considerably smaller than in wired networks. Moreover, resources such as node-processing power and node energy are limited. Protocols should use them very efficiently.
- *High dynamicity of a mobile ad hoc network*: the quality of a radio link is highly versatile, because of changes in the conditions of radio propagation (e.g. fading with distance, multiple paths, obstacles, weather conditions, radio interferences with other flows). Moreover, the topology of a MANET changes very frequently: new links are created whereas others are broken. New nodes join the network and others leave it.
- *Support of node mobility*: it partly explains the success of MANETs. Multicast protocols should preserve this property.

These specificities make the multicast protocols designed for wired networks inefficient in wireless networks. New multicast protocols have been designed for MANETs. This is the topic of this chapter.

This chapter is organized as follows. Section 5.2 deals with the design considerations of a multicast protocol and identifies the evaluation criteria of such a protocol. Section 5.3 describes the state of the art. It first establishes a classification based on two independent criteria distinguishing between:

1. mesh-based and tree-based multicast protocols. In case of a tree, the distinction can be refined between source-based tree versus shared tree,
2. flat structure versus overlay structure.

To illustrate this classification, four examples of multicast protocols are overviewed. Three protocols use a flat structure: ODMRP as mesh-based, MOLSR as tree-based and MAODV as shared tree–based. A fourth protocol, MOST, based on an overlay structure, is then described. Section 5.4 summarizes some thoughts for practitioners. Reliable multicast protocols have been introduced to provide a delivery guarantee, provided that all receivers are reachable. Such protocols can be used to build distributed fault-tolerant systems. They are briefly described in Section 5.5. Directions for future research are given in Section 5.6. These directions concern the quality of service (QoS) support in multicast protocols and a very new topic, network coding. Finally, Section 5.7 summarizes this chapter and concludes. The interested reader can refer to Section "Terminologies" for a definition of the terminology used throughout this chapter. The Section "Questions" provides a list of questions/ answers on the main concepts and mechanisms presented here. The Section "References" gives all the references used.

## 5.2 Multicast Protocol Design Considerations

Designing a new multicast protocol or selecting an existing one is not an easy task. The first thing to do is to define precisely the multicast problem that must be solved. For this purpose, we propose here a list of criteria organized as follows:

- *Size of the multicast problem*: these criteria aim at quantifying the multicast groups and their members. If, for instance, near all nodes belong to the considered multicast group, an optimized broadcast (i.e., a protocol minimizing the number of retransmissions in the network, see SMOLSR [6] for an example) can provide good performance. Moreover, a high number of multicast groups and sources are not in favor of multicast protocols maintaining a structure per pair (multicast source, multicast group). These criteria are:

  o Number of multicast groups,
  o Client density per multicast group: this is defined as the ratio between the number of clients and the total number of nodes within the network,
  o Number of sources per multicast group.
  o Traffic rate that must be sustained by the multicast protocol.

- *QoS (Quality of Service) requirement*: QoS requirement is generally expressed in terms of throughput or delay. With a QoS requirement, interferences can no longer be neglected: the transmission of a packet consumes bandwidth not only on the sender and its one-hop neighbors but also on all nodes up to a distance of two transmission ranges from the sender.
- *Adaptativity and reactiveness*: how does the considered multicast protocol, running in a wireless ad hoc network, adapt to frequent topology changes? How long does it take to recover? Similarly, new sources can be added in the multicast group, others can depart. Multicast group members can join or leave. How does the multicast protocol react? With which latency?
- *Reliability*: does the application expect from the multicast protocol a delivery guarantee with regard to all multicast clients, provided they are reachable? If the answer is yes absolutely or yes with a high probability, a reliable multicast protocol is required: a deterministic one in the first case and a probabilistic one in the second case.
- *Mobility support*: is the network static or do some nodes move? In case of mobility, what is the pertinent mobility model (i.e., pause time, speed, direction). Mobile nodes considerably increase the frequency of topology changes that the multicast protocol must be able to cope with. Most multicast protocols are able to support a pedestrian mobility, but nodes embedded in moving vehicles require specific consideration.
- *Control overhead and scalability*: a multicast protocol based on retransmissions by the sender of the lost messages is adapted to small multicast groups but is unable to scale to large groups. The control overhead induced by the multicast protocol must be kept small because of the limited resources in wireless ad hoc networks.
- *Requirement on the underlying unicast routing protocol:* some multicast routing protocols are totally independent from the unicast routing protocol; others rely on it and have specific requirement on its type (e.g., a reactive protocol like AODV [7] or DSR, or on the contrary a proactive protocol like OLSR [8])

In order to select the best multicast protocol among several candidates or to improve the design of a new multicast protocol by choosing the best variant, we can compare their performance according to the following criteria:

- Delivery rate,
- Sustained throughput,
- End-to-end transmission delay,
- Control overhead: it can be evaluated:

    o At the bandwidth level, by the messages exchanged per second to maintain the multicast structure or by the redundant messages that would be useless in the absence of message loss.

  o At the memory level, by the amount of memory needed to store the control information maintained by the multicast protocol.
  o At the processing level, by the complexity of the multicast algorithm used.
- Time needed by the multicast protocol to:
  o Add a new client or source joining the multicast group,
  o Remove a client or source leaving the multicast group,
  o Recover from topology changes (i.e., breakage of an existing link, creation of a new link, appearance of a new node, disappearance of a neighbor node).

These comparisons will be made on scenario representative of the environment in which the multicast protocol will run:

- Number of multicast groups and multicast group size
- Client density,
- Various traffic rates,
- Various mobility scenario,
- Frequencies of group membership and topology changes, etc.

The goal is to determine the best candidate for the multicast problem that has been previously defined.

## 5.3  Background

### 5.3.1  Multicast Protocols Classification

Recall that the main objective of a multicast structure remains to connect multicast group receivers and multicast sources, in order to deliver data to each participant, as well as saving bandwidth by preventing unnecessary retransmissions of the same data. In wired networks, data replication is delayed as much as possible as we have seen in the first example (Fig. 5.1). In wireless ad hoc networks, and due to the nature of the radio medium, this strategy may be not well suited for wireless context and even may weaken the global structure, as it will be explained further. Nevertheless, the main target of reducing the number of retransmissions remains the same. Numerous multicast protocols have been proposed during the last years for wireless mobile ad hoc networks. In the literature we can find several types of classification such as proactive/reactive protocols, which is actually inherited from the classification of the unicast protocols for MANET. Another classification could be based on the ability of the multicast protocol to work on a standalone manner without an underlying unicast protocol. The classification that we adopt here is more related to the routing structure built by the multicast protocol. Basically, the protocols can be grouped into two types: tree-based and mesh-based.

Moreover, these structures can also be flat or overlay ones. In the following we detail the classification we adopt:

- *Multicast structures*: Multicast protocols can be classified into two categories: tree-based and mesh-based. The tree-based family can also be divided into two sub-categories: the source tree–based and the shared tree–based.

  - *Shared tree* In the shared tree–based family only one tree is built for each multicast group. Sources are not required to be a part of the multicast structure; they need an entry point to send their data to (the root of the tree, for example, or the nearest tree member).
  - *Source based*: In the source–based family, a tree is built for each tuple source multicast group. For each multicast group we have several trees. Notice that IGMPv3 [9] enables multicast source selection, which is straightforward with this kind of multicast tree.
  - *Mesh-based protocols*: These maintain a structure containing all the participants to the multicast group: all the multicast sources and the multicast receivers. The target is to have several paths from one sender to each destination. Data are relayed and delivered through different paths to the receivers. Hence, it increases the robustness against link breakages. This robustness against the topology changes in mesh-based protocols are however more demanding in terms of bandwidth consumption compared to the tree-based protocols, which are more efficient in terms of resource usage.

The major inconvenience of the first two categories is the vulnerability of the tree. In fact, few link breakages due to mobility or simply due to topology changes may affect a large part of the multicast tree and prevent from delivering the data to the multicast receivers. In contrast, meshes offer more redundancy to overcome this weakness.

- *Flat/overlay structure*: The protocols can also be categorized into two categories, flat and Overlay. In the first one fall the protocols where all the nodes are assumed to handle multicast data and can participate in the multicast structure building and maintenance (tree, mesh). In the second one, multicast nodes of a same group build and maintain a virtual structure on top of physical structure that links all the participants using unicast tunnels. In this case, not all nodes within the network are supposed to know about the multicast protocol routing, they only have to forward the encapsulated multicast data that flow inside the unicast tunnels. It means that only nodes interested in the multicast communication have to take part to the protocol operations. The advantages of this approach are high robustness and reliability, thanks to the use of unicast tunnels. Indeed, unicast communications are acknowledged, which is not the case when using the broadcast to multicast data.

Table 5.1 summarizes the proposed classification with different examples of wireless multicast protocols.

**Table 5.1** Multicast protocol classification

| Protocol | Structure | Flat/Overlay | Standalone |
|---|---|---|---|
| MOLSR [10] | Source tree | Flat | No |
| MAODV [11] | Shared tree | Flat | No |
| ODMRP [12] | Mesh | Flat | Yes |
| MOST [13] | Shared tree | Overlay | No |
| FGMP [14] | Mesh | Flat | No |
| MCEDAR [15] | Mesh | Flat | No |
| AMRoute [16] | Shared tree | Overlay | Yes |
| DDM [17] | Source tree | Flat | No |

## 5.3.2 Protocols Description

### 5.3.2.1 Mesh-Based Example: ODMRP

On-Demand Multicast Routing Protocol (ODMRP) [12] is a mesh-based multicast routing protocol. It is a standalone protocol and does not need any unicast underlying protocol to run on top of it. Moreover, although it was designed as a multicast protocol, ODMRP offers unicast routing capability. It is based on the forwarding group concept where only a subset of nodes forwards the multicast packets. The forwarding group connects through a mesh the sources and the clients of a given multicast group.

In ODMRP, group membership and multicast routes are established and updated by the source on demand. The mesh creation is basically composed of two steps: a request phase and a reply phase. In fact, when a multicast source node has packets to send without any knowledge of the group members, it starts flooding to the entire network a "Join Query" message. This message is broadcast periodically to update the mesh routes and refresh group membership.

When an intermediate node receives a non-duplicate "Join Query" message (duplicates are simply discarded), it stores the upstream node address or updates it as the next hop to reach the source node in its "Routing Table". This message is then re-broadcast if the Time-To-Live value is greater than zero.

When a "Join Query" message reaches a multicast receiver, it creates and broadcasts a "Join Reply" to its neighbors. This message includes among other information, a list of the IP addresses of the sources of the multicast group which the receiver is interested in, and a list of the IP addresses of the next hops (corresponding to a subset of neighbor nodes of this multicast receiver) to reach each source of this multicast group. The neighboring nodes receiving this message process this message only if their address matches one of the entries in the list of the next hop addresses. If so, it means that this node is on the reverse path of one of the multicast sources, and thus is a part of the forwarding group. It then broadcasts its own "Join Reply" built upon the matched entries. The next node address field is filled in by extracting the information from its routing table. This way, the "Join Reply" is propagated by each forward group member until it reaches the multicast sources via the selected paths. This process constructs (or updates) the routes from sources to receivers and builds a mesh of nodes, the forwarding group (see Fig. 5.2).

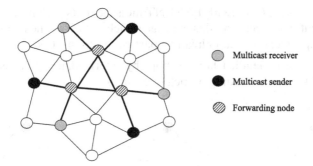

**Fig. 5.2** ODMRP mesh example

In ODMRP, there is no specific message for the leave process. When a multicast source wants to leave a group, it simply stops sending the periodic message "Join Query". Whereas multicast receivers that want to leave the multicast group have only to stop replying to these messages (i.e., stop sending "Join Reply" messages). Consequently, the nodes in the forwarding group being not refreshed disconnect themselves from the mesh.

### 5.3.2.2 Tree-Based Example: MOLSR

The Multicast Optimized Link State routing protocol (MOLSR) [10] belongs to the source tree–based family. The MOLSR protocol takes benefit of the topology knowledge gathered by the OLSR unicast routing protocol with its Topology Control messages exchange to build multicast trees. MOLSR is developed as an extension to OLSR. A multicast tree is built and maintained for any tuple (source, multicast group) in a distributed manner without any central entity and provides shortest routes from the source to the multicast group members. The trees are updated whenever a topology change is detected.

Three steps are distinguished: the tree building, the tree maintenance, and the tree detachment.

Once a source wants to send data to a specific multicast group G, it sends a SOURCE_CLAIM message enabling nodes, which are members of this group, to detect its presence and to attach themselves to the associated multicast tree.

This message is flooded within the ad hoc network using the optimized flooding technique of OLSR. Branches are built in a backward manner: group members that do not know yet about this source try to attach themselves to the corresponding tree.

More specifically, when a group member receives a SOURCE_CLAIM message and it is not already a participant of this (source, multicast group) tree, it attaches itself to the (source, multicast group) tree:

- It looks into the multicast routing table for the next hop to reach the source (the multicast routing table provides shortest routes to all the multicast capable nodes). This next hop becomes its parent in the multicast tree.

- Then it sends a CONFIRM_PARENT message to its parent node.
- The parent node receiving this message attaches itself to the (source, multicast group) tree, if it is not already a participant to this tree.

This message is handled hop by hop, by intermediate multicast routers, which build the corresponding branch (see Fig. 5.3).

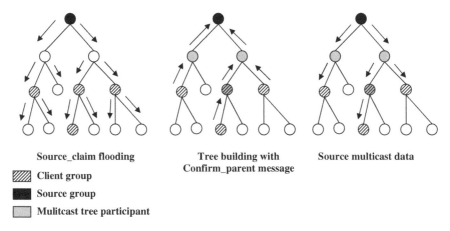

Source_claim flooding            Tree building with            Source multicast data
                                 Confirm_parent message
▨ Client group

■ Source group

▢ Mulitcast tree participant

**Fig. 5.3** MOLSR multicast tree building example

The trees are periodically refreshed, by means of the SOURCE_CLAIM message and the CONFIRM_PARENT message. Notice that topology changes are still detected by the exchange of topology control messages, which is done naturally by OLSR. Thus, trees updates are triggered by the detection of topology changes.

If a node wants to leave the multicast tree and it is a leaf, it detaches itself from the tree: it just sends a LEAVE message to its parent in this multicast tree. If its parent becomes a leaf, and this parent is not a group member, it detaches itself from the tree on its turn.

This message is processed hop by hop and unused branches are deleted automatically.

### 5.3.2.3 Shared Tree Example: MAODV

Multicast Ad hoc On-Demand Distance Vector (MAODV) [11] protocol belongs to the shared-based tree protocols family. MAODV protocol is associated with the Ad hoc On-Demand Distance Vector (AODV) unicast routing protocol, and as such it shares many similarities and packet formats with it. A shared tree is built for each multicast group whenever there are multicast receivers in the network. A multicast group leader is associated with each multicast group. The group leader has the responsibility of the initialization and maintenance of the group sequence number, which is broadcast periodically to the entire network. The tree maintenance and branches repairing is assured by the inter-mediate nodes once they detect link breakages in a reactive manner.

The shared multicast tree is set up on demand by a series of "Route Request (RREQ)/Route Reply (RREP)/Multicast activation (MACT)" messages exchanges. In fact, when a node either wishes to join a multicast group (i.e., the node is interested in receiving multicast data of that group) or find a route to a multicast group, it broadcasts a RREQ message to discover a route to the multicast tree associated with that group. A node receiving a RREQ message first updates or creates a reverse route to the originator of this message. After that, the node checks whether the RREQ is a join request or not. If it is a join request, the node must reply only if it is a member of the requested multicast tree and its multicast sequence number is at least as great as the one contained in the RREQ message. If the RREQ is not a join request, the node can respond only if it has a fresh route to the multicast tree. If the node is not able to reply to the RREQ solicitation, it must re-broadcast this request to the rest of the network. The route reply is unicast back to the source using the reverse path, following the pointers set during the RREQ broadcast process. Once the source node has waited the discovery period to receive RREPs, it selects the best route (in terms of number of hops, for example) to the multicast tree and unicasts the next hop along that route a MACT message. This message activates the route.

Link breakages are repaired by means of expanding ring search started by the downstream node in order to reconnect the downstream branches to the rest of the multicast tree. The downstream node sends a fresh join request contain-ing the distance in terms of the number of hops to the group leader and the last known sequence number. Only tree members having smaller distance to the group leader, which means that they are closer, and having a greater sequence number can respond to this request (see Fig. 5.4).

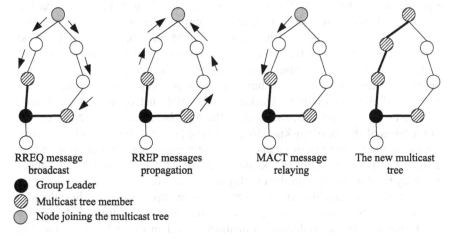

RREQ message          RREP messages          MACT message          The new multicast
broadcast             propagation            relaying              tree

● Group Leader
▨ Multicast tree member
◯ Node joining the multicast tree

**Fig. 5.4** MAODV multicast tree building example

Nodes that wish to leave the group has only to prune themselves by unicasting a MACT message to the next hop along the path to the multicast tree with a prune flag set. Only leaf nodes can prune themselves at anytime. They cannot leave the tree if they are playing the role of a router for any other nodes in the multicast group.

Notice that if the node broadcasts a join request to a multicast group and does not receive any reply after a predefined number of attempts, it takes the lead of this multicast group, initializes the multicast group sequence number, and starts broadcasting a Group Hello message (GRPH) periodically to announce to the network that it is now the leader of that group.

### 5.3.2.4 Overlay Example: MOST

The Multicast Overlay Spanning Tree protocol (MOST) [13] as its name indicates is an overlay protocol, where only nodes that have joined a multicast group (generally a subset of nodes in the network) participate in the overlay tree building. MOST needs an underlying protocol that provides link state information. It is developed as an extension to the OLSR unicast routing protocol. It is a fully distributed protocol where each node computes a minimum shared spanning tree for each multicast group autonomously. A minimum shared tree that connects all the multicast nodes can be calculated by applying the well-known Prim's algorithm. The links of the resulting overlay tree are the unicast tunnels used to forward multicast data among the members. The tree is recalculated whenever the topology changes.

When a node wants to join or to leave a group it broadcasts a join or a leave message. These messages are relayed to the whole network using the Multi-Point Relay (MPR) technique of the OLSR protocol. These messages enable the nodes to learn about the group membership. When a node starts broadcasting a join message periodically, it starts declaring also periodically to the rest of the network its neighbor set. The reason behind that is in OLSR only partial topology is known, but still each node can compute the shortest paths between itself and the other nodes in the network. In MOST, multicast nodes must calculate the same shared tree based on the shortest paths between each pair of multicast nodes in a distributed manner. So, in order to apply the Prim's algorithm, they must be able to have the same topology vision on each node in terms of number of hops. Declaring the whole list of neighbors of each multicast node ensures this, and each of them is able to compute the same tree on the basis of the topology knowledge and the group membership knowledge. For each group, each multicast client computes periodically the corresponding shared overlay tree in order to take into account the eventual topology changes that may have an impact on the overlay tree structure.

When the node decides to leave a multicast group, it stops sending the join messages and broadcast a leave message. Meanwhile, it continues acting as a group member for a predefined transition period in order to avoid the tree breakage and give enough time to the rest of the multicast nodes to recalculate a new overlay tree. During this transition period, the node must continue multicast data relaying in order to guarantee a smooth handover.

## 5.4 Thoughts for Practitioners

Practitioners could be from different work fields and communities: students, researchers, network managers, engineers, etc. Hence, their needs and their use of the multicast could be different. In fact, network managers and engineers would simply use an existing protocol, and their needs are essentially limited to the choice and deployment of a solution off-the-shelf. In contrast, students or researchers may have to test, study, and simulate the behavior of a multicast protocol deeply. They may need to modify or design a new protocol to answer a specific problem. In this section we present some issues in order to face a situation where multicast is needed.

The paradigm for considering a multicast protocol that will be used in a real environment includes three steps, detailed hereafter:

1. *Observation and context description*: The goal of this step is to define the multicast needs and define problem statement, taking into account the specificities and the criteria defined in Section 5.2. For that purpose, one has to answer the following questions:

   - Is multicast needed for a research project or for real deployment (demonstration or industrial purpose)?
   - What are the characteristics of our network: How large are the multicast groups? What kind of mobility should we support? Do we need high reliability? What about the group and topology dynamicity? Do we have requirements in throughput and delay? Etc.

2. *State of the art and off-the-shelf solutions*: At this step, one has to review the existing solutions, learn from experience feedback, and use the know-how.

   Notice that there is no standardized solution. In fact, wireless multicast for ad hoc networks are still at a research step, and there is no IETF document that recommends a particular protocol. All the existing solutions are basically originated from research communities without any large deployment nor feedback from different parts.

3. *Actions*: The actions to be taken as stated before depend on the practitioner type: decider, network manager, end user in a community, or researcher. The actions to be taken depend also on the nature of the project (research, or real deployment) and on the amount of time dedicated to this task. The actions are the following:

   - While a network manager will select or adapt an existing solution to fulfill the requirements, the researcher will design a new solution from scratch to meet the specificities of the project. At this step, the practitioner has to make tradeoffs between the different criteria (e.g., reliability versus mobility/scalability) if needed.

- Evaluate the performance by simulation or analytical modeling to validate the design principles in case of a complete design. And in case of a real deployment, the performance evaluation gives an idea about the behavior of the future network, and allows choosing the best values of different parameters of the multicast protocol.
- Implement on a real platform: the multicast stack has been designed for wired networks and it is not adapted to route multicast data within a MANET. Thus, we need some hacking to emulate the multicast routing when implementing a new protocol, and must patch the operating system when we want to install the binary code on a machine.
- Test and measure on a real platform: Simulations provide interesting guidelines to improve the solutions. However, they are not sufficient in a wireless environment because of the randomness in the air transmissions (obstacles, fading, multipath, interferences, etc). That is why it is required to test an implementation on a real platform, to obtain a better accuracy of the protocol behaviour in real conditions and validate the parameters values chosen during the design step. In order to get tests and measurements, representative of the real environment, the practitioner, has to cope with different aspects related to:
  - ○ The distributed nature of the network,
  - ○ The dynamicity of the topology and group membership,
  - ○ Non-uniform traffic rates,
  - ○ Mobility scenario.

Once the experimentation scenarios are chosen, and due to the distributed nature of the mobile ad hoc networks, we should not forget that we need the help of several persons to play these scenarios. In fact, mobility scenarios require a strong coordination and synchronization between all people moving around or driving the vehicles with embedded wireless nodes. During these experimentations, we must plan for an adequate tool to collect the measures in a distributed manner. After measurement campaigns, the large amount of collected data must be analyzed and exploited in order to improve the multicast protocol. A performance evaluation by measurements conducted on a real platform represents a very long task.

## 5.5 Reliable Multicast Protocols

Due to frequent and unpredictable topology changes in MANETs, the packet delivery rate of multicast messages can be low, lower than required by the applications such as software distribution and access to distributed databases. Reliable multicast protocols have been introduced to improve the delivery rate of messages multicast to the multicast group members. Their goal is to ensure that all multicast group members receive, certainly or with a high probability, the data multicast in this group, provided they are reachable.

## *5.5.1 Classification of Reliable Multicast Protocols*

This delivery guarantee is brought by the use of recovery mechanisms. Four mechanisms can be identified, leading to four reliable multicast families [18]:

- *ARQ-based* (Automatic Retransmission reQuest): receivers send positive (ACK) or negative (NACK) acknowledgments to the source that retransmits lost packets. This process is repeated until all packets are correctly received by all members. This mechanism is subject to feedback implosion, when the size of the multicast group increases. To limit this phenomenon, the source can be assisted by other designated nodes. Examples studied in the next subsection are RMA [19] and RALM [20].
- *FEC-based* (Forward Error Correction): lost packets can be rebuilt from redundant information contained in the received packets. More precisely, the source encodes $m$ packets into $n>m$ packets and multicasts them. For a multicast group member, it suffices to receive any $m$ out of $n$ packets to be able to rebuild the original $m$ packets. This mechanism tolerates packet losses as long as the loss rate is lower than $1-m/n$. Otherwise, it must be combined with another recovery mechanism. In the RMDP protocol [21], the additional mechanism is ARQ with NACK.
- *Gossip-based*: receivers recover lost messages by gossiping with their neighbors. They periodically exchange their list of missing messages and their list of most recently received messages. This mechanism allows them to recover their missing messages by neighboring retransmission (see, for instance, AG [22] and RDG [23] detailed in the next subsection).
- *Epidemic protocols* can be seen as a specific case of gossip protocols, where the gossip message, instead of being sent to a neighbor, is sent to all neighbors of the sending node. The EraMobile protocol [24] is an example of this family.

The type of guarantee [25] provided by a reliable multicast protocol depends on the recovery mechanism used. It is deterministic for an ARQ-based or FEC-based protocol: all packets are delivered to all group members, provided that they are reachable. It is only probabilistic for a gossip-based or epidemic-based protocol: packets are delivered with a high probability.

Some reliable multicast protocols, such as those belonging to the ARQ-based family, require the sources to know the group membership. Gossip-based protocols have no such requirement.

The recovery mechanism used (i.e., retransmission of the lost message or transmission of redundant information enabling to rebuild the lost message) relies on:

- multicast for ARQ-based, FEC-based, and epidemic protocols,
- unicast for strict gossip-based protocols.

We can also distinguish between the reliable multicast protocols integrating a multicast routing protocol (e.g., RMA, RMDP, AG, and RDG), and those running on top of an underlying routing protocol (e.g., RALM). This latter, unlike the others, includes a congestion control based on a window mechanism similar to the one used by TCP.

We then obtain the classification illustrated in Table 5.2. In the next subsection, we describe a protocol representative of each family.

**Table 5.2** Classification of reliable multicast protocols

| Family | Delivery guarantee | Requires the group membership knowledge | Protocol | Feedback | Recovery | Requires an underlying routing protocol |
|--------|--------------------|-----------------------------------------|----------|----------|----------|-----------------------------------------|
| ARQ-based | Deterministic | Yes | RMA | Ack | Multicast | No |
|  |  |  | RALM | Nack/ Ack | Multicast | multicast routing |
| FEC-based | Deterministic | No in general, but yes if combined with ARQ | RMDP | Nack | Multicast | Multicast routing |
| Gossip-based | Probabilistic | No | AG | Gossip | Unicast | Multicast routing |
|  |  | Partial | RG | Gossip | Unicast | On-demand unicast routing |
| Epidemic | Probabilistic | No | EraMobile | Gossip | Multicast | No |

## 5.5.2 Description of Reliable Multicast Protocols

### 5.5.2.1 ARQ-Based Example: RMA and RALM

- *RMA* (Reliable Multicast Algorithm) [19] belongs to the ARQ family of reliable multicast protocols. As such, it assumes that each source knows the group membership. For any group member, if the source knows a route to that member, the information to multicast is routed along that path, using the MKNOWN message. Otherwise, it is broadcast using the MUNKNOWN message. In order to reduce the overhead, multicast messages with the same next hop are grouped together and similarly for broadcast messages. A destination acknowledges the received message back to the source through the MACK message if it is the first receipt, and the BMACK message otherwise. If the source detects that a receiver has not acknowledged a message, it retransmits it. The originality of this protocol resides in the criterion used by the source to select the best path to a destination; it is no longer the shortest path, as usual, but the path with the highest lifetime. Each node maintains a routing table including the usual information (destination IP address, next hop IP address) with in addition some quality of service parameters (path bandwidth, path lifetime, and for each active multicast session: the number of members and non-members on the path). The aim is to favor paths with the highest lifetime and involving the smallest number of non-members.

A new node can join a multicast session by broadcasting a JOIN message through the network. Similarly, any node that wants to leave a multicast session broadcasts a LEAVE message through the whole network.

The main advantage of this protocol resides in its reliability, improved by the use of link lifetime. The overhead is reduced by the use of message grouping. However, all receivers have to acknowledge the received packets to the sender, causing feedback implosion. This drawback vanishes with RALM.

- *RALM* (Reliable Adaptive Lightweight Multicast) [20] runs on top of an underlying multicast routing protocol and provides a reliable congestion controlled multicast similar to this used in TCP. The control overhead is reduced by selecting the multicast receiver, called feedback receiver, in charge of sending its positive or negative acknowledgments back to the source. The selection is round robin. The sender window size is adjusted according to the feedback receiver response. It is halved upon a loss detection and increases linearly otherwise. Lost packets are requested one at a time to the source by the feedback receiver. The source multicasts them again, one at a time.

### 5.5.2.2 FEC-Based Example: RMDP

- *RMDP* (Reliable Multicast Distribution Protocol) [21] belongs to the FEC-based family of reliable multicast protocols; If the FEC mechanism fails to rebuild the lost messages, an ARQ mechanism is used in addition. In the latter case, the feedback storm is limited by spreading the generation of feedback from receivers over a suitable time interval and avoiding duplicate feedback. The source splits the information to multicast in groups of $m$ packets. These $m$ packets are encoded in $n>m$ packets, which are multicast. A receiver must wait the receipt of $m$ packets before starting to decode them. The main advantage of this protocol is that if the redundancy degree has been correctly dimensioned, no feedback is generated. However, the response time, time elapsed between the multicast transmission request and the multicast delivery at the receivers, can be large. The control overhead (i.e., the transmission of $n>m$ packets instead of $m$) is not negligible.

### 5.5.2.3 Gossip-Based Example: AG and RDG

- *AG* (Anonymous Gossip) [22] belongs to the gossip family of reliable multi-cast protocols. It proceeds in two phases. In the first phase, any multicast routing protocol is used to multicast the message to the group. In the second phase, lost messages are recovered by periodic gossiping with other group members. (The closest neighbors are chosen with high probability.) More precisely, each node randomly selects one of its neighbors to which it sends a gossip message. A node receiving the gossip message either accepts the gossip request and sends back the lost message to the initiating node or propagates the gossip request to another node.

  This protocol has the advantage of not requiring the knowledge of the group membership. However, the delivery guarantee is only probabilistic: some messages cannot be recovered.

- *RDG* (Route-Driven Gossip) [23] is another example of gossip-based protocol, where the nodes use their partial view of the group membership provided by the underlying unicast routing protocol. (An on-demand routing protocol is requested.)

  A node wanting to join the group floods the network with a GROUP-REQUEST message. Upon receipt of this message, a multicast group member updates its local view of the group membership and replies with a probability $P_{reply}$. The new joining node updates its local view from the responses received. The underlying unicast on-demand routing protocol records the route of each incoming packet.

  RDG is a pure gossip-based protocol insofar as messages are multicast only by means of gossip messages. Periodically, each member of the group sends a gossip message to $F$ other nodes randomly chosen in its local network. Each message is gossiped $t$ times. The gossip message includes the new received packets, the identifiers of the most recent missing packets, the local view of the sender, and a flag indicating whether the sender wants to leave the group. Upon receipt of a gossip, a member updates its local view of the group membership, stores the new packets, and replies to the gossip requests.

  In the topology-aware variant, close members are selected with a high probability in the gossip process.

### 5.5.2.4 Epidemic Protocol Example: EraMobile

- *EraMobile* [24] is a reliable multicast protocol representative of the epidemic family. The source broadcasts the new packet once, and all nodes participate in the dissemination through periodic gossips with their neighbors. No group membership knowledge is requested.

  A gossip message is sent to all neighbor nodes. It includes the list of the most recent messages that are not yet stable (i.e., gossiped a number of times less than the stability threshold).

Upon receipt of a gossip message, a node finding the identifier of a missing message requests it to the gossip sender, which will transmit it if it has not exceeded the maximum number of messages in this gossip period. In order to reduce the control overhead in high-density networks, the gossip period is extended, the number of messages sent in gossip period is decreased, and a random jitter on message transmission time is used.

## 5.6  Directions for Future Research

### 5.6.1  QoS Multicast Routing

Multicast algorithm design for ad hoc networks is a complex problem in itself. One must take into account the characteristics of these networks such as the versatility of the shared wireless medium and the dynamicity of the nodes within the network. Most multicast protocols deal with multicast structure building; some of them try to improve the reliability, but only few of them are addressing the QoS requirements like QAMNET [26], QMR [27].

- *QAMNET*: the basic idea of QAMNET [26] is to extend the existing approaches of mesh-based multicasting and unicast QoS provisioning (like SWAN [28]). The source floods periodically the network with a message containing the required bandwidth and a special field called bottleneck field, which indicates the bottleneck bandwidth found along the route (BB). The intermediate nodes relay the first non-duplicate copy of this message after updating the BB field if this latter is greater than the local available bandwidth (the local capacity of a node is calculated as in SWAN). The intermediate nodes must also set a pointer to the upstream node. The receiver (multicast destination) waits a small period in order to collect all the request messages coming from different branches. Hence it can select the branch containing the largest available bandwidth, which must be greater than the required one. Consequently, it sends a reply packet toward the source, following the pointers set during the relaying of the request message.
- *QMR* (QoS Multicasting Routing) [27] is also a multicast mesh-based protocol coupled with its own mechanism to evaluate the residual capacity locally on each node. In QMR, the sources flood the network periodically with the bandwidth requirements. The intermediate nodes forward the request only if they have enough bandwidth locally; otherwise the request is rejected. The nodes have to interact with their MAC layer to estimate the available bandwidth. The multicast receivers send back to the source their replies to establish the mesh.

These protocols build a mesh structure in a reactive manner. Thus, there is neither control nor knowledge of the constructed mesh. In this case, the use of the overall network capacity is not optimized, and new nodes may be prevented from joining the multicast communicating groups. Both protocols try to estimate the residual bandwidth with different mechanisms, but they do not consider the interferences problem efficiently during the bandwidth reservation. In fact, in QAMNET and QMR, bandwidth reservation is made locally on multicast mesh nodes. The neighboring nodes have to evaluate continuously the available bandwidth on their own; hence, their residual capacity is updated only when the multicast data flows start later on. Nodes transmissions from a same multicast group mesh may interfere in that case.

## 5.6.2 *Network Coding*

In classical wired or wireless networks, coding is restricted to the sources and the end receivers, whereas intermediate nodes are only in charge of routing and copying packets. In their seminal work [29], Ahlswede et al. have introduced the concept of network coding where coding is performed by the intermediate nodes also. They have shown that classical routing solutions are generally unable to achieve the multicast capacity, but network coding is. The multicast capacity is defined as the maximum rate that a source S can send information to a group of receivers. It is equal to the minimum for each receiver T, of the maximum flow rate sustained between S and T.

Chou et al. [30] have proposed a distributed scheme for practical network coding in real packet networks. This scheme achieves a throughput close to the theoretical optimum, the multicast capacity, while being robust to packet loss, random delays, and topology changes.

Work is in progress to apply this concept to wireless ad hoc networks. In [31], Adjih et al. proved the optimality of network coding in case of dense and homogeneous networks (e.g., grid networks). They are now designing a practical protocol, called DRAGONCAST, for wireless ad hoc networks that are not dense.

## 5.7 Conclusions and Synthesis

Wireless ad hoc networks have limited resources, and are subject to radio interferences. They also cope with frequent topology changes, increased by mobility. These specificities make the multicast protocols designed for wired networks inefficient in MANETs. The main goal of the multicast protocols is to save bandwidth by avoiding unnecessary data transmissions. But maintaining a multicast structure implies additional overhead. Using the same tree to forward multicast data for all the group members means that when a multicast packet is lost (notice that a multicast packet is generally not acknowledged by the receiver(s)) or when a link fails, a subset of the tree is prevented from receiving the multicast data. Mesh structures may offer more redundancy in some situations to cope with this problem. Multicast overlay structures use unicast tunnels and are less sensitive to packet loss than the basic multicast trees or meshes.

If a deterministic or probabilistic delivery guarantee is requested, a reliable multicast protocol must be used. It ensures that all members of the multicast group receive, certainly or with a high probability, the data multicast in the group, provided they are reachable. In the ARQ-based protocols, the members of the multicast group send a positive or negative acknowledgement to the source that is in charge to retransmit the lost packets. Such protocols, subject to feedback implosion, do not scale well and exhibit good performance only in small multicast groups. FEC-based multicast protocols rely on redundant coding of packets to recover from losses. The redundant coding is determined according to the bit error rate on the wireless links. It can be combined with an ARQ mechanism. Gossip-based and epidemic protocols have the advantage of not requiring the knowledge of the group membership. In both protocols, lost messages are recovered by gossiping with the neighbors, gossiping appears as a background task. In epidemic protocols, the gossip message is sent to all one-hop neighbors of the sending node, whereas in gossip-based protocols it is sent to one neighbor only, this neighbor can forward it. Furthermore, some reliable multicast protocols enhance an existing multicast routing protocol, and others integrate it. Some rely on the topology information brought by the underlying unicast routing protocol to improve their performance. The choice of a reliable multicast protocol is clearly a tradeoff between reliability and scalability/ mobility.

In this chapter, we have identified two research directions related to multi-cast protocols: QoS support and network coding. QoS support in multicast protocol is needed for multimedia applications that have strong requirements on the end-to-end throughput and sometimes on the end-to-end delays. Each node must estimate the bandwidth available in its neighborhood. This bandwidth is then used to select the routes. Radio interferences must be taken into account to ensure that enough bandwidth is granted to the multicast stream. Network coding is a promising area; it achieves the theoretical optimum throughput. The design of practical protocols that can be applied to real wireless networks is in progress.

## Terminologies

*Multicast* is a transmission technique that allows the sender to send only one packet to a group of destinations. The packet is replicated only when needed to reach each destination.

*Multicast group* contains all the destinations interested in receiving data multicast within this group.

*Multicast client* is a member of the multicast group.

*Multicast receiver* is a multicast client.

*Group membership* determines which nodes belong to the group.

*Tree-based structure is* a structure built using a tree.

*Source-based structure is* a multicast tree structure having one multicast source as a root and connecting all members of a same multicast group.

*Shared tree structure is* a single multicast tree structure that connects all members of a same multicast group without considering multicast sources.

*Mesh-based structure is* a structure containing all the participants to the multicast group (the multicast sources and the multicast receivers). This structure offers several paths from one sender to each destination.

*Overlay structure* is a virtual structure built on top of a physical structure.

*Reliable multicast protocol* provides a delivery guarantee (probabilistic or deterministic) to multicast receivers, provided they are reachable.

*ARQ-based (Automatic Retransmission request)*: with such a protocol, receivers send positive or negative acknowledgments to the source in charge of retransmitting lost packets.

*FEC-based (Forward Error Correction):* with such a protocol, sources split the information to multicast in $m$ packets that are encoded in $n>m$ packets. The receipt of $n$ packets among the $m$ allows to rebuild the original information.

*Gossip-based protocol* recovers from lost packets by gossiping with neighbors. Each node gossips with a selected neighbor.

*Epidemic multicast protocol:* a node periodically gossips with all its one hop neighbors, taking advantage of the semi broadcast nature of the wireless interface.

## Questions

1. Goal of a multicast protocol? Why is it different from a broadcast one?
2. Does a wireless ad hoc network increases the complexity of designing a multicast protocol, and why?
3. Give a possible classification of wireless multicast protocols
4. What does a reliable multicast protocol ensure? Which type of delivery guarantee does it provide?
5. What is the problem of the ARQ-based family of reliable multicast protocols? How can it be alleviated?
6. Why the epidemic family can be interesting in a wireless ad hoc network?
7. Give a possible classification of the reliable multicast protocols?
8. What is network coding? Why is it a promising research area?
9. How the multicast mesh structure and the overlay structure can increase the robustness?
10. What are the main steps to select a multicast protocol?

# References

1. A. Adams, J. Nicholas, W. Siadak, "Protocol Independent Multicast – Dense Mode (PIM-DM: Protocol Specification (Revised)", IETF RFC 3973, January 2005.
2. B. Fenner, M. Handley, H. Holbrook, I. Kouvelas, "Protocol Independent Multicast – Sparse Mode (PIM-SM): Protocol Specification (Revised)", IETF RFC 4601, August 2006.
3. D. Waitzman, C. Partridge, S. Deering, "Distance Vector Multicast Routing Protocol", IETF RFC 1075, November 1988.
4. A. Ballardie, "Core Based Trees (CBT version 2) multicast routing: protocol specification", IETF RFC 2189, September 1997.
5. J. Moy, "Multicast extensions to OSPF", IETF RFC 1584, March 1994.
6. Hipercom project, "Simple Multicast OLSR (SMOLSR)", http://hipercom.inria.fr/SMOLSR-MOLSR/
7. C. Perkins, E. Belding-Royer, S. Das, "Ad hoc on-demand distance vector (AODV) routing", RFC 3561, 2003.
8. T. Clausen, P. Jacquet, C. Adjih, A. Laouiti, P. Muhletaler, P.Minet, A. Qayyum, L. Viennot, "Optimized link state routing protocol", RFC 3626, 2003.
9. H. Holbrook, B. Cain, B. Haberman, "Using Internet Group Management Protocol Version 3 (IGMPv3) and Multicast Listener Discovery Protocol Version 2 (MLDv2) for Source-Specific Multicast", IETF RFC 4604, 2006.
10. A. Laouiti, P. Jacquet, P. Minet, L. Viennot, T. Clausen, C. Adjih, "Multicast Optimized Link State Routing", INRIA research report RR-4721, 2003.
11. E. Royer, C. Perkins, "Multicast Ad hoc On-Demand Distance Vector (MAODV) routing", IETF, Internet Draft: draft-ietf-manet-maodv-00.txt, 2000.
12. S. Lee, W. Su, M. Gerla., "On demand multicast routing protocol in multihop wireless mobile networks", ACM/Baltzer Mobile Networks and Applications, 2000.
13. G. Rodolakis, A. Meraihi Naimi, A. Laouiti, "Multicast overlay spanning tree protocol for Ad Hoc networks", WWIC, Coimbra, Portugal, 2007.
14. C.-C. Chiang, M. Gerla, L. Zhang, "Forwarding group multicast protocol (FGMP) for multihop, mobile wireless networks", ACM/Baltzer Journal of Cluster Computing, Vol. 1 (2), pp. 187–196, November 1998.
15. P. Sinha, R. Sivakumar, V. Bharghavan, "MCEDAR: Multicast core-extraction distributed Ad Hoc routing", Proceedings of WCNC '99 – the IEEE Wireless Communications and Networking Conference, pp. 1313–1317, New Orleans, USA, 21–24 September 1999.
16. M. Liu, R. R. Talpade, A. McAuley, E. Bommaiah, "AMRoute: Ad Hoc multicast routing protocol", Center for Satellite and Hybrid Communication Networks Technical Research Report, University of Maryland, August 1999.
17. L. Ji, M. S. Corson, "Explicit multicasting for mobile ad hoc networks", ACM Mobile Networks and Applications, Vol. 8 (5), pp. 535–549, October 2003.
18. B. Ouyang, X. Hong, Y. Yi, "A comparison of reliable multicast protocols for mobile ad hoc networks", IEEE Southeast Conference, 2005.
19. T. Gopalsamy, M. Singhal, D. Panda, P. Sadayappan, "A reliable multicast algorithm for mobile ad hoc networks", ICDCS, July 2002.
20. K. Tang, K. Obraczka, S.J. Lee, M. Gerla, "A reliable congestion-controlled multicast transport protocol in multimedia multi-hop networks", IEEE WPMC, Honolulu, October 2002.
21. L. Rizzo, L. Vicisano, "RMDP: a FEC-based reliable multicast transport protocol for wireless environments", ACM Mobile Computing and Communications Review, Vol. 2(2), April 1998.
22. R. Chandra, V. Ramasubramanian, K. Birman, "Anonymous gossip: improving multicast reliability in mobile ad hoc networks", ICDCS, April 2001.

23. J. Luo, P. Eugster, J.P. Hubaux, "Route driven gossip: probabilistic reliable multicast in ad hoc networks", INFOCOM'03, San Francisco, March 2003.
24. O. Ozkasap, Z. Genc, E. Atsan, "Epidemic-based approaches for reliable multicast in mobile ad hoc networks", ACM SIGCOM Operating Systems Review, special issue on self-organizing systems, Vol. 40 (3), July 2006.
25. E. Vollset, P. Ezhilchelvan, "A survey of reliable broadcast protocols for mobile ad hoc networks", Technical report, CS-TR792, University of NewCastle upon Tyne, 2003.
26. H. Tebbe, A. Kassler., "QAMNet: Providing Quality of Service to Ad-hoc Multicast Enabled Networks", ISWPC, Thailand, 2006.
27. M. Saghir, T. C. Wan, R. Budiarto, "Load Balancing QoS Multicast Routing Protocol in Mobile Ad hoc Networks", AINTEC, Bangkok, Thailand, 2005.
28. G. Ahn, A. T. Campbell, A. Veres, L. Sun, "Supporting Service Differentiation for Real-Time and Best Effort Traffic in Stateless Wireless Ad hoc Networks (SWAN)", IEEE Transactions on Mobile Computing, September 2002.
29. R. Ahlswede, N. Cai, S-Y Li, R. Yeung, "Network information flow", IEEE Transactions on Information Theory, Vol. 46 (4), July 2000.
30. P. Chou, Y. Wu, K. Jain, "Practical network coding", Allerton Conference on Communication, Control and Computing, 2003.
31. C. Adjih, S-Y. Cho, P. Jacquet, "Near optimal broadcast with network coding in large sensor networks", 1st International Workshop on Information Theory for Sensor Networks, Santa Fe, June 2007.

# Chapter 6
# Broadcast in Ad Hoc Networks

Justin Lipman, Hai Liu, and Ivan Stojmenovic

**Abstract** Broadcast is the process of sending a message from one node to all other nodes in an ad hoc network. It is a fundamental operation for communication in ad hoc networks as it allows for the update of network information and route discovery as well as other operations. The chapter presents a comprehensive review and analysis of existing localized solutions on broadcast, where only local knowledge is required. The techniques reviewed include optimized broadcast techniques, such as multipoint relay and dominating set-based broadcasting with fixed transmission radii, resource awareness, localized minimum energy broadcasting with adjustable transmission radii, and solutions for increasing reliability of broadcasting are also reviewed. Further, the chapter highlights the use of broadcast in route discovery and new approaches to route discovery based upon self-selecting search techniques as opposed to traditional broadcast approaches.

## 6.1 Introduction

### 6.1.1 Background

Broadcast forms the basis of all communications in ad hoc networks. The simplest form of broadcast in an ad hoc network is referred to as *blind flooding*. In blind flooding, a node transmits a packet, which is received by all neighboring nodes that are within the transmission range. Upon receiving a broadcast packet, each node determines if it has transmitted the packet before. If not, then the packet is retransmitted. This process allows for a broadcast packet to be disseminated throughout the ad hoc network. Blind flooding terminates when all nodes have received and transmitted the packet being broadcast at least

J. Lipman (✉)
Intel Asia Pacific, Research and Development Ltd., 2nd floor, Block 4, 555 Dong Chuan Road, Shanghai, Zizhu Science Park, 200241, PRC
e-mail: justin.lipman@gmail.com

S. Misra et al. (eds.), *Guide to Wireless Ad Hoc Networks*,
Computer Communications and Networks, DOI 10.1007/978-1-84800-328-6_6,
© Springer-Verlag London Limited 2009

once. As all nodes participate in the broadcast, blind flooding suffers from the *Broadcast Storm Problem* [38]. The broadcast storm problem states that, in a CSMA/CA network, blind flooding is extremely costly and may result in the following:

- *Redundant rebroadcasts* – occur when a node decides to rebroadcast a message to its neighbors; however, all neighbors have already received the message. Thus the transmission is redundant and useless.
- *Medium contention* – occurs when neighboring nodes receive a broadcast message and decide to rebroadcast the message. These nodes must contend with each other for the broadcast medium.
- *Packet collision* – because of the lack of the back-off mechanism, RTS/CTS dialog, and the absence of CD, collisions are more likely to occur and result in lost or corrupted messages.

Figure 6.1 shows redundant broadcast and contention of the broadcast storm problem when performing a blind flood. In Fig. 6.1, node *A* initiates a broadcast of a message and the message is received by nodes *B* and *C*. According to blind flooding, *B* and *C* rebroadcast the message if they had not broadcasted it before. Therefore, *D* will receive the message and also rebroadcast the message if there is no collision. It may cause the following problems:

- Since node *A* is within the transmission range of nodes *B* and *C*, it will receive two redundant copies of the message from nodes *B* and *C*. This is also the case with nodes *B* and *C*, which receive the message from node *D* and from each other.
- Nodes *B* and *C* may contend for the broadcast medium in the shadow area as shown in Fig. 6.1. If there are more nodes in the shadow area, there will be an increase in contention for the broadcast medium.

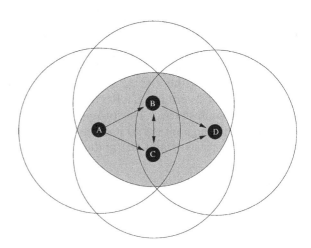

**Fig. 6.1** Broadcast storm problem

- If nodes $B$ and $C$ broadcast at approximately the same time, there is a possibility of a packet collision at node $D$. Even if nodes $B$ and $C$ broadcast the message at a different time, node $D$ will receive a total of two broadcast messages (one each from nodes $B$ and $C$).

From Fig. 6.1 only two rebroadcasts (nodes $A$ and $B$ or nodes $A$ and $C$) are actually necessary for all nodes to receive the broadcast from node $A$.

## 6.1.2 Overview of the Chapter

There exists significant literature describing various approaches to alleviate the broadcast storm problem. These approaches are *localized* or *globalized* methods, depending upon the degree of neighbor topology information, which is required to make broadcast decisions. Globalized approaches are centralized in nature. They require global topology information and attempt to determine an optimal, in terms of energy efficiency, broadcast tree. The computing of such optimal broadcast trees is NP-Hard [33].

However, given the dynamic nature of ad hoc networks, a centralized approach is neither desirable nor feasible as the cost of obtaining global topology information and maintaining the broadcast tree is restrictive in terms of overhead. Localized approaches are distributed in nature and require only localized neighbor topology information. Compared with globalized approaches, localized approaches are able to adapt to the topology change in ad hoc networks. Thus localized approaches are more appropriate to be applied in ad hoc network environments and are the focus of this chapter.

This chapter is organized as follows:

Section 6.2 describes and analyses various approaches of optimized broadcast methods that utilize a fixed transmission radius. These approaches include Multipoint Relay (MPR), Connected Dominating Sets (CDS), and Active/Passive Clustering. Given the resource constrained nature of ad hoc networks, the section also describes a novel resource-aware approach that accounts for a node's resource constraints (such as battery power) and potential device or user constraints.

Mechanisms in Section 6.2 are optimized to alleviate the broadcast storm problem by reducing the number of redundant rebroadcasts. This is achieved by limiting the number of participating nodes. These approaches do not scale well as their performance is degraded when node density increases. Section 6.3 describes the optimized graph theoretic broadcast approaches, which are based upon minimum spanning tree (MST) and relative neighborhood graphs (RNG). MST and RNG utilize radio transmission power control to limit the transmission radius, thereby significantly reducing energy consumption and alleviating the broadcast storm problem.

Optimized broadcast mechanisms alleviate the broadcast storm problem by reducing redundant broadcasts or varying transmission power. However, the very nature of blind flooding and the large number of redundant transmissions is able to achieve higher reliability than optimized broadcast. It has been shown in literature that the reliability of optimized broadcast is drastically affected by background traffic when compared to the blind flooding. Section 6.4 describes new approaches that address reliability issues while providing optimized broadcasting.

Section 6.5 describes the application of broadcast in many reactive and proactive ad hoc routing protocols. The section explores current research that utilizes self-selection as the basis of a search strategy to enhance reactive route discovery. We conclude the chapter in Section 6.6.

## 6.2 Optimized Broadcast with Fixed Radius

This section explores the optimized broadcast mechanisms that utilize a fixed transmission range and focus on the reduction of unnecessary rebroadcasts, which lead to the broadcast storm problem.

### 6.2.1 Heuristic-Based Broadcasting

Heuristic-based broadcasting methods require careful selection of parameters and thresholds, which are closely related to ad hoc network environments. Their performance is highly dependent on the selected parameters and thresholds in the heuristic.

Ni et al. [38] and Tseng et al. [56] proposed several heuristics on broadcast:

- *Counter-based* – the decision of rebroadcast is based upon a threshold value for the number of duplicate packets received by the broadcasting node. If the number of duplicate packets is less than the threshold value, then the node will rebroadcast. Otherwise, it will not rebroadcast. An expected additional coverage function may be defined, which shows that the more times a host has heard the same broadcast packet, the less additional coverage the host contributes if it rebroadcasts the packet.
- *Distance/location-based* – the heuristic may involve distance in a relative sense – physical distance between nodes or the transmission power required. Each node is equipped with a GPS device or is able to determine signal strength of a neighboring node. Given the distance or location of broadcasting nodes, it is possible to calculate the expected additional coverage (in terms of area) a node may contribute by rebroadcast.
- *Probability-based* – the decision of rebroadcast is based upon a random probability. This probability may be as simple as flipping a coin or it may be more complex involving probabilities that include parameters such as node density, duplicate packets received, battery power, or a node's participation/benevolence within the network.

## 6.2.2 Neighbor Coverage–Based Broadcast

In Neighbor Coverage–Based (NCB) broadcast, nodes periodically or dynami-cally broadcast beacon messages to advertise their own existence and also discover the existence of neighboring nodes within the transmission range (one hop). Beacon messages may typically contain the broadcasting node's address and the neighboring nodes that the node may be aware of. Thus, the information of neighbor topology within two hops can to be obtained. The use of neighbor information allows the link state topology of nodes to be deter-mined. It is also useful in situations where GPS may not work, such as indoor applications. The exchange of beacon messages allows for attaching additional information about neighboring nodes. The additional information may include a node's remaining battery power, any user-based constraint, physical coordinates acquired through a GPS device, signal-to-noise ratio (SNR) measurements (acquired from the MAC layer), and possible device characteristics such as maximum broadcast power.

However, the use of beacons for neighbor discovery may suffer from various problems: (i) Consider a $n$-node degree network: a node wishing to discover its local two-hop topology must first wait for each of its $n$ neighbors to receive beacons from their neighbors. Thus, at least $n^2$ messages are required to discover two-hop topology; (ii) Nodes do not transmit beacons simulta-neously. Thus multiple exchanges of beacons over an extended period of time may be necessary to discover two-hop topology; (iii) As beacons are sent using broadcast packets, there is a possibility of packet collisions resulting in loss of the beacons. It affects the accuracy of two-hop knowledge; (iv) Node mobility may result in link state errors; (v) Exchange of link state information will increase the packet size of the beacon, which may increase the probability of collision of broadcast packets; (vi) Frequent movement of nodes may require more frequent exchange of beacons, thus introducing additional overhead and packet collisions. Therefore, neighbor discovery over two or more hops using beacon messages becomes less reliable.

The simplest NCB mechanisms are "Self-Pruning" [32] and "Neighbor Coverage" [56]. Both mechanisms are equivalent. Two neighbor sets are main-tained at each node. Suppose node $i$ broadcasts a message to node $j$. Set $N_i$ and $N_j$ denote the neighbors of node $i$ and $j$, respectively. When node $j$ receives a broadcast packet from a node $i$ for the first time, it determines its coverage set as follows:

$$C_j = N_j - N_i - \{i\}. \tag{1.1}$$

The resulting coverage set $C_j$ is the set of neighbors of node $j$, which are not covered by node $i$ yet. This keeps track of pending hosts in $j$'s neighborhood, which have not received a direct broadcast from node $i$ as they are outside node $i$'s broadcast range. Node $j$ does not rebroadcast the packet if $C_j$ is an empty set.

An empty set implies that all neighbors of node $j$ are also neighbors of node $i$. This calculation is performed on each node that receives a broadcast packet prior to rebroadcasting.

The "Scalable Broadcast Algorithm" (SBA) [43] utilizes two-hop neighbor knowledge and a broadcast delay timer to determine whether or not to rebroadcast. Upon receiving a broadcast message from node $i$, node $j$ utilizes Equation 1.1 to determine if it has any neighbors that are not covered by node $i$. If the result is an empty set, then the node will not rebroadcast. Otherwise, node $j$ will schedule a broadcast with a specific delay. The delay is calculated based on the node $j$'s degree ($D_j$) and its neighbor's maximum node degree ($D_{N\max}$) as shown in Equation 1.2. It implies that nodes with the maximum number of neighbors have higher priority to broadcast than those nodes with less neighbors.

$$T_{\text{delay}} = \frac{D_{N\max}}{D_j}. \tag{1.2}$$

"Dominant pruning" [32] makes use of two-hop neighbor knowledge and a greedy set cover algorithm to alleviate the broadcast storm problem. Unlike previous mechanisms, the sender in the dominant pruning specifies a set of nodes in a forward list (attached to the broadcast packet), which are responsible for rebroadcasting the packet so that the packet reaches all nodes within two hops. Finding the minimum forwarding list is equivalent to the minimum set cover problem, which is NP-complete. In [34], dominant pruning is analyzed and two new algorithms, "Total Dominant Pruning" and "Partial Dominant Pruning" are proposed. The algorithms utilize two-hop neighbor knowledge to achieve further reductions in redundant broadcasts.

"Multipoint Relaying" (MPR) [47] is the broadcast scheme in which a sending node selects adjacent nodes as relay nodes to complete the broadcast. The IDs of selected adjacent nodes are appended to the packet as a forward list. An adjacent node that is requested to relay the packet must determine its own forward list. This process is iterated until the broadcast is completed. MPR makes use of two-hop neighbor knowledge and is employed in the OLSR [21] routing protocol for the optimized dissemination of link state information. MPR aims to reduce the number of redundant retransmissions during broadcast by restricting the number of retransmitters to a small set of relay nodes. The method proposed in [47] is based on a "greedy set cover" heuristic that selects the minimal subset of neighbors of a given node $A$, which will "cover" all two-hop neighbors of node $A$. A node is called "covered by node $A$" if it receives (directly or via retransmissions by other nodes) a message originating from $A$. Relay nodes of node $A$ are $A$'s one-hop neighbors that cover all two-hop neighbors of node $A$. That is, after all relay nodes of node $A$ retransmit the message, all two hop neighbors of node $A$ will receive the message. The goal is to minimize the number of relay nodes of node $A$. The proposed greedy set cover heuristic is as follows:

1. Find all two-hop neighbors reachable from only a single one-hop neighbor. Assign the one-hop neighbors as MPRs.
2. Determine the resultant cover set – the set of two-hop neighbors that will receive the packet from the current MPR set.
3. From the remaining one-hop neighbors not in the MPR set, find the ones that cover the most two-hop neighbors not in the cover set.
4. Repeat from Step 2 until all two-hop neighbors are covered.

Figure 6.2 shows the process of selecting multipoint relays in MPR. Nodes $B$, $D$, and $E$ are one-hop neighbors of node $A$. Nodes $C$, $G$, and $F$ are two-hop neighbors of node $A$. A broadcast is initiated by node $A$. According to the proposed MPR heuristic Step 1, node $B$ is selected as a MPR as node $C$ may only be reached by node $B$. The remaining nodes $G$ and $F$ are similarly covered by node $D$ that is then added to the MPR list (node $E$ covers only node $F$, not node $G$). Node $E$ is not added to the MPR list as its neighboring node $F$ is already covered by node $D$.

The mechanisms described so far rely upon explicit reasoning to determine whether or not to rebroadcast. In [49], the Lightweight and Efficient Network-Wide Broadcast (LENWB) mechanism is proposed. LENWB utilizes implicit reasoning based upon the information provided by neighboring nodes to implicitly determine which nodes received a broadcast packet. LENWB utilizes two-hop neighbor knowledge to determine the node degree of all neighboring nodes. Each neighboring node is assigned a priority that is proportional to its node degree.

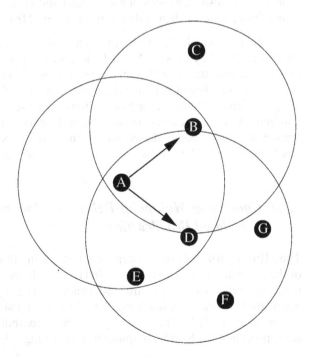

**Fig. 6.2** Multipoint relay selection

A node selects its neighboring nodes with higher priority to perform rebroadcasts. Thus, LENWB can proactively determine which neighboring nodes will rebroadcast and also which neighboring nodes will receive the broadcast. If the node determines that some of its neighbors will not receive a broadcast, then it rebroadcasts the message.

### 6.2.3 Dominating Sets–Based Broadcasting

In [58, 59], the authors describe a simple and efficient distributed algorithm for determining a connected dominating set (CDS). CDS may be used to limit the broadcast storm problem, by limiting broadcasting nodes to those gateway nodes. A dominating set is the set that any node in the network either belongs to the set or is the direct neighbor of some node in the set. The authors define a node $i$ as an "intermediate" node if there exist two neighbors $j$ and $k$ of $i$, which are not direct neighbors of each other. Two rules are applied:

- *Rule 1* – Given two intermediate neighboring nodes $u$ and $v$. If neighbors of $u$ are also neighbors of $v$ and the node identifier of node $u$ is less than the node identifier of node $v$, then node $u$ is not an "inter-gateway" node. Therefore, node $u$ is covered by node $v$.
- *Rule 2* – Assume three inter-gateway nodes $u$, $v$, and $w$ with shared neighbors. If the neighbors of node $u$ are contained within the neighbors of nodes $v$ and $w$ (that are also neighbors of each other) and node $u$'s identifier is less than both node $v$ and $w$, then node $u$ may be removed from the gateway set (CDS).

Stojmenovic et al. [52] propose a method that replaces the use of node identifier's as a key in Rule 1 and Rule 2 with a node's neighbor degree and its $(x,y)$ coordinates as additional keys. The neighbor degree is defined as a node's total number of neighboring nodes. The use of neighbor degree allows for a significant reduction in size of the dominating set. Nodes that belong to the dominating set are referred to as "internal" nodes. Broadcasting nodes are limited to those nodes selected as internal nodes. Nodes that have unique neighbors as with MPR are always selected as internal nodes.

### 6.2.4 Combining Multipoint Relay and Dominating Set–Based Broadcasting

The MPR algorithm is source-dependent, requiring that a relay node be aware of the preceding broadcasting node. In [1], a localized algorithm is proposed to make the relay selection source-independent. The algorithm also improves upon MPR by determining a smaller relay set, yet still providing equivalent performance to MPR. The authors proposed to combine MPR and dominating set approaches. Each node computes its forwarding neighbors set and transmits this to its neighbors. Each node then determines whether it belongs to the "MPR-dominating set", if:

- it has the smallest ID in its neighborhood;
- the node is a forwarding neighbor of the neighbor with the smallest ID.

In [57], the authors extend this work to further reduce the size of the relay set without introducing additional cost. The definition of the first condition is enhanced as follows: It has the smallest ID in its neighborhood and it has two unconnected neighbors, or the node is a forwarding neighbor of the neighbor with the smallest ID. In [57], forward node selection is as follows: Free neighbors do not need to be covered (*u* is a free neighbor of *v*, if *v* is not the smallest ID neighbor of *u*) at all, if a two-hop neighbor is covered by only a single one-hop neighbor, then that neighbor is taken immediately, then after that continue with the explained greedy set coverage heuristic.

## 6.2.5 Hexagonal and Dominating Set–Based Broadcasting

In [42], a broadcast protocol based on the hexagonal tiling of a plane is proposed, with transmission radius as the edge length of hexagons. The source chooses six nodes, which are closest to points that best approximate a regular hexagon, for retransmitting the message. Each designated node continues the process in a similar manner. The reliability (the broadcast message reaches all nodes) is not proven, but it is possible to provide examples showing that the opposite can be constructed. As observed in [23], the protocol may repeat broadcasting if neighbors do not agree on the choice of node near the common ideal point. Work in [23] introduced a stopping rule to prevent repeated broadcast and applied this type of broadcast as the basis of route discovery.

## 6.2.6 Cluster-Based Broadcasting

Clustering [15] is the process of grouping nodes together into clusters (groups) as shown in Fig. 6.3. A representative of each cluster is called the *clusterhead* (nodes *B* and *D*). A cluster encompasses all nodes within a clusterhead's transmission range. Nodes that belong to a cluster, but are not the clusterhead, are called *ordinary* nodes. Often nodes may belong to more than one cluster. These nodes are called *gateway* nodes (node *C*). Only clusterhead nodes and gateway nodes are responsible for propagating messages. The process of forming clusters may be either active or passive. In Fig. 6.3, ordinary node *A* broadcasts a message. The message is received by node *B* and is relayed to all nodes within node *B*'s broadcast range. Node *C* is a gateway node that receives the message from node *B* and rebroadcasts the message. Clusterhead node *D* receives the message and rebroadcasts it to its neighboring nodes. The

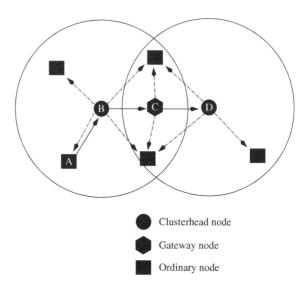

**Fig. 6.3** Example of a cluster-based flood initiated from node A with only clusterhead and gateway nodes rebroadcasting

⬤ Clusterhead node

⬢ Gateway node

◼ Ordinary node

directed solid lines show the propagation of the message among those nodes that are allowed to rebroadcast. Dashed directed lines show the propagation of the message from the clusterheads and the gateway nodes to ordinary nodes.

In *active clustering*, nodes must cooperate in order to elect clusterheads. This is achieved through periodic exchange of control information. The formation of clusters in active clustering is independent of the background data traffic. The selection of a clusterhead may be based upon Lowest ID algorithm or Highest ID algorithm [29]. In [44] and [45], clustering is used as an optimized flooding mechanism, whereby only clusterheads and gateways rebroadcast messages. Additionally, the clusterheads in the mechanism ensure reliable delivery of the message to those nodes belonging to their cluster. In [34], a mechanism that builds a cluster-based backbone for the dissemination of information is proposed. They propose the creation of a static and a dynamic backbone. The static backbone is created using a source-independent connected dominating set. The dynamic backbone is created using a source-dependent connected dominating set.

In *passive clustering* [61, 62], cluster formation is dependent on background data traffic. Therefore, passive clustering will not form clusters until there is background traffic. This is because, in passive clustering, the flow of data traffic is used to propagate cluster control information and collect neighbor information through promiscuous packet reception. Promiscuous packet reception is achieved by allowing the MAC layer to pass all received packets up the TCP/IP stack irrespective of MAC address. Passive clustering is beneficial in that it utilizes the existing data traffic to form clusters. However, without existing data traffic, it is unable to form clusters and provide the benefits of an optimized flood. Active clustering requires that cluster control information be exchanged between nodes and clusterheads. Thus, it requires more overhead than passive clustering or non-clustered flooding mechanisms for the formation of clusters. However, unlike passive clustering, there is no delay involved as it does not require background traffic.

## 6.2.7 Resource-Aware Broadcasting

Resource-aware broadcasting is the process of disseminating information in such a way that mechanisms are aware of and utilize available resources within the network in an efficient and aware manner. It makes sense for mechanisms to strive to extend lifetime of devices in the network. This can be achieved through optimization in utilizing devices that are most suitable based upon their resources and constraints. Optimized broadcast mechanisms are proposed to alleviate the broadcast storm problem. However, they do not address the need for "Resource Awareness". These two requirements may possibly oppose each other, as an optimal broadcast is not necessarily the most efficient resource-aware broadcast.

In *Activity Scheduling* [54], nodes must actively determine if they are in an active or passive state in order that the network remains connected and the lifetime of both the network and the nodes are maximized. Nodes in a passive state (sleeping) do not consume constant energy. They are not involved in the reception of packets for which they are not specified receivers. In [51], a topology maintenance scheme is proposed with the aim of extending lifetime of the network while preserving network connectivity. A node is either active or has an active neighboring node. Thus, flooding (and routing) activities are restricted to those active nodes. The active nodes create a connected dominating set. Nodes update their activity status periodically during short transition periods when all nodes are active and packets to passive nodes are delivered. It is possible for nodes that have greater remaining energy remain active longer than the nodes with less remaining energy – which may enter a passive state more often and on awakening collect packets from their neighbors, which could not be received in the passive state. In [51], the authors further propose metrics for determining activity status, which are based on the combinations of node-degree and remaining battery power.

In [57], the author extends MPR flooding to reduce size of the relay node set without introducing additional overhead. The process of selecting relay nodes may also be done in a resource-aware manner, which accounts for the remaining battery power of nodes. This mechanism still utilizes "Step 1" of the MPR algorithm, which is to select those nodes with unique neighbors. However, the majority of relay nodes that could be selected to relay a message are selected only because they have unique neighbors. Thus the selection of remaining relay nodes based solely on their resources (battery power) is limited in its results, as these relay nodes only constitute a fraction of all relays selected.

$$U_f = BU_p U_n, \tag{1.3}$$

$$U_p(i) = \frac{1}{1 + e^{(-P_i + S)}}, \tag{1.4}$$

$$U_n(i) = \frac{\text{unallocated two hop neighbours of node } i}{\text{total two hop neighbours of node } i}. \tag{1.5}$$

Lipman et al. [27] describe a distributed optimized broadcast mechanism for ad hoc networks called Utility Based Flooding (UBF). Unlike the existing optimized broadcast mechanisms, UBF is fully resource-aware. UBF selects relay nodes based solely on a forwarding utility $U_f$ (Equation 1.3). The utility $U_p$ (Equation 1.4) is a sigmoid function where $P$ is the remaining energy at a node while $s = \text{MaxEnergy}/2$ is half the maximal energy. $U_n$ (Equation 1.5) is the ratio of unallocated (uncovered) local two-hop neighbors to the total number of two-hop neighbors. The significant point in UBF is that the selection of nodes with unique neighbors is not a priority as in MPR or [57]. In this way the coverage provided by those best nodes (in terms of resources) is accounted for irrespective of unique neighbors as is done in MPR. $U_n$ increases the utility of possible relay nodes that may have unique neighbor. However, if a node with a unique neighbor is not suitable because of constraints or low resources, then its utility will remain low such that it may only be selected after all other possible relays are selected. The benevolence (B) is intended to capture or represent any constraints imposed upon a device. A user may allow a device attached to a reliable power source to fully participate in network activities. However, if the device is mobile and the battery power drops below a specified threshold, the user may not wish the device to participate. Existing broadcasting mechanisms do not account for this type of behavior. Thus their performance will be degraded in such a network as they may select restricted nodes as relay nodes that will not rebroadcast messages. UBF is shown to significantly improve broadcast reachability over successive broadcasts in a resource-constrained environment while not adversely affecting performance. Moreover, UBF extends the lifetime of both nodes participating in the ad hoc network and lifetime of the network itself.

### 6.2.8 Distributed and Efficient Flooding

To solve the broadcast storm problem, Liu et al. [31] propose a distributed and efficient flooding scheme. The authors first study the sufficient and necessary condition of 100% deliverability for flooding schemes that are based on only one-hop neighbor information. It proves that a one-hop flooding scheme achieves 100% deliverability if and only if for each node $s$ the union of coverage disks of $s$'s neighbors is fully covered by the forwarding node set determined by $s$. The proposed flooding algorithm achieves the local optimality in two senses: 1) the number of forwarding nodes in each step is the minimal; 2) the time complexity for computing forwarding nodes is the lowest, which is $O(n \log n)$, where $n$ is the number of neighbors of a node.

The basic idea of the flooding scheme is as follows. When a node (called the source) has a message to be flooded out, it computes a subset of its neighbors as forwarding nodes and attaches the list of the forwarding nodes to the message. Then it transmits (broadcasts) the message out. According to the sufficient and necessary condition of 100% deliverability, for each node $s$, the forwarding nodes computed by $s$, denoted by $F(s)$, should fully cover the union of coverage disks of $s$'s neighbors. After that, every node in the network does the same as follows. Upon receiving a flooding message, if the message has been received before, it is discarded; otherwise the message is delivered to the application layer, and the receiver checks if itself is in the forwarding list. If yes, it computes the next hop forwarding nodes among its neighbors and transmits the message out in the same way as the source. The message will eventually reach all the nodes.

To minimize the unnecessary rebroadcast messages, the set of forwarding nodes should be minimized. Liu et al. proposed an efficient method to compute the minimum forwarding set. For each node $s$, the strategy of this method is to compute the neighbor's boundary of $s$, and thus the nodes that contribute to this boundary are the nodes in $F(s)$. The pair-wise boundary merging method is adopted to compute the boundary efficiently. Initially, each node is arbitrarily paired with another node to merge their coverage boundaries. Then, the merged pair's boundary is further merged with another pair's boundary. This merge operation is repeated until eventually there is only one big merged boundary, which is the neighbor's boundary of $s$. The minimal $F(s)$ consists of the nodes that contribute to this boundary. In the flooding operation, each node computes its minimum forwarding node set based on its one-hop information. After a flooding message is initiated from a source node, only the nodes in the forwarding node set will relay the message. It greatly reduces the unnecessary rebroadcast messages in the network while guaranteeing 100% delivery.

## 6.3 Variable Radius Optimized Broadcasting

In variable radius optimized broadcasting, nodes utilize transmission power control when broadcasting packets. The use of transmission power control allows for the isolation of broadcasts through reduction of transmission range and is beneficial for the following reasons. The required power for a transmission distance of $d$ between two nodes is proportional to $d^\lambda$. Typically $\lambda$ takes a value between 2 and 6, depending on the characteristics of the communications medium [60]. Isolating a broadcast increases the probability of only necessary nodes hearing a broadcast. This helps to both reduce duplicate packet reception and the power consumed with packet reception at receivers. Limiting the nodes that will hear a broadcast reduces medium contention between nodes, increases medium utilization, and reduces the probability of packet collisions. The use of transmission power control may result in one high-power transmission being replaced with two or more low-power transmissions. A common analogy would be, "In a room full of people, it would be better for people to whisper, rather than yell at one another".

Optimized flooding mechanisms that utilize transmission power control require a node's location coordinates in order to determine the required transmission power. These coordinates may be obtained via a positioning system like GPS and shared via periodic exchange of beacon messages. If a positioning system is not available, distances may be determined through received signal strength of beacon messages.

## 6.3.1 Relative Neighborhood Graph

The Relative Neighborhood Graph (RNG) [55] shown in Fig. 6.5 is formed when two nodes are connected with an edge if their *lune* contains no other nodes of the graph. The lune of two nodes $u$ and $v$, shown in Fig. 6.4 (in gray), is defined as the intersection of two spheres of radius $d(u,v)$, one centered at node $u$ and the other at node $v$. Use of a localized RNG was first proposed in [7] as a topology control algorithm to minimize node degrees, hop diameter, and maximum transmission range and ensure connectivity. The resulting RNG graph is the same irrespective of whether it is calculated in a distributed or centralized manner.

In [11], the authors propose a distributed flooding protocol based upon the RNG called RNG Relay Subset (RRS). RRS allows for self-selection of forwarding neighbors. In RRS a node $v$ will select itself as a relay for a node $u$ if and only if node $v$ is also neighbor of node $u$. Node $v$ must also have a RNG neighbor that is not covered by node $u$. RRS alleviates the broadcast storm problem by reducing the transmission range of a broadcasting node to include

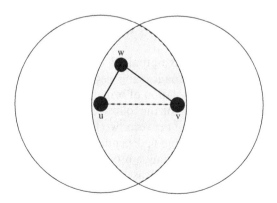

**Fig. 6.4** Formation of relative neighborhood graph using a lune

**Fig. 6.5** Relative
neighborhood graph

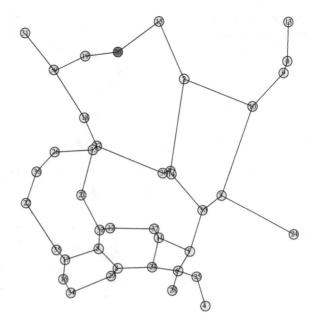

only those RNG neighbors that must receive the broadcast, thereby ensuring
the flood propagates. The use of self-selection by nodes using RRS allows nodes
to determine if they need to rebroadcast, without the need for additional
information attached to the broadcast packet.

## 6.3.2 Minimum Spanning Tree Graph

The Minimum Spanning Tree (MST) [55] graph, as shown in Fig. 6.6, is a
connected graph (path of edges between any two vertices) that uses the minimum
sum of edge weight. This results in a graph with one less edge than the number of
vertices. The MST has traditionally been used in networks for determining
broadcast trees using global topology information. The MST is a subgraph of
RNG and may be computed from the RNG by removing edges that create a cycle
in the graph. This results in the formation of a tree or directed acyclic graph from
all nodes back to the broadcasting node. Thus the MST generates a more optimal
broadcast path than RNG, but suffers as there is no fault tolerance in the
resulting graph [7]. Fault tolerance refers to the number of alternative paths a
message may travel toward a node, thus improving the probability of delivery.

Lipman et al. [26] and Cartigny et al. [12] proposed to apply the MST
algorithm in a distributed manner to improve the performance of flooding in
ad hoc networks. In the localized MST approach, the topology available to the
MST algorithm is restricted to one hop, yet still allows for an optimal broadcast
set of nodes with minimal transmission range to be determined as with the

**Fig. 6.6** Centralized
minimum spanning tree

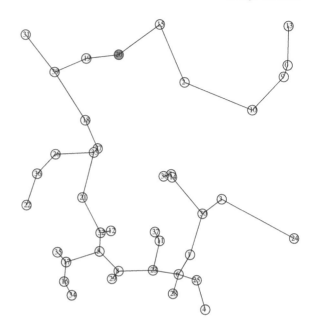

centralized approach within that one hop. The resulting localized MST graph
does not exhibit the tree-like structure of the centralized MST with global
topology knowledge. It can be seen by comparing Figs. 6.5, 6.6, and 6.7 that
MST ⊆ Localized MST ⊆ RNG as described in [30]. Thus many of the benefits
of MST are maintained with the addition of fault tolerance not found in the
centralized approach.

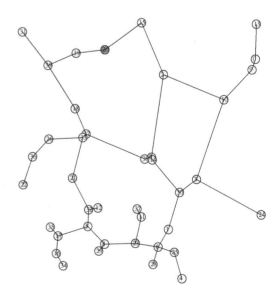

**Fig. 6.7** Localized minimum
spanning tree

### 6.3.3 Determining Common Transmission Radius and Constructing MST

To maintain network connectivity in an ad hoc network, the minimum transmission range $R$ that may be utilized by nodes is equal to the longest edge in the
minimum spanning tree [40]. Existing algorithms for determining $R$ require
globalized network knowledge or use distributed approaches that are based
upon straightforward adaptations of centralized algorithms. Ovalle et al. [39]
propose to use the longest edge as determined by localized MST (LMST) to
approximate $R$. This is achieved by using a wave propagation quazi-localized
algorithm, which allows nodes to make decisions based on both local knowledge and additional information obtained by wave propagation. Each node
maintains a record of the longest edge it knows so far (initially its own longest
edge in its LMST). In each round, each node receiving a larger edge in the
previous round will broadcast its new longest edge. At the end, all nodes will
receive the same longest edge, which is then used as the transmission radius.
Thus, the wave propagation allows for the longest $R$ to be determined locally by
all nodes and disseminated globally. It was observed that, although LMST has
less than 5% additional edges than MST, the edges tend to be longer, hence the
longest LMST edge selected as transmission radius may double the energy
consumption with respect to selecting the longest MST edge.

Furthermore, Ovalle et al. [39] propose a scheme for converting the distributed LMST graph into the equivalent centralized MST graph. They first prove
that MST is a subset of LMST. The conversion is achieved through a two-step
iteration that traverses and eliminates dangling (tree) edges and breaks loops
found in LMST that differentiate it from the MST. The process terminates at a
node (the network leader), which through the two-step process has learned
the longest MST and LMST edges. The value of the longest MST edge may
then be broadcast to other nodes in the ad hoc network. Additionally, a simple
algorithm is proposed, which allows for the determined MST to be updated if a
node is added or deleted.

### 6.3.4 RNG- and LMST-Based Broadcasting

Cartigny et al. [12] propose the use of RNG and LMST as the basis of a
localized minimum energy broadcast protocol. Lipman et al., [26] also applied
LMST to describe a minimum energy-broadcasting scheme. In this protocol,
each node will apply neighbor elimination scheme after receiving the first copy
of a message. When timeout expires, the node will not retransmit packet if all its
RNG (or LMST) neighbors are eliminated. Otherwise, the transmission radius
is selected to be the distance to the furthest RNG (or LMST) neighbor that is
not eliminated.

Work in [18] showed that, if $c>0$ in the power energy consumption model $d^\lambda+c$, the algorithm may produce energy-inefficient solution (compared to existing globalized protocols). The reason is that, for dense networks, the protocol tends to select short transmission radius. Therefore, many nodes need to retransmit the message. However, each of these nodes spends constant amount $c$ of energy at least, which can accumulate. It may be better, for dense networks, to apply a minimal "target" transmission radius, which can be derived by considering an ideal hexagonal network and the radius that minimizes the two energy consumption terms in the expression. For $c>0$ and $\lambda>2$, the target radius, found theoretically, is $[2c/(\lambda-2)]^{1/\lambda}$. The transmission radius computed by the above procedure is then increased to that target radius, so that a number of nodes are covered with a single transmission.

Work in [9] considered the impact of energy needed to overhear transmissions in dense networks on the construction of minimum energy broadcast trees. In effect, they consider the power energy consumption model $d^\lambda+c$, with three values of $c$: $c1=0$ when only transmitting energy is considered, $c2>0$ when electronic cost of transmission is also considered, and $c3>c2$ when also electronic cost of intended receiver and nodes overhearing transmission are added. They conclude that minimum energy tree becomes "bushier" with increasing $c$ value. It corresponds to the observations made by [18] about the choice of target radius $[2c/(\lambda-2)]^{1/\lambda}$, with consequence of reducing the number of transmitting nodes when $c$ increases.

### 6.3.5 Neighbor-Aware Adaptive Power Broadcasting

Lipman et al. [28] describe the following efficient broadcast mechanism: Upon receiving a broadcast message from a node, say $h$, each node, say $i$ (that was determined by $h$ as a forwarding node) determines which of its one-hop neighbors also received the same message. For each of its remaining neighbors $j$ (which did not receive a message yet, based on $i$'s knowledge), node $i$ determines whether $j$ is closer to $i$ than any one-hop neighbors of $i$ (which are also forwarding nodes of $h$) who received the message already. If so, $i$ is responsible for message transmission to $j$. Otherwise it is not. Node $i$ then determines a transmission range equal to that of the farthest neighbor it is responsible for. A node $i$ may decide to perform local optimization whereby it determines a reduced set of its closest neighbors that still provide coverage of its remaining further neighbors. Thus node $i$ may adapt the transmission power (limiting the transmission range) to include only those closest neighbors.

### 6.3.6 Incremental Power Broadcasting

In ad hoc networks, the radios installed in nodes generally use omni directional antennae. When a node transmits a message with a given transmission power, each node with the transmission range will receive the message. This characteristic of broadcast radio transmission is referred to as the *wireless multicast advantage* [60]. To take advantage of this characteristic, they proposed the *Broadcast Incremental Power* (BIP) protocol. BIP constructs an efficient broadcasting tree from a source node to all other nodes in the ad hoc network while considering the multicast advantage. Although BIP is energy-efficient, it requires global knowledge of the ad hoc network topology, which is counterproductive as this would then introduce significantly more overhead to acquire first and thus reduces the energy efficiency. To resolve this global knowledge problem, some distributed versions have been proposed. However, these distributed extensions usually result in significant message exchange and overhead to gather topology information, and hence also suffer from energy efficiency issues. In [17], the authors propose an incremental localized protocol that locally applies the BIP heuristic method. Given an initial connected graph, it enables a node to broadcast a message with high energy efficiency. The localized protocol allows each node to calculate its own BIP within its *k*-hop neighborhood based on prior knowledge from the preceding node. The localized tree is then forwarded to the next nodes with the broadcast packet. In this way, the global broadcasting structure is incrementally constructed.

## 6.4 Reliable Broadcasting

Blind flooding in ad hoc networks may be used as a "fall back" mechanism that provides more reliable broadcasting in situations of increased mobility, channel noise, or packet traffic where optimized broadcasting mechanisms may fail. This reliability is due to the inherently high degree of redundancy present in blind flooding – whereby all nodes retransmit received broadcast packets at least once. However, as stated earlier, blind flooding results in the broadcast storm problem. Optimized broadcast mechanisms reduce the level of redundancy during a broadcast, thereby reducing the broadcast storm problem. However, there exists a significant problem in broadcast environments where a broadcast transmission may be lost due to packet corruption, packet collision, or hidden node transmissions. Therefore, it is possible that nodes may not receive a broadcast transmission. Furthermore, those nodes that do not receive a broadcast transmission may be required to receive a transmission. This is especially true in the case of optimized broadcast mechanisms, where selected nodes are responsible for retransmission. Given that optimized broadcast mechanisms greatly reduce the redundancy found in blind flooding, there may be situations where a packet may be lost and a broadcast may not propagate due to reduced redundancy.

A reliable flooding mechanism described in [44] and [45] is based on the use of a clustering. Each cluster consists of a single clusterhead that is responsible for nodes within its cluster. Clusterheads are responsible for ensuring messages flooded are received by nodes they are responsible for. The clusterhead will wait for acknowledgements from each node within its cluster. The gateway nodes will then forward the message to the clusterheads of other clusters that may also belong to. In this way a message is reliably propagated from cluster to cluster. The mechanism ensures reliability by utilizing unicast messages between cluster heads, and the collection of unicast acknowledgements from nodes belonging to a cluster. Gateway nodes will delay acknowledgement of a received message from the preceding clusterhead while they transmit the message to another clusterhead. Once the last cluster is reached, then acknowledgements will start flowing back toward the originating clusterhead and ultimately the source of the flood. In this way the source of the flood is able to determine which nodes the flood was received by. The problem with a cluster-based approach is the formation and maintenance of the clusters, which is costly especially in the presence of mobility. The formation of the cluster tree does not ensure that all nodes are covered by a clusterhead as nodes may leave a cluster. Therefore, it is possible for some nodes to be excluded from receiving a broadcast. Additionally, given node mobility, the reverse path back to the source node may be destroyed. To solve this problem, nodes may flood acknowledgements back to the clusterhead of the originating node.

In [16], a flooding mechanism is proposed to limit the broadcast storm problem and also to provide reliability. The mechanism consists of three phases. The first phase is the *scattering phase*, in which the source of node initiates a flood that utilizes the counter-based [38] flooding mechanism. The idea is to disseminate the message to as many nodes as possible. A handshake procedure as described in [6] is utilized to ensure neighboring nodes have received the same messages. During the scattering phase, a tree graph is formed from all nodes back to the original source of the flood. The second phase is *gathering phase*, in which acknowledgements are collected from all nodes. Acknowledgements travel back toward the source of the flood via the acyclic graph formed during the scattering phase. Unicast packet transmission is used for the transfer of acknowledgement. The third stage is the *purging phase* and is initiated by the source node, which floods a request for all data structures maintained during the reliable flood at each node to be deleted.

In [53], a reliable flooding mechanism is described. The mechanism consists of two schemes: Duplicate Broadcast Scheme (DBS) and Broadcast Acknowledgement Scheme (BAS). In DBS, a node maintains its local set of one-hop neighboring nodes in a table called Local Connectivity Table (LCT). When a node broadcasts a message, it relies upon the BAS to determine which neighbors receive the message. Given the number of successfully received messages and the number of nodes in the LCT, the authors propose to determine whether or not it is necessary to perform an additional broadcast, thus attempt to reach those nodes that did not receive the message yet. The BAS is a positive acknowledgement scheme that involves modifying the IEEE 802.11 MAC while maintaining compatibility. The BAS requires that all nodes successfully receiving a broadcast message respond with an acknowledgement. The scheme allows receiving nodes to utilize the DIFS period after receiving the data frame to transmit an acknowledgement. The DIFS period is divided into mini-slots, and nodes select a mini-slot in which to send their acknowledgement to the broadcasting node.

In [35], the authors propose a simple broadcast algorithm that provides a high delivery ratio for packets being flooded in an ad hoc network and provides limited reduction of redundant broadcasts. The algorithm allows for only selected forward nodes (one hop neighbors) of a broadcasting node to send acknowledgements, confirming reception of a broadcast packet. Forward nodes are selected so as to ensure that all two-hop neighbors of the broadcasting node are covered. Moreover, no acknowledgment is needed from one-hop neighbors that are covered by at least two forwarding neighbors. The broadcasting node waits for acknowledgements from its entire forwarding one-hop neighbor nodes. If not all acknowledgments are received, the broadcast node will rebroadcast the packet until a maximum number of retries is reached.

In [22], the authors proposed to apply the concept of double domination for providing more reliable service in ad hoc networks. A node in a double cluster structure is said to be covered if it has at least two clusterheads in its neighborhood. This means that each node that traditionally needs to be covered by one clusterhead is no longer covered by a unique clusterhead, but at least two clusterheads. They describe several double-clustering schemes based on some early clustering schemes that were already reported to not always work correctly. Work in [22] generalized the double cluster definition to a set of nodes that are up to $k$ hops away from a clusterhead, i.e., a $k$-hop double dominating set. This means that, in a $k$-hop double dominating set, a node is covered by at least two clusterheads that are at a distance up to $k$ hops.

In [35] and later improved in [36], the authors proposed a simple but reliable broadcast algorithm, called double-covered broadcast (DCB), which takes advantage of broadcast redundancy to improve the delivery ratio in environments that have a high transmission error rate. Among the one-hop neighbors of the sender, only selected forward nodes retransmit the broadcast message. Forward nodes are selected in such a way that (1) the sender's two-hop neighbors are covered and (2) the sender's one-hop neighbors are either a forward node or a non-forward node, but covered by at least two forwarding neighbors. The retransmissions of the forward nodes are received by the sender as confirmation of their receiving the packet. The non-forward one-hop neighbors of the sender do not acknowledge the reception of the broadcast. If the sender does not detect all its forward nodes' retransmissions, it will resend the packet until the maximum times of retry is reached. Simulation results show that the algorithm provides good performance for a broadcast operation under high transmission error rate environment. However, the channel errors are assumed to follow uniform distribution with fixed probability, which may not be realistic. A realistic physical layer model shows that the error rate depends on the distances between nodes.

In [25], a reliable and optimized broadcast mechanism called Reliable Minimum Spanning Tree (RMST) broadcasting is proposed. RMST utilizes a combination of the unique properties of the localized MST and unicast packet transmission to improve the reliability of optimized broadcasting. Each node in RMST uses one-hop topology information to calculate its local MST and determine those closest neighboring nodes that must be included within any transmissions to ensure a broadcast propagates throughout the ad hoc network. The distributed calculation of the localized MST results in a connected graph with a neighbor degree greater than one but less than six, and an average neighbor degree of less than 2.04 nodes [30]. Therefore, if the prior transmitting node is removed, the average neighbor degree is reduced to 1.04 nodes. This low neighbor degree results in a reduced set of neighboring nodes to which a broadcasting node must transmit a message, and allows for IEEE 802.11 broadcast transmission (as used by existing broadcast mechanisms) to be replaced with IEEE 802.11 unicast transmission. Unicast transmission provides a more reliable transport mechanism than broadcast transmission, as it may implement a RTS/CTS exchange at the MAC layer prior to transmission in order to reduce the problems associated with the hidden node problem. More importantly, unicast transmission utilizes a frame retransmission mechanism at the MAC layer based on a positive acknowledgement scheme (ARQ). Thus, a transmitting node will retransmit a frame if it does not receive a positive acknowledgement from the destination node. The IEEE 802.11 ARQ is not completely reliable and packet loss may occur. However, it provides a more reliable transport mechanism than broadcast and requires no modifications of the IEEE 802.11 MAC layer. The number of retransmissions before a timeout occurs is adjustable and is generally 4–7 retransmissions. If a node fails to retransmit a message to a destination node, it is able to detect the failure and may utilize an alternative approach to continue dissemination.

## 6.5 Broadcasting in Routing

Ad hoc network routing protocols allow for point-to-point communication in ad hoc networks. Routing protocols are responsible for delivering packets between nodes not within the transmission range. This requires the use of cooperative intermediate nodes that are able to act as routers in a distributed manner, thus allowing for data packets to be forwarded toward their destination. Ad hoc network routing protocols may be classified as proactive or reactive (on-demand) depending on how they determine routes. In this section we explore how broadcasting is utilized in reactive routing for route discovery and in proactive routing for link dissemination.

### 6.5.1 Proactive Routing Protocols

Proactive routing protocols [8] require that each node maintains route information to every other node in the network. Route tables are periodically or dynamically updated if the network topology changes. Proactive routing protocols differ in how they detect changes in network topology, how they maintain route tables, and how they disseminate this information to other nodes in the network. Proactive routing protocols experience minimal delay when routing packets as routes are available immediately from constantly maintained route tables. Although there is no initial penalty when a route to a destination is required, there is a constant overhead associated with disseminating link state or route table information throughout the network. This results in a reduction in network capacity due to constant and possibly heavy control traffic delivery. This is made worse in the presence of node mobility. Additionally, proactive protocols do not scale effectively as node density and node numbers increase [50]. Constant dissemination of control information throughout the ad hoc network also results in increased power consumption.

The majority of proactive routing protocols disseminate control information throughout the ad hoc network using blind flooding. Examples of proactive routing protocols that utilize blind flooding are Destination-Sequenced Distance Vector (DSDV) [41] and Wireless Routing Protocol (WRP) [37]. Other proactive routing protocols such as Fisheye State Routing (FSR) [13] limit the rate at which they update route information depending on the distance. Routes to closer nodes are maintained more regularly, whereas routes to remote nodes are maintained less regularly. Source-Tree Adaptive Routing (STAR) [14] eliminates periodic dissemination of control information in favor of conditional dissemination, thus reducing the constant overhead. However, blind flooding is still required. In Cluster-head Gateway Switch Routing (CGSR) [10], a hierarchy is created based on node clustering. Clusterheads control the flow of route information within or among their clusters, thus reducing the amount of route information and limiting the dissemination of the route information. The Optimized Link State Routing (OLSR) [21] protocol attempts to reduce the problems associated with blind flooding by utilizing an optimized flooding algorithm called Multipoint Relay (MPR) flooding [47]. The use of an optimized flooding algorithm reduces the problems associated with blind flooding and allows OLSR to scale more effectively, given an increased number of nodes.

## 6.5.2 Reactive Routing Protocols

Reactive routing protocols [8] are designed to reduce the overheads associated with proactive routing protocols. They do this by only maintaining information for active routes. Reactive routing protocols do not proactively maintain routes to all nodes; therefore, they must perform route discovery when a route to a destination node is required. Route discovery requires that a "route request" (RREQ) packet be blind flooded throughout the network. When the destination (or a node with an active route to the intended destination) receives the RREQ a "route reply" (RREP) is sent back to the source of the route request. The RREP may either be blind flooded back to the source or it may be unicast back along the path followed by the RREQ. The inherent nature of blind flooding is that it always chooses the shortest path, as the broadcast packet follows all possible paths in parallel. This is one reason why most reactive routing protocols use blind flooding to perform route discovery.

As routes are not immediately available, reactive protocols have a much higher initial delay at the start of communication than proactive routing protocols. Given that flooding forms the basis of route discovery, reactive routing protocols suffer from the broadcast storm problem. This is made worse by increasing node density, heightening node mobility and the number nodes of performing route requests for peer-to-peer communications. It is important for mobile devices that there is no constant power usage due to periodic flooding of link state or route table information as with proactive routing protocols.

Both Dynamic Source Routing (DSR) [20] and Ad hoc On-Demand Distance Vector Routing (AODV) [46] protocols utilize blind flooding as a means of performing route discovery. However, they differ in the way they maintain routes to destination nodes and also in the amount of information required to route packets. To reduce the effects of blind flooding, these protocols use route caching as well as limiting the number of hops for route discovery. The Routing On-demand Acyclic Multi-path (ROAM) [48] protocol limits the effects of flooding by using directed acyclic subgraphs based on distance between the source and the destination for the propagation of a flood. This eliminates the propagation of a flood in a direction along a subgraph if the destination is not reachable along that subgraph. In Relative Distance Micro-discovery Ad-hoc Routing (RDMAR) [5], overhead associated with route discovery is reduced and localized by limiting each RREQ packet to a certain number of hops. However, this localization of route requests can occur only if the source and destination node have communicated before and exchanged position information. If the nodes have not communicated before, then the route request is not localized. Location Aided Routing (LAR) [24] requires that each node is equipped with a GPS device and therefore is aware of its location. Overhead associated with route discovery is reduced by limiting the direction and scope of flooding. This protocol defines zones specifying which direction a RREQ packet may travel toward. Route request packets therefore only travel in the approximate direction of the intended destination. Cluster-Based Routing Protocol (CBRP) [19] is a hierarchal routing protocol based on clustering. Clusterheads are defined and responsible for the nodes within each cluster. To reduce the effects of route discovery, only clusterheads exchange and propagate RREQ packets.

## 6.5.3 Self-selecting Route Discovery

Existing research in ad hoc network routing has contributed significantly to improve routing through maximizing the usage of prior knowledge of nodes, improving stability of routes, and creating a collaborative environment between nodes. However, little work has been done in improving the process of route discovery when no prior node or topology knowledge is available. In the case of reactive routing, improving the efficiency of route discovery is one key to providing higher scalability as network density increases. Moreover, if only a blind flooding is performed, then the route determined is generally the shortest path (as all routes are searched in parallel during a blind flood) and is not necessarily the best route in terms of efficient usage of resources.

In [4], the authors propose novel distributed search strategies that may be incorporated into reactive route discovery in heterogeneous ad hoc networks. The proposed search strategies are self-selecting and aim to be resource-aware, avoid unidirectional links, and reduce control packet overhead associated with traditional route discovery caused by blind flooding. The first proposed search strategy allows for efficient resource-aware route discovery in ad hoc networks where heterogeneous nodes may have varying transmission ranges. The second proposed search strategy addresses route discovery issues associated with the existence of unidirectional links experienced when nodes have varying transmission ranges or as a result of a varying wireless propagation environment.

In [2] and [3], authors present a number of different self-selecting route discovery strategies, which allow for intermediate nodes to selectively participate in route discovery. The aim of these strategies is to reduce the broadcast storm problem in terms of the number of control packets exchanged and the level of medium contention in the network, thereby achieving higher levels of scalability. Additionally, such strategies are able to provide more control to individual nodes to better manage their limited resources (such as battery power) and to determine more effective routes between end nodes.

## 6.6 Thoughts for Practitioners

Broadcast/flooding is widely utilized in the applications of wireless ad hoc networks. To improve energy efficiency and shorten communication latency, cross-layer design could be adopted for specific applications of ad hoc networks. For example, improved performance can be obtained by jointly considering physical and network layer issues. Such novel approaches incorporated protocol layer functions will provide advantages over traditional network architectures.

In some broadcast protocols, there are several parameters that should be determined to optimize the performance before applying the protocols. For example, it may be required to decide predetermined thresholds and whether or not to employ the acknowledgement scheme in the broadcast protocols. A suggestion is to set the simplest environment and apply the protocol to the real applications. Based on the feedback of the parameters, the system can iteratively adjust the parameters to achieve a stable and good performance of the broadcast protocols. It is better than directly setting the parameters as required in the protocols. The reason is that there is always a gap between research work and real applications. Even for the works based on real testbed, performance of the protocols is significantly affected by different environments.

## 6.7 Conclusions and Directions for Future Research

As can be seen from this chapter, a significant amount of research on designing efficient broadcast schemes in ad hoc networks has been carried out in the past few years. This has resulted in significant advances in the state of the art. The most notable are the developments of efficient localized broadcast schemes where each node requires only one or possibly two-hop neighbor knowledge about the network. The performance of the protocols is close to the performance of protocols that require global network knowledge. This chapter also identifies how broadcast is used and optimized in routing protocols and introduces new research on route discovery and the transition away from standard broadcast and optimized broadcast techniques to more search-oriented route discovery for routing protocols.

Although broadcast in ad hoc networks has been well studied, there are challenges for future research. Most existing solutions for efficient broadcasting are usually designed for static networks or the networks where nodes do not move fast. It is because that these efficient broadcast protocols are normally based on the information of neighbors, which is hard to maintain in mobile environment. Therefore, design of efficient broadcast protocol in highly mobile networks is expected in future work.

Moreover, we observe that there are more and more applications in which ad hoc networks are integrated with other wireless networks. For instance, wireless sensor actuator networks consist of actuator networks (ad hoc networks) and wireless sensor networks. Broadcast in wireless sensor actuator networks is normally from an actuator to all sensors or the sensors in a specific area. Another example is the integration of ad hoc networks and wireless mesh networks.

**Acknowledgments** This research is supported by NSERC Collaborative Research and Development Grant CRDPJ 319848-04, and the UK Royal Society Wolfson Research Merit Award.

## Terminologies

*Broadcast* – Sending a message from one node to all other nodes in the network.

*Blind flooding* – Each node in the network retransmits the flooding message at the first time it receives the message.

*Broadcast storm problem* – Blind flooding may cause serious problems of redundant retransmissions, medium contention and packet lost.

*Covering with disks* – Given a set of points in the plane, the problem is to identify the minimum set of disks with prescribed radius to cover all the points.

*Localized method* – Each node uses the information of only its neighbors or nature information.

*Globalized method* – Some node or all nodes have entire network information.

*Dominating Set (DS)* – A subset of the vertices of a graph if every vertex in the graph is either in the subset or is adjacent to at least one vertex in the subset.

*Connected Dominating Set (CDS)* – A connected DS.

*Clusterhead* – A representative of each cluster.

*Gateway nodes* – The nodes that belong to several clusters.

*Active clustering* – Nodes cooperate with each other to elect clusterheads. This is achieved through periodic exchange of control information.

*Passive clustering* –Cluster formation is dependent on background data traffic.

*Activity scheduling* – Nodes must actively determine if they are in an active or passive state in order that the network remains connected and the lifetime of both the network and the nodes are maximized.

## Questions

1. What is the difference between broadcast and multicast?
2. What is blind flooding? Why it may cause broadcast storm problem?
3. Describe the basic idea of Multipoint Relaying (MPR) [47]?
4. What is the basic idea of self-pruning [32]?
5. What are the basic steps of dominating sets–based broadcasting?
6. What is resource-aware broadcasting?
7. What are the advantages and disadvantages of passive clustering and active clustering?
8. What is the sufficient and necessary condition to guarantee 100% deliverability for the flooding algorithms based only on one-hop information?
9. Descript Broadcast Incremental Power (BIP) protocol.
10. What is the difference between proactive routing and reactive routing?

## References

1. Adjih, C., Jacquet, P., and Viennot, L. (2002). *Computing connected dominated sets with multipoint relays*. Technical Report 4597, INRIA, October 2002.
2. Abolhasan, M. and Lipman, J. (2005). An Efficient and Highly Scalable Route Discovery for On-demand Routing Protocols in Ad hoc Networks. In *proceedings of the IEEE 30th Conference of Local Computer Networks (LCN)*, Sydney, Australia.
3. Abolhasan, M. and Lipman, J. (2006). Self-Selecting Route Discovery Strategies for Reactive Routing in Ad hoc Networks. In *the International Conference on Integrated Internet Ad hoc and Sensor Networks (InterSense2006)*, Nice, France.

4. Abolhasan, M., Lipman, J., and Chicharo, J. (2004). A routing Strategy for Heterogeneous Mobile Ad hoc Networks. In *IEEE 6th CAS Symposium on Emerging Technologies: Frontiers of Mobile and Wireless Communications* (MWC), Shanghai, China, pp. 13–16.

5. Aggelou, G. and Tafazolli, R. (1999). RDMAR: A bandwidth-efficient routing protocol for mobile ad hoc networks. In *ACM International Workshop on Wireless Mobile Multimedia (WoWMoM)*, Seattle, WA, pp. 26–33.

6. Alagar, S. and Venkatesan, S. (1995). Reliable broadcast in mobile wireless networks. In *Proceedings Military Communications Conference*, vol. 1, pp. 236–240.

7. Borbash, S. A. and Jennings, E. H. (2002). Distributed Topology Control Algorithm for Multihop Wireless Networks. In *Proceedings 2002 World Congress on Computational Intelligence (WCCI 2002)*, Honolulu, Hawaii.

8. Broch, J., Maltz, D. A., Johnson, D. B., Hu, Y.-C., and Jetcheva, J. (1998). A Performance Comparison of Multi-Hop Wireless Ad Hoc Network Routing Protocols. In *Mobile Computing and Networking*, Dallas, Texas, US, pp 85–97.

9. Basu, P. and Redi, J. (2004), *Effect of Overhearing Transmissions on Energy Efficiency in Dense Sensor Networks*, ACM IPSN, Berkeley.

10. Chiang, C.-C., Gerla, M., and Zhang, L. (1997). Routing in Clustered Multihop Mobile Wireless Networks with Fading Channel. In *proceedings of IEEE SICON*, Kent Ridge, Singapore, pp. 197–211.

11. Cartigny, J., Ingelrest, F., and Simplot, D. (2003). RNG relay subset flooding protocols in mobile ad-hoc networks. In *International Journal of Foundations of Computer Science*, Vol. 14, no. 2, pp. 253–265.

12. Cartigny, J., Ingelrest, F., Simplot-Ryl, D., and Stojmenovic, I. (2003). Localized LMST and RNG based minimum energy broadcast protocols in ad hoc networks. *Proc. IEEE INFOCOM*, San Francisco, CA, USA; Ad Hoc Networks.

13. Gerla, M. (2002). *Fisheye State Routing Protocol (FSR) for Ad hoc Networks.* Internet Draft, draft-ietf-manet-fsr-03.txt, http://www.ietf.org.

14. Garcia-Luna-Aceves, J., and Spohn, C. M. (1999). Source-Tree Routing in Wireless Networks. In *proceedings of the 7th Annual International Conference on Network Protocols*, Toronto, Canada, pp. 273–283.

15. Gerla, M. and Tsai, J. T. C. (1995). Multicluster, Mobile Multimedia Radio Network. In *Journal of Wireless Networks*, Vol. 1, pp. 255–265.

16. Hsu, C.-S. and Tseng, Y.-C. (2002). An Efficient Reliable Broadcasting Protocol for Wireless Mobile Ad Hoc Networks. In *IASTED Networks, Parallel and Distributed Processing, and Applications (NPDPA)*, Japan.

17. Ingelrest, F. and Simplot-Ryl, D. (2008), Localized Broadcast Incremental Power Protocol for Wireless Ad Hoc Networks, *Wireless Networks*, Vol. 14, no. 3, pp. 309–319.

18. Ingelrest, F., Simplot-Ryl, D., Stojmenovic I. (2004), Target transmission radius over LMST for energy-efficient broadcast protocol in ad hoc networks, *IEEE International Conference on Communications* ICC, Paris.

19. Jiang, M., Ji, J., and Tay, Y. (1999). *Cluster Based Routing Protocol.* Internet Draft, draft-ietf-manet-cbrp-spec-01.txt, work in progress, http://www.ietf.org.

20. Johnson, D. and Maltz, D. (1996). Dynamic Source Routing in Ad hoc Wireless Networks. In *Mobile Computing*, Kulwer Academic, pp. 153–181.

21. Jacquet, P., Muhlethaler, P., Qayyum, A., Laouitim, A., and Viennot, L. (2000). *Optimized Link State Routing.* draft-ietf-manet-olsr-06.txt, http://www.ietf.org/.

22. Koubaa, H. and Fleury, E. (2003). On the performance of double domination in ad hoc networks. In *Proceedings IFIP Medhoc 2003*, Tunisia, Mahdia, Tunisia.

23. Kim, D. and Maxemchuk, N. (2003). A comparison of flooding and random routing in mobile ad hoc network. In *Proceedings of 3rd New York Metro Area Networking Workshop*, New York, USA.

24. Ko, Y.-B. and Vaidya, N. H. (1998). Location Aided Routing (LAR) in Mobile Ad hoc Networks. In *Proceedings of the 4th Annual ACM/IEEE International Conference on Mobile Computing and Networking (Mobicom'98)*, Dallas, Texas.

25. Lipman, J., Boustead, P., and Chicharo, J. (2004). Reliable minimum spanning tree flooding in ad hoc networks. *Proceedings of the IEEE 6th CAS Symposium on Emerging Technologies: Frontiers of Mobile and Wireless Communication*, Shanghai, China.

26. Lipman, J., Boustead, P., Chicharo, J., and Judge, J. (2003). Optimized flooding algorithms for ad hoc networks. In *Proceedings of the 2nd Workshop on the Internet, Telecommunications and Signal Processing (WITSP'03)*, Coolangatta, Gold Coast, Australia.

27. Lipman, J., Boustead, P., Chicharo, J., and Judge, J. (2003b). Resource aware information dissemination in ad hoc networks. In *Proceedings of the 11th IEEE International Conference on Networks (ICON 2003)*, Sydney, Australia.

28. Lipman, J., Boustead, P., and Judge, J. (2003). Neighbor aware adaptive power flooding in mobile ad hoc networks. *International Journal of Foundations of Computer Science*, Vol. 14, no. 2, 237–252.

29. Lin, C. and Gerla, M. (1997). Adaptive Clustering for MobileWireless Networks. In *IEEE Journal on Selected Areas in Communications*, Vol. 15, no. 7, pp. 1265–1275.

30. Li, N., Hou, J. C., and Sha, L. (2003). Design and analysis of an mst-based topology control algorithm. In *Proceedings of IEEE Infocom 2003*, San Francisco California, USA.

31. Liu, H., Jia, X., Wan, P., Liu, X., and Yao, F. (2007). A Distributed and Efficient Flooding Scheme Using 1-hop Information in Mobile Ad Hoc Networks, *IEEE Transactions on Parallel and Distributed Systems*, vol. 18, no. 5.

32. Lim, H. and Kim, C. (2000). Multicast tree construction and flooding in wireless ad hoc networks. In *Proceedings of the 3rd ACM international workshop on modeling*, analysis and simulation of wireless and mobile systems, ACM Press, pp 61–68.

33. Li, F. and Nikolaidis, I. (2001). On minimum energy broadcasting in all wireless networks. In *Proceedings of IEEE Annual Conference on Local Computer Networks (LCN 2001)*, Tampa, Florida, USA.

34. Lou, W. and Wu, J. (2002). On Reducing Broadcast Redundancy in Ad hoc Wireless Networks. In *IEEE Transactions on Mobile Computing*, Vol. 1, no. 2, pp.111–122.

35. Lou, W. and Wu, J. (2003). A reliable broadcast algorithm with selected acknowledgements in mobile ad hoc networks. In *Proceedings of IEEE GLOBECOM'03*, San Francisco, USA, San Francisco, USA.

36. Lou, W. and Wu, J. (2004). Double-covered broadcast (dcb): a simple reliable broadcast algorithm in manets. In *Proceedings of IEEE INFOCOM'04*, Hong Kong, China.

37. Murthy, S. and Garcia-Luna-Aceves, J. (1995). A routing protocol for packet radio networks. In *proceedings Mobile Computing and Networking*, Berkeley, California, US, pp. 86–95.

38. Ni, S.-Y., Tseng, Y.-C., Chen, Y.-S., and Sheu, J.-P. (1999). The broadcast storm problem in a mobile ad hoc network. In *Proceedings of the 5th annual ACM/IEEE international conference on Mobile computing and networking*, ACM Press, pp. 151–162.

39. Ovalle-Martinez, F. J., Stojmenovic, I., Garciea-Nocetti, F., and Solano-Gonzalez, J. (2004). Finding minimum transmission radii for preserving connectivity and constructing minimal spanning trees in ad hoc and sensor networks, *the third Workshop on Efficient and Experimental Algorithms*, Angra dos Reis, Brazil.

40. Penrose M. (1997), The longest edge of the random minimal spanning tree, *The Annals of Applied Probability*, vol. 7, no. 2, pp.340–361.

41. Perkins, C. and Bhagwat, P. (1994). Highly Dynamic destination-sequenced distance vector routing (DSDV) for mobile computers. In *Proceedings Computer Communications Review*, vol. 24, no. 4, pp. 234–244.

42. Paruchuri, V., Durresi, A., Dash, D., and Jain, R. (2003). Optimal flooding protocol for routing in ad hoc networks. In *proceedings of IEEE Wireless Communications and Networking Conference*, New Orleans, Louisiana.

43. Peng, W. and Lu, X. C. (2000). On the reduction of broadcast redundancy in mobile ad hoc networks. In *Proceedings 1st Annual Workshop on Mobile and Ad Hoc Networking and Computing*, Boston, USA, pp. 129–130.

44. Pagani, E. and Rossi, G. P. (1997). Reliable Broadcast in Mobile Multihop Packet Networks. In *Mobicom 97*, Budapest, Hungary, pp.34–42.

45. Pagani, E. and Rossi, G. P. (1999). Providing reliable and fault tolerant broadcast delivery in mobile ad hoc networks. In *Mobile Networks and Applications*, vol. 4, pp. 175–192.

46. Perkins, C. E. and Royer, E. M. (1999). Ad hoc On-Demand Distance Vector (aodv) routing. In *Proceedings of the 2nd Annual IEEE Workshop on Mobile Computing Systems and Applications*, New Orleans, La, USA, pp. 90–100.

47. Qayyum, A., Viennot, L., and Laouiti, A. (2001). Multipoint relaying: An efficient technique for flooding in mobile wireless networks. In *proceedings of the 35th Annual Hawaii International Conference on System Sciences*, Maui.

48. Raju, J. and Garcia-Luna-Aceves, J. (1999). A new approach to on-demand loop free multipath routing. In *Proceedings of the 8th Annual IEEE International Conference on Computer Communications and Networks (ICCCN)*, Boston, MA, pp. 522–527.

49. Sucec, J. and Marsic, I. (2000). An efficient distributed network-wide broadcast algorithm for mobile ad hoc networks. *Technical Report 248*, Rutgers University, CAIP.

50. Santivez, C. A., Ramanathan, R., and Stavrakakis, I. (2001). Making link-state routing scale for ad hoc networks. In *Proceedings of the 2nd ACM international symposium on Mobile ad hoc networking and computing, Long Beach*, CA, USA, pp. 22–32. ACM Press.

51. Shaikh, J., Solano, J., Stojmenovic, I., and Wu, J. (2003). New metrics for dominating set based energy efficient activity scheduling in ad hoc networks. In *Proceedings of WLN Workshop at IEEE Conf. on Local Computer Networks*, Bonn, Germany.

52. Stojmenovic, I., Seddigh, M., and Zunic, J. (2002). Dominating sets and neighbor elimination based broadcasting algorithms in wireless networks. *IEEE Trans. On Parallel and Distributed Systems*, vol. 13, no. 1, pp. 14–25.

53. Sheu, S.-T., Tsai, Y., and Chen, J. (2002). A Highly Reliable Broadcast Scheme for IEEE 802.11 Multi-hop Ad Hoc Networks. In *Proceedings IEEE ICC*, vol. 1, pp. 610–615.

54. Stojmenovic, I. and Wu, J. (2004). Broadcasting and activity scheduling in ad hoc networks, in: *Ad hoc Networking*, S. Basagni, et al., eds. IEEE Press, 2004.

55. Toussaint, G. (1980). The relative neighborhood graph of finite planar set. *Pattern Recognition*, vol. 12, no. 4, pp. 261–268.

56. Tseng, Y.-C., Ni, S.-Y., and Shih, E.-Y. (2001). Adaptive Approaches to Relieving Broadcast Storms in a Wireless Multihop Mobile Ad Hoc Network. In *Proceedings of the International Conference on Distributed Systems*, Washington, DC, USA, pp. 481–488.

57. Wu, J. (2003). An enhanced approach to determine a small forward node set based on multipoint relays. In *Proceedings of 2003 IEEE Semiannual Vehicular Technology Conference (VTC2003-fall)*, Orlando, USA.

58. Wu J. and Li, H., (August 1999) A dominating set based routing scheme in ad hoc wireless networks, *Proceedings DIAL M*, Seattle, pp. 7–14.

59. Wu J. and Li, H., (2001) A dominating set based routing scheme in ad hoc wireless networks, *Telecommunication Systems*, vol. 18, no. 1–2, pp.13–36.

60. Wisielthier, J., Nguyen, G. and Ephremides, A. (2000). On the construction of energy-efficient broadcast and multicast trees in wireless networks. In *Proceedings IEEE INFOCOM*, Tel Aviv, Isreal.

61. Yi, Y., Gerla, M., and Kwon, T. J. (2003). Efficient Flooding in Ad hoc Networks using On-Demand (Passive) Cluster Formation. In *2nd Proceedings of the Mediterranean Workshop on Ad-hoc Networks*, Lausanne, Switzerland.

62. Yi, Y., Kwon, T. J., and Gerla, M. (2001). *Passive Clustering (PC) in Ad hoc Networks*. Internet Draft, draft-ietf-yi-manet-pac-00.txt, http://www.ietf.org.

# Chapter 7
# Geographic Routing in Wireless Ad Hoc Networks

Dazhi Chen and Pramod K. Varshney

**Abstract** Geographic routing has become an efficient solution for communications and information delivery in wireless ad hoc networks where the position information of nodes is available. This chapter provides a comprehensive overview of basic principles, classical techniques, as well as latest advances in geographic routing. The chapter first presents in detail the topic of geographic unicast routing, where the presentation is focused on two operation modes of geographic forwarding, that is, greedy forwarding and void handling. The chapter also briefly introduces three advanced topics in geographic routing: geographic multicast, geocast, and trajectory-based forwarding. Finally, the chapter makes some comments on the practical aspects of geographic routing for practitioners and discusses the directions for further research with a list of open issues in the area of geographic routing.

## 7.1 Introduction

The idea of geographic routing,[1] simply described as heuristic greedy forwarding algorithms, was originally proposed for an early example of wireless ad hoc networks – packet radio networks – in the 1980s [1, 2]. In recent years, with the availability of small, inexpensive, and low-power Global Positioning System (GPS) receivers and the advances in the area of self-configuring ad hoc localization techniques [3–6], geographic routing has regained considerable attention,

---

This work was supported in part by the Air Force Office of Scientific Research (AFOSR) under grant FA-9550-06-1-0277.

[1] Geographic routing is also referred to as geo-routing, position-based routing, location-based routing in the literature.

D. Chen (✉)
Department of Electrical Engineering and Computer Science, Syracuse University, Syracuse, NY, 13244, USA
e-mail: dchen02@syr.edu

S. Misra et al. (eds.), *Guide to Wireless Ad Hoc Networks*,
Computer Communications and Networks, DOI 10.1007/978-1-84800-328-6_7,
© Springer-Verlag London Limited 2009

as it promises a scalable and efficient solution for information delivery in emerging wireless ad hoc networks such as Mobile Ad hoc Networks (MANETs) [7], Vehicular Ad hoc Networks (VANETs) [8], Wireless Mesh Networks (WMNs) [9], and Wireless Sensor Networks (WSNs) [10].

Geographic routing takes advantage of geographic information of nodes (i.e., actual geographic coordinates or virtual relative coordinates), instead of topological connectivity information in the network to move data packets to gradually approach and eventually reach their intended destinations. In the literature, most of the research effort on geographic routing is focused on geographic unicast routing that enables the routing of data packets from a single source to a single destination. In geographic unicast routing, geographic forwarding plays a major role and it makes the forwarding decision to move a data packet from the current sender to a next-hop node at every hop on a source–destination path. Geographic forwarding mainly operates in a greedy forwarding mode that primarily uses the positions of the sender, its neighboring nodes, and the destination to move the packet further toward the destination at every hop, if possible. However, greedy forwarding may not always be possible. For example, distance to the destination is a natural metric for greedy forwarding, because it provides an inherent gradient toward the intended destination. What if all the neighboring nodes of a sender are farther away from the destination than the sender itself? In this scenario, greedy forwarding fails and an alternate mode is needed for geographic forwarding. Otherwise, the packet has to be discarded and the delivery fails. This problem is called *communication void* [11], *local maximum phenomenon* [12], or *local minimum phenomenon* [13] while the solution to this problem is called *void handling* [11], *backup* [14], or *recovery* [15]. This chapter presents an extensive overview of geographic forwarding techniques in wireless ad hoc networks and, in particular, it focuses on the presentation of two operation modes of geographic forwarding, i.e., greedy forwarding and void handling. For each mode, we introduce the basic principles involved and describe classical techniques as well as latest advances in the area.

In addition to geographic unicast routing, we also introduce three advanced topics in geographic routing, including geographic multicast (i.e., the routing of data packets to multiple destinations) [16], geocast (i.e., a special type of geographic multicast that enables the routing of data packets to multiple destinations within a given geographic region) [17], and trajectory-based forwarding (i.e., routing on a curve [18]). The chapter provides a general description of these topics as well as some relevant techniques, allowing readers to learn the basic principles and further investigate geographic routing research issues if desired.

Our goal in this chapter is to present a comprehensive overview of geographic routing in wireless ad hoc networks in a tutorial manner while keeping sufficient depth. Hence, we make every effort to provide a balanced treatment in terms of general descriptions and necessary technical details for each geographic routing technique. The rest of the chapter is organized as follows. In Section 7.2, we provide some background information for geographic routing.

We introduce geographic unicast routing in detail in Section 7.3. Section 7.4 presents three advanced topics in geographic routing. We describe some practical issues relevant for practitioners in Section 7.5 and discuss some possible directions for further research in Section 7.6. Section 7.7 concludes this chapter.

## 7.2 Background

Wireless ad hoc networks are formed by the cooperation of an arbitrary set of independent wireless nodes that communicate with each other without using a fixed network infrastructure or centralized administration. They have a wide range of potential applications in both civilian and military contexts such as disaster relief and battlefield activities. In a wireless ad hoc network, each node can only communicate directly with other nodes in its vicinity, because of limited transmission range. Thus, if two nodes are not within the transmission range of each other, they have to communicate over multiple wireless hops. This necessitates the development of efficient and scalable routing protocols specifically optimized for multihop wireless ad hoc networks.

Since the early 1980s, there have been a large number of routing protocols proposed for wireless ad hoc networks, covering a broad range of design choices and approaches. From the perspective of information exploited, the existing routing protocols can mainly be classified into two categories: topology-based routing and geographic routing.

Topology-based routing exploits topological connectivity information about the links in the network to establish and maintain source–destination paths. It can further be divided into proactive (also called table-driven), reactive (also called on-demand), and hybrid approaches. If geographic information of nodes is available to use, topology-based routing can also incorporate the additional information to improve its efficiency. For example, if the position of the destination is known by the source, the use of the position information of nodes can prevent network-wide searches for the destination by limiting the network flooding range, as either control or data packets can be sent in the general direction of the destination. A review and comparison of topology-based routing protocols are available in [19, 20].

Geographic routing relies solely on the geographic information of nodes in the network, i.e., actual geographic or virtual relative coordinate information, to move a data packet to its intended destination. Since most geographic routing techniques only use the local knowledge of one-hop neighboring nodes to make forwarding decisions, they do not require the establishment and maintenance of end-to-end paths. Thus, nodes do not have to store routing tables and there is no need to transmit control packets to update path states either [12]. Since topology changes do not disturb forwarding decisions in geographic routing unless they affect the local knowledge of a sender, topology changes have less impact on geographic routing than on topology-based routing. The operation of geographic routing is highly localized and thus it is a scalable solution for routing in wireless ad hoc networks.

Although geographic routing, when compared to topology-based routing, has some advantages such as good scalability, it has its disadvantages. First, geographic routing relies on the existence of GPS receivers or complex localization techniques to calculate the positions of nodes, which increases cost and complexity of wireless ad hoc networks. In addition, geographic routing is inefficient when the network topology is not well captured by the geographic coordinates of nodes, due to obstacles, wireless channels, or localization errors. Second, geographic routing usually requires a location service to look up the positions of destinations. The problem of designing efficient schemes for updating the positions of nodes in a location service might even be more difficult than geographic forwarding itself in wireless ad hoc networks with high node mobility [15]. The overhead introduced by the location service may also be very costly. Finally, the advantages of geographic routing mostly come from the operation mode of greedy forwarding. However, greedy forwarding may fail in the presence of communication voids and the operation mode of void handling is necessary for geographic routing to get stuck packets around voids. Whether geographic routing can still retain its advantages also depends on how communication voids are handled.

Geographic routing in the context of wireless ad hoc networks is the theme of this chapter. The chapter covers most of the existing geographic routing techniques in a systematic manner. Before starting a formal presentation of geographic routing, we make the following two general assumptions in this chapter, and other important assumptions will be mentioned later when necessary.

- All nodes in a wireless ad hoc network know their own positions. The positions of nodes can be in the form of either actual geographic coordinates, as obtained through the GPS system, or virtual relative coordinates, as obtained through reference points on some fixed coordinate systems [3]. When powerful positioning devices such as GPS receivers are available, the positions of nodes can be configured manually by the network installer, provided that all nodes are physically accessible and remain stationary after configuration, or nodes can be equipped with positioning devices themselves to determine their positions directly. In most wireless ad hoc networks and applications, due to the considerations of cost, power consumption, device form factor, or indoor applications, only some nodes in the network are equipped with positioning devices and such nodes are called *anchors*, *beacons*, or *landmarks*. In this case, advanced ad hoc localization techniques, including range-based and range-free localization, can be deployed for nodes to infer their positions, based on signal measurements or network connectivity relative to anchors [3–6]. For example, in a range-based localization technique called lateration, determination of a node's 2D position, as shown in Fig. 7.1, requires distance measurements from three non-collinear anchors (four for 3D position). Several measurement techniques such as relative signal strength and time difference of arrival are used to estimate distances to anchors. When powerful

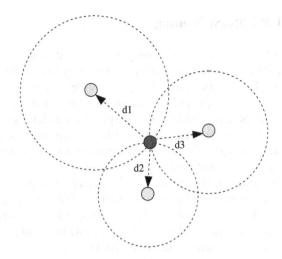

**Fig. 7.1** Localization for a 2D position using lateration

positioning devices are unavailable (i.e., in the absence of anchors), some localization techniques that construct local coordinate systems, by using trigonometric relationships (sine or cosine rules) among nodes, can be deployed for nodes to infer their virtual relative coordinates. A detailed review of the techniques that determine the positions of nodes is available in [3].

- Before a packet can be sent, the source knows the geographic information of the routing destination, such as the position of a destination node, the positions of destination nodes, or a given geographic region. Geographic information of the destination is then carried in the header of the packet so that other intermediate nodes can learn where the packet is destined for. Typically, a *location service*, which maintains the up-to-date positions of nodes in the network, is used to complete this task. In [12], existing location services were classified into some-for-some, some-for-all, all-for-some, all-for-all services, according to how many nodes operate the location service and how many positions of other nodes are maintained in one of such nodes. A viable location service can be either a centralized service external to the wireless ad hoc network or a decentralized service that is part of the network. A survey of existing location services is available in [12], while a survey of location update schemes is available in [21]. In the presence of a location service in the network, a source node directly asks the location service for geographic information of the destination. If a location service is unavailable in the network, a source node can initiate a destination search, which floods the network with a search packet. The destination then reports back to the source node with its position information. In some scenarios, the destination can also announce its position to the source directly. For instance, in sensor networks, a fixed sink can broadcast its position to the whole network via network-wide flooding or can piggyback its geographic information in querying packets.

## 7.3 Geographic Unicast Routing

Geographic unicast routing only exploits geographic information of nodes in the network to deliver data packets from a single source to a single destination. Since we assume that the position of a packet's destination can always be determined before sending the packet, geographic unicast routing is actually equivalent to geographic forwarding. Geographic forwarding is a forwarding mechanism that moves a data packet to gradually approach and eventually reach the intended destination. It usually operates in one of the following two modes: greedy forwarding and void handling. In the greedy forwarding mode, the packet is forwarded from the current sender to its next-hop node in a locally optimal and greedy manner. If greedy forwarding fails to move the packet further due to the presence of communication voids, the void handling mode is invoked. In this mode, the sender attempts to route the data packet around the void, because a topologically valid path from the source to the destination may still exist. In this section, we introduce these two modes in detail. For each mode, we present both classical and newly proposed techniques. Most of these techniques are extracted from their respective geographic forwarding protocols proposed for wireless ad hoc networks. In other words, we introduce basic principles, original ideas, and inherent characteristics of these techniques in a cohesive manner, independent of any individual geographic forwarding protocol as well as of any specific wireless ad hoc network with unique characteristics. This will enable the readers to have a clearer understanding of geographic forwarding as a whole, which will help them to better analyze a specific geographic forwarding protocol themselves in the future.

### 7.3.1 Greedy Forwarding

Greedy forwarding makes packet forwarding decisions based on locally available information, such as the positions of the sender, its neighbors, and the destination, and thus it is a simple and scalable forwarding strategy. This greedy strategy is applied wherever possible until the destination, if reachable, is eventually reached. In greedy forwarding, a next-hop node, with respect to the packet destination, needs to be established for the sender at every hop to move the data packet further toward the destination, leading to the research issue: how and based on which criterion to select a next-hop node?

In the literature, a number of greedy forwarding algorithms have been proposed. Most of the algorithms use a centralized selection of the next-hop node at the sender. In other words, the sender employs a specific criterion to select the next-hop node before it actually forwards the data packet. For this, the sender must know the positions of its neighbors before it makes the choice of the next-hop node. A beaconing scheme is typically responsible for this task. Each node periodically transmits a beacon, which includes its own position and maybe other information such as node residual energy, to its one-hop neighbors, so that each node maintains a list of positions and other state information of its neighbors in its neighborhood table. Since the beaconing rate is closely related to the accuracy of positions of neighbors as well as communication overhead incurred, it needs to adapt to local network dynamics in order to achieve optimized performance. In recent years, some algorithms [14, 22–26] have been proposed, which use a distributed selection to determine the next-hop node among the neighbors of the sender. In other words, the neighbors of the sender contend to establish the next-hop node in a distributed fashion for the sender to forward its data packet. In this way, the beaconing scheme is no longer needed. The positions and other necessary state information of neighbors are directly obtained from neighbors in a distributed fashion. There also exist some hybrid algorithms that use instant access of information of the neighbors while still employing the centralized selection of the next-hop node at the sender [27].

A next-hop node selection criterion depends on the performance optimization objectives and network assumptions made for a wireless ad hoc network. In the following, we review several of the currently available next-hop node selection criteria, taken from existing greedy forwarding algorithms in the literature.

Some early proposed criteria of next-hop node selection only use geometric calculations to determine the next-hop node. Most such criteria can be illustrated by means of Fig. 7.2, where S and D denote the sender and the destination, respectively. The transmission range of S is modeled as a fixed circle with a radius of $r$.

- *Most Forward Within Radius* (MFR): *Progress* is defined as the orthogonal projection of the line connecting the sender and the next-hop node onto the line connecting the sender and the destination. A neighbor is in *forward direction* if the progress is positive. Otherwise, it is said to be in *backward direction*. The MFR criterion [1] is to select a neighbor that has greatest positive progress toward the destination as the next-hop node. For example, in Fig. 7.2, node A is chosen. This criterion tries to minimize the number of hops a packet has to traverse in order to reach D. The number of hops is related to performance objectives such as packet delay.
- *Nearest with Forward Progress* (NFP): Using the NFP criterion [2], the next-hop node is chosen as the nearest neighbor with positive progress. For example, node B in Fig. 7.2 is chosen. Under the assumption that the sender can adjust its transmission power to the distance between the two nodes, this criterion tries to reduce the probability of packet collisions in the region around the sender. The probability of packet collisions is closely related to link transmission reliability performance.
- *Random selection*: In the random selection criterion [28], one node from the set of neighbors with positive progress is randomly (with equal probability) chosen as the next-hop node. For example, a node from the set of shaded nodes A, B, C, E, F, and G, shown in Fig. 7.2 is randomly chosen. This criterion tries to trade off progress and transmission reliability performance,

**Fig. 7.2** Next-hop node selection criteria using geometric calculations

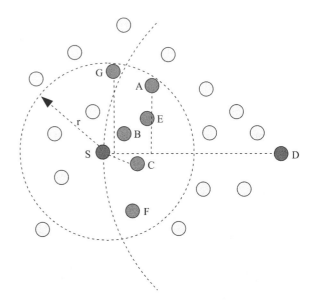

with the assumption that the transmission power can be adjusted to the distance between the two nodes.

- *Closest-direction-based*: This criterion is defined in compass routing [29], and the next-hop node is selected as the neighbor that has the direction closest to the line between the sender and the destination, such as node C shown in Fig. 7.2. This criterion tries to minimize the spatial distance that a packet has to travel.

A desirable property for a routing protocol in wireless ad hoc networks is that it is free of loops. Otherwise, a routing protocol cannot guarantee[2] that a packet will always be successfully delivered to its destination. The aforementioned criteria for selection of the next-hop node cannot guarantee that a loop-free path can always be formed. An example [30] is illustrated in Fig. 7.3, where nodes A and B are neighbors and node D is the destination. Since both nodes A and B have positive progress with respect to each other, a possible loop may form (A-B-A-B...), where A selects B as its next-hop node and B selects A as its next-hop node. Another example is presented in [31], which shows that the closest-direction-based criterion in compass routing is not loop-free either. Thus, a geographic routing protocol that uses any of the aforementioned four criteria may generate loops on routing paths and cannot guarantee packet delivery.

---

[2] The guarantee of packet delivery is ensured only at the topology level. Other factors such as packet collisions at the MAC layer or packet loss due to network congestion are not considered here.

**Fig. 7.3** An example where progress-based criteria generate a loop (A-B-A)

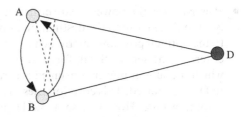

Since distance to the destination provides an inherent gradient toward the intended destination, a criterion using distance is a natural criterion for greedy forwarding to select the next-hop node. Two such criteria are described as follows:

- *Most Advance Within Radius* (MAR):[3] *Advance* is defined as the distance between the sender and the destination minus the distance between a neighbor and the destination [30]. The MAR criterion in Cartesian Routing [32] is to select as the next-hop node the neighbor closest to the destination among a set of neighbors which are closer to the destination than the sender, i.e., a neighbor that has greatest positive advance toward the destination. In the example of Fig. 7.2, node E is chosen. Note that although E has less progress than A, E is closest to the destination D and thus has more advance than A.
- *Nearest Closer* (NC): The NC criterion [33] is to select the nearest neighbor among the set of neighbors closer to the destination than the sender, i.e., the nearest neighbor with positive advance is selected as the next-hop node. In Fig. 7.2, the NC criteria still selects node B, as chosen using the NFP criterion.

In practice, MFR and MAR, in most network topologies, discover the same path to the destination. Similarly, NFP and NC also discover the same path in most network topologies. Unlike any progress-based criterion, an advance-based criterion can guarantee that a loop-free path can always be formed. In other words, if a next-hop node is selected only from a set of neighbors with positive advance, loops can always be avoided.[4]

In addition to geometric criteria that use only the geographic coordinates of nodes to select the next-hop node, non-geographic local knowledge such as power consumption, node energy reserve, and node queue size can also be included in the next-hop node selection criteria for different optimization purposes, so that the desired network performance can be attained. Note that, in most such criteria, only those neighbors with positive advance, i.e., closer to destination, are considered as next-hop candidates, as whichever node from this set is selected, the loop-free property can be guaranteed.

---

[3] This name is introduced for our convenience.

[4] Mobility-induced loops [15] are not considered here.

- *Power-aware*: the power-aware criterion in [33] is to select the neighbor that is not only as close to the destination as possible but also as close as possible to the optimal position determined by the theorem proved in [33]. Using this criterion, the sender S will forward the packet to the next-hop node B for which the sum of power consumed for transmission from S to B and from B to D is minimized. This criterion tries to minimize the energy consumption of a routing task. However, some nodes might be selected by many routing tasks, which will deplete their energy very fast and nodes might die very quickly. In order to maximize the number of successful routing tasks in the network, other criteria proposed in [33] also consider the remaining battery power information (i.e., node residual energy) along with the transmission power factor.
- *Maximum Weighted Progress* (MWP): *weighted progress* in [27] is defined as the advance of a neighbor to the destination divided by the transmission power consumed by the sender for the delivery of the packet to the neighbor, i.e., the advance toward the destination achieved per unit transmitted power. The MWP criterion is to select the neighbor that has maximum weighted progress as the next-hop node. This criterion tries to provide a trade-off between advance and transmission power.
- *Energy-advance-random*: A heuristic function, which incorporates information regarding advance and residual energy of nodes as well as a random value, was proposed in [25] to select the next-hop node. The introduction of a random value is to further disperse the system workload. This energy-advance-random criterion tries to prolong the lifetime of the network by balancing the energy consumption among nodes.

Recent experimental studies on wireless ad hoc networks have shown that wireless links can be extremely unreliable [34, 35]. For example, [36] demonstrated the existence of a large *transitional region* where link quality has high variance. Thus, link reliability should be explicitly taken into account when selecting the next-hop node in greedy forwarding. Two criteria that consider the more realistic lossy wireless channels are introduced as follows:

- *Best Packet Reception Rate* (PRR) × *Distance*: *PRR× distance* is defined as the product of the packet reception rate and the advance achieved by forwarding the packet to this neighbor [37]. The best PRR × distance criterion is to select the neighbor with the highest value of the product as the next-hop node. This criterion was concluded in [37] to be the optimal criterion for achieving the distance-hop energy trade-off for geographic routing in lossy wireless networks with Automatic Repeat reQuest (ARQ) mechanisms. Nodes using this criterion often take advantage of the neighbors in the transitional region that has links with high variance. Note that a similar criterion was independently proposed in [38].
- *Maximum Normalized Advance* (MNA): *Normalized Advance* (NADV) [39] is defined as the advance of a neighbor divided by the link cost incurred by the sender for the delivery of the packet to this neighbor, i.e., the advance

toward the destination achieved per unit cost. The MNA criterion is to select the neighbor with largest NADV as the next-hop node. Since various types of link costs such as packet error rate, link delay, and energy consumption can be included in NADV, this criterion is a generalized criterion. The MNA criterion balances the advance against the link cost and it tries to provide the best trade-off between link cost and geographic proximity.

A greedy forwarding algorithm that considers link reliability in its next-hop node selection has to access the MAC layer to obtain the necessary information such as the packet reception rate or packet error rate. Thus, a geographic routing protocol that uses such a greedy forwarding algorithm is actually a cross-layer protocol, as routing is originally designed as a task at the network layer, which cannot utilize the MAC layer information directly in a traditional layered protocol design. Table 7.1 presents a summary of the main characteristics of greedy forwarding algorithms that use the next-hop node selection criteria we discussed in this section.

## 7.3.2  Void Handling

The void handing mode is invoked when the greedy forwarding mode fails to forward the packet further due to the presence of communication voids. This occurs when none of the neighbors of the sender can be selected as the next-hop node. For example, as illustrated in Fig. 7.4, node S is closer to the destination D than any of its neighbors. Thus, no advance-based greedy forwarding algorithm can move a data packet at S further toward D. However, there exists a topologically valid path from S to D: S-A- B-C-E-D. In this case, the packet is said to have encountered a *communication void* with respect to D and gets stuck at S, which is called a *void node*.

Before geographic forwarding switches from the greedy forwarding mode to the void handling mode, it should first determine whether the void is temporary due to temporary unavailability of next-hop nodes or packet collisions. The problem of temporary voids will get resolved as nodes wake up, mobile nodes move in, or collisions are resolved. Temporary voids only affect packet delivery delay. Only if it is determined that the void is not temporary, the void handling mode is invoked and a void handling solution is employed for geographic forwarding to handle the void. Geographic forwarding switches back to the greedy forwarding mode if the stuck packet reaches a node closer to the destination than the node where geographic forwarding switched to the void handling mode.

Network-wide flooding is the simplest void handling solution, which certainly enables the stuck packet to reach the destination if at least a topologically valid path exists. However, this solution is effective but inefficient in terms of network resource utilization. Every node in the network has to forward the packet once and too many redundant copies of the stuck packet are received by

**Table 7.1** Characteristics of greedy forwarding algorithms

| Name | Transmission range | Link reliability | Optimization objective | Criterion | Loop-free |
|---|---|---|---|---|---|
| MFR | Fixed circle | Lossless | To minimize the number of hops | Maximum progress | No |
| NFP | Adjustable circle | Lossless | To reduce the probability of packet collisions | Nearest with positive progress | No |
| Random | Adjustable circle | Lossless | To trade off progress and transmission reliability | Random with equal probability | No |
| Compass routing | Fixed circle | Lossless | To minimize the spatial travel distance | Closest direction | No |
| MAR | Fixed circle | Lossless | To minimize the number of hops | Maximum advance | Yes |
| NC | Adjustable circle | Lossless | To reduce the probability of packet collisions | Nearest with positive advance | Yes |
| Power-aware | Adjustable circle | Lossless | To minimize energy consumption of a routing task or to maximize the number of successful routing tasks | Distance, transmission power, and node residual energy | Yes |
| MWP | Adjustable circle | Lossless | To maximum advance per unit transmission power | Distance, transmission power | Yes |
| Energy-advance-random | Fixed circle | Lossless | To balance the energy consumption of nodes | Distance, node residual energy, and random value | Yes |
| PRR × distance | Statistic | Lossy | To minimize energy consumption | Distance, packet reception rate | Yes |
| MNA | Statistic | Lossy | To maximize advance per unit cost | Distance, cost such as packet error rate | Yes |

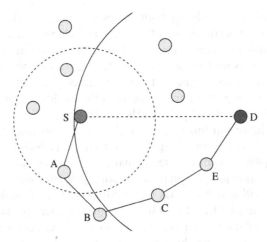

**Fig. 7.4** A communication void, with respect to the destination D, occurs at node S where advance-based greedy forwarding fails

the same destination. Thus, some advanced solutions are desired for geographic forwarding to handle voids.

In the following, we present six categories of void handling solutions [11], each designed with a different approach, that is, planar-graph-based, topology-based, link-reversal-based, geometric, heuristic, and hybrid. For each category, we present its basic principle and describe one representative solution. More details about the void problem and existing void handling solutions are available in [11].

1. Planar-graph-based void handling
   Planar-graph-based solutions to the void problem, which exploit the properties of planar graphs, have attracted a major amount of research effort. In graph theory, a planar graph is defined as a graph that can be embedded in the plane with no intersecting edges. Based on the ancient idea of the *right-hand rule*, a planar graph traversal algorithm can be used to find a path from the source to the destination on an *embedding* of the planar graph. The rule states that it is likely to traverse every wall in a maze by keeping one's right hand against the wall while moving forward. In a wireless ad hoc network, two nodes are neighbors if the distance between them is at most $r$, where $r$ is the transmission range that is equal for all nodes in the network. Using this model, the network becomes a *unit disk graph*. However, the underlying graph of a wireless ad hoc network is usually not planar. An additional algorithm is required to extract a planar subgraph from the original graph, called *planarization*. Otherwise, routing paths may contain loops. Thus, a complete planar-graph-based void handling solution should include a planarization algorithm and a planar graph traversal algorithm. In addition, these algorithms should be distributed due to the distributed nature of wireless ad hoc networks. The performance of a planar-graph-based void

handling solution depends on both the performance of the planarization algorithm and the performance of the traversal algorithm. For the former, we need to consider whether planarization is performed in an efficient and effective manner as well as the quality of topologically optimal paths in the planar subgraph, when compared to those in the original network graph. For the latter, the main performance concern is the quality of paths discovered, when compared to optimal paths in the subgraph.

Existing solutions in this category include some distributed planarization algorithms such as Relative Neighborhood Graph (RNG) [40] and Gabriel Graph (GG) [40]; some distributed planar graph traversal algorithms such as convex face routing [41], original face routing [29], the face-2 algorithm [42], Other Face Routing (OFR), and Other Adaptive Face Routing (OAFR) in GOAFR [43], and GOAFR+ [44]; and some complete planar-graph-based void handling solutions such as perimeter routing in Greedy Perimeter State- less Routing (GPSR) [40], bypass in Priority-based Stateless Geo-Routing (PSGR) [26], and Request Response (RR) in Beacon-Less Routing (BLR) [14].

For example, Fig. 7.5 illustrates the RNG distributed planarization algo- rithm, where an edge remains in the planar subgraph if the intersection of the circles centered at two nodes x and y is free of other nodes such as node w. After that, a network graph becomes a planar graph. Figure 7.6 illustrates how the face2 traversal algorithm is carried out [42]. Starting in the face of the graph just beyond the source S along the line SD and the algorithm moves around the face counterclockwise. When the line SD is about to be crossed, the algorithm crosses over into the next face along the line SD and continues to move. The algorithm proceeds over multiple faces along the line SD until D is eventually reached. Sometimes the line SD may intersect a face more than twice, the algorithm needs to determine all face boundary cross- ings with SD and selects the one farthest from S. As shown in Fig. 7.6, when the algorithm traverses to node X, it encounters two face boundary crossings (i.e., the edge XZ and the edge XY), the algorithm selects XY to cross because it is farther from S than XZ.

Perimeter routing is a complete void handling solution proposed in GPSR [40], consisting of a distributed planarization algorithm that uses either

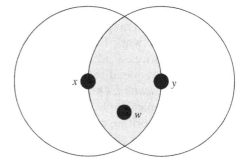

**Fig. 7.5** The RNG planarization algorithm. For edge (x,y) to be included in the planar subgraph, no witness node w is located within the shaded lung area

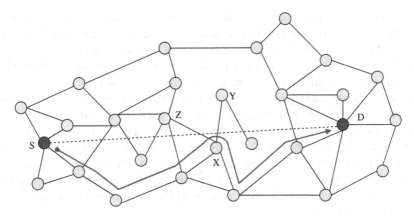

**Fig. 7.6** The face2 planar traversal algorithm

RNG or GG and a planar traversal algorithm similar to the face2 algorithm. All nodes execute the distributed planarization algorithm periodically so that a planar subgraph of the original network graph is always maintained in the network. When a packet encounters a void node in greedy forwarding, the planar traversal algorithm is used for GPSR to handle the void. The header of the stuck packet usually carries information regarding the position of the void node, the first edge traversed on the current face, and the position of the last intersection that a face change occurs. Such information helps each node that receives the packet make routing decisions locally. For example, the information regarding the first edge traversed is used to judge if the stuck packet traverses the first edge on the current face for the second time. When the stuck packet encounters a node that is closer to the destination than the void node, greedy forwarding is reactivated to forward the stuck packet.

2. Topology-based void handling[5]

Topology-based solutions to the void problem rely on topological connectivity information to overcome voids. So in a strict sense, a geographic forwarding protocol that uses a topology-based void handling solution is a hybrid routing protocol, instead of a pure geographic routing protocol.

When a packet encounters a void node, the simplest way to use topological information is the flooding of the stuck packet at the void node to the whole network in order to exploit all possible connectivity to the destination, called *network-wide flooding* or *full flooding*. As we know, full flooding is inefficient in terms of network resource utilization. Thus, topology-based void handling solutions attempt to control the network range of flooding as well as the frequency of occurrence of flooding at void nodes, called

---

[5] We presented this category as flooding-based void handling in [11] and here we add some latest work to this category.

*restricted flooding* or *partial flooding*, so that the flooding cost is minimized while handling voids effectively. Flooding can be used to forward the stuck packet directly if this is a one-time delivery of the stuck packet of small size. If there is a long-lived session of packet delivery from the void node to the destination, flooding can first be used to discover a path to another node, which is closer than the void node, where greedy forwarding can be reactivated, or a path to the destination directly. Then the actual data packet is routed on the discovered path, similar to topology-based unicast routing. Note that the discovered path is only a part of the end-to-end path between the source and the destination. In addition, some routing discovery techniques other than flooding which is a breadth-first search technique, for example, the depth-first search technique [45, 46], can also be used to find a desired partial path.

Restricted flooding or partial path discovery is a reactive topology-based approach to handle voids, which is initiated on demand only when a packet encounters a void node. If some efficient topology/routing structures can proactively be built, the pre-computed structures can be exploited to route stuck packets in the presence of voids. This approach is a proactive topology-based approach and the efficiency (e.g., low overhead) and scalability (e.g. localized operation) of structures is the key for such a solution to be viable. Currently, two efficient structures, *grids* [47] and *hull trees* [48], have been designed for geographic routing to handle voids.

Topology-based void handling solutions include some algorithms that directly flood the data packet at the void node such as limited flooding in Cartesian Routing [32] and one-hop flooding [31]; some algorithms that first discover a partial path for the delivery of the stuck data packet such as Partial Source Routing (PSR) in On-demand Geographic Forwarding (OGF) [49] and Partial Hop-by-hop Routing (PHR)[6] in Geographic Routing Algorithm (GRA) [46]; and some algorithms that proactively build some efficient topology structures for void handling such as Grid-based Recovery in Robotic Routing Protocol (RRP) [47] and Greedy Distributed Spanning Tree Routing (GDSTR) [48]. We only describe the PSR solution in detail in the following.

PSR is comprised of two phases: partial path discovery and source packet forwarding. In the first phase, a void node uses an algorithm similar to expanded ring search [20] to discover a partial path to get around the void. For example, in Fig. 7.7, the void node T tries to discover a partial path by flooding a search packet to its two-hop neighbors and this discovery fails. T then initiates another discovery to its three-hop neighbors and a partial path is found from T to node C, which is closer to the destination D. In practice, the search range is increased until a targeted node is found or the maximum number of runs (defined as a protocol parameter) is reached. In the latter case, the struck packet is discarded. In the second phase, the void node

---

[6] This name is introduced for our convenience.

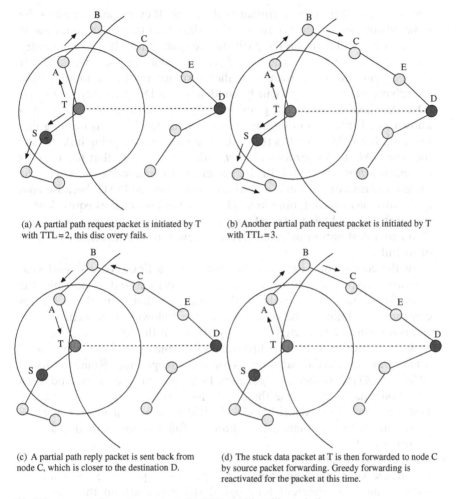

(a) A partial path request packet is initiated by T with TTL = 2, this disc overy fails.

(b) Another partial path request packet is initiated by T with TTL = 3.

(c) A partial path reply packet is sent back from node C, which is closer to the destination D.

(d) The stuck data packet at T is then forwarded to node C by source packet forwarding. Greedy forwarding is reactivated for the packet at this time.

**Fig. 7.7** An example to demonstrate the PSR void handling solution

includes the discovered partial path in the data packet's header and forwards the packet to the given next-hop node. All immediate nodes will follow the given partial source path and move the packet further in-between hops. When the packet arrives at the targeted node, for example, node C in Fig. 7.7, greedy forwarding is then reactivated at the targeted node.

3. Link-reversal-based void handling[7]

In some types of wireless ad hoc networks such as wireless sensor networks, often all nodes are required to deliver their data packets to a fixed

---

[7] We presented this category as cost-based void handling in [11]. After an in-depth study, we believe link reversal is the fundamental idea behind all solutions in this category.

destination, called the *base station* or *data sink*. If every node employs the same advance-based greedy forwarding algorithm to forward data packets, no loops will form on the paths to the destination. If we assign every link between any two neighbors a direction from the node with a larger distance and the node with a smaller distance to the destination, the underlying network graph can be modeled as a Directed Acyclic Graph (DAG), which is a directed graph without loops. In graph theory, it is known as a finite DAG that has at least one node which has no outgoing link [50]. If a DAG only has the destination with no outgoing link, we say that the DAG is *destination-oriented*. Otherwise, we say that the DAG is *destination-disoriented* [51]. In the presence of communication voids, the underlying network graph is a destination-disoriented DAG, because void nodes also have no outgoing link. Thus, the void problem is equivalent to the problem of how to transform a connected destination-disoriented DAG to a destination-oriented DAG by reversing the directions of some of its links.

In the context of wireless ad hoc networks, a link-reversal-based void handling solution should be a distributed algorithm and, if possible, the algorithm should exploit only local information that is available. In this category, a solution is based on one of the following two methods: full reversal method and partial reversal method. In the full reversal method, each void node reverses the directions of all incoming links. Cost-based forwarding in Partial-partition Avoiding Geographic Routing-Mobile (PAGER-M) [52] is such an algorithm. In the partial reversal method, each void node may only reverse the directions of some of its incoming links. Distance Upgrading Algorithm (DUA) [53] is such an algorithm. Fig. 7.8 shows an example to demonstrate how the full reversal method is able to resolve a void.

4. Geometric void handling

Geometric void handling exploits the geometric properties of voids by considering some inherent topological structures behind the seemingly unorganized nodes in the network. For example, *holes*, which are defined as the regions of the network with boundaries consisting of all the nodes that can possibly become void nodes [13], are such useful topological structures. Thus, when a packet encounters a void node, it must be at the boundary of at least one of the holes. Distributed algorithms can be designed to identify holes as well as to discover hole-surrounding paths. These paths then can be exploited by geographic forwarding to handle voids. We describe BOUNDHOLE [13], the seminal work in this category, in the following.

In BOUNDHOLE, each node first uses the TENT rule [13], which exploits the geometric properties of neighbors, to detect if it can possibly be a void node. After being identified as a void node, the void node can initiate the BOUNDHOLE algorithm that uses greedy sweeping to set up a path containing a hole inside. The basic idea is shown in Fig. 7.9(a). The

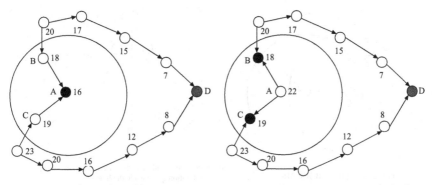

(a) Due to the void node A, the graph becomes a destination disoriented DAG.

(b) Node A reverses all its incoming links by increasing its distance value from 16 to 22, node B and node C become void nodes.

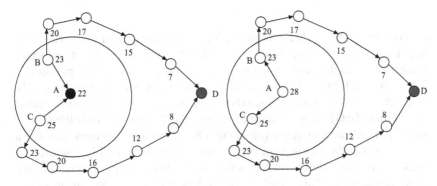

(c) Node B and Node C reverses all their incoming links by increasing their distance values from 18 to 23 and from 19 to 25, respectively. Node A becomes a void node again.

(d) Node A again reverses all its incoming links by increasing its distance value from 22 to 28 and the void is resolved. The graph becomes a destination oriented graph.

**Fig. 7.8** An example to show how the full reversal method resolves a void

algorithm starts from node p and sweeps over the stuck direction by sending a packet to node t1 in a counterclockwise direction. Node t1 then passes the packet to node t2. The above process is repeated until the packet moves around the boundary of the hole and returns to p. Now that the boundary of a hole has been identified and cached locally, it provides a conduit for geographic forwarding to handle voids. For example, as shown in Fig. 7.9(b), when a packet encounters the void node p, the stuck packet is forwarded along the boundary of the hole. When the packet reaches a node closer to the destination q than p, greedy forwarding is reactivated to forward the packet. If the destination is outside the hole, the stuck packet can always reach it.

5. Heuristic void handling
   Heuristic void handling solutions are based on heuristics, i.e., intuitive ideas that are not subject to a strict theoretical analysis on their efficiency and

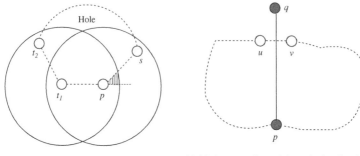

(a) Greedy sweeping in BOUNDHOLE.

(b) A hole-surrounding path is used to handle voids.
The destination q is outside the hole.

**Fig. 7.9** The BOUNDHOLE algorithm to handle voids

effectiveness. They either exploit the inherent properties of network topology itself and the geographic properties of void areas or utilize extra resources to handle voids. Such solutions include passive participation [22, 25], active exploration [27, 54], void avoidance [55], alternate network [31], Intermediate Node Forwarding (INF) [56], and Anchored Geodesic Packet Forwarding (AGPF) in terminode routing [57]. We describe only the passive participation solution in the following. Once a node determines that it is a void node for the destination of a received packet, it simply discards the packet and keeps itself from forwarding any subsequent packets toward the same destination. This simple strategy has a reverse-propagation effect, which greedily informs other upstream nodes at previous hops to utilize other potential paths in the network. Figure 7.10 shows an example. Node

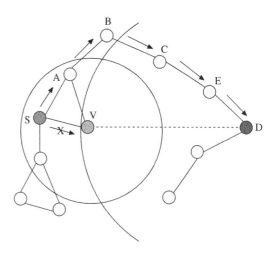

**Fig. 7.10** An example to
demonstrate how passive
participation handles a void

V finds itself to be a void node with respect to the destination D, after receiving a packet from node S. V discards the packet and passively participates in the forwarding of the subsequent packets destined for the same destination. The packet will automatically follow another path from S to D: S-A-B-C-E-D.

6. Hybrid void handling

A hybrid void handling solution combines at least two individual void handling solutions together to handle voids, in order to achieve better performance. These solutions may belong to the same category or different categories. There are two situations where a hybrid void handling solution is desired. The first is when one void handling solution, combined with another solution, can enhance the resource efficiency of handling voids. The other is when a single void handling solution cannot handle voids effectively for all possible network topologies. The hybrid void handling solutions include BOUNDHOLE plus restricted flooding [13], PSR plus passive participation [49], active exploration plus passive participation [54], and AGPF plus perimeter routing [57]. For example, a geometric void handling solution, BOUNDHOLE [13], cannot guarantee packet delivery if the destination is inside the hole, because it is likely that all nodes in the hole are not closer to the destination than the void node. As shown in Fig. 7.11, the void node p is closer to the destination q than all other nodes on the boundary of the hole. In this situation, the hole-surrounding path cannot be used to handle the void, because the stuck packet will return to node p without being able to locate a node closer to the destination in the boundary of the hole. Restricted flooding, a topology-based void handling solution, can be initiated to allow each node on the boundary of the hole to broadcast the struck packet to all its one-hop neighbors. In Fig. 7.11, node $v_1$ will receive the stuck packet and then use greedy forwarding to deliver the packet to the destination q inside the hole.

Table 7.2 provides a summary of the main characteristics of six representative void handling solutions we introduced in this section. A detailed comparison of the existing void handling solutions from six categories is provided in [11].

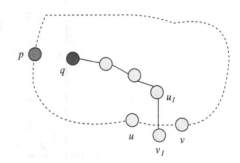

**Fig. 7.11** If the destination q lies within the hole, restricted flooding is used to handle the void

**Table 7.2** Characteristics of six representative void handling solutions

| Solution | Category | Guaranteed delivery | Optimal path | Reactive | Distributed | States | Overhead |
|---|---|---|---|---|---|---|---|
| Face2 | Planar-graph-based | Yes | No | Proactive | Localized | State-less | Medium |
| PSR | Topology-based | No | Yes | Reactive | Global | Partial-path-state | High |
| Full reversal | Link-reversal-based | Yes | No | Proactive | Localized | State-less | Medium |
| B.H. | Geometric | No | No | Either | Localized | State-less | Medium |
| Passive | Heuristic | No | No | Reactive | Localized | State-free | Low |
| B.H + Flooding | Hybrid | Yes | No | Either | Localized | State-less | High |

## 7.4 Advanced Topics in Geographic Routing

In this section, we introduce three advanced topics in geographic routing, i.e., geographic multicast, geocast, and trajectory-based forwarding. Since these topics are relatively new, we only present a general description of the main research problems associated with these topics and some relevant techniques. This will provide readers with a more complete view of advances in the area of geographic routing. If readers are interested in a specific research problem and would like to further study it, they can refer to the related references for more details.

### *7.4.1 Geographic Multicast*

Multicasting is used to deliver packets from a single source to a set of destinations. For this purpose, two trivial solutions can be used readily: one is to employ network-range flooding and the other is to use unicast routing to each destination. Both solutions are effective but inefficient in terms of network resources, because paths from the source to each different destination may share many links. Thus, multicast routing tries to minimize the consumption of network resources by reducing redundant links, based on a specific metric such as number of hops or cost. Usually a multicast tree is built. A multicast routing protocol is used to find an optimum multicast tree, which is rooted at the source and spans all destinations.

There have been a lot of topology-based multicast routing protocols proposed for wireless ad hoc networks in the literature. Most of them are designed for MANETs to handle node mobility. They can be further divided into tree-based and mesh-based protocols. In tree-based protocols such as Multicast Ad hoc On Demand Distance Vector Routing (MAODV) [58], data packets are forwarded on a single-path to a specific destination and the union of the paths to all destinations forms a multicast tree. However, due to frequent broken links, maintaining multicast trees is very difficult and tree-based multicasting may not be reliable. Mesh-based protocols such as On-Demand Multicast Routing Protocol (ODMRP) [59] expand a multicast tree with some redundant paths, which can be used to forward multicast data packets when some of the links break, thus enhancing the robustness at the cost of higher forwarding overhead. Both tree-based and mesh-based multicast protocols need to maintain state information about multicast topology structures and rely on periodic flooding to update them. Thus, these protocols generally do not have good scalability.

There have also been some protocols taking advantage of geographic information of nodes to perform more efficient topology-based multicast routing for wireless ad hoc networks. Geographic information can be utilized to efficiently construct the multicast topology structure, to reduce the topology maintenance overhead, and to improve mobility prediction [60, 61].

Compared to topology-based multicast routing, geographic multicast routing is still in an early stage. Since a geographic multicast protocol does not require the maintenance of states about multicast topology structures such as tree structures, it does not use flooding to update them. In order to use the idea of geographic unicast routing in geographic multicast routing, two key problems have to be solved [62]. The first problem is that multiple copies of a multicast packet should be created at some intermediate nodes, in order to reach all destinations. The technical challenge here is to decide when such a copy should be created, using only the local information available. For example, in Fig. 7.12, two copies of a packet are created at the source S and one copy is further copied into three more copies at node C. The second problem is that a void handling solution needs to be reconsidered for multiple destinations. For instance, as illustrated in Fig. 7.12, S is a void node with respect to the destination D2, but not with respect to D1, D3, and D4. How the void handling mode interacts with the normal greedy forwarding mode for multiple destinations is an interesting and challenging problem.

Position-Based Multicast Routing (PBM), proposed in [62], is a representative geographic multicast protocol for wireless ad hoc networks. PBM is a generalization of geographic unicast routing (e.g., Greedy-Face-Greedy (GFG) [42] or GPSR [40]) for multicast. Similar to GPSR, PBM consists of a greedy multicast forwarding mode and a perimeter multicast forwarding mode. In the greedy multicast forwarding mode, next-hop nodes are selected based on local geographic information, i.e., on the positions of the sender, its neighbors, and the destinations. Perimeter multicast forwarding mode is invoked when greedy multicast forwarding fails. The rule for splitting a multicast packet is determined by a heuristic function, parameterized by $\lambda$, which provides a trade-off between the total number of nodes forwarding the multicast packet and the optimality of individual paths toward the destinations, so that the latency and bandwidth performance can be controlled. Note that a multicast tree in geographic multicast does not exist proactively before a multicast packet is forwarded. Instead, the tree is formed gradually as the packet propagates toward

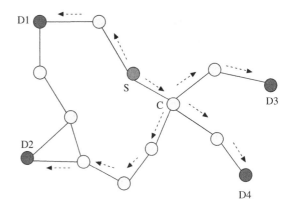

**Fig. 7.12** An example of a multicast group with four destinations

its destinations. However, determining an optimal value of $\lambda$ for a good performance trade-off is not trivial and the algorithm used in PBM to select next-hop nodes at every hop is computationally complex. Geographic Multicast Routing (GMR) [16] tries to solve the shortcomings of PBM. In GMR, a heuristic greedy neighbor selection scheme, which requires a low computational cost and is able to generate very efficient multicast paths in terms of a cost over progress metric, is proposed. GMR is able to provide a good trade-off between the efficiency of multicast packet delivery and the optimality of the multicast tree.

In both PBM and GMR, a multicast packet has to include all destinations' positions in the packet header for intermediate nodes to learn geographic information of destinations. Thus, for a multicast group with a very large number of destinations, these protocols may not scale well and thus they are applicable for a small multicast group only. Scalable Position-Based Multicast (SPBM) [63] was designed to improve the scalability with group size. It uses the positions of nodes to provide a scalable multicast group membership scheme with a hierarchical organization of nodes for membership. The multicast forwarding decision in SPBM is based on the information from the group management scheme. However, since SPBM uses one separate unicast routing for each destination, it fails to provide efficient multicast routing. In addition, SPBM requires periodic local or network-wide flooding to update membership, which makes the membership management inefficient due to significant control overhead. In Efficient Geographic Multicast Protocol (EGMP) [64], geographic information of nodes is used in all aspects of multicasting, including the hierarchical structure of group membership management, zone-based multicast tree construction, and multicast packet forwarding. EGMP, similar to SPBM, uses geographic unicast routing to forward multicast packets and is thus not efficient. In addition, EGMP, as a zone-based protocol, has to solve the challenging empty zone problem, which may incur additional cost.

## 7.4.2 Geocast

Geocast, which enables the delivery of data packets to a set of nodes that are within a specified geographic region, was originally proposed for the Internet [17]. Geocast is regarded as a more natural and economic routing service for location-based applications than traditional IP address-based multicast [17]. Actually, geocast can be regarded as a special type of multicast with its own unique characteristics. In a geocasting protocol, the members of a multicast group are determined by their physical positions. The source specifies a region, called *the geocast region*, for a geocast packet to be delivered and a geocasting protocol tries to deliver data packets only to the nodes in that region. Since the source does not need to include all destinations' positions and only needs to encode the geographic coordinates of the region in a geocast packet header, geocast is scalable to the destination group size. Obviously, network-wide flooding, topology-based multicast, or geographic multicast can be used to implement a geocast service. However, they are effective but may not be efficient in terms of resource utilization, especially in the context of wireless ad hoc networks, because they do not exploit the unique characteristics of geocast such as the geographic relation of destinations.

A geocasting protocol usually consists of two main components, as illustrated in Fig. 7.13. The first component is responsible for forwarding data packets from a source to one or more nodes in the geocast region, if the source is not within the region. The second one is responsible for forwarding data packets from one or more nodes in the region to all nodes in the region. Based on different approaches in the first component, most of the existing geocasting protocols for mobile ad hoc networks can be classified into two categories: flooding-based and unicast-routing-based [65]. Flooding-based protocols, such as Location-Based Multicast algorithm (LBM) [66], Voronoi diagram based geocasting [67], and GeoGRID [68], use restricted flooding to forward geocast packets from the source to the geocast region. For example, LBM restricts the flooding range within a forwarding zone with respect to the geocast region [66]. However, restricted flooding may still incur serious redundancy, contention, and collision problems. Unicast-routing-based protocols, including GeoTORA [69], Mesh-based Geocast Routing Protocol (MGRP) [70], and Geocast Adaptive Mesh Environment for Routing (GAMER) [71], first use control packets to create one or multiple paths from the source to the geocast region, and then deliver data packets along a path to the geocast region. For example, GeoTORA uses the unicast routing protocol TORA (i.e., Temporally Ordered Routing Algorithm) to forward geocast packets to the geocast region. The advantage of unicast-routing-based protocols is the reduced overhead for transmitting data packets while the disadvantage is that the control overhead is required to create and maintain paths, which may also incur some packet delivery delay [65].

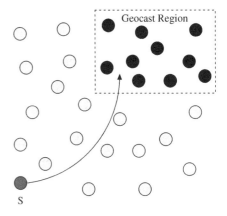

**Fig. 7.13** A geocasting example

Geographic unicast routing, which does not need to create and maintain paths, can be used for delivering data packets from the source to the geocast region. For example, Geographic-Forwarding-Geocast (GFG) and Geographic-Forwarding-Perimeter-Geocast (GFPG) were proposed in [72] to trade off the overhead incurred by the geocast packet and the proportion of nodes in the geocast region that are able to receive the geocast packet. Interestingly, the face routing, which we introduced previously as a planar-graph-based void handling solution, can also be used to guarantee packet delivery for geocast. Such a geocasting algorithm was proposed in [42]. The algorithm enumerates all faces, vertices, edges for a planar graph intersecting a geocast region, and then defines a total order on the edges of the planar graph and traverses these edges in a depth-first manner.

The aforementioned geocasting protocols (except the planar-graph-based geocast) have a similar second component. That is, whenever a node in the geocast region receives a geocast packet, it will flood the packet to its neighbors so that all nodes in the region are able to receive it, called *region flooding*. If nodes within the region are mobile, as common in MANETs, flooding is a good choice to deliver data packets to all current nodes in the region. Further, some intelligent flooding approaches [73] could be used to reduce the flooding overhead.

Geocast has a very important application in sensor networks: query dissemination. Geocast can route a query to sensors in a region of interest to collect the desired information from the physical space. Since in most of the sensor networks nodes are stationary and energy consumption is of main concern, geocast in the context of sensor networks has special considerations in its two components: one is to efficiently deliver the query to the sensor closest to a given position within the queried region while the other is to efficiently deliver the query from the sensor to all other sensors within the region. A representative protocol on this subject is Geographic and Energy-Aware Routing (GEAR) [74], where it uses geographically informed and energy-aware neighbor selection heuristics to route a packet to the sensor closest to the centroid of the region while it uses a recursive geographic forwarding technique to deliver the packet to every sensor within the region. Some other protocols proposed in [75–78] also try to solve similar problems for sensor networks. These protocols differ in energy efficiency metrics, the position in the queried region used for the sink to unicast the query, as well as the topology structures built in two components.

### 7.4.3  Trajectory-Based Forwarding

Trajectory-Based Forwarding (TBF) [79] is a generalization of traditional geographic routing we introduced throughout this chapter as well as traditional source routing. Similar to traditional source routing, a packet in TBF carries its path information so that intermediate nodes can follow it to forward the packet.

However, traditional source routing specifies the information in terms of a discrete number of intermediate nodes in the packet header so that the packet can be routed along a given path on a hop-by-hop basis to its destination. In TBF, an ideal curve is specified analytically and its description is encoded in the packet header, so that intermediate nodes can forward packets to those nodes that lie more or less on the curve. For example, a curve representing a sine wave can be encoded with a sine function ($x = t, y = \sin t$) in the packet header at the source. Representing paths as curves is an efficient and scalable encoding technique in large-scale and densely deployed wireless ad hoc networks such as sensor networks. As a result, instead of routing to a given destination, TBF is *routing on a specified curve* [18]. As illustrated in Fig. 7.14, when a curve is given in the form ($x(t), y(t)$) in a packet header, nodes A-H are selected in turn to forward the packet along the trajectory.

Similar to traditional geographic routing, TBF makes a greedy decision at every hop to select the next-hop node, based on locally available geographic information such as the positions of neighbors. Along the same line, different criteria can be used in TBF to select the next-hop node [79]. For example, a sender may select the neighbor that is geographically closest to the given curve as its next-hop node. Different from traditional geographic routing that makes forwarding decisions with respect to a destination, a sender in TBF bases its forwarding decisions on a curve. Similar to traditional geographic routing, TBF needs to handle the void problem. This occurs when obstacles in the graph are non-monotonous along the direction of a given curve. Note that if the position of the destination is known to the source and the curve specified is a line between the source and the destination, TBF degenerates to traditional geographic routing.

The idea of TBF can be used to implement all routing functions including unicast, multicast, and broadcast, where the main problems are trajectory/ curve specification and encoding, as well as the next-hop node selection criteria. Since TBF completely decouples the path description from the actual forwarding nodes in the path, it is very robust to topology changes caused by network

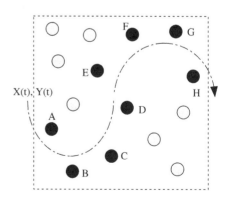

**Fig. 7.14** An example for Trajectory-Based Forwarding

dynamics. Note that TBF is quite suitable for situations where the topology of the network matches the topography of the physically surroundings in which is it deployed, e.g., large-scale and densely deployed wireless ad hoc networks of sensors in the physical space, as an appropriate level of node density can be maintained in the network along the specified packet trajectory.

## 7.5  Thoughts for Practitioners

It is important for a practitioner to know that geographic routing is yet to have a successful deployment for real-world wireless ad hoc networks, although it has been extensively studied in theory and via simulation. One main reason is that most of the existing geographic routing protocols made unrealistic assumptions about underlying wireless ad hoc networks. For instance, most of the proposed geographic routing protocols assume the communication graph of an underlying wireless ad hoc network as a unit-disk graph, where every node has an equal radio transmission/reception range in a perfect circle around the node and all links are thus bidirectional. Most protocols also assume perfect link reliability, independent of distance within the radio range, as well as assume accurate localization without errors. Recent experimental studies on wireless ad hoc networks have shown that these assumptions are not valid in real-world scenarios [34–35, 81]. Radio irregularities due to obstacles and time-varying wireless channels pose a situation that bears little resemblance to ideal wireless ad hoc networks, and the situation also varies with time, space, and different types of wireless ad hoc networks.

Due to the fact that idealistic assumptions do not hold true in practical situations, many of the proposed protocols suffer from severe performance degradation or they may even completely fail. For example, those geographic routing protocols that rely on planar-graph-based solutions to handle voids, e.g., GPSR, need to use a distributed planarization algorithm such as GG to make the underlying graph planar. It has been experimentally shown [81, 82] that GG does not work well in real-world wireless ad hoc networks. Radio irregularities can lead to the existence of crossed links despite the use of the GG technique. Radio irregularities also make the void problem very common in the network. When a geographic routing uses face routing in the presence of crossed links to overcome voids, data packets may enter loops and never be delivered to their destinations. Therefore, more studies under realistic assumptions are required to revisit existing geographic routing protocols, algorithms, and techniques, before they can successfully be deployed for real-world wireless ad hoc networks. Some preliminary work for more practical geographic routing techniques has been conducted in [37, 39, 80–85].

It is also important for a practitioner to know that for a given performance objective, a geographic forwarding algorithm never achieves better performance than a centralized routing algorithm with the availability of global information, because it only exploits localized information to make greedy forwarding decisions at each hop. But it is possible for a geographic forwarding algorithm to achieve performance that is close to that of the relevant centralized algorithm. For example, using the number of hops (i.e., hop count) as a performance metric, a greedy algorithm with the MFR criterion or the MAR criterion, if successful, nearly matches the performance of a shortest path algorithm such as Dijkstra's algorithm. In the presence of communication voids, the performance of a geographic forwarding algorithm also depends on the performance of a void handling solution. Usually, the combination of greedy forwarding and void handling achieves worse performance than greedy forwarding alone. This is why the greedy forwarding mode is required to be used wherever possible in a geographic forwarding protocol. In real-world wireless ad hoc networks, all algorithms need to be implemented as distributed algorithms executed in each node without global information. If a node wants to use any information in the network to make its routing decisions, it has to obtain such information first, which consumes network resources. In addition, information received by the node may already be stale due to network dynamics. Clearly, localized algorithms such as greedy forwarding algorithms are more scalable to network size and dynamics than those algorithms requiring more than local information. Note that the scalability of a geographic forwarding protocol also depends on the scalability of its void handling solution.

Finally, it is important for a practitioner to know that each geographic routing technique introduced in this chapter has its advantages and its disadvantages. This chapter attempts to help practitioners understand and compare the existing geographic routing techniques in a systematic manner. In a real-world deployment, selecting an existing technique or designing a new technique depends on the desirable features of a geographic routing protocol, as well as the unique characteristics of a targeted wireless ad hoc network and its applications. Practitioners have to decide on engineering trade-offs themselves when designing and implementing their own geographic routing protocols, so that the desired network performance can be achieved in their targeted wireless ad hoc networks.

## 7.6 Directions for Future Research

In this chapter it has been shown that there are quite a number of geographic routing techniques proposed for wireless ad hoc networks. However, research in this area is far from over and there still exist many interesting problems that need to be addressed in the future research, as elaborated in the following.

Assessing the robustness of existing geographic routing techniques to non-ideal conditions corresponding to real-world environments and designing new techniques that take these conditions into account are vital for the successful deployment of geographic routing for wireless ad hoc networks in the near future. As a result, a lot of experimental studies are needed to investigate if and how existing geographic routing techniques work in real-world wireless ad hoc networks. Based on the results of real-world experiments, researchers should make realistic assumptions and develop new models to revisit existing geographic routing techniques and accordingly design practical geographic routing techniques.

In greedy forwarding, new next-hop node selection criteria are desired. For instance, application requirements, network traffic conditions, and wireless channel status can also be taken into account in next-hop node selection to make adaptive packet forwarding decisions. Such cross-layer information may enable novel QoS-aware geographic routing techniques. Other issues such as the way in which a next-hop node is established and how to obtain information of the neighbors greatly affect the properties of greedy forwarding and need more detailed studies.

Most of the research effort in void handling is focused on the planar-graph-based approach. This approach assumes that routing takes place at a much faster rate than topology changes due to network dynamics such as node movement, thus a network can be viewed as a static wireless ad hoc network at the routing layer. While this approach is very mature for static wireless ad hoc networks, it would be quite interesting to understand how this approach behaves in mobile wireless ad hoc networks. Further, the effectiveness of this approach greatly relies on whether or not a perfect planar graph can be extracted from the underlying wireless network graph. As pointed out in [81], current distributed planarization algorithms such as GG cannot function well in real-world scenarios; new practical planarization algorithms are needed to complete this task in an effective and efficient manner. Another important issue is how to guarantee the quality of the discovered paths, when compared to topologically optimal paths in the original graph. Planar-graph-based void handling is not the only approach to handle voids, as we have reviewed six categories of void handling, each with a different approach. A quantitative comparison of existing void handling techniques from different categories and the discovery of more novel void handling approaches are highly desired. A summary of open issues in each category of void handling is available in [11].

Whether geographic multicast routing can compete with topology-based multicast routing as well as face multicast routing is still an open issue. In addition, current geographic multicast routing protocols are not scalable to the multicast group size. New next-hop node selection algorithms with low computation complexity are desired for targeted performance objectives. How the greedy forwarding mode interacts with the void handling mode (when using different void handling techniques) for multiple destinations in an effective and efficient manner is an interesting problem. Geocast in the context of wireless sensor network not only needs to enable the efficient delivery of information packets (e.g., queries or commands) to sensors in a given geocast region but also needs to build an efficient aggregate routing structure for sensors in the region to send back the desired information. This still needs further study. Trajectory-based routing is still at an early stage and many open issues exist. For example, the design of efficient representation methods for complex trajectories in routing functions such as multicast is still under consideration. More research efforts on greedy next-hop node selection criteria, the void problem, and new applications are needed to advance this topic.

Security issues in geographic routing are attracting more attention. In addition to similar security issues in topology-based routing such as dropping of packets at a misbehaving node, geographic routing has its unique security issues. For example, the sender trusts nodes to provide their position information and uses it to make forwarding decisions. What if some nodes falsify this information [86]? On the other hand, location privacy of nodes in geographic routing needs protection, which is another security issue that needs careful consideration.

Other interesting topics include cross-layer design of geographic routing, transport layer protocols such as TCP over geographic routing, geographic routing over physical layer with new capabilities such as multiple channels, multiple radios, or cognitive radios, geographic routing in three-dimensional space (i.e., 3D geographic routing [87]), and geographic QoS routing. These topics are largely unexplored and they need a lot of research effort.

## 7.7 Conclusion

In this chapter we have provided an extensive overview of geographic routing in wireless ad hoc networks. Given a significant amount of work done in this area, we have attempted to organize them in a logical and systematic manner. Our effort was mostly focused on two operation modes of geographic forwarding in geographic unicast routing: greedy forwarding and void handling. In greedy forwarding, we have concentrated on the introduction of various next-hop node selection criteria. In void handling, we have presented six different categories of approaches, each with a representative void handling solution described. In addition to geographic unicast routing, we have also briefly introduced three advanced topics in geographic routing: geographic multicast, geocast, and trajectory-based forwarding. In order for practitioners to better understand geographic routing in practice, we have provided some comments on the practical implementation of geographic routing protocols. Finally, we have discussed open research issues in the area of geographic routing in order to stimulate more creative research in the future.

## Terminologies

*Geographic routing*: A routing approach that exploits only the geographic information of nodes in the network to enable communications among nodes.

*Geographic unicast routing*: A geographic routing technique that enables the routing of data packets from a single source to a single destination.

*Geographic forwarding*: The forwarding function of geographic unicast routing that forwards a packet from the sender to its next-hop node.

*Greedy forwarding*: A forwarding strategy that uses locally available information, usually the positions of the sender, its neighbors, and the destination, to make packet-forwarding decisions in a greedy manner. Greedy forwarding is used wherever possible in geographic forwarding.

*Void handling*: A communication void occurs when a sender cannot locate a next-hop node among its neighbors, which is closer to the destination than itself. Void handling is a forwarding mode used to handle voids in geographic forwarding.

*Ad hoc localization*: A technique for determining the location of a node in a wireless ad hoc network.

*Planarization*: A planar graph is a graph that can be embedded in the plane with no intersecting edges. Planarization is a technique used to extract a planar subgraph from the original graph. Distributed planarization algorithms are required in wireless ad hoc networks.

*Geographic multicast*: A geographic routing technique that enables the routing of data packets from a single source to multiple destinations.

*Geocast*: A routing service that enables the routing of data packets from a single source to multiple destinations within a given geographic region. Geocast can be regarded as a special type of geographic multicast.

*Trajectory-based forwarding*: A geographic routing technique that enables routing on a given curve/trajectory, instead of routing to a given destination.

## Questions

1. Describe the advantages and disadvantages of geographic routing, when compared to topology-based routing.
2. What are two modes of geographic forwarding? How are these two modes used in geographic forwarding?
3. Why is greedy forwarding required to be used wherever possible in geographic forwarding?
4. In the following six next-hop node selection criteria in different greedy forwarding algorithms: MFR, NFP, random selection, closest direction, MAR, and NC, which of them are guaranteed to be loop-free? Why?
5. What is the communication void problem? Why does the network need the ability to handle voids in geographic routing?
6. How many categories can we classify the existing void handling solutions into? Describe their respective approaches in brief.
7. What determines the performance of planar-graph-based void handling?

8. What are the two general methods in link-reversal-based void handling? Describe them in brief.
9. Can BOUNDHOLE in geometric void handling guarantee packet delivery? If not, how can it guarantee packet delivery?
10. What are the unrealistic assumptions made in most of the existing geographic routing techniques? How do these techniques work in the real world?

## References

1. H. Takagi and L. Kleinrock, "Optimal Transmission Ranges for Randomly Distributed Packet Radio Terminals," *IEEE Transactions on Communications*, Vol. 32, No. 3, pp. 246–257, 1984.
2. T. Hou and V. Li, "Transmission Range Control in Multihop Packet Radio Networks," *IEEE Transactions on Communications*, Vol. 34, No. 1, pp. 38–44, January 1986.
3. A. Savvides and M. B. Srivastava, "Location Discovery," Book Chapter in *Mobile Ad Hoc Networking*, John Wiley and Son Inc., August 2004.
4. N. Bulusu, J. Heidemann, and D. Estrin, "GPS-less Low Cost Outdoor Localization for Very Small Devices," *IEEE Personal Communications Magazine*, Vol. 7, No. 5, pp. 28–34, October 2000.
5. J. Hightower and G. Borriello, "Location Systems for Ubiquitous Computing," *IEEE Computer*, Vol. 34, No. 8, pp. 57–66, August 2001.
6. T. He, C. Huang, B. M. Blum, J. A. Stankovic, and T. Abdelzaher, "Range-free Localization Schemes for Large Scale Sensor Networks," *Proc. of ACM MobiCom*, San Diego, CA, September 2003.
7. S. Basagni, M. Conti, I. Stojmenovic, and S. Giordano, *Mobile Ad Hoc Networking*, John Wiley and Son Inc., August 2004.
8. M. Torrent-Moreno, D. Jiang, and H. Hartenstein, "Broadcast Reception Rates and Effects of Priority Access in 802.11-Based Vehicular Ad-Hoc Networks", *Proc. Of ACM VANET*, Philadelphia, PA, October, 2004.
9. I. F. Akyildiz, X. Wang, and W. Wang, "Wireless Mesh Networks: A Survey," *Elsevier Computer Networks Journal*, Vol. 47, pp. 445–487, March 2005.
10. I. F. Akyildiz, W. Su, Y. Sankarasubramaniam, and E. Cayirci, "A Survey on Sensor Networks," *IEEE Communications Magazine*, Vol. 40, No. 8, pp. 102–114, August 2002.
11. D. Chen and P. K. Varshney, "A Survey of Void Handling Techniques for Geographic Routing in Wireless Networks," *IEEE Communications Surveys and Tutorials*, Vol. 9, pp. 50–67, First Quarter, 2007.
12. M. Mauve, J. Widmer, and H. Hartenstein, "A Survey on Position-based Routing in Mobile Ad Hoc Networks," *IEEE Network Magazine*, Vol. 15, No. 6, pp. 30–39, November 2001.
13. Q. Fang, J. Gao, and L. J. Guibas, "Locating and Bypassing Routing Holes in Sensor Networks," *Proc. of IEEE Infocom*, Hong Kong, March 2004.
14. M. Heissenbüttel, T. Braun, T. Bernoulli, and M. Wälchli, "BLR: Beacon-Less Routing Algorithm for Mobile Ad-Hoc Networks," *Elsevier Computer Communications*, Vol. 27, No. 11, pp. 1076–1086, 2004.
15. I. Stojmenovic, "Position-Based Routing in Ad Hoc Networks," *IEEE Communications Magazine*, Vol. 40, pp. 128–134, July 2002.
16. J. A. Sanchez, P. M. Ruiz, and I. Stojmenovic, "GMR: Geographic Multicast Routing for Wireless Sensor Networks," *Proc. of IEEE SECON 2006*, Sept., 2006.

17. J. C. Navas and T. Imielinski, "Geographic Addressing and Routing," *Proc. of ACM MobiCom*, Sept. 1997.
18. B. Nath and D. Niculescu, "Routing on a Curve," *ACM SIGCOMM Computer Communication Review*, pp. 155–160, October 2002.
19. J. Broch, D. A. Maltz, D. B. Johnson, Y. –C. Hu, and J. Jetcheva, "A Performance Comparison of Multi-hop Wireless Ad Hoc Network Routing Protocols," *Proc. of ACM MobiCom*, Dallas, Texas, August 1998.
20. E. Royer and C. –K. Toh, "A Review of Current Routing Protocols for Ad Hoc Wireless Networks," *IEEE Personal Communications*, April 1999.
21. I. Stojmenovic, "Location Updates for Efficient Routing in Ad Hoc Networks," Book Chapter in *Handbook of Wireless Networks and Mobile Computing*, Wiley, New York, 2002.
22. M. Zorzi and R. R. Rao, "Geographic Random Forwarding (GeRaF) for Ad Hoc and Sensor Networks: Multihop Performance," *IEEE Transactions on Mobile Computing*, Vol. 2, No. 4, 2003.
23. B. M. Blum, T. He, S. Son, and J. A. Stankovic, "IGF: A Robust State-Free Communication Protocol for Sensor Networks," *Technical Report CS-2003-11*, CS Department, University of Virginia, 2003.
24. H. Füßler, J. Widmer, M. Käsemann, M. Mauve, and H. Hartenstein, "Contention-Based Forwarding for Mobile Ad Hoc Networks," *Elsevier Ad Hoc Networks*, Vol. 1, No. 4, pp. 351–369, 2003.
25. D. Chen, J. Deng, and P. K. Varshney, "A State-Free Data Delivery Protocol for Wireless Sensor Networks," *Proc. of IEEE WCNC*, New Orleans, LA, March 2005.
26. Y. Xu, W. Lee, J. Xu, and G. Mitchell, "PSGR: Priority-based Stateless Geo-Routing in Wireless Sensor Networks," *Proc. of IEEE MASS*, November, Washington DC, 2005.
27. D. Ferrara, L. Galluccio, A. Leonardi, G. Morabito, and S. Palazzo, "MACRO: An Integrated MAC/Routing Protocol for Geographic Forwarding in Wireless Sensor Networks," *Proc. of IEEE Infocom*, Miami, FL, March 2005.
28. R. Nelson and L. Kleinrock, "The Spatial Capacity of a Slotted Aloha Multihop Packet Radio Network with Capture," *IEEE Transactions on Communications*, Vol. 32, No. 6, pp. 684–694, June 1984.
29. E. Kranakis, H. Singh, and J. Urrutia, "Compass Routing on Geometric Networks," *Proc. of 11th Canadian Conference on Computational Geometry*, Canada, 1999.
30. T. Melodia, D. Pompili, and I. F. Akyildiz, "Optimal Local Topology Knowledge for Energy Efficient Geographical Routing in Sensor Networks," *Proc. of IEEE Infocom*, Hong Kong, March 2004.
31. I. Stojmenovic and X. Lin, "Loop-free Hybrid Single-Path/Flooding Routing Algorithms with Guaranteed Delivery for Wireless Networks," *IEEE Trans. Parallel Dist. Sys.*, Vol. 12, No. 10, pp. 1023–1033, 2001.
32. G. G. Finn, "Routing and Addressing Problems in Large Metropolitan-Scale Internetworks," *ISI Research Report,* ISU/RR-87-180, March 1987.
33. I. Stojmenovic and X. Lin, "Power-Aware Localized Routing in Wireless Networks," *IEEE Transactions on Parallel and Distribution System*, Vol. 12, No. 11, pp. 1122–1133, 2001.
34. D. Couto, D. Aguayo, J. Bicket, and R. Morris, "A High-Throughput Path Metric for Multi-hop Wireless Routing," *Proc. of ACM MobiCom*, San Diego, CA, Sept. 2003.
35. J. Zhao and R. Govindan, "Understanding Packet Delivery Performance in Dense Wireless Sensor Networks," *Proc. of ACM Sensys*, California, Nov. 2003.
36. M. Zuniga and B. Krishnamachari, "Analyzing the Transitional Region in Low Power Wireless Links," *Proc. of IEEE SECON*, Santa Clara, CA, 2004.
37. K. Seada, M. Zuniga, A. Helmy, and B. Krishnamachari, "Energy-Efficient Forwarding Strategies for Geographic Routing in Lossy Wireless Sensor Networks," *Proc. of ACM SenSys*, Baltimore, Maryland, USA, November, 2004.

38. M. Zorzi and A. Armaroli, "Advancement Optimization in Multihop Wireless Networks," *Proc. of IEEE VTC*, Orlando, October, 2003.

39. S. Lee, B. Bhattacharjee, and S. Banerjee, "Efficient Geographic Routing in Multihop Wireless Networks," *Proc. of ACM MobiHoc*, Champaign, IL, May 2005.

40. B. Karp and H. T. Kung, "Greedy Perimeter Stateless Routing for Wireless Networks," *Proc. of ACM MobiCom*, Boston, MA, August 2000.

41. F. Zhao and L. J. Guibas, *Wireless Sensor Networks: An Information Processing Approach*, Morgan Kaufmann Publishers, July 2004.

42. P. Bose, P. Morin, I. Stojmenovic, and J. Urrutia, "Routing with Guaranteed Delivery in Ad Hoc Wireless Networks," *Wireless Networks*, Vol. 7, No. 6, pp. 609–616, 2001.

43. F. Kuhn, R. Wattenhofer, and A. Zollinger, "Worst-case Optimal and Average-case Efficient Geometric Ad-Hoc Routing," *Proc. of ACM MobiHoc*, Annapolis, MD, USA, June 2003.

44. F. Kuhn, R. Wattenhofer, Y. Zhang, and A. Zollinger, "Geometric Ad-Hoc Routing: Of Theory and Practice," *Proc. Of ACM PODC*, Boston, MA, July 13–16, 2003.

45. R. Jain, A. Puri, and R. Sengupta, "Geographical Routing Using Partial Information for Wireless Ad Hoc Networks," *IEEE Personal Communications*, Feb. 2001.

46. I. Stojmenvoic, M. Russell, and B. Vukojevic, "Depth First Search and Location Based Localized Routing and QoS Routing in Wireless Networks," *Computers and Informatics*, Vol. 21, No. 2, pp. 149–165, 2002.

47. D. Kim and N. Maxemchuk, "Simple Robotic Routing in Ad-Hoc Networks," *Proc. of ICNP*, 2004.

48. B. Leong, B. Liskov, and R. Morris, "Geographic Routing without Planarization," *Proc. of USENIX Symposium on Networked Systems Design and Implementation*, April 2006.

49. D. Chen and P. K. Varshney, "On Demand Geographic Forwarding for Data Delivery in Wireless Sensor Networks," *Elsevier Computer Communications, Special Issue on Network Coverage and Routing Schemes for Wireless Sensor Networks*, 2007.

50. G. Agnarsson and R. Greenlaw, *Graph Theory: Modeling, Applications, and Algorithms*, Prentice Hall, 2006.

51. E. M. Gafni and D. P. Bertsekas, "Distributed Algorithms for Generating Loop-Free Routes in Networks with Frequently Changing Topology," *IEEE Transactions on Communications*, Vol. 29, No. 1, pp. 11–18, 1981.

52. L. Zou, M. Lu, and Z. Xiong, "PAGER-M: A Novel Location-based Routing Protocol for Mobile Sensor Networks," *Proc. of Broadwise*, California, October 2004.

53. S. Chen, G. Fan, and J. Cui, "Avoid 'Void' in Geographic Routing for Data Aggregation in Sensor Networks," *International Journal of Ad Hoc and Ubiquitous Computing (IJA-HUC), Special Issue on Wireless Sensor Networks*, Vol. 2, No. 1, pp. 169–178, 2006.

54. D. Chen, J. Deng, and P. K. Varshney, "On the Forwarding Area of Contention-Based Geographic Forwarding for Ad Hoc and Sensor Networks," *Proc. of IEEE SECON*, Santa Clara, California, September, 2005.

55. T. He, J. A. Stankovic, C. Lu, and T. Abdelzaher, "SPEED: A Stateless Protocol for Real-time Communication in Sensor Networks," in *Proc. of ICDCS*, May 2003.

56. D. S. J. De Couto and R. Morris, "Location Proxies and Intermediate Node Forwarding for Practical Geographic Forwarding," *Technical Report*, MIT-LCS-TR-824, MIT Laboratory for Computer Science, June 2001.

57. L. Blazevic, S. Giordano, and J. Y. Le Boudec, "Self Organized Terminode Routing," *J. Cluster Computing*, Vol. 5, No. 2, pp. 205–218, April 2002.

58. E. M. Belding-Royer and C. E. Perkins, "Multicast Operation of the Ad-Hoc On-Demand Distance Vector Routing Protocol," *Proc. of ACM MobiCom*, 1999.

59. S. J. Lee, W. Su, and M. Gerla, "On-Demand Multicast Routing Protocol in Multihop Wireless Mobile Networks," *ACM/Kluwer Mobile Networks and Applications*, Vol. 7, No. 6, pp. 441–452, Dec. 2002.

60. I. Chlamtac, S. Basagni, and V. Syrotiuk, "Location-aware, Dependable Multicast for Mobile Ad Hoc Networks," *Computer Networks*, Vol. 36, No. 5–6, pp. 659–670, August 2001.

61. S. Su and M. Gerla, "Wireless Ad Hoc Multicast Routing With Mobility Prediction," *ACM/Kluwer Mobile Networks and Applications*, Vol. 6, No. 4, pp. 351–360, 2001.

62. M. Mauve, H. FÜβler, J. Widmer, F. Lang, "Position-Based Multicast Routing for Mobile Ad-Hoc Networks," *Technical Report*, TR-03–004, Department of Computer Science, University of Mannheim, March 2003.

63. M. Transier, H. FÜβler, J. Widmer, M. Mauve, and W. Effelsberg, "Scalable Position-Based Multicast for Mobile Ad-Hoc Networks," *Proc. of IEEE BroadWim*, San Jose, CA, Oct. 2004.

64. X. Xiang, X. Wang, Z. Zhou, "An Efficient Geographic Multicast Protocol for Mobile Ad Hoc Networks," *Proc. of IEEE WoWMoM*, Buffalo, NY, June, 2006.

65. X. Jiang and T. Camp, "A Review of Geocasting Protocols for a Mobile Ad Hoc Network," *Proc. of the Grace Hopper Celebration (GHC)*, 2002.

66. Y. Ko and N. H. Vaidya, "Geocasting in Mobile Ad Hoc Networks: Location-Based Multicast Algorithms," *Proc. of IEEE WMCSA*, 1999.

67. I. Stojmenovic, "Voronoi Diagram and Convex Hull Based Geocasting and Routing in Wireless Networks," *Technical Report*, TR-99–11, University of Ottawa, December, 1999.

68. W.-H. Liao, Y.-C. Tseng, K.-L. Lo, and J.-P. Sheu, "Geogrid: A Geocasting Protocol for Mobile Ad Hoc Networks Based on Grid," *Journal of Internet Technology*, Vol. 1, No. 2, pp. 23–32, 2000.

69. K. Yo and N. H. Vaidya, "GeoTORA: A Protocol for Geocasting in Mobile Ad Hoc Networks," *Proc. of IEEE ICNP*, Nov. 2000.

70. J. Boleng, T. Camp, and V. Tolety, "Mesh-Based Geocast Routing Protocols in an Ad Hoc Network," *Proc. of IEEE IPDPS*, April 2001.

71. T. Camp and Y. Liu, "An Adaptive Mesh-Based Protocol for Geocast Routing," *Journal of Parallel and Distributed Computing*, Vol. 62, No. 2, pp. 196–213, 2003.

72. K. Seada and A. Helmy, "Efficient Geocasting with Perfect Delivery in Wireless Networks," *Proc. of IEEE WCNC*, Atlanta, Georgia, March 2004.

73. S. Ni, Y. Tseng, Y. Chen, and J. Sheu, "The Broadcast Storm Problem in a Mobile Ad Hoc Network," *Proc. of ACM MobiCom*, 1999.

74. Y. Yu, R. Govindan, and D. Estrin, "Geographical and Energy Aware Routing: A Recursive Data Dissemination Protocol for Wireless Sensor Networks," *Technical Report*, UCLA/CSD-TR-01-023, UCLA, Computer Science Department, 2001.

75. A. Mizumoto, H. Yamaguchi, and K. Taniguchi, "Cost-Conscious Geographic Multicast on MANET," *Proc. of IEEE SECON*, Santa Clara, CA, 2004.

76. J. Lian, L. Chen, K. Naik, M. T. özsu, and G. Agnew, "Localized Routing Trees for Query Processing in Sensor Networks," *Technical Report*, CS2005-15, University of Waterloo, 2005.

77. A. Coman, M. A. Nascimento, and J. Sander, "A Framework for Spatio-Temporal Query Processing over Wireless Sensor Networks," *Proc. of Int. Workshop on Data Management for Sensor Networks, in conjunction with VLDB 2004*, Canada, 2004.

78. W. Zhang, X. Jia, and C. Huang, "Distributed Energy-Efficient Geographic Multicast for Wireless Sensor Networks," *Int. J. Wireless and Mobile Computing*, Vol. 1, No. 2, pp. 141–147, 2006.

79. D. Niculescu and B. Nath, "Trajectory Based Forwarding and Its Applications," *Proc. of ACM MobiCom*, San Diego, CA, 2003.

80. L. Barrière, P. Fraigniaud, and L. Narayanan, "Robust Position-Based Routing in Wireless Ad Hoc Networks with Unstable Transmission Ranges," *Proc. of 5th ACM Int. Wksp. Discrete Algorithms Methods for Mobile Comp. and Commun.*, 2001.

81. Y. J. Kim, R. Govindan, B. Karp, and S. Shenker, "Geographic Routing Made Practical," *Proc. of USENIX Symposium on Networked Systems Design and Implementation*, Boston, MA, April 2005.

82. Y. J. Kim, R. Govindan, B. Karp, and S. Shenker, "Lazy Cross-Link Removal for Geographic Routing," *Proc. of ACM Sensys*, Colorado, November, 2006.
83. K. Seada, A. Helmy, and R. Govindan, "On the Effect of Localization Errors On Geographic Face Routing in Sensor Networks," *Proc. of IPSN*, California, 2004.
84. R. C. Shah, A. Wolisz, and J. M. Rabaey, "On the Performance of Geographical Routing in the Presence of Localization Errors," *Proc. of IEEE ICC*, 2005.
85. S. Funke and N. Milosavljevic, "Guaranteed-Delivery Geographic Routing Under Uncertain Node Locations," *Proc. of IEEE Infocom*, 2007.
86. N. Abu-Ghazaleh, K.-D. Kang, and K. Liu, "Towards Resilient Geographic Routing in WSNs," *Proc. of ACM MSWiM*, Canada, 2005.
87. R. Flury and R. Wattenhofer, "Randomized 3D Geographic Routing," *Proc. of IEEE Infocom*, 2008.

# Chapter 8
# Formal Verification of Routing Protocols for Wireless Ad Hoc Networks

**Daniel Câmara, Antonio A.F. Loureiro, and Fethi Filali**

**Abstract** Routing is one of the most basic and important tasks in a collaborative computer network. Having a correct, robust, and efficient routing protocol is fundamental to any wireless network. However, a difficult problem is how to guarantee these desirable qualities. Neither simulations nor testbed implementations can ensure the quality required for these protocols. As an alternative to these methods, some researchers have successfully investigated the use of formal verification as a mean to guarantee the quality of routing protocols. Formal verification is a technique that assures a system has, or has not, a given property, based on a formal specification of the system under evaluation. This technique has proved to be a valuable tool, even contradicting some authors' claims and informal proofs. This chapter presents the main tools, proposals, and techniques available to perform formal verification of routing algorithms for wireless ad hoc networks.

## 8.1 Introduction

This chapter discusses the importance of applying formal verification in the development of routing algorithms for wireless ad hoc networks. It also presents a concise description of some of the most important proposals on this field.

We start by answering the following two questions: "What is a formal verification technique and why should it be applied to routing protocols for wireless ad hoc networks?" In short, the term formal method refers to mathematical-based techniques used in specification, development, and verification of software and hardware systems. The use of formal methods intends to increase the rigor on the design and development of systems, leading to more reliable products.

D. Câmara (✉)
Department of Computer Science, Federal University of Minas Gerais,
Caixa Postal 702, 30123-970, Belo Horizonte, Minas Gerais, Brazil
e-mail: danielc@dcc.ufmg.br

S. Misra et al. (eds.), *Guide to Wireless Ad Hoc Networks*,
Computer Communications and Networks, DOI 10.1007/978-1-84800-328-6_8,
© Springer-Verlag London Limited 2009

Looking at this definition, and mainly keeping in mind the mathematics involved in the process, some people tend to believe that the use of formal methods, and manly formal verification, is hard and worthy only for safety-critical systems. However, the fact is that formal methods may help the development of any system and the mathematics involved is quite easy and straightforward [1]. Formal methods, especially formal verification, can help the protocol designers to decrease the development time [1], find design errors and validate the proposed solutions. Thus the use of such methods tends to improve the final quality of the verified pieces of software. Following this line, this chapter focuses on formal verification as a tool to increase the quality of routing algorithms for wireless networks. Formal verification is the mathematical proof that the formal specified system, and hopefully the developed system, has, or has not, a given property. Such verification can be done manually or automatically.

Normally, designers perform a manual verification of a system when they want to understand better the system they are developing. Such proofs aim human readability and, sometimes, lack the required precision and formalism. Usually, manual proofs are done in high level and, not rarely, in natural language. Unfortunately, the ambiguity, inherent to the natural language, may lead to subtle errors that can be neglected. Another point to observe is that the continuous improvement in computing capacity has increased the complexity of hardware and software systems. Given such scenario, it is virtually impossible for humans to manually check all aspects of the system.

Automatic verification, on the other hand, presents a more accurate method to check the correctness of a system. The use of verification tools also requires a simpler, and more common, mathematical background than the one required to perform a manual verification. This makes this technique accessible to a wider audience and applicable to a broader range of cases.

Next section discusses the main formal verification techniques and their main variants and problems. After that Section 8.3 presents some of the main available tools to perform formal verification of computer systems. Section 8.4 presents and exemplifies a simple and interesting way to perform formal verification. Then Section 8.5 presents a concise view of some of the most important works on the field. Sections 8.6 and 8.7 present some thoughts about future research on the field and final comments about the chapter.

## 8.2 Background

Formal verification is the process of verifying, through a series of formal proofs, if a system has or has not a given property. The US Department of Defense DOD 5200.28-STD standard [2], the orange book, states that "a formal proof is a complete and convincing mathematical argument, presenting the full logical justification for each proof step, for the truth of a theorem, or set of theorems, composed as a series of inference steps. This process is machine checkable and each step follows the results of one or more previous steps".

It is important to notice that formal verification is not a substitute for testing or simulation. These three quality assurance techniques are much more complementary rather than competitive approaches. They should be used together to improve the system reliability once each one has a different approach and objective. Test is a way to think how the system works, trying to find situations where it may fail. Simulation offers the possibility to run a large battery of tests under identical circumstances where some parameters can be varied and the effect studied [3]. Formal verification is used to prove the correctness of the system, according to some properties. However, even the most enthusiastic supporters of formal methods recognize that other approaches are sometimes better [4].

Notice also that neither formal verification nor testing can guarantee that the system is perfect [1]. As Edsger W. Dijkstra said once about testing, "Program testing can best show the presence of errors but never their absence". In the same way, formal verification can prove that a system presents, or not, a characteristic we can think of. However, this does not guarantee, by no means, the system is perfect. Even further, the truth is that formal systems are also fallible. The fallibility is the most fundamental limitation of formal verification methods, and it arises from two facts: first, some properties can never be proved and, second, we can make mistakes in the proofs of those aspects we want to prove [1].

## 8.2.1 Formal Verification Techniques

There are basically three kinds of automated formal verification techniques, namely model checking, theorem proving and equivalence checking. Model checking is a method to verify if a formally modeled system satisfies a given property [5]. Theorem proving technique uses mathematical methods, such as axioms and rules, to prove the correctness of a system [8]. Equivalence checking formally checks if two models, at different abstraction levels, are equivalent [8]. This section will discuss these techniques, but it is worth to remember that even though they propose automated solutions, neither approach works without some degree of human assistance. For example, theorem proving sometimes requires advice of which properties worth to verify. Model checkers, on the other hand, can quickly get stuck when checking millions of useless states and human guidance can be handy.

- *Model checking*: Model checking verifies, using an algorithm, if a given model is in accordance with the specification. The model is normally programmed in a special purpose language and it is based on the system specification. Given the complexity of the current systems, the models

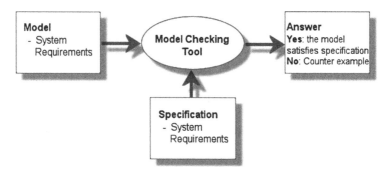

**Fig. 8.1** The model checking approach

often represent a simplified version of the target systems. Some tools express the properties to be verified using temporal logic formulae. Temporal logic allows the programmer to express system properties and verify them against the model. Figure 8.1 presents the model checking approach. The tool receives as input the system model and the desired/undesired property to be checked. The output is the answer whether the system holds or not the requested property. In the last case, it is common to provide a counter-example showing why the property is not satisfied.In model checking, all the valid inputs and possibilities are verified to guarantee the correctness of the system. To accomplish this task, a model checking tool uses a combinatorial amount of states to represent the system. In another words, the number of states required to represent a system increases exponentially with its size, leading to a problem known as state explosion, discussed later in Section 8.2.2. See [34] for a good explanation about model checking and its roots.

The success of the model checking verification depends on the user's expertise. Building a good model is a tradeoff between representing the important points of the system and decreasing the size of the model to avoid the state space problem. In this process, the model designer must be really careful to not remove the fundamental system characteristics and, at the same time, reduce the system complexity so the model is feasible to be verified.

- *Theorem proving*: Theorem proving involves verifying the truth of mathematical theorems postulated about the design. Theorem proving is like any traditional proof; it starts with axioms and using rules of inference the designer tries to prove the truth of a conclusion. The specification of the system is done in first-order or higher-order logic. From this precise formulation of the system, the designer can infer relations to prove its correctness. Often the theorem proving tool requires some guidance from the user, and the proof itself can be almost obtained by an interactive process. This technique requires highly trained and experienced people able to guide the tool through the right path.

- *Equivalence checking*: Equivalence checking is the process of verifying whether two implementations of the same system, in different abstract levels, are identical. Equivalence checking is very popular in the industry, and it is commonly used in the development of digital integrated circuits to formally prove that two representations of a circuit present exactly the same behavior. In this case, typically, the gate-level implementation is compared with its representation at a higher level, Register Transfer Level (RTL). However, in general, equivalence checking can work well for two structurally similar designs.

  Notice that equivalence checking does not verify if the design is error-free. In addition, when a difference between two design implementations is found, the error diagnosis capability of an equivalence checking tool is, often, limited and so it is difficult to determine the exact cause of the difference.

## 8.2.2  The State Explosion Problem and Remedies

Reachability analysis has been proved to be one of the most effective techniques to formally verify a broad range of systems. It consists of the analysis of which states the system can reach in the next steps, giving the current state. Although it is a powerful technique, it has its application severely restricted due to the "state explosion problem" [9]. This term refers to the situation in which the state space storage grows exponentially with the size of the model. The state space explosion problem occurs because of the large number of possible interleaving between processes in a reactive concurrent system. In this case the verification may fail, not because the model is wrong but simply because there is not enough memory to verify the target system [3]. A number of proposals have been made to minimize this problem, and, thus, enable its application to the verification of real systems. In the following, we list some of the main techniques, according to the description provided by Clarke et al. [5]:

- *Symbolic representation*: This technique refers to the use of compact data structures to represent the state space [10], i.e., encoding the transition relations of a Kripke structure as a Binary Decision Diagram (BDD). In this case it is possible to save storage space by exploiting the often inherent regularity of a hardware or software system. Other example of symbolic representation is the Constraint system representation of continuous parameters such as clock ranges, i.e., UPPAAL [11]. In this case it would not even be possible to store explicitly all time points, regardless the available amount of memory [3].
- *Partial order reduction*: It is based on the principle [12] that if two, or more, processes do not exchange information during their lifetime, it does not matter if they run in parallel or in any sequential order. This makes the verification easier since these processes can be verified isolated from each other. Partial order reduction is about to analyze the processes execution and

exploit the commutativity of concurrently executed transitions, which result in the same state when executed in different orders. Notice that the verification property must also be taken into account since it might introduce additional data dependencies between processes. Partial order reduction has been successfully applied to a series of tools, such as SPIN [13].

- *Compositional reasoning*: In short, it is the decomposition of the system into components, which are verified apart from the other components [15]. Composing over these parts, global properties can then be inferred. Even if mutual dependencies between components exist, the components can be verified separately, assuming that the other components work as expected.
- *Abstraction*: Abstraction is a way to decrease the complexity of system models [16]. Normally, when modeling a system, one may use abstractions in many ways. For example, instead of verifying the behavior of the system for all possible floating inputs, different classes of values can be modeled and used. When modeling network protocols, normally, other stack layers and protocols are abstracted from the problem to decrease the complexity of the model. Automatic abstraction methods are also available and can help in the formal verification of a broad range of systems.
- *Symmetry*: Many systems are symmetric in their design and implementation. Sometimes this symmetry can be seen as a form of redundancy [17]. Symmetry reduction [18] is a technique that combines states, which are similar, into equivalence classes. From these a new reduced model is built, choosing one representative of each equivalence class. Hopefully, the new model will be smaller than the original one, preserving the state transition graph. Using this technique it is possible to reach a substantial, often exponential, savings in terms of states.

## 8.3 Tools

The automatic verification of protocols is intrinsically linked to the software tools. This section briefly presents some of the most used formal verification tools available. All these are general tools and can be applied to a number of different applications to verify a broad range of systems.

- *HOL – Higher-Order Logic*: HOL [19] is a powerful and widely used interactive theorem proving tool. It is used to construct formal specifications and proofs in higher-order logic. HOL is used in a broad range of areas and problems, being successful in both industry and academia. HOL is a complete programming environment in which theorems can be proved and proof tools implemented. An important characteristic of this tool is its high degree of programmability based on the ML meta-language. To help the developers HOL has some built-in decision procedures and theorem proofs. An oracle mechanism also gives access to external programs such as SAT and BDD engines. Obradovic et al. [20, 21] used HOL to verify the AODV protocol.

- **SPIN** – SPIN is an efficient verification system for modeling distributed software systems. It provides a powerful and concise notation for expressing general correctness requirements [22]. SPIN accepts the design specifications written in the verification language PROMELA (Process Meta Language) [23], and the specification or correctness properties are expressed in linear temporal logic (LTL). The description of a concurrent system in PROMELA consists of one or more user-defined process templates and at least one process instantiation. The templates define the behavior of different types of processes. Any running process can instantiate further asynchronous processes, using the process templates [22]. SPIN translates each process template into a finite automaton.

  Instead of doing the verification directly on the PROMELA code, to improve its performance, SPIN generates C code from the model. This saves memory, improves performance, and allows the insertion of C code directly into the model. SPIN was used in a number of proposals [24, 25, 21, 26] to verify different routing protocols for wireless ad hoc networks, such as AODV, WARP, LAR, DREAM, LUNAR.

- *CPN – Colored Petri Nets*: Petri nets [27] is a tool that allows the creation of mathematical representations of discrete distributed systems in an intuitive and graphical way. The systems are modeled as graphs consisting of place nodes, transition nodes, and directed arcs connected by transitions. In Petri nets, modules interact using a set of well-defined interfaces. The graphical representation makes it easier to see the basic structure of a complex model.

  An important characteristic of Petri nets it is that they can handle more than one data stream at each time. This provides great expressiveness, especially when modeling distributed and parallel systems. There are two main variants of Petri nets, the standard one and Colored Petri Net (CPN) [28]. Unlike standard Petri nets where tokens are indistinguishable, in a Coloured Petri Net, every token has a value. This make easier for the designer to express different events and actions. Colored Petri Nets have also a formal, mathematical representation with a well-defined syntax and semantics. Petri nets have been used to prove the correctness of AODV [29], DSDV [30], and ERDP [31].

- *UPPAAL*: UPPAAL [11] is a tool suited for modeling, simulating, and verifying a broad range of systems, but mainly real-time systems. To do so it uses a collection of non-deterministic processes with finite control structure and real-valued clocks, communicating through channels and/or shared variables.

  UPPAAL has three main parts: a description language, a simulator, and a model checker. The description language is non-deterministic and serves to describe the system behavior using a network of timed automata. The simulator is used for interactive and automate analysis of the model, and for the verification of the correctness of the programmed model examining specific executions of the model. The model checker can also be used in an interactive way to find failures in the modeled system. However, its full power is shown when it automatically covers the exhaustive dynamic behavior of the modeled system. UPPAL automatically generates a diagnostic trace that explains why a property is, or is not, satisfied by a system description.

## 8.4 Thoughts for Practitioners

The use of formal verification techniques does not need to be difficult or extremely complex. Even simple approaches can lead to good and useful results. This section presents and exemplifies a simple, yet powerful, way to formally verify routing protocols for wireless networks. The technique initially presented in [32] and improved in [25] was used to find errors in DREAM [6] LAR [7] and OLSR [35]. In a nutshell, the technique is basically a list of procedures to be followed when verifying protocol. The method encourages the utilization of standard formal verification techniques, some of them presented in Section 8.2.2, and serves as a guide for newbies. Algorithm 8.1 presents the basic steps of the methodology described in [25].

The method is quite general and its steps may be useful to verify different protocols for wireless networks. In contrast to the work presented in [33], this specific method focuses on the qualitative aspects of the protocols rather than on quantitative ones [25]. It is indicated to those who are interested in identifying specification problems such as routing loops, packet delivery failures, unexpected reception of messages, and other "pathological" cases not treated in the protocol specifications. For those interested in quantitative aspects, such as max/min number of messages, the particular behavior of a given topology or timing aspects, other methods such as [20, 33], discussed later in this chapter, are more appropriated.

As general advices, when implementing this technique, we can highlight two points: first, build a simpler model and improve over it. Second, make sure to model all possible relations of the system under consideration. Starting with a simple verifiable code makes the work easier and with faster results. This also helps in filtering false positive results. Whenever a property is verified, the presented result must be analyzed in order to determine whether it is a failure in the algorithm or in the model. Unfortunately, up to now, this task cannot be done automatically and following traces of complex models can be very hard.

Regarding the second advice, when using a model checker tool, one of the main concerns is to be careful enough to model all relations that can happen in the real world. Model checkers normally work performing a search through all possible states, trying to find a condition that satisfies, or denies the requested property. For these tools it does not really matter if the verified property occurs in 99% of the cases or just in one single case. When the property does not hold, the tool finds this case and, often, presents the scenario where it occurs. After that, it is up to the system designer to analyze the scenario and figure out if the presented counterexample is a real problem or if it is a failure in the model.

**Algorithm 8.1** High-level algorithm of the proposed method [25]

| | |
|---|---|
| 1: | Acquire the information needed to model the protocol; |
| 2: | **Repeat** |
| 3: | Create a detailed pseudo-code or finite state machine of the protocol; |
| 4: | Compare carefully all cases described in the protocol with the pseudo code; |
| 5: | **Until** (Pseudo code is consistent with the protocol) |
| 6: | **For** (each kind of message) |
| 7: | **For** (each kind of node) { |
| 8: | Specify the semantics of the message to the node; |
| 9: | } // For |
| 10: | }// For |
| 11: | Divide the protocol into internal and external behaviors |
| 12: | **Internal behavior**: describes the message flows and behaviors for the node; |
| 13: | **External behavior**: describes the behaviors related to the node interactions; |
| 14: | **For** (each behavior) |
| 15: | Creates an algorithm or an state machine representation to understand it better; |
| 16: | } // For |
| 17: | Model the External vs. Internal interactions |
| | // The internal behavior should be modeled as if it was a routine call. In this |
| | // way the external behavior becomes independent of the internal behavior. |
| | // Ideally, the external and internal behaviors should be independent. |
| 18: | Build a simple model based on the internal and external behavior machines |
| 19: | **While** (the desired model complexity was not reached) |
| 20: | Analyses and verify the model |
| 21: | **For** (each error found in the model) |
| 22: | Verify whether the error is due to a protocol failure or a modeling failure; |
| 23: | Find a solution for the problem; |
| 24: | Model the solution; |
| 25: | Test the solution; |
| 26: | } // For |
| 27: | Increase the model complexity; |
| 28: | } // While |
| 29: | Identify and isolate verified procedures to be used in other protocols. |

## 8.4.1 Case Study

This section presents the methodology applied to OLSR protocol [35], using SPIN model checker. Notice, however, that any model checker which allows the modeling of non-deterministic channels can be used. The examples presented here are in PROMELA, the SPIN language. Another important observation is that even though OLSR has newer and more precise descriptions [14, 36], in order to meet our didactical objectives, given its simplicity, we use the original version [35].

### 8.4.1.1 Understanding the Protocol

The Optimized Link State Routing Protocol (OLSR) [35] is a proactive protocol. In order to disseminate routing information, a node sends both hello messages to its neighbors and topology control (TC) messages to a set of selected nodes that in turn rebroadcast them to other MPR (multipoint relay) nodes. The broadcasted information includes the node address and a list of distance information of nodes' neighbors. With this information each node builds its routing table, using a shortest path algorithm. In this way, the node can create a route to every other node in the network. If a node receives a duplicated packet, it discards it rather than retransmitting it. However, the key concept of OLSR is the multipoint relay (MPR) nodes. OLSR relies on MPR nodes to retransmit information in an organized and smart way. The MPR nodes are chosen among the one-hop neighbors in such a way that they are the minimum set that covers all the two-hop neighbors.

### 8.4.1.2 OLSR State Machine

To better understand how the protocol works, it is often useful to rewrite it in an algorithmic form or translate it to a state machine. Just doing this sometimes one can find errors or flaws in the protocol description. Some of the most important errors found in AODV by Obradovic et al. [20] were exactly specification errors. These specification problems occur, in some extent, often because the protocol designers are much more concerned about the main protocol ideas and are somehow more careless describing the details and smaller, but important, design decisions. Figure 8.2 shows a simplified diagram for the OLSR intermediate node, suitable for an early verification model. In the

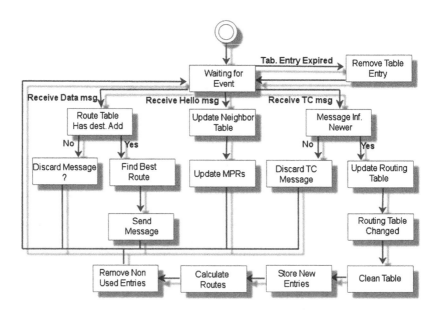

**Fig. 8.2** Simplified OLSR intermediate node diagram

diagram the discard message box, when there is no route for data packets, has the symbol "?" because the protocol description has no explicit reference about this specific case, so we will need to infer what to do.

### 8.4.1.3 Messages and Kinds of Nodes

OLSR defines three different kinds of messages: hello, topology control (TC), and data message. Each one has its own purpose and semantics for each different node. Table 8.1 presents these messages and their meaning for each kind of node.

### 8.4.1.4 Dividing into Internal and External Behaviors

The main part of the verification is done by the internal model behavior. This is expected once if we model the behavior of each node in respect to each different message, it is exactly how the distributed protocol should behave in the network. However, sometimes some behaviors can only be captured globally and, in this case, the external view is quite useful. For the OLSR, this is clear when we observe the creation of the tree of MPRs and the transmission of messages through it.

For the OLSR protocol, the modeling of the external behavior was worthy because it showed that the source node may receive the same TC packet it has generated. This may occur because the source node may be the MPR, or may be in the vicinity of an MPR of the node that is retransmitting its TC message. We cannot characterize this as an error, but at least for us this is somehow surprising, once the whole structure is intended to decrease the number of messages sent and it is always represented as a tree without loops.

### 8.4.1.5 Modeling the Channel

One of the key aspects of the methodology is doubtlessly the channel modeling. Modeling the channel in a non-deterministic way makes the results more general, and indeed it is what makes possible for the intermediate node to act as a cloud of nodes. This simple observation greatly decreases the complexity of the

**Table 8.1** OLSR messages and their semantics for each node

| Node | Hello | TC | Data |
|------|-------|-----|------|
| Origin | Spread information of links among neighbors | Spread information about the network table | Share upper-layer protocol information |
| Intermediate | Update its own neighbor table, find MPRs | Update routing tables and rebroadcast | Find new route and rebroadcast |
| Destination | This role does not exist; act as intermediate | This role does not exist; act as intermediate | Send message to upper layers |

models and allows their verification independently of any specific topology. The channel could be modeled in PROMELA as follows:

```
mtype = {RouteRequest, RouteReply, DataPacket};
/* type, origID, destID, TTL, inf timestamp */
chan medium = [nnodes] of {mtype, byte, byte, bool, bool};
```

However, the most important aspect of the channel is the way the nodes receive messages from it. The choice must be random and any node should be able to receive the message at any time. The protocol behavior will be in charge of handling the messages according to its specification.

Another important channel's characteristic is that it should be able to hold more than one message. OLSR is heavily based on a controlled flooding, but flooding nevertheless. If all relations are modeled, putting at least two messages in the channel, it enables the verification of all possible node states, even concurrency.

### 8.4.1.6 Creating the Model

To model a new protocol, it is better starting with the simplest model as possible; Algorithm 8.2 shows a first version of the OLSR model. After that the model complexity can be increased slowly until the desired level. Building the model in this way has several advantages. First, it helps a designer to better understand the protocol. Second, if one finds an error in the protocol, it is easier to isolate the problem and find a solution for it. Third, building the model in this way makes it easier to debug and finding out errors on it. However, the most important advantage is to have results sooner. Starting the process by building a complex model is probably a much more difficult task and, sometimes, people simply give up in the middle of the work. It is better to have results sooner and being able to stop at some comfortable point than to have a complex model, which most surely will even drive the model to the state explosion problem and, worse, without any result.

The model should be as simple as possible, while this does not compromise the protocol verification. For example, the condition of the OLSR timer to send a hello message can be modeled as a Boolean variable, i.e., either it is time to send the hello message or not. In this case there is no need for modeling a real timer. This trigger could be programmed in PROMELA as:

```
bool sendHello;
if
:: skip -> sendHello = true
:: skip -> sendHello = false
fi;
```

**Algorithm 8.2**  Example of a simple first version of the OLSR model

```
#define nnodes 3 /* Interm,Origin,Dest */        if
                                                 :: skip $->$ type = RouteRequest
chan medium = [nnodes] of { mtype, byte, byte }; :: skip $->$ type = RouteReply
mtype = { RouteRequest, RouteReply,              :: skip $->$ type = DataPacket
  DataPacket };
mtype = { Origin, Destination, Intermediate };   fi;
bool packt = false;                              medium!type(ttl); /* sends the packet */
                                                 fi;
proctype nodes() {                               if /*threat packet*/
  bool ttl;                                      :: skip $->$ node = Origin
  byte type;                                              /*play origin node*/
  if /* Creates random packet */                 :: skip $->$ node = Destination
  :: packt = = false $->$ packt = true $->$              /*play dest*/
  If                                             :: skip $->$ node = Intermediate
    :: skip $->$ ttl = true                               /*play inter*/
    :: skip $->$ ttl = false                     fi;
  fi;                                            }
                                                 init { run nodes(); }
```

Every variable, as much as possible, should be initialized with a random value. This increases the number of verified cases and makes the verification more independent and broader. Once a variable is initialized randomly, all cases related to that variable will be verified automatically. Of course, in the hello message example above, we are assuming that the message-sending procedure represents an independent event. It is neither triggered nor affected by other events. The drawback of this approach is that the model may become so broad that even cases that can never occur in the real world may be present in the model, and, thus, leading to false positives that need to be analyzed and discarded. Although this is probably better than missing a real-world case in the model.

### 8.4.1.7 Verifying the Model

In SPIN a verification is done based on the propositions represented in linear time temporal logic (LTL). To make the verification easier, each protocol scenario, specially the ones that can lead to errors, should be identified with a different variable, normally a Boolean one. Building the model in this way enables the creation of simpler and straightforward LTL formulae. To verify a proposition becomes just a matter of verifying the state of a variable. For example, the LTL formula to verify if the protocol fails to deliver a message could be:

```
[] ( !failDeliveryMessage && PathExists)
```

### 8.4.1.8 Case Study Results

Using the technique and incrementing slowly the complexity of the model presented in Algorithm 8.2, one can find a series of errors in the OLSR protocol, some presented here. It is important to highlight again that neither formal verification nor testing can guarantee that the system is perfect [1]. Indeed, we do not intend, by no means, to claim that the errors presented here are the only ones present in the protocol specification. However, what we do claim is that, for sure, at least these ones exist.

As proposed in [37], the algorithm to recalculate the routing table first cleans the entire routing table prior rebuilding it. It is not clear in the original work how this procedure is done; actually, this can be seen as an incompleteness of the protocol description. However, if a data packet arrives at this time, OLSR may raise an error because there will be no route available and the packet may even be discarded.

Other problem that occurs in OLSR is that when a message arrives in a node, just after the link is marked as unidirectional instead of bidirectional, the control messages may be discarded. With this, possibly, not all two-hop neighbors will receive such message. The problem here is authors argue that the MPR nodes are enough to guarantee that *all* two-hop neighbors will receive the control messages, once they represent the minimum set to cover all two-hop away nodes. This statement may not hold if, for any reason, a node stops to act as MPR. In this case, part of the network may be uncovered until another node takes its place.

Authors also argue that OLSR is resilient to a message loss. However, the protocol removes old entries from its tables if they are not refreshed in a defined amount of time. If the update message is lost, even if the route is still valid and able to deliver messages, the entry may be removed.

The last two OLSR problems presented here are the following. It has no explicit control for counter overflow. Thus, whenever a counter overflow occurs, the older information is kept on the routing tables instead of newer ones. This situation holds at least until the information entries are discarded by aging. This apparently is not an important problem, although it can lead to another more serious problem that is a routing loop, at least for a short amount of time. The loop case may befall if a conjunction of overflow and message lost situations occurs leading to inconsistent routing tables among nodes.

## 8.5 Proposals for Routing Verification

This section will present some important proposals on formal verification for routing in wireless networks, giving special attention to their strong and weak points. Figure 8.3 presents a schematic representation of the relations between the techniques presented, protocols verified, main concerns and tools used by the proposals.

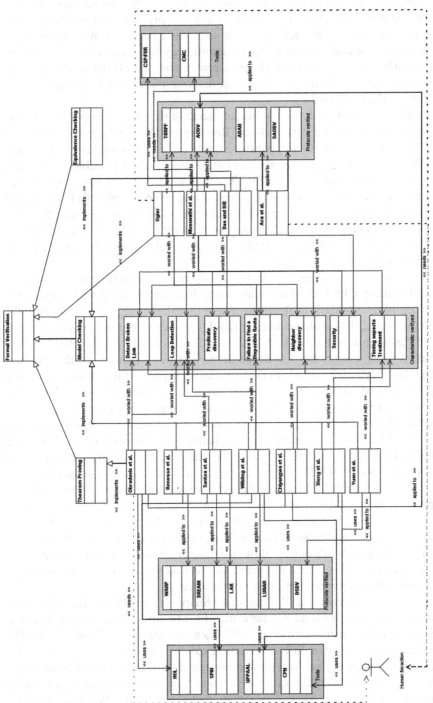

**Fig. 8.3** Relations among the formal verification proposals, protocols, and main issues

In [20] and [21], Obradovic et al. show how to use the theorem prover HOL and the model checker SPIN to prove the key properties of distance-vector routing protocols. The technique is powerful and one of the most cited in the literature. The main disadvantage of this strategy is its intense user interaction [38]. HOL, as semi-automatic theorem proving tool, needs the user to guide it. Another problem is the complexity in defining the theorems and lemmas to perform the real proof.

In [3], Wibling et al. use the model checking technique to verify the Lightweight Underlay Network Ad-hoc Routing (LUNAR) Protocol. They use SPIN to verify the data and control aspects of the LUNAR protocol and the UPPAAL tool to verify the protocol timing properties. A possible drawback is that the authors only verify LUNAR, which was designed by the same group. Furthermore, the work is based on some strong assumptions: only bidirectional links are allowed; messages must be delivered in order; and each node in the network can only receive and handle one message at a time. Such assumptions, in some cases, may even prevent the whole protocol verification, if it is based on any of these points.

Chiyangwa and Kwiatkowska [33] focus their work on the timing aspects of AODV using UPPAAL. They build a timed-automata model and evaluate the effects of the standard protocol parameters on the timing behavior of AODV. In that work, they evaluate a linear topology in which the source is node 0 and the destination is node $n-1$. All other nodes involved are sequentially placed between the source and the destination. The work focuses on this peculiar topology because it intends to evaluate timing aspects of the routing discovery problem and to find the maximum possible network diameter. The work reaches its purpose, but it only verifies the timing aspects of the protocol and does not consider qualitative aspects, such as loops and other routing problems.

In [38] Yuan et al. illustrate the dynamic operations of a MANET using CPN. They show a simple way to model the dynamic topology changes of ad hoc networks with CPN. The great strength of the proposal is the simplicity and elegance of the model. However, because of the simplifications in the modeling process, the work does not really handle the process of sending messages. Thus, for example, there is no difference between full and incremental routing table updates. This simplification may hide important errors that are not verifiable. The technique also does not allow two different nodes to receive and process, simultaneously, broadcast messages. In this case errors caused by concurrent sending/receiving messages cannot be detected.

Renesse and Aghvami [4] present a technique to use SPIN to formally verify routing algorithms for ad hoc networks. In their work, Renesse and Aghvami argue that the supertrace mode of SPIN is more suitable for large models. The supertrace mode of a SPIN validation can be performed in much smaller amount of memory, and still present reasonable coverage. They present simple examples in PROMELA of how to implement timers, mobility, and other needful procedures. They apply their technique to the Wireless Adaptive

Routing Protocol (WARP) using a five-node network. No strong justification or proof is given for using this number of nodes.

Building models to use a model checker tool is a hard and error-prone task. When verifying an abstraction, rather than the code itself, it is easy to miss possible implementation errors. Observing this, Musuvathi et al. [39] suggest a new way to perform formal verification. They propose a new model checker, CMC (C Model Checker), which checks C and C++ code implementations directly, eliminating the need for a model to abstract the system behavior. Performing the verification on the real code, one neither misses the errors that would be omitted from a model nor wastes time evaluating bugs that appear in the model but not in the real implementation. CMC is an interesting and promising tool, but it misses the point of evaluating profound design errors in protocols. Even the authors arguing in contrary [39], it becomes harder to verify if the protocol specification has design errors considering a real C/C++ implementation. For example, Obradovic et al. [20] found a number of flaws in AODV specification exactly because they were not bounded by a real implementation. Specifically for routing protocols, another point completely missed by this technique is the interaction among nodes. In another words, how to verify the protocol's dynamic behavior.

Zakiuddin et al. [40] propose a methodology to verify ad hoc networks protocols through model checking. Their approach is limited to a small number of nodes, typically about five. The authors argue that this is enough to characterize undesirable behaviors. They also argue that given the characteristics of the data and the tool they use, CSP and FDR, the results are applicable for an unbound number of nodes. Although the authors claim about the specification of a methodology, the proposed technique heavily relies on specific characteristics of the used tool. Another point to notice is that the application of the methodology depends on the proficiency of the designer with the tool. Once the procedure to apply the methodology is not fully specified, easily two people applying the same technique would arrive at different implementations and possibly results.

Xiong et al. [29] propose a timed model for AODV protocol, based on the idea of topology approximation mechanism. This mechanism describes the aggregate behavior of nodes where their long-term average behaviors are of interest. With this technique the nodes and their relationships are modeled as a graph where nodes become the vertices and the links become the edges of the graph. With this, the vertex degree shows the number of neighbors of the node. This structure is then translated into CPN. To perform the verification this work uses five nodes, but, again, there is no explanation about why to use such number. The verification also is, partially, bounded by the computational power available to the user, i.e., if there are more resources, it is possible to add more nodes. This is, by no means, not a good characteristic of any technique. The protocol verification should be independent of any particular scenario.

Ács et al. [41] propose a framework model to verify security of on demand routing algorithms. Basically the authors propose the creation of two distinct models, a real world and an ideal world model. The real-world model should describe the real operation of the system, and an ideal-world model should capture what the system wants to achieve in terms of security. Then, in order to prove the security of the system, the outputs of these two models must be indistinguishable [41]. The ideal world is secured by construction. It is what one wants to achieve, so no attack can be successful on it. On the other hand, the attacks can be successful at the other model, once no precautions are made in the sense of avoiding such attacks. The proposal has some drawbacks, but the main one is that it is still theoretical, and no automated proof is presented.

Das and Dill [42] propose a way to discover quantified predicates automatically from the model. They use this technique to prove the absence of loops in a simplified version of AODV. The initial predicate set is formulated in a manual step where conditions on next-node pointers, hop counters, and existence of routes are constructed. The method successfully discovers all required predicates for the version of AODV considered [24]. Unfortunately, for the general case, the problem of finding predicates to an unbounded system is intractable. However, the authors claim that the presented technique, Predicate Abstraction, is an efficient way of reducing infinite state systems into more tractable finite state systems.

As a manual verification, we can refer to the work of Ogier [43], which proves the correctness of the Topology Dissemination Based on Reverse-Path Forwarding (TBRPF) routing protocol. Since TBRPF consists of two modules, the routing module and the neighbor discovery module, the work presents the correctness proof for both modules separately. Even though this kind of proof is not easy, its results and procedures stands for TBRPF and only for it. Another point to observe is that when verifying a protocol, all cases must be considered and doing so manually it can be even harder for other protocols.

## 8.6 Directions for Future Research

The formal verification technique applied to routing algorithms for wireless networks is a quite unexplored field yet, and therefore there are lots of opportunities for new research. Indeed, the field is in need of more specific techniques and tools.

Until this moment, at the best of our knowledge, no attempt was made trying to apply equivalence checking techniques in the verification of routing for wireless networks. Equivalence checking is a powerful technique and may be extremely helpful in the development and mainly in the evolution of wireless routing algorithms.

Every new routing algorithm is a target for the techniques already developed. The verification of newer algorithms often reveals crucial failures that, if corrected earlier, can lead to more stable and trustable algorithms.

In terms of individual proposals, the work of Musuvathi et al. [39] has a huge merit in the sense it presents a different and, in some terms, more practical approach. Verifying directly the algorithm code, instead models, may be an interesting and promising path to follow in the verification field. Advances in this kind of verification technique would have a wide applicability.

The work of Chiyangwa and Kwiatkowska [33] has also a remarkable value in the way it limits the verification scope and targets very specific limit problems. Such kind of approach may be interesting and applicable for other problems and situations. A good and valuable work, apart from expanding the existing one and applying it to other algorithms, could define a list of general situations and limits where one can use this kind of technique.

A good path for research on this field could be the mixing of different existent tools and approaches. For example, Obradovic et al. [20, 21], uses both: SPIN and HOL in their work. A good advice, for any field, is always try to use the right tools to solve the problems. If the problems have different characteristics, the use of different tools and techniques should be considered.

Routing is a key aspect for the network and, mainly for wireless networks, security is a key aspect, and the work of Ács et al. [41] points this need. The verification of security aspects of routing protocols for wireless networks is also a promising research field.

## 8.7 Conclusions

Formal verification is a promising technique to validate algorithms for wireless networks. Different from which someone can think the application of formal verification techniques in the development of new routing algorithms can be easy and presents an expressive increase in the quality of the protocol. The techniques presented here are a good start point for people who want to follow the research on this field, or at least apply formal verification on their own algorithms.

## Terminologies and Keywords

*Formal verification:* The mathematical proof that a formal specified system has, or has not, a given property.

*Model checking:* Technique to verify if a defined property stands against a formal specified model.

*Model:* An abstraction of the target system, normally defined in a special purpose language.

*Theorem proving:* A formal verification technique that employs axioms, theorems, and rules of inference to prove the truth of a conclusion.

*Equivalence checking:* The process of verifying whether two implementations of the same system are identical or not.

*Reachability:* The analysis of which states the system can reach in the next steps, giving the current state.

*State explosion:* One of the main problems in the formal verification field. Reachability analysis creates an exponential number of states. The term *State Explosion* refers to the situation in which the state space storage grows exponentially with the size of the model.

*Colored Petri Nets (CPN):* Graphical language for specification and verification of computational systems.

*Symbolic representation:* The use of encoded structures to represent complex concepts trying, in this way, to decrease the need of computational resources.

*Ambiguity:* Situation where a definition may have more than one meaning.

*Routing:* The process of deciding which is the best node sequence to send a message through the network.

## Questions

1. What are the main purposes of using formal verification to validate algorithms for wireless networks?
2. Explain, with your own words and based on other references, the most fundamental limitation of the formal verification technique.
3. Why problems such as hidden and exposing nodes are the problems for routing algorithms?
4. Explain the differences among HOL, SPIN, and CPN tools.
5. When using model checking to formally verify protocols, which are the main points one should have in mind?
6. Search for new proposals on this field and insert them in the diagram of Fig. 8.3.
7. Create a simple new routing algorithm and try to prove it is correct using SPIN.
8. Find a routing algorithm for sensor or ad hoc network and, without any specific technique, try to spot three weak points of it.
9. Get the same protocol and use the technique described in [25] and exemplified here, and try to formally verify the protocol using SPIN.
10. Verify the protocol using HOL and CPN.

## References

1. A. Hall. Seven myths of formal methods. IEEE Software, 7(5):11–19 Sep/Oct, 1990
2. Department Of Defense Standard: Department Of Defense Trusted Computer System Evaluation Criteria (Aka. The Orange Book). DoD 5200.28-STD; Supersedes; CSC-STD-00 l-83, dtd l5 Aug 83; Library No. S225,7ll.

3. O. Wibling. Ad hoc routing protocol validation, Licentiate Thesis 2005-004, Department of Info Technology, Uppsala University, Sweden, 2005.
4. J. P. Bowen, and M. G. Hinchey. Seven more myths of formal methods. IEEE Software, 12(4):34–41, Jul, 1995.
5. E. M. Clarke, O. Grumberg, and D. A. Peled. Model Checking. MIT Press, Cambridge, MA, 1999.
6. S. Basagni, I. Chlamtac, V. Syrotiuk, and B. Woodward. A distance routing effect algorithm for mobility, MobiCom'98, Dallas, TX, 1998.
7. Y. Ko, and N. H. Vaidya. Location-aided routing (LAR) mobile ad hoc networks, MobiCom'98, Dallas, TX, 1998.
8. C. Kern and M. R. Greenstreet. Formal verification in hardware design: a survey. ACM Transactions on Design Automation of Electronic Systems, 4(2):123–193, April 1999.
9. F. J. Lin, P. M. Chu, and M. T. Liu, Protocol verification using reachability analysis: the state space explosion problem and relief strategies, SIGCOMM '87, ACM Press, New York, NY, 1988.
10. S. Hendriex, and L. Claesen, A symbolic core approach to the formal verification of integrated mixed-mode applications, EDTC '97 European Conference on Design and Test, 1997.
11. J. Bengtsson, K. G. Larsen, F. Larsson, P. Pettersson, and W. Yi. Uppaal – a tool suite for automatic verification of real-time systems, In Proceedings of the 4th DIMACS Workshop on Verification and Control of Hybrid Systems, New Jersey, October 1995.
12. P. Godefroid, An approach to the state-explosion problem, PhD. thesis, University of Liege, Computer Science Department, 1994.
13. G.J. Holzmann, and D. Peled. An improvement in formal verification. In Proc. 7th IFIP WG 6.1 International Conference on Formal Description Techniques, October 1994.
14. T. Clausen, and P. Jacquet, Optimized Link State Routing Protocol (OLSR), Request for Comments: 3626, October 2003.
15. S. Berezin, S. Campos, and E. M. Clarke. Compositional reasoning in model checking. In Compositionality: The Significant Difference: International Symposium, V. 1536 of Lecture Notes in Computer Science, Springer-Verlag, September 1997.
16. E. M. Clarke, O. Grumberg, and D. E. Long. Model checking and abstraction. ACM Transactions on Programming Languages and Systems, 16(5):1512–1542, September 1994.
17. E. A. Emerson, and R. J. Trefler. From asymmetry to full symmetry: new techniques for symmetry reduction in model checking, Conference on Correct Hardware Design and Verification Methods, 142–156, 1999.
18. E. M. Clarke, E. A. Emerson, S. Jha, and A. S. Sistla. Symmetry reductions in model checking. V. 1427 of Lecture Notes in Computer Science, Springer-Verlag, June/July 1998.
19. M. Aagaard, M. E. Leeser, and P. J. Windley. Toward a super duper hardware tactic, 6th International Workshop, HUG'93, Vancouver, B.C., August 11–13, 1993.
20. K. Bhargavan, D. Obradovic, and C. A. Gunter. Formal verification of standards for distance vector routing protocols, Journal of the ACM, 49(4): 538–576, July 2002.
21. D. Obradovic. Formal analysis of convergence of routing protocols, Ph.D. Thesis Proposal, Department of Computer and Information Science, University of Pennsylvania, November 2000.
22. G. J. Holzmann, The model checker SPIN. IEEE Trans. on Software Eng., 23(5), May 1997.
23. G. J. Holzmann, Design and Validation of Computer Protocols. Englewood Cliffs, N.J.: Prentice Hall, 1991.
24. O. Wibling, J. Parrow, and A. Pears. Ad hoc routing protocol verification through broadcast abstraction, 25th IFIP FORTE, Taiwan, 2005.
25. D. Câmara, A. A. F. Loureiro, and F. Filali. Methodology for formal verification of routing protocols for ad hoc wireless networks, IEEE GLOBECOM 2007, Washington DC, November 2007.

26. R. de Renesse, and A. H. Aghvami, Formal verification of ad-hoc routing protocols using SPIN model checker, 12th Mediterranean Electrotechnical Conference, Croatia, 2004.
27. T. Murata, Petri nets: properties, analysis and applications, Proceedings of the IEEE, pp. 541–580, Vol. 77, No 4, April, 1989.
28. K. Jensen, Coloured Petri Nets. Basic Concepts, Analysis Methods and Practical Use. Volume 1, Basic Concepts, Monographs in Theoretical Computer Science, Springer-Verlag, 1997.
29. C. Xiong, T. Murata, and J. Tsai, Modeling and simulation of routing protocol for mobile ad hoc networks using colored petri nets, Research and Practice in Information Technology, 12: 145–153, Australian Computer Society, 2002.
30. C. Yuan, and J. Billington, An abstract model of routing in mobile ad hoc networks, Sixth Workshop and Tutorial on Practical Use of Coloured Petri Nets and the CPN Tools, Aarhus, Denmark, 26 October 2005.
31. L. M. Kristensen, and K. Jensen, Specification and validation of an edge router discovery protocol for mobile ad hoc networks, Integration of Software Specification Techniques for Applications in Engineering, V. 3147 of Lecture Notes in Computer Science, Springer-Verlag, September 2004.
32. D, Câmara, C. F. Santos, and A. A. F. Loureiro. Formal verification of routing protocols for ad hoc networks, Brazilian Symposium on Computer Networks, SC, Brazil, 2001. (In Portuguese)
33. S. Chiyangwa and M. Kwiatkowska. A timing analysis of AODV, 7th IFIP FMOODS, June 2005.
34. E. Clarke. Model checking: My 25 year quest to overcome the state-explosion problem, 25 Years of Model Checking Symposium, The 2006 Federated Logic Conference, Seattle, Washington, August 10–22, 2006.
35. T. Clausen, P. Jacquet, A. Laouiti, P. Muhlethaler, A. Qayyum, and L. Viennot. Optimized link state routing protocol for ad hoc networks, IEEE INMIC Pakistan 2001.
36. T. Clausen, and P. Jacquet. Optimized Link State Routing Protocol (OLSR), Request for Comments: 3626, October 2003.
37. T. Clausen, P. Jacquet, A. Laouiti, P. Muhlethaler, A. Qayyum, and L. Viennot. Optimized link state routing protocol for ad hoc networks, IEEE INMIC Pakistan 2001.
38. C. Yuan, and J. Billington. An abstract model of routing in mobile ad hoc networks, Sixth Workshop and Tutorial on Practical Use of CPN and the CPN Tools, Aarhus, Denmark, 2005.
39. M. Musuvathi, D. Y. W. Park, A. Chou, D. R. Engler, and D. L. Dill. CMC: A pragmatic approach to model checking real code, 5th Symposium on Operating System Design and Implementation, USENIX Association, Massachusetts, December 2002.
40. I. Zakiuddin, M. Goldsmith, P. Whittaker, and P. H. B. Gardiner. A methodology for model-checking ad-hoc networks, Lecture Notes in Computer Science, Volume 2648, Springer Verlag, May 2003.
41. G. Ács, L. Buttyán, and I. Vajda. Provable security of on-demand distance vector routing in wireless ad hoc networks, Second European Workshop on Security and Privacy in Ad Hoc and Sensor Networks (ESAS 2005) Visegrád, Hungary, July 13–14, 2005.
42. S. Das, and D. L. Dill. Counter-example based, predicate discovery in predicate abstraction. Formal Methods in Computer-Aided Design, Portland, Oregon, November, 2002.
43. R. Ogier. Topology dissemination based on reverse-path forwarding (TBRPF): Correctness and simulation evaluation, Technical report, SRI International, October 2003.

# Chapter 9
# Mobility Management in MANETs:
# Exploit the Positive Impacts of Mobility

Feng Li, Yinying Yang, and Jie Wu

**Abstract** The question of whether mobility is a blessing (Burleigh et al., *IEEE Communications Magazine,* 41:128–136, 2003; Capkun et al., *Proc. of ACM MobiHoc,* 2003) or a curse (Zhang et al., *Proc. of ACM MobiHoc,* 2005) to ad hoc networks has attracted a significant amount of research interest. Some researchers argue that mobility is a hurdle, as it makes the routing, naming and addressing, and location services more challenging. Some new mechanisms such as Zhang et al. (*Proc. of ACM MobiHoc,* 2005) have been proposed to tackle the problems caused by node mobility in MANETs. Others argue that far?from being a hurdle, mobility can be exploited to increase the system performance. Carefully designed protocols may exploit the mobility to obtain advantages in many important aspects of ad hoc networks, such as network capacity, security, and information dissemination.

This chapter surveys the impact of mobility in ad hoc networks from a wide perspective. We refrain from going into minute details of mobility, and instead?head for giving a broader picture. The goal of this chapter is to endorse new approaches to employ mobility in ad hoc networks based on the current situation and show why mobility can help in many different aspects (Cooper et?al., *Proc. of IEEE MASCOTS,* 2005; Camp et al., *Wireless Communications and Mobile Computing,* 2(5):483–502, 2002).

## 9.1 Introduction

Mobility is an inherent character of ad hoc networks. Mobile ad hoc networks (MANETs) are characterized by their node mobility and lack of infrastructure. Nodes' movements are usually irrelevant to the application. However, the mobility patterns are usually crucial to the networks' performance. Although

F. Li (✉)
Department of Computer Science and Engineering, Florida Atlantic University,
Boca Raton, FL 33431
e-mail: fli4@fau.edu

S. Misra et al. (eds.), *Guide to Wireless Ad Hoc Networks,*
Computer Communications and Networks, DOI 10.1007/978-1-84800-328-6_9,
© Springer-Verlag London Limited 2009

**Fig. 9.1** Overview of the
positive impacts of mobility

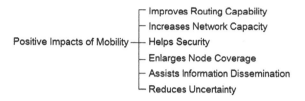

each node's movement is random, there are still some underlining disciplines in their mobility model. To design and select a realistic mobility model that truly depicts and predicts nodes' mobility in a MANET is the first step of mobility management.

Based on the mobility model, protocols that take mobility into consideration at?the design phase should be used. By doing so, these protocols, i.e., the mobility management schemes, can fully exploit the positive impacts of mobility in ?MANETs. Mobility management in MANETs is still relatively understudied. The main issue centers on whether mobility should be treated as a foe (undesirable) or a friend (desirable).

The traditional connection-based model used in MANETs, including the existing protocols (DSR [6], AODV [4], and ZRP [14]), is built on the premise that the underlying network is connected and views node mobility as undesirable. More recently, mobility has been identified to cause asynchronous sampling of Hello messages and various protocol delays that result in an inconsistent global state. Several tolerant schemes have been proposed [1, 3] as the first attempt to mask the effect of node movement and to construct a consistent global state for various applications.

Several recent research works examine mobility from new angles. They show that far from being a hurdle, mobility can be exploited to increase the ad hoc network's performance. By revising the traditional connection-based models and designing protocols that take mobility into consideration at the beginning, we can exploit mobility to improve routing capability, increase network ?capacity, improve security, and reduce uncertainty. This chapter summarizes the positive impacts of mobility and provides a foundation for readers to capture the essentiality of these favorable impacts. Figure 9.1 shows some positive impacts of mobility that we will discuss in this chapter.

## 9.2 Overview

In this chapter, we will introduce the widely used mobility models, survey the possible positive impacts of mobility, and summarize how to exploit mobility management schemes to benefit from these positive impacts.

In traditional protocols for networks, links are considered to be permanent. For one round of communication, the source and destination are connected through a path in a connected graph representing the network. Mobility violates this underline assumption. With mobility, links are temporary and time-variant. Therefore, mobility is treated as a side issue in these traditional protocols, and the effect of mobility is usually counteracted through a simple recovery scheme. For example, a route disruption caused by node movement is dealt with either by route rediscovery or by a local fix in a typical reactive approach.

If we take mobility into consideration in the first place, and design network protocols based on the inherent nodes' mobility in MANETs, the results seem to be quite different. If routing protocols are based on temporary connections instead of permanent links, the opportunities to set up a route between source and destination will be increased. By exploiting mobility to reduce interference, network capacity can be increased. Mobility also helps nodes to disseminate information, buildup security associations, spread trust, and reduce uncertainty. Through movement, the nodes' coverage area will be enlarged. Figure 9.1 ?summarizes these possible positive impacts, and we discuss them in detail in the following sections.

## 9.3  Thoughts for Practitioners: Positive Impacts of Mobility

### 9.3.1  Mobility Models

Using a realistic mobility model to depict nodes' movement is the first step to conduct mobility management. Different mobility models have different focuses and different application scenarios [8, 12, 16, 18, 30]. Besides random mobility models, to achieve better performance, some recent mobility research papers have adapted a method to control the movement of a small portion of designated nodes and exploit this movement to improve the network's overall performance. Figure 9.2 summarizes the most widely used mobility models in recent research papers.

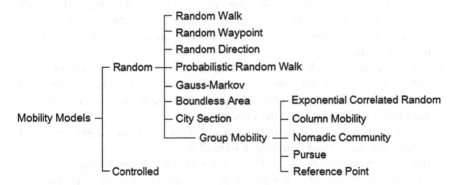

**Fig. 9.2** Classification of the existing mobility models

Therefore, the mobility models in recent papers can be classified as uncontrolled models (reactive schemes) or controlled models (proactive schemes). In?uncontrolled models such as epidemic routing [2], applications rely on ?movement that is inherent in the devices themselves to help deliver messages. When disconnected, nodes passively wait for their own mobility to allow them to reconnect. Since encounters between nodes can be unpredictable and rare, these approaches suffer potentially low data delivery rates and large delays. To increase delivery rate and reduce delay, nodes typically propagate messages throughout the network, which exacerbates contention for limited buffers in nodes and drains the nodes' limited energy. In controlled models, nodes modify their trajectories proactively for communication purposes. Li and Rus propose an optimal algorithm [20] to compute the trajectories of nodes in an effort to?minimize message transmission delay. In [15], Wu and Yang propose a *trajectory planning* scheme that gives an up-bound of the expected total moving distance while controlling the number of relays through a hierarchical structure of trajectory.

For the uncontrolled random mobility model, researchers have proposed many different schemes to model the inherent mobility of nodes. These models [25] attempt to realistically represent the behaviors of mobile nodes without the use of traces. Changes in speed and direction must occur, and they must occur in reasonable time slots. For example, we would not want mobile nodes to travel in straight lines at constant speeds throughout the course of the entire simulation, because real mobile nodes would not travel in such a restricted manner.

1. *Random walk mobility model*: Each node moves from its current location to a new location by randomly choosing an arbitrary direction and speed from a given range. Such a move is performed for either a constant time or a constant distance traveled. Then a new speed and direction are chosen. At the boundaries, nodes bounce off like billiard balls on a pool table. The random walk mobility model is described as a memoryless mobility pattern, because it retains no knowledge concerning its past locations and speed values.

2. *Random waypoint mobility model*: This model is equivalent to the random walk model except that before any change of speed and direction, a predetermined pause time is performed. This model is widely used for evaluating ad hoc network routing protocols.

3. *Random direction mobility model*: Here, the node must travel to the edge of the simulation area (or some other condition must be met) at a constant speed and direction. Then, the nodes pause and a new direction and velocity are chosen randomly. Then the process repeats.

4. *A boundless simulation area mobility model:* This model exchanges the planar rectangular simulation field with a boundless torus.

5. *Gauss-Markovmobility model:* This is a model that uses one tuning parameter to vary the degree of randomness in the mobility pattern. The random Gauss-Markov mobility model is introduced as an improvement over the

smooth random mobility model. A node's next location is generated by its past location and velocity. Depending upon parameters set, this allows modeling along a spectrum from random walk to fluid-flow.

6. *A probabilistic version of the random walk mobility model*: In this model the last step made by the random walk influences the next one. Under the condition that a node has moved to the right, the probability that it continues to move in this direction is higher than the probability that movement will cease. This leads to a walk that leaves the starting point much faster than the original random walk model.

7. *City Section Mobility Model*: Here the random waypoint movement is ?combined with a street map of a virtual city. The paths of the mobile nodes are limited to these streets in the field. In a related model, the streets are replaced by Voronoi graphs. Furthermore, obstacles are used, which obstruct radio signals.

The group-mobility models [29] are usually an extension of the above models, where either a function describes the group behavior or the nodes are somehow associated with a group leader or a target. We list the following group mobility models here:

(1) Exponential Correlated Random Mobility Model: Here a motion function creates a group behavior. (2) Column Mobility Model: The set of mobile nodes form a line and move forward in a particular direction. (3) Nomadic Community Mobility Model: A group mobility model where a set of mobile nodes move together from one location to another. (4) Pursue Mobility Model: For each group the group members follow a target node moving over the simulation area. (5). Reference Point Group Mobility Model: The group movement is based upon the path traveled by a logical center. Again the logical center moves according to an individual mobility model.

## 9.3.2 Different Levels of Mobility

In static networks, the mobility of nodes, users, and the monitored phenomenon itself is minimal or ignored. For example, sun and temperature sensors in a sunroom may collect relevant information and use it to control motorized shades in order to maintain these parameters within preset limits. This static paradigm may be expanded by introducing mobility in one or more of the below-mentioned three levels of the ad hoc networks:

- *Node level mobility*: the ad hoc nodes themselves may be moving. Examples include nodes mounted on moving cars or flying unmanned aerial vehicles, collecting information as their carriers constantly change their location and/ or orientation.
- *Information level mobility*: the event (source) monitored by or occurring in?the network is mobile [7]. For example, the smog generated by a poorly

maintained truck is moving along with the truck. Another example may be the evolution of an oil spill that we try to model through measurements at distinct buoy locations.

- *User level mobility*: users (destination) accessing the information collected by the network may themselves be moving, and thus the information that is pertinent to them may change over time. For example, monitoring the traffic conditions on the way to the nearest hospital changes as the user is changing his/her position.

### 9.3.3 Mobility Improves Routing Capability

Routing [9, 10, 21, 22] in ad hoc networks has been an active research field in recent years,?producing many routing algorithms such as DSR, DSDV, and AODV. However, most of the existing work focuses on connected networks where an end-to-end path exists between any two nodes in the network. In sparse ?networks, where partitions are not exceptional events, these routing algorithms will fail to deliver packets because no route is found to reach their destinations. To overcome partitions in sparse networks, a straightforward approach is to use radios with longer transmission ranges and maintain persistent network connectivity. However, since many mobile nodes use batteries for power supply, the use of a long-range radio leads to excessive energy consumption. In addition, the availability of such devices in critical scenarios would be questionable. Mobility becomes the natural choice to help nodes set up connections in these scenarios.

#### 9.3.3.1 Main Target: Connection-Based Routing

The common assumption behind existing ad hoc routing techniques is that there is always a connected path from the source to the destination. However, the advent of short-range wireless communication environments and the wide physical range and circumstances over which such networks are deployed means that this assumption is not always valid in realistic scenarios. Unfortunately, with original ad hoc routing protocols, packets are not delivered if a network partition exists between the source and the destination when a message is originated. Certain applications, such as real-time, constant bit rate communication, may require a connected path for meaningful communication. However, a number of other application classes benefit from the eventual and timely delivery of messages, especially in the case where frequent and numerous network partitions would prevent messages from ever being delivered end to end.

In the context of such applications, the goal of this work is to develop techniques for delivering application data with high probability, even when there is never a fully connected path between source and destination.

Thus, some works, such as epidemic routing [2] and message ferrying [27, 28], make minimal assumptions about the connectivity of the underlying ad hoc network: (1) the sender is never in range of any receivers, (2) the sender does not know where the receiver is currently located or the best route to follow, (3) the receiver may also be a roaming wireless host.

The intermediate nodes of the routing in these methods are called carriers. They have the ability to store messages before forwarding them to the next intermediate node. The goals of using these carriers to improve routing ?capability are to: (1) efficiently distribute messages through partially connected ad hoc networks in a probabilistic fashion, (2) minimize the amount of resources consumed in delivering any single message, and (3) maximize the percentage of messages that are eventually delivered to their destination.

There are some common concerns in all of those methods, as they let the intermediate nodes *store-and-carry* the messages: (1) routing under Uncertainty: Message senders have inexact knowledge of the location of nodes throughout the system. Thus, a key issue is determining whether to transmit a message when a carrier comes into the range. The problem is partly solved when we exploit the controlled movement of the carrier. However, hosts in the controlled movement scenarios still need to decide whether they have enough messages to send and approach the carrier. (2) Buffer Overflow: The carriers' buffer is limited. The system must balance the conflicting goals of maximizing message delivery and?minimizing resource consumption. Determining when a message should be?dropped is a critical issue in both controlled and uncontrolled schemes. (3)?Performance: A given message exchange and routing protocol can be ?evaluated along a number of different axes. Performance metrics include the average latency in delivering messages, the average amount of system storage and communication bandwidth consumed in delivering a message, and the amount of energy consumed in transmitting the message to its destination. (4) Reliability: Given the probabilistic delivery of messages in these schemes, how to guarantee delivery or raise the delivery ratio is considered to be one critical issue.

### 9.3.3.2  Mobility Scheme: Controlled vs. Uncontrolled

With regard to the aforementioned design goals, different schemes with different mobility models have been proposed to increase the *routing capability*. One?is to exploit the nodes' inherent random movement. These methods use a per-connection-based flooding method. The most representative method is epidemic routing. In the other category of schemes, nodes modify their trajectories proactively for communication purposes. The most representative method is message ferrying.

Epidemic routing supports the eventual delivery of messages and only requires periodic pair-wise connectivity. The epidemic routing protocol, as show in Fig. 9.3(a), works as follows. The protocol relies upon the transitive distribution of messages through ad hoc networks, with messages eventually reaching their destination. Each host maintains a buffer consisting of messages that it has originated as well as messages that it is buffering on behalf of other hosts. For efficiency, a hash table indexes this list of messages, keyed by a unique identifier associated with each message. Each host stores a bit vector called the summary vector that indicates which entries in their local hash tables are set. When two hosts come into communication range of one another, the host with the smaller identifier initiates an anti-entropy session with the host with the larger identifier. To avoid redundant connections, each host maintains a cache of hosts that it has spoken with recently. Anti-entropy is not re-initiated with remote hosts that have been contacted within a configurable time period.

During anti-entropy, the two hosts exchange their summary vectors to determine which messages stored remotely have not been seen by the local host. In turn, each host then requests copies of messages that it has not yet seen. The receiving host maintains total autonomy in deciding whether it will accept a message. For example, it may determine that it is unwilling to carry messages larger than a given size or destined for certain hosts. Epidemic routing associates a unique message identifier, a hop count, and an optional ack (acknowledge) request with each message. The hop count field determines the maximum number of epidemic exchanges that a particular message is subject to. As epidemic routing can be regarded as a temporary connection-based flooding, the hop count field can be regarded as the TTL of each message and therefore control the scale of the flooding.

Message ferrying, as illustrated in Fig. 9.3(b), is a proactive mobility-assisted approach, which utilizes a set of special mobile nodes called message ferries (or ferries for short) to provide communication services for nodes in the network. Similar to their real-life analog, message ferries move around the deployment area and take responsibility for carrying data between nodes. The main idea

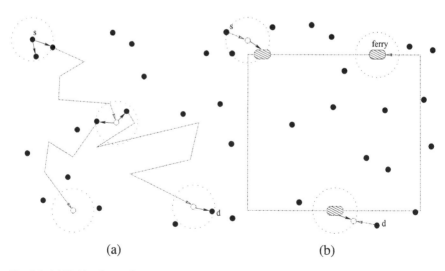

**Fig. 9.3** (a) Epidemic routing, (b) Ferry-based routing

behind the message ferrying approach is to introduce non-randomness in the movement of nodes and exploit such non-randomness to help deliver data. Message ferrying can be used effectively in a variety of applications including battlefields, disaster relief, wide area?sensing, non-interactive Internet access, and anonymous communication. For example, in an earthquake disaster scenario, unmanned aerial vehicles or ground vehicles that are equipped with large-storage and short-range radios can be used as message ferries to gather and carry data among disconnected areas. This enables rescue participants and victims to use available devices such as cell phones, PDAs, or smart tags for communication.

Two variations of the message ferrying schemes, depending on whether ferries or nodes initiate non-random proactive movement, are proposed. In the node-initiated message ferrying scheme, ferries move around the deployed area according to known routes and communicate with the other nodes they meet. With the knowledge of ferry routes, nodes periodically move close to a ferry and communicate with the ferry. In the ferry-initiated message ferrying scheme, ferries move proactively to meet nodes. When a node wants to send packets to other nodes or receive packets, it generates a service request and transmits it to a chosen ferry using a long-range radio. Upon reception of a service request, the ferry will adjust its trajectory to meet up with the node and exchange packets using short-range radios. In both schemes, nodes can ?communicate with distant nodes that are out of range by using ferries as relays.

The message ferrying design is distinguished from other epidemic routing-like mechanisms by its explicit exploitation of non-random node mobility and the use of message ferries, which improves data delivery and energy efficiency. By using ferries as relays, routing is efficient without the energy cost and the network load burden involved in other mobility-assisted schemes that use flooding.

### 9.3.3.3  Why Mobility Helps: Enlarge Routing Probability

The reason that mobility can help to increase routing capacity in these scenarios is that the dynamic connection is now considered to be the base of routing. In traditional networks, the base of routing is a permanent link. When a link in the selected path is broken, the routing process should restart and find a new path. However, in ad hoc networks, such permanent link-based paths may never occur or persist. The ad hoc network is highly dynamic, which differentiates it?from the traditional network. Network partitions may exist all the time. Permanent link-based routing schemes seem especially unsuitable in these dynamic environments. Temporary connection-based routing schemes can be used to solve this problem.

Mobility, whether controlled or uncontrolled, can always increase the probability that two nodes meet with each other. If we use a logic link between two nodes to represent that they meet each other at a particular point in time, the nodes' movement actually largely increases the number of links in this dynamic topology graph. With these additional links, the chance that there is a logical path between two nodes largely increases. Therefore, the routing probability increases because of the mobility.

## 9.3.4 Mobility Increases Network Capacity

The capacity of ad hoc wireless networks is constrained by the mutual interference of concurrent transmissions between nodes. These nodes are assumed to be mobile. We examine the per-session throughput for applications with loose?-delay constraints, such that the topology changes over the time-scale of packet delivery. Under this assumption, the per-user throughput can increase dramatically when nodes are mobile rather than fixed. This improvement can be achieved by exploiting a form of multi-user diversity via packet relaying [17].

### 9.3.4.1 Main Target: Increase Throughput with Loose Delay Constraints

There is some theoretical work on the capacity of mobile ad hoc networks. In a seminal paper, Gupta and Kumar [19] study a model of ad hoc networks with fixed nodes and show that when the number of nodes per unit area $n$ increases, the per-node throughput decreases. Grossglauser and Tse [17] show that with?-loose delay constraints, node mobility can dramatically improve *network capacity*. They prove that the per-node throughput can be kept constant as the number of nodes per unit area increases. The improvement of throughput comes at the price of increased delay.

### 9.3.4.2 Mobility Scheme: Random Movement

It is shown in Gupta's paper that the average available throughput per node decreases as the square root of the number of nodes $n$, in a static ad hoc network. Equivalently, the total network capacity can be increased to an amount equal to the value of the square root of $n$, at most. This result holds quite generally. In particular, it holds irrespective of the network topology, power control policy, or any transmission scheduling strategy. Given this limitation on the achievable throughput, a natural question that arises is whether the average throughput available per node can be increased. There are two approaches discussed in literature.

1) Add relay-only nodes in the network: This increases the total network capacity, thus increasing the share available to each sender. However, a major drawback of this scheme is that the required number of relay nodes is huge. For example, in a network with 100 senders, at least 4476 relay nodes are needed to increase the capacity fivefold.
2) Add mobility: In a network where nodes move randomly in a circular disk?such that their steady-state distribution is uniform, Grossglauser and Tse [17] showed that it is possible for each sender–receiver pair to obtain a constant fraction of the total available bandwidth. This constant remains independent of the number of sender–receiver pairs.

Various mobility models have been considered in literature to evaluate the effect of the node mobility on the performance of algorithms and protocols. The most widely used of these is probably the random waypoint model. Other models include Random Gauss-Markov and Fluid flow models.

### 9.3.4.3 Why Mobility Helps: Short-Range Transmission

The mechanisms in this category aim to achieve close-to optimal capacity; some of them also try to keep the delay small. These algorithms exploit the patterns in the mobility of nodes to provide guarantees on the delay. Moreover, the throughput achieved by the algorithm is only a poly-logarithmic factor off from the optimal.

Intuitively, if a node s transmits a message to some node at a distance, $d$, then due to the nature of wireless transmission, this causes interference to all the nodes within a distance of approximately $d$ from s. Hence, if the average distance of transmission is about $d$, then at most $n/d^2$ users can transmit simultaneously.

This forms the basis of the result of Gupta et al. [5]. If a node transmits a packet to another node $d$ steps away, then in a disk topology the number of hops between source to destination will be $\sqrt{n}/d$ on the average. This implies that the total throughput can be at most $(n/d^2)/(\sqrt{n}/d) = \sqrt{n}/d$. Thus, it helps to have short-range transmissions (i.e., $d = 1$) and hence the total capacity can increase an amount equal to at most $\sqrt{n}$.

Obtaining an $O(1)$ average throughput per node is a very stringent requirement and it implies several things. First, this means that each node must be sending packets to its destination for a constant fraction of time. Second, each packet traveling from the source to the destination must involve at most a constant number of relays. The idea of Grossglauser et al. [17] is that each node hands over a packet to its nearby mobile node at all times. When the mobile node is close to the destination node, it hands over the packet to the destination. Note that this does not provide any guarantees on how long the packet will take to reach the destination.

Some restrictions in the mobility model are needed to realize the request for delay boundary: (1) providing good delay guarantees the need to assume that the position of the destination is fixed. Indeed, if the destination is a mobile node, it will be impossible to provide any guarantees on the delay. (2) Second, to obtain a constant throughput per sender, senders need to be able to transmit most of the time. This requires that the number of mobile nodes must be at least $O(n)$, since otherwise the throughput is bounded by the number of static nodes $O(n)$. (3) The number of relays per packet should not be too large. To do this, we will need to exploit the patterns in the mobility of nodes. (4) Finally, to ensure a small delay, we must ensure that a packet does not stray along the path. This requires new ideas and we do not know of any previous work that considers these issues. Note that, at any time, a relay node will have several packets corresponding to various destinations. However, when it meets another relay node along its way, it can hand over very few of these packets, since the duration during which they are nearest neighbors (hence are in communicating range) is quite small. Their algorithm exploits the patterns in the mobility of nodes to provide guarantees on the delay. Moreover, the throughput achieved by the algorithm is only a poly-logarithmic factor off from the optimal.

### 9.3.5 Mobility Helps Security

Security and mobility seem to be at odds with each other. *Security* is usually enforced by a static, central authority that is generally in charge of securing the system under consideration, be it a communication network, an operating system, or the access system to the vault of a bank. In this case, because users are static as well, their locations are predictable, they are more likely to be available, and the system can more easily perform appropriate controls.

However, this intuition can be misleading: mobility, far from being a hurdle, can be useful to establish the security associations between any two mobile nodes of a given network [13, 23, 24, 26]. The idea that mobility can help security is extremely straightforward, as it simply mimics human behavior: if people want to communicate securely, they just get close to each other in order to exchange information and to establish (or reinforce) mutual credentials. In spite of its simplicity, this idea is very powerful, as it can be applied to virtually any mobile ad hoc network at any layer (from the MAC up to the application layer).

#### 9.3.5.1 Main Target: Building Security Association

As the ad hoc networks are self-organized, there is no infrastructure (hence no PKI), no central authority, no centralized trusted third party, no central server, and no secret share dealer, even in the initialization phase. Each node is able to?generate cryptographic keys, to check signatures and, more generally, to accomplish any task required to secure its communications (including to agree on cryptographic protocols with other nodes).

A *Security Association* means two nodes have verified each other's identities and set up an association. If a user i can relate the name (or the face) of another user j to his (j's) public key, we will say that there is a one-way security association from i to j. Two one-way security associations between i and j (one in each direction) constitute a two-way security association between i?and j. A (two-way) security association between two nodes i and j can be represented by the triplet $(u_i, k_i, a_i)$ at the side of j and the triplet $(u_j, k_j, a_j)$ at the side of i, where $u_i$ and $u_j$ are the names of the users that are associated with nodes i and j; $k_i$ and $k_j$ are the public keys of nodes i and j; and $a_i$ and $a_j$ are the node addresses of i and j, respectively.

Given a security association between nodes i and j, they can verify that the node addresses match the public keys, and they can set up a secure communication channel between themselves, which protects the integrity and confidentiality of the exchanged messages. In fact, for efficiency reasons, i and j may want to use symmetric key cryptography for the protection of their messages; in this case, the symmetric keys are established using the public keys in the security association.

### 9.3.5.2 Mobility Scheme: Uncontrolled

When they meet, users are naturally given the possibility to visually identify each other. The decision to set up a security association between two nodes is based on this physical encounter. Assume that each device is equipped with a short-range connectivity system (e.g., infrared or wire). A channel established by this mechanism is called a secure side channel. A secure side channel can only be point to point and works only when the nodes are within a secure range of each other. The secure side channel is used to set up security associations between nodes by exchanging cryptographic material.

The users are given the opportunity to associate a unique identifier to the established security association. This operation is very similar to the exchange of business cards; in fact, it can even be transparently combined with the exchange of electronic business cards. If a user wants to establish a security association with a user-independent device (e.g., a printer), he/she will identify the device visually and bind its identity to the context in which the device operates.

Assume that an adversary can eavesdrop on all radio links and can manipulate messages in all kinds of ways. Therefore, having two nodes verify each other and build a security association over radio links is insecure. However, it is much harder for the adversary to modify the messages transmitted over the secure side channel. Note that we do not require the secure side channel to protect the confidentiality of exchanged information.

The basic mechanism for setting up security associations relies on the physical encounters of users and the activation of the secure side channel. However, in order to expedite the process, we assume that nodes can also rely on friends. Two?nodes, i and j, are said to be friends if (1) they trust each other to always provide correct information about themselves and about other nodes they have previously encountered, and (2) they have already established a security associa-tion between each other (typically, they know each others' public keys).

When nodes randomly move around, they may enter the secure communication range of other nodes. In this case, they will set up the secure communication channel, and try to find if they have at least one common trusted friend. If so, they will authenticate each other by using the secret from the commonly trusted friend and build up security association. Random walk and restricted random waypoint have been selected to depict the move pattern of the nodes.

### 9.3.5.3 Why Mobility Helps: Encounters and Help from Common Friends

The idea underlining the above solution is extremely straightforward, as it simply mimics human behavior: if people want to communicate securely, they just get close to each other in order to exchange information and to establish (or?reinforce) mutual credentials. In spite of its simplicity, this idea is very powerful, as it can be applied to virtually any mobile ad hoc network at any layer (from the MAC up to the application layer). It makes it possible to provide security either without any kind of central authority or with an authority the role of which is limited to the initial delivery of certificates. There are two important factors that makes mobility useful in security association setup.

The first is the secure side channel. If two nodes are far away from each other, their communication may travel many intermediate nodes. If they did not set up a security association, it would be hard for them to verify each other, as the intermediate nodes can launch attacks and a third party may eavesdrop on their communication. However, a privileged side channel may be extremely useful. Nodes can easily establish security associations when they are in the vicinity of each other. The location-limited channels are assumed to be secure against active attacks.

Second, a common trusted friend is used as the introducer between two newly encountered nodes. These two nodes will verify each other by using the secret given by their common friend. Therefore, they can authenticate each other and set up security association.

After explaining these two factors, the effect of mobility on security becomes clear. Nodes' mobility creates more chances that two nodes meet with each other. When close enough, these two nodes can set up a secure side channel and authenticate each other if they have a common trusted friend. A great chance of encounters leads to lower convergence times.

## 9.3.6 Mobility Enlarges Node Coverage

Many works on the coverage of mobile node networks focus on algorithms to reposition nodes in order to achieve a static configuration with an enlarged covered area. In [3], the authors study the dynamic aspects of the coverage of a mobile node network, which depends on the process of node movement. As time goes by, a position is more likely to be covered; targets that might never be?detected in a stationary node network can now be detected by moving nodes. The main metrics to measure *node coverage* could be the area coverage at specific time instants and during time intervals, as well as the time it takes to detect a randomly located stationary target. Exploiting mobility, both metrics can be improved.

### 9.3.6.1 Main Target: Increase Node Coverage and Detection Rate

Coverage issues are mainly discussed in the scenario of wireless sensor networks. Coverage can be considered as the measure of quality of service of a sensor network. For example, how well the network can observe a given area and what the chances are that a fire starting in a specific location will be detected in a given time frame. The coverage of a mobile sensor network now depends not only on the initial network configurations but also on the mobility behavior of the sensors.

There are algorithms to reposition sensors in desired positions in order to enhance network coverage. Three types of coverage are the main focus of many papers:

1) Blanket coverage: to achieve a static arrangement of elements that maximizes the detection rate of targets appearing within the coverage area. Sensors deploy themselves so that the resulting configuration maximizes the net sensor coverage of the network with the constraint that each node has at least $k$ neighbors.
2) Barrier coverage: the objective is to achieve a static arrangement of elements, which minimizes the probability of undetected penetration of the barrier.
3) Sweep coverage: the objective is to move a number of elements across a coverage area in a manner that addresses a specified balance between ?maximizing the number of detections per time and minimizing the number of missed detections per area.

### 9.3.6.2 Why Mobility Helps: Dynamic Coverage

Among these three kinds of coverage, many research papers discussed the first one. Most of this work focuses on algorithms to reposition sensors in desired positions in order to enhance network coverage. More specifically, these ?proposed algorithms strive to spread sensors in the field so as to maximize the covered area. The main differences among these works are how exactly the desired positions of sensors are computed. Although the algorithms can adapt to changing environments and recompute the sensor locations accordingly, sensor mobility is exploited essentially to obtain a new stationary configuration that improves coverage after the sensors move to their desired locations.

There are also some papers that try to identify and characterize the dynamic aspects of network coverage that depend on the movement of sensors, that is, the coverage provided by the sensor movement.

So the above second and third category of dynamic coverage have also?attracted much research interest. Some metrics to measure the dynamic coverage have also been raised, i.e., area coverage over a time interval and detection time. These algorithms focus on the coverage resulting from the continuous movement of sensors. This coverage is not available if the sensors stop moving.

### 9.3.6.3 Why Mobility Helps: After-Deployment Movement

If we observe the moving sceneries, we will find that previously uncovered areas become covered as sensors move through them and covered areas become uncovered as sensors move away. As a result, the locations covered by sensors change over time, and a greater area will be covered over time than in the case where sensors are stationary. Also, a location is now not always covered. It alternates between being covered and not being covered.

Second, note that an initially undetected intruder or event will never be detected in a stationary sensor network if the intruder remains stationary or moves along an uncovered path. In a mobile sensor network, an intruder is more likely to be detected as the moving sensors patrol the field. Thus, sensor mobility provides a time-varying coverage not available in a sensor network with stationary sensors. This can significantly improve the intrusion detection capability of a sensor network.

## 9.3.7 Mobility Assists Information Dissemination

Ad-hoc wireless networks can get partitioned. Therefore, we need to develop schemes that helps to disseminate information. One possible approach for information dissemination in such networks is to replicate information at multiple nodes acting as repositories, and utilize these nodes' movement to disseminate information.

### 9.3.7.1 Main Target: Information Dissemination

Existing solutions for the dissemination of information in static or cellular networks may not be applicable to ad hoc networks. First, these solutions usually do not consider changing topology of the network backbone that contains the information servers. Second, the possibility of partitioning means that some nodes may not be able to communicate updates to other nodes and/or may be unable to retrieve the latest information on queries. Third, earlier solutions that do consider network partitioning approach the problems from the direction of replica consistency in distributed databases. While the problems are similar, several instances of information dissemination problem in ad-hoc networks are simpler in nature. Employing the sophisticated replica consistency solutions would result in reduced availability of data, while also incurring unacceptably high communication overheads. Fourth, unlike traditional distributed database systems where the timing of updates is ?independent of network topology, location sensitive information should be updated as a function of network topology in ad hoc networks.

Unlike the approaches for routing protocols, route or delay optimization is?not the primary design goal of our architecture, reducing communication overhead is. This is especially true in energy-constrained environments, which includes many sensor networks, particularly for one-shot queries, where no path is established to be used for further communication. We design our protocol to be scalable, self-configuring, and highly adaptive to mobility.

### 9.3.7.2  Mobility Scheme: Zone and Contact

It is not appropriate to expect an exact location query semantic for queries in mobile sensor networks. Thus, the notion of queries in mobile sensor networks needs to be further developed to clarify the semantic of queries and to specify validity and imprecision of queries. If the nodes' movement is random, it is important to determine answers to the following questions: When should the information be updated? Where the updates should be sent? Which nodes should be queried for information?

One way to utilize node mobility to help information dissemination is to divide the network into zones. Each zone will elect a selector. Nodes on the boundary of the zone may move out of the zone and therefore be selected as the contacts. The selector will keep in touch with these contacts, thereby enlarging the information dissemination area.

Zone establishment is performed by each node independently by sending link state messages $R$ hops away. $R$ is called the zone radius. Contacts are defined as?short cuts to the outside world (i.e., out-of-zone), which provide useful information when needed. To reduce the discovery delay, these contacts are established in anticipation of queries. With network dynamics and mobility, it?may be quite expensive to establish and maintain routes to all far away contacts. Instead, candidate contacts are established from within the zone. As these candidates move out of the zone, they become contacts and can be used in the query process, thus taking advantage of mobility.

Not all nodes in the network need to establish contacts. In fact, if all nodes?establish contacts, this may constitute a large overhead for large-scale networks. Only a small subset of nodes, called selectors, independently chooses to establish contacts. Selectors are not fixed, but are dynamic and may be chosen (in a distributed manner, without extra overhead) in a way that achieves load balancing. A selector keeps a list of (a subset of) its zone borders, and chooses its contacts from those border nodes that move out of the zone. This choice takes advantage of zone information in an attempt to reduce overlap between contact zones. Once the contact is out-of-zone, a simple contact ?discovery mechanism is invoked to keep track of it.

Routes maintained to contacts are loose (perhaps suboptimal) routes. Since each node knows about neighboring nodes up to $R$ hops away, the contact route (that has initial length of $R$ hops) may be extended up to $R^2$ hops, without any extra overhead as the nodes en-route move away. Once a contact (or one of its en-route nodes) moves too far away, then the contact is dropped and another is chosen.

Once these contacts are chosen, they may be used in the query process for resource discovery. A querying node sends messages to its contacts, and their contacts, and their contacts' contacts, and so on, up to the maximum contact level or until the object is found. Mechanisms to prevent loops and re-visits of already-searched zones are also introduced.

### 9.3.7.3 Why Mobility Helps: Query and Information on the Fly

Node mobility results in dynamic networks with frequent topology transients. We can make use of mobility to distribute messages to another island of nodes. We can exploit different types of contacts, such as scheduled, opportunistic, and predicted, to disseminate information. Scheduled contacts can exist, for instance, between a base station somewhere on earth and a low earth-orbiting relay satellite. Opportunistic contacts are created simply by the presence of two entities at the same place, in a meeting that was neither scheduled nor predicted. Finally, predicted contacts are also not scheduled, but predictions of their existence can be made by analyzing previous observations.

We can also take advantage of mobility to increase the efficiency of query resolution, which provides an efficient query resolution for one-shot, frequent, simple queries. Here, we can introduce the concept of contacts that act as short cuts to reduce the degrees of separation between the sources of the query and the targeted objects. We can reduce communication overhead by focusing on contact selection and maintenance.

## 9.3.8 Mobility Reduces Uncertainty

*Uncertainty* increases the transaction cost and decreases the acceptance of communication and cooperation [11]. Our objective is to reduce the trustor's perceived uncertainty so that transaction cost is lowered and a long-term exchange relationship is sustained. One key way to efficiently reduce uncertainty is to exploit one important property of MANETs: mobility. Node movement can increase the scope of direct interaction and recommendation propagation, hence speeding-up trust convergence. We study this effect under different mobility models and analyze several factors that will strongly influence the convergence speed and cost. We present a detailed design of a two-level Mobility-Assisted Uncertainty Reduction Scheme (MAURS). It exploits configurable level partition and ?movement schemes to provide a range of trade-offs between convergence time, cost, and uncertainty level. MAURS offers flexibility for users to achieve their application objectives.

### 9.3.8.1 Main Target: Reducing Uncertainty while Building Trust

Uncertainty is an important factor in trust evaluation. The highly dynamic environment and self-organizing nature of wireless ad hoc networks makes uncertainty unavoidable. However, high uncertainty is unfavorable as it may lead to unstable or incorrect decisions.

Neighbor monitoring is a unique mechanism that helps to evaluate the trustworthiness of a node. Exploiting the promiscuous nature of broadcast communication in wireless media, nodes are able to track the outgoing packets of their one-hop neighbors through passive observation. An observation is classified as either a success or a failure and, accordingly, the corresponding variable, $\alpha$ for successful forwarding and $\beta$ for failed forwarding, is incremented. Each node can then estimate its neighbor's reliability based on its accumulated observations using Bayesian inference. Bayesian inference is a statistical model in?which evidence or observations are used to update or to newly infer the probability that a hypothesis is true. The beta distribution, Beta($\alpha$, $\beta$), is used in the Bayesian inference. The beta distribution is a family of continuous probability distributions defined on [0, 1] differing in the values of their two nonnegative shape parameters, $\alpha$ and $\beta$. The examples of beta distributions are illustrated in Fig. 9.4.

We introduce the concept of uncertainty and use a triplet to represent the node's opinion toward reliability: $(b, d, u) \in [0, 1]^3$ and $b + d + u = 1$ where $b$, $d$, and $u$ designate belief, disbelief, and uncertainty, respectively. The values of $(b, d, u)$ will be derived from Beta($\alpha$, $\beta$) using the method below.

Two important attributes can be observed from the general understanding of?the concept of uncertainty. First, when there is more evidence, which implies $(\alpha, \beta)$ is higher in our reliability estimation model, it consequently lowers uncertainty $u$. Second, when the evidence for success or failure dominates, there will be less uncertainty when compared to the situation in which there is?equal evidence for both success and failure. After examining the major

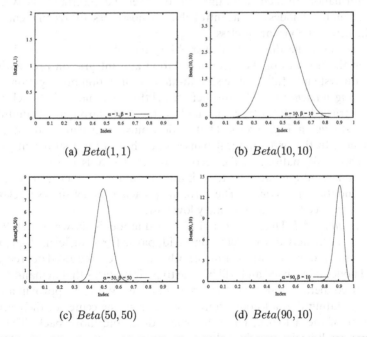

(a) $Beta(1, 1)$                          (b) $Beta(10, 10)$

(c) $Beta(50, 50)$                          (d) $Beta(90, 10)$

**Fig. 9.4** The Beta distributions. (**a**) $Beta(1,\psi1)$ ,$\psi$ (**b**) $Beta(10,\psi10)$ ,$\psi$ (**c**) $Beta(50,\psi50)$ ,$\psi$ (**d**) $Beta(90,\psi10)$

statistical metrics of the beta distribution, we find that the normalized variance satisfies these observations. Therefore, we define $u$ as follows:

$$u = \frac{12 \cdot a \cdot \beta}{(\alpha + \beta) \cdot (\alpha + \beta + 1)}.$$ (9.1)

In social life, if people want to raise their confidence in the evaluation of someone, they just get closer to that person and create chances for direct contact, or take the recommendations from someone they trust who knows the subject better. In MANETs, mobility increases the chance that two ?separated nodes meet and come into directly contact. It also allows each node?to have more evidence to verify future recommendation. Intuitively, we consider mobility to be a good method for reducing uncertainty.

### 9.3.8.2 Mobility Scheme: Hierarchical Movement

First, we analyze the effect of mobility based on the random waypoint model. Using this model, nodes will have a new neighborhood during each pause time. A?node can contact and observe its new neighbors directly. The results of these direct contacts increase the $\alpha$ or $\beta$ in both nodes' first-hand opinion, therefore reducing uncertainty. However, the randomness also restricts the use of second-hand information. In each pause time, the disbelief and uncertainty between the newly encountered nodes are uncontrollable. In most cases, the recommendations from the new neighbors are useless.

We now examine the controlled mobility models, which can be designed based on the features of the recommendation and integration process in the reputation system to fully utilize second-hand information propagation.

*Traveling preacher model:* Another straightforward model is to select one common trusted node to travel around all the grids through a Hamiltonian path, as shown in Fig. 9.5(b). That node's movement can be divided into two?rounds. In the first round, it pauses in each grid for a sufficient time to collect trust information. In the second round, it travels to each grid again to?disseminate all the gathered trust information about other grids using the recommendation mechanism. The traveling preacher model shows a relatively long convergence time, but extremely low cost.

*Town hall model:* The first straightforward model is shown in Fig. 9.5(a). All nodes in the network travel to one grid, pause for a sufficient time, build up trust, and reduce the uncertainty of other nodes to a required degree. After that, all nodes move back and will be able to perform tasks that demand remote nodes to cooperate and have trust requirements. We can approximate this model by stating that all nodes start moving from the center of their grid, to the center of the network, pause for some time, and move back. The town hall?model will lead to a relatively short convergence time with an extremely high cost.

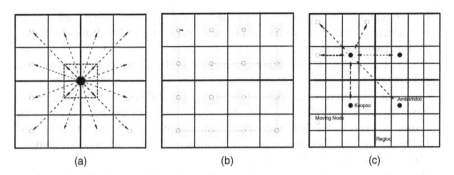

**Fig. 9.5** (a) Town hall model, (b) Traveling preacher model, (c) MAURS

*Mobility-assisted uncertainty reduction scheme*: When the requirement is a short convergence time to quickly start a trust-based application, or a controllable cost, the above two mobility models will offer extreme options. However, these two methods are not flexible enough and we lack a way to find a trade-off between convergence time and cost to satisfy different application objectives. A two-level controlled mobility model called MAURS could be used here. In MAURS, we divide the whole network into several regions, allowing each region to contain a specified number of grids, and choose mobility models for intra- and inter-region movement. MAURS combines the advantages of the above two models and offers more options for MANET implementation. There are three phases in the MAURS: moving node election; region partition; and trajectory planning.

### 9.3.8.3 Why Mobility Helps: Recommendation from Trusted Moving Nodes

Node movement increases the chance for potential contactors to gather more trust information and evidence, thus enlarging the scope of reputation-qualified candidate nodes for future tasks. The controlled heretical mobility model offers many ways to adjust the convergence time and total cost related to a specific certainty goal. Under a certain requirement $U_{max}$, the user can achieve a desired convergence time, cost, and general trust decay.

## 9.4  Directions for Future Research

Most research studies on mobility-assisted schemes assume simplistic mobility models, such as the random walk (in general, i.i.d. models), a priori knowledge of future mobility, or using controlled mobility models. These assumptions provide scenarios amenable to mathematical analysis, which provides good insight to system performance. However, these simple mobility models do not address the complexity of node mobility in real-life settings. The differences between these models and real-life models can be summed up as follows:

1. Heterogeneous or homogenous: In the existing mobility models, all mobile nodes behave statistically identical to each other. In real life, nodes in an ad-?hoc network are usually heterogeneous. They may have different moving ability, e.g., a vehicle may have a moving speed that is not attainable by a pedestrian.
2. Location/time preference: Most of the existing mobility models assume nodes have no location preference, and that their behaviors do not change with respect to time. In real life, nodes do have location preferences that are related to their attributes. The behavior also changes over time and usually a pattern can be formed for this behavior change.
3. Boundary: In the existing mobility models, the inter-meeting time (the time between two independent mobile nodes meet with each other twice) has been?assumed to be exponentially distributed so as to make their analysis tractable. However, recent studies on real mobility tracing data are contradictory to this claim. This is because the existing mobility models assume there is a finite boundary for the possible moving area.

As the underlying mobility model is an important factor in the performance of all of the above and future mobility-assisted schemes, there is an increasing need for mobility models that capture the realistic mobility characteristics and remain mathematically manageable. For these new and usually more complex mobility models, researchers also need to re-analyze the performance of those mobility-assisted schemes and make some changes in the detailed schemes.

## 9.5 Conclusion

Before these recent works, one might think that mobility has only a negative impact on the behavior of wireless networks. However, these works have shown that this is not the case. Mobility shows many positive impacts in ad hoc networks. Protocols that take mobility into consideration could utilize these positive impacts to increase routing capability and network capacity, improve security and reduce uncertainty, enlarge node coverage, and assist with information dissemination.

However, more research efforts in these aspects are needed, and traditional networking protocols, such as the routing schemes and security protocols, need to be revised to fully utilize these positive impacts. New and more realistic mobility models also influence the design of those protocols, which aim to utilize utility.

**Acknowledgments** This work was supported in part by NSF grants CCR 0329741, CNS 0422762, CNS 0434533, CNS 0531410, and CNS 0626240. Contact information: Jie Wu, (561)297-3855, Fax: (561)297-2800, jie@cse.fau.edu.

## Terminologies

*Mobility management:* *Mobility management* describes the design requirement that protocols need to take mobility into consideration. This is because mobility is an inherent characteristic of MANETs.

*Mobile ad hoc networks:* A *Mobile Ad Hoc Network* (MANET) is a kind of wireless ad hoc network, and is a self-configuring network of mobile routers (and associated hosts) connected by wireless links – the union of which forms an arbitrary topology.

*Routing:* Routing is the process of selecting paths in computer networking along which to send data or physical traffic.

*Network capacity:* Network capacity is one of the measurements for a network. *Network capacity* is defined as the per-session throughput for applications with loose delay constraints in this chapter.

*Security association:* A *Security association* means two nodes have verified the identity of each other and set up an association.

*Node coverage:* Node coverage is defined by the area coverage at specific time instants and during time intervals, as well as the time it takes to detect a randomly located stationary target.

*Information dissemination:* One-way communication flow providing information.

*Uncertainty:* A state of having limited knowledge where it is impossible to exactly describe existing trust state or predict result.

*Trajectory planning:* Design of the mobile nodes' mobility model to achieve certain desirable properties.

*Store-carry-forward:* A mobile node first stores the routing message from the source, carries it while in motion, and then forwards it to an intermediate node or the destination. This model supports routing in an unconnected graph by virtual connectivity through node movement.

## Questions

1. In a simulation, each node randomly chooses an arbitrary direction and speed from [1.0, 10.0] at the beginning of every time slot. On moving at the selected direction and speed for 10.0 s, the node will pause at the destination before the end of the current time slot. What kind of mobility model are we using in this simulation?
2. Why we need to develop different mobility models?
3. Use your own words to explain the impact of node buffer size in epidemic routing and message ferrying. Why delivery ratio and message delay will be affected by node buffer size?
4. Why mobility is treated as a threat for traditional routing schemes?

5. Explain why mobility reduces interference and increase network capacity? What is the price for the capacity increase?
6. Explain Store-carry-forward model, and why should we use such model?
7. What is the difference between static and dynamic coverage? How can mobility assist the node coverage?
8. Why do we want to reduce uncertainty? Why mobility helps?
9. Why does the hierarchical movement scheme outperform other one-level schemes?
10. Summarize the aspects that mobility can help. Besides the examples in the chapter, find one aspect that you think mobility can also help.

# References

1. A. Agarwal and P. R. Kumar. Capacity bounds for ad-hoc and hybrid wireless networks. In *Proc. of ACM SIGCOMM*, 2004.
2. A. Vahdat and D. Becker, Epidemic routing for partially-connected ad hoc networks, Technical Report, Duke University, 2002.
3. B. Liu, P. Brass, O. Dousse, P. Nain, and D. Towsley. Mobility improves coverage of sensor networks. In *Proc. of ACM MobiHoc*, 2005.
4. C. E. Perkins and E. M. Royer. Ad-hoc on-demand distance vector routing. In *Proc. of IEEE WMCSA*, 1999.
5. D. Austin, W. Bowen, and J. McMillan. Intraspecific variation in movement patterns: modeling individual behaviour in a large marine predator. In *Proc. of ACM SIGCOMM*, 2004.
6. D. B. Johnson. Routing in ad hoc networks of mobile hosts. In *Proc. of the Workshop on Mobile Computing Systems and Applications*, 1994.
7. D. Cooper, P. Ezhilchelvan, I. Mitrani, and E. Vollset. Optimization of encounter gossip propagation in mobile ad-hoc networks. In *Proc. of IEEE MASCOTS*, 2005.
8. DTN research group. In http://www.dtnrg.org/.
9. F. Bai, N. Sadagopan, and A. Helmy. Important: a framework to systematically analyze the impact of mobility on performance of routing protocols for ad hoc networks. In *Proc. of IEEE INFOCOM*, 2003.
10. F. Bai, N. Sadagopan, and A. Helmy. Brics: a building-block approach for analyzing routing protocols in ad hoc networks – a case study of reactive routing protocols. In *Proc. of IEEE ICC*, 2004.
11. F. Li and J. Wu. Mobility reduces uncertainty in manets. In *Proc. of IEEE INFOCOM*, 2007.
12. Garmin website. In http://www.garmin.com/.
13. J. Chiang and Y. Hu. Extended abstract: cross-layer jamming detection in wireless broadcast networks. In *Proc. of ACM MobiCom*, 2007.
14. J. Haas. A new routing protocol for the reconfigurable wireless networks. In Proc. of *IEEE 6th International Conference on Universal Personal Communications,* 1997.
15. J. Wu, S. Yang, and F. Dai. Logarithmic store-carry-forward routing in mobile Ad Hoc networks. *IEEE Transactions on Parallel and Distributed Systems*, 18(6):735–748, 2007.
16. M. Grossglauser and M. Vetterli. Locating nodes with EASE: Mobility diffusion of last encounters in Ad Hoc networks, In *Proc. of IEEE Infocom*, 2003.
17. M. Grossglauser and D. Tse. Mobility increases the capacity of ad-hoc wireless networks. In *Proc. of IEEE INFOCOM*, 2001.
18. Mit trace. In http://nms.lcs.mit.edu/mbalazin/wireless/.

19. P. Gupta and P. Kumar. The capacity of wireless networks. *IEEE Transactions on Information Theory*, 46(2):388–404, 2000.
20. Q. Li and D. Rus, Sending messages to mobile users in disconnected ad-hoc wireless networks. In *Proc. ACM MobiCom*, 2000.
21. R. Atkinson, C. Rhodes, D. Macdonald, and R. Anderson. Scale-free dynamics in the movement patterns of jackals. In *OIKIS*, 98:134–140, 2002.
22. S. Burleigh, A. Hooke, L. Torgerson, K. Fall, V. Cerf, B. Durst, K. Scott, and H. Weiss. Delay-tolerant networking: an approach to interplanetary internet. In *IEEE Communications Magazine*, 41:128–136, 2003.
23. S. Capkun, J. Hubaux, and L. Buttyán. Mobility helps security in ad hoc networks. In *Proc. of ACM MobiHoc*, 2003.
24. S. Capkun, M. Cagalj, and M. Srivastava. Securing localization with hidden and mobile base stations. In *Proc. of IEEE INFOCOM*, 2006.
25. T. Camp, J. Boleng, and V. Davies. A survey of mobility models for ad hoc network research. *Wireless Communications and Mobile Computing*, 2(5):483–502, 2002.
26. W. Zhang, H. Song, S. Zhu, and G. Cao. Least privilege and privilege deprivation: towards tolerating mobile sink compromises in wireless sensor networks. In *Proc. of ACM MobiHoc*, 2005.
27. W. Zhao, M. Ammar, and E. Zegura, A message ferrying approach for data delivery in sparse mobile ad hoc networks, In *Proc. ACM MobiHoc*, 2004.
28. W. Zhao, M. Ammar, and E. Zegura, Controlling the mobility of multiple data transport ferries in a delay-tolerant network, In *Proc. IEEE INFOCOM*, 2005.
29. X. Hong, M. Gerla, G. Pei, and C. Chiang. A group mobility model for ad hoc wireless networks. In *Proc. of ACM/IEEE MSWiM*, 1999.
30. Zebranet website. In http://www.princeton.edu/mrm/zebranet.html.

# Chapter 10
# Mobility Models for Ad Hoc Networks

Mihail L. Sichitiu

**Abstract** Network simulators emerged as the most common method of evaluating the performance of large and complex networking systems. However, for systems involving mobile nodes, the movement of the mobile nodes has a significant influence on the results of the simulation. Therefore, in the past decade, a significant amount of research was devoted to develop mobility models suitable for evaluating the performance of wireless networks. Existing mobility models vary widely in their realism, from completely artificial to very realistic as well as in their statistical properties. In this chapter we provide an overview of those mobility models and their most important properties.

## 10.1 Introduction

Computer networks are, arguably, some of the most complex systems ever designed. The Internet has become so large and complex that researchers publish surprising facts on regular basis. Generally speaking, there are four different methods for evaluating the performance of a networking system: analytical, simulation, emulation, and testbed experiments.

Analytical methods, however, are limited to very simple network configurations, or have to rely on abstraction models of varying accuracy. Therefore, they are of limited use for complex networks and protocols.

Network simulation relies on detailed replication of the networking protocols in a network simulator that is often implemented as an event-based simulator. Different simulators often trade-off speed for accuracy: the more detailed the models (especially at the lower layers), the slower the simulation. While using network simulators for performance evaluation is far from perfect, their

M.L. Sichitiu (✉)
Department of Electrical and Computer Engineering, Campus Box 7911, NC State University, Raleigh, NC, 27695, USA
e-mail: mlsichit@ncsu.edu

S. Misra et al. (eds.), *Guide to Wireless Ad Hoc Networks*,
Computer Communications and Networks, DOI 10.1007/978-1-84800-328-6_10,
© Springer-Verlag London Limited 2009

convenience for quickly exploring a large design space makes them, by far, the most common method of performance evaluation in the literature.

Testbeds are, by far, the most realistic method of evaluating the performance, as they practically use the protocol implementations and the hardware that is the same, or very similar, to the one used for the production networks. However, large testbeds are expensive to build and manage, and for very large (or highly mobile) networks, practically impossible to implement. Furthermore, the degrees of freedom in a testbed are significantly reduced by comparison with a network simulation (e.g., if the wireless cards in the testbed only have one transmission power, it would be impossible to test the effect of different transmission powers on the network's performance).

Network emulation is typically somewhere between simulations and testbeds in terms of convenience and accuracy. Some emulators use the output of the wireless cards as inputs and emulate the wireless transmissions between the nodes, creating the nodes the illusion that they move relative to each other although in reality the nodes are fixed. Others also emulate the wireless interfaces, allowing multiple instances of the same networking stack to run in parallel on the same physical hardware.

Practically, for all methods of performance evaluation (but especially for simulation and emulation), it is important to have good models for the offered traffic, networking protocols, wireless propagation (if the transmissions are wireless), and, if the network is mobile, network mobility. It was shown time and again that, for all the components mentioned above, unrealistic models result in unrealistic results. When using unrealistic models, the simulation results do not translate in real life when the system is eventually deployed.

While wireless propagation models and traffic models and their impact on the performance of network simulations [1–3] have been thoroughly studied, only recently researchers have considered the realism and influence of mobility model on network performance [4–7]. Results show that, indeed, the mobility model has a significant influence on the performance of the network. This is true for all wireless networks, but is especially true for ad hoc networks that rely on intermediate nodes for relying data between the source and the destination.

## 10.2 Classification

There is currently a large number of mobility models used in the literature. These models have different properties, each with its advantages and disadvantages. We consider the following properties of mobility models:

*Realism* is the degree of accuracy of the mobility model, with respect to the movements of mobile nodes in a real scenario. The higher the realism, the better the mobility models' ability to predicting the performance of the network system in a real scenario (assuming all other models are also accurate).

*Diversification* qualifies the ability of a model to diversify to a large number of different scenarios, including different types of mobile nodes (e.g., vehicles, pedestrians, zebras) and different types of environments (e.g., campus, conference, city).

*Complexity* is a measure of the computational resources required to produce the traces for the simulation. Just as more detailed wireless propagation models may significantly slow down a network simulation, an overly complicated mobility model may take a significant fraction of the total simulation time.

Figure 10.1 shows a loose classification of existing mobility models and their relative strengths. The classification in Fig. 10.1 is not absolute – in reality there is a continuum of models from stochastic to highly detailed models; however, the classification helps the discussion of the characteristics of each model.

*Stochastic* models rely primarily on random movements, without imposing any constraints on the movement of the nodes. Classical examples include the random waypoint, random direction (Section 10.3.1). By construction, the models are not tied to any particular scenario.

*Detailed* models, in contrast, are custom-built for a particular scenario. For example, for a detailed mobility model of students in a campus, the student's schedules can be considered, means of arriving on campus (bus, bike, car), location of parking lots and classroom, library, lunch in cafeteria, etc.

*Hybrid* models aim to balance the realism of detailed models with the diversification convenience of stochastic models. Depending on the particular target, the model can lean more on one side or the other of the two extremes. We subdivide hybrid models into group, obstacle, and trace-based mobility models.

*Real traces* are, as the name implies, collections of trajectories of real users in a particular scenario. CRAWDAD [8] is making several such collections available to researchers.

In a way, the classification of mobility models mirrors the classification of the methods for evaluating the performance of a network: the stochastic models

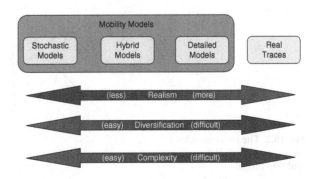

**Fig. 10.1** A loose classification of mobility models and their relative advantages and disadvantages

are similar to theoretical analysis in their simplicity and lack of realism, however being well understood and possible to analyze; testbeds and detailed models are both expensive and realistic, but lack diversification convenience; network simulators and hybrid models strike various middle-ground positions in terms of both realism and ease of diversification.

## 10.3 Mobility Models

In this section we will review several common mobility models used in the literature, as well as comment on their relative merits. For an excellent survey of stochastic mobility models, see [4]. We will start with a few stochastic mobility models and continue toward more realistic mobility models.

### *10.3.1 Stochastic Mobility Models*

In this section we focus on stochastic models that are very simple, but bear little or no resemblance to reality. On the up-side they are relatively easy to study and diversify. The models presented in this section are often encountered with various modifications in the literature (e.g., with no pause times, using the surface of a sphere or of a torus instead of a rectangle, bouncing off the walls or wrapping-up).

#### 10.3.1.1 Random Waypoint (RWP) Mobility Model

The random waypoint model (shown in Fig. 10.2) [9] is, by far, the most widely used model in the literature, in part due to its simplicity and convenience (it is readily available in practically all network simulators) and in part due to the lack of a clearly better alternative.

The RWP assumes a fixed number of nodes in a fixed size rectangle. The simulation starts with the nodes uniformly distributed in the rectangle. Each node chooses a random destination and chooses a random speed distributed

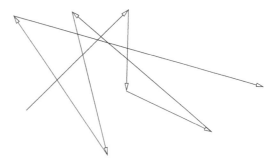

**Fig. 10.2** The movement of one node with an RWP mobility model

uniformly in the interval $[v_{min}, v_{max}]$. Once it arrives at the destination, it pauses for a random time uniformly distributed in $[P_{min}, P_{max}]$, then it chooses a new speed and destination and repeats the process.

It is hard to imagine a scenario for which RWP represents a realistic mobility model. Also, due to its simplicity, it is fairly easy to diversify – if one wants to change from student in a campus to taxis in a city, all it has to do is change the speed and pause time intervals. If a larger campus is needed, just increase the size of the bounding rectangle.

Different researchers use RWP with different choices of parameters, most often with $v_{min} = 0$ and $P_{min} = 0$, and, sometimes, even with $P_{max} = 0$, i.e., without any pause at the destination. Due to its simplicity and widespread usage, RWP is well studied. Over time, several shortcomings of RWP have surfaced.

Although the destination is chosen uniformly over the area of the deployment, the density of nodes is non-uniform. In particular, nodes are more likely to be closer to the center of the rectangle than to the edges [10–12]. Thus, initial distribution (uniform) and the stationary one (weighted in the center) do not match, thus creating a transient period at the beginning of the simulation. The brute-force solution is to wait until the stationary regime is achieved and discard the transition period. The problem is that it may take a long time to achieve the stationary regime. A better solution is to sample the initial distribution from the stationary regime instead of a uniform distribution [10–12].

If the minimum speed $v_{min} = 0$, in the course of the simulation, many nodes will, eventually, choose a very small minimum speed and proceed at this snail pace toward a distant destination [13]. The net effect is that the average speed of the nodes continuously decreases throughout the simulation (and does not have an average). The obvious fix is to limit the minimum speed $v_{min}$ at a reasonable value.

Finally, it was also shown [14] that because RWP uses a bounding rectangle, the inter-meeting times (i.e., time between two nodes will be in wireless range of each other) are decaying at least exponentially. The problem is that analysis of real traces reveals that, for real systems, the inter-meeting times have a power-law distribution (which decreases far slower than the exponential). This should not come as a surprise, as RWP is clearly not meant to be a realistic mobility model. However, what is surprising is that by removing RWP's boundary (or making it "sufficiently large"), the inter-meeting time becomes power-law (at least for the time-scales of interest), thus taking a first step toward making RWP more realistic.

### 10.3.1.2 Random Walk Mobility Model (RWM)

In a random walk (shown in Fig. 10.3), each node chooses a random direction (uniformly distributed in $[0, 2\pi]$) and a random speed (also uniformly distributed in $[v_{min}, v_{max}]$); it then moves for a time period (or over a fixed distance) with this speed, then it repeats its choice. This model is often referred to as Brownian motion, as it resembles the movement of particles suspended in a fluid.

**Fig. 10.3** The movement of
three nodes with an RWM
mobility model

In an equivalent view of this model, the world is divided into cells (e.g., squares) and at each step a node can jump into any of the neighboring cells (up to several steps away).

When the nodes reach the edges of the bounding rectangle, they can "bounce" (or "reflect") from the edges or "wrap-around", by assuming that the world is a torus. The movement can also take place on a sphere.

Similar to the RWP, RWM is highly unrealistic for most scenarios. Furthermore, in the long term, the nodes tend to stay close to their origin; thus the overall mobility is limited. Also, when bounded, RWM has the same exponential inter-meeting times as RWP (also due to the existence of the bounds [14]).

### 10.3.1.3 Random Direction Mobility Model (RDM)

In a RDM (shown in Fig. 10.4), a node chooses a direction and travels on in that direction with a random speed until it encounters an edge. It then chooses another direction and repeats the algorithm. The advantage of this strategy is that it results in a uniform stationary distribution in the rectangle. The

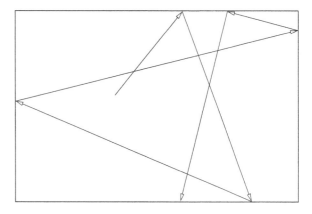

**Fig. 10.4** The movement of
a node with an RDM
mobility model

disadvantage is that it is just as unrealistic as RWP and RWM, with sudden changes in speed and direction.

### 10.3.1.4 Smooth Mobility Model (SM)

To avoid the unrealistic and sudden changes as well as the edge effects of RWP, RWM, and RDM, Haas proposed a smooth mobility model (shown in Fig. 10.5), where the mobile nodes only change the speed gradually and the world is a torus [15].

In SM, each node is characterized by a motion vector $(v, \theta)$, where $v$ is the speed of the node and $\theta$ is the direction. The position $(x,y)$ of a node and its motion vector are updated periodically (every $\Delta t$ seconds) as follows:

$$v(t + \Delta t) = \min[\max(v(t) + \Delta v, 0, V_{max})] \tag{10.1}$$

$$\theta(t + \Delta t) = \theta(t) + \Delta\theta \tag{10.2}$$

$$x(t + \Delta t) = x(t) + v(t)\cos(\theta(t)) \tag{10.3}$$

$$y(t + \Delta t) = y(t) + v(t)\sin(\theta(t)), \tag{10.4}$$

where $V_{max}$ is the maximum speed, (the minimum speed is zero), and $\Delta v$ and $\Delta\theta$ are random variables denoting the change of speed and direction at each step.

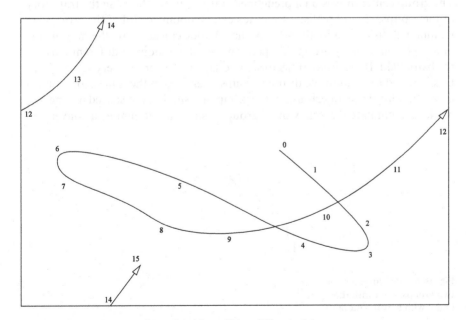

**Fig. 10.5** The movement of a node with an SM mobility model

The uniform intervals for $\Delta v$ and $\Delta \theta$ can be chosen relatively small to force a smooth trajectory of the mobile nodes.

## 10.3.2 Group Mobility Models

A first step toward more realistic models is acknowledging that humans (who are often the ones that carry the mobile nodes) as well as other animals (e.g., zebras, red wolfs) tend to operate in groups. Classical examples include military, search and rescue, and public safety, but also students or employees in a campus. The size and the movements of these groups (and within the group) vary from scenario to scenario, but several characteristics are common to all these scenarios: the nodes are split in several smaller groups, and each group acts seemingly independently of the other groups. Also, within each group, each user has its own liberty to move with respect to the center of the group or with respect to the other members of the group. In this respect, the group mobility models have two sub-models – the *group model* describing the movements of the groups and the *individual model* describing the movement of a node within the group. Any of the "flat", stochastic models, presented in Section 10.3.1 can be used for any of the two sub-models.

In the Reference Point Group Mobility (RPGM), shown in Fig. 10.6 [16], the individual model uses RWP (random waypoint – Section 10.3.1.1) uniformly distributed in a disk around the group center (reference point) and no pauses. The group center moves on a predefined trajectory. By choosing the trajectory of the groups, several realistic scenarios can be obtained [16]: if the groups are maintained disjoint (and almost stationary), they can simulate the movements of a deployed military group that performs similar actions in different parts of the battlefield. If the group trajectories overlap, a disaster recovery scenario can be simulated (with multiple distinct groups operating in the same geographical area). Finally, if the trajectories of the groups are similar, but spaced in time, the model can emulate the behavior of groups visiting a museum or a convention

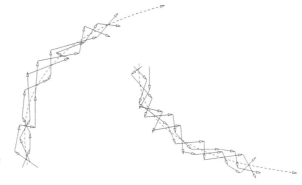

**Fig. 10.6** The movement of two groups, each with three nodes with a group mobility model

center and stopping to admire the exhibits or discuss one project or another. Also, different types of nodes (e.g., infantry and UAVs) can be simultaneously modeled by assigning them to different groups with different characteristics (i.e., group trajectories, speeds, pause times, etc.).

Another approach for modeling group behavior is based on social networks [17, 18]. The assumption is the users with strong relationships tend to stay together during the day. Given a matrix of relationships between the nodes, the nodes will have a higher preference toward a goal where "friends" are located, rather than a random location. The model also allows nodes to move (with a low probability) to locations without any other nodes.

### 10.3.3  Obstacle Models

The mobility models presented above do not take into consideration obstacles or, more generally, constraints on the trajectories of the mobile nodes. While this may be an appropriate assumption for UAVs or ships at sea, it is highly unrealistic for most other scenarios.

In [5, 19], the authors consider the effect of buildings in a campus both on the mobility model as well as on the wireless propagation model. They construct "paths" between the buildings by considering the Voronoi diagram: each building is defined by a polygon, and each vertex of the polygon is a location point in the Voronoi diagram. The mobile nodes move between buildings by walking only on paths and using the shortest possible route. The authors also consider the effect the obstacles have on shadowing the wireless transmissions between nodes. The authors show that using this model results in significantly different performance results for AODV [20] (in comparison with RWP and RDM).

One of the simplest ways to restrict the movement of mobile nodes is to provide a list of allowable trajectories. In [7], the Manhattan and the Freeway mobility models (Fig. 10.7) are introduced. In the Manhattan mobility model the nodes are only allowed to move on a predefined grid. At each intersection, a

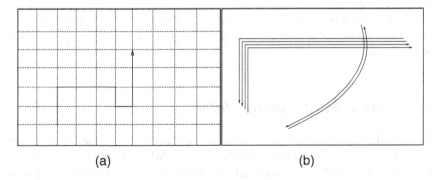

(a)                                           (b)

**Fig. 10.7** The Manhattan (a) and freeway (b) mobility models

node will continue to go straight with a probability of 0.5 or turn left or right
with a probability of 0.25. In the freeway mobility model, the nodes are not only
restricted to follow the existing roads but also restricted to follow their lanes
without passing the vehicles in front and have limited acceleration and breaking
capabilities.

### 10.3.4 Detailed Mobility Models

Both the group models as well as the obstacle models are clear steps toward
increased realism and decreased flexibility. While a freeway model is clearly a
better model for a scenario involving vehicles on a highway than RWP, it is not
well suited for a scenario involving students in a campus or participants at a
conference.

In the extreme, the models increase in realism and complexity while focusing
on a single scenario (or a narrow class of scenarios). For example, the STreet
RAndom Waypoint (STRAW) [21] uses real street maps (in the United States
the maps are free to download for any region), and their associated maximum
speeds to simulate vehicle movements. In this model a vehicle chooses a random
destination and then it computes (taking into account the maximum speed on a
particular road) the shortest path to arrive at that destination. Once it arrives, it
chooses another destination and it repeats. Another supported mode is similar
to the Manhattan mobility model, where vehicles reaching an intersection turn
with a specified probability. While the model is not perfect (e.g., it does not
implement lane changes or stop lights and vehicles can do U-turns in the middle
of the highway), it is a far more realistic model for vehicular networks than any
of the stochastic models (and even the group or obstacle models).

Numerous [22–28] other vehicular traffic simulators offer different degrees
of realism in the areas of car-following, lane changes, stopping at intersections,
etc. An especially accurate traffic simulator that is well-respected by civil
engineers is CORSIM [29], which can simulate highly detailed driving behaviors
like changing lanes to allow others to pass, traffic light timings, and aggressive-
ness of the drivers. Several research groups used traces from CORSIM for
VANET simulations [30–34]. TRANSIMS [35] is another very detailed simulator that simulates movements of both vehicles and pedestrians in urban
environments.

### 10.3.5 Trace-Based Mobility Models

The detailed simulators, while potentially very accurate for the scenario they
have been crafted for, cannot diversify easily to different scenarios. For example, CORSIM, while being an excellent vehicular simulator, would be a poor
choice for simulating the movements of zebra herds.

Attempting to walk the fine line between realism and diversification, hybrid mobility models try to "generalize" mobility trends based on the characteristics of real traces. In [36], the authors study the common characteristics of mobility of WLAN users in school campuses. By analyzing both the individual movements and the encounters between the nodes in the real traces, the authors point out the characteristics not available in any existing network simulator (for example, that some students never encounter other students, and that the "friendship" relationships are highly asymmetric). In [37], both user mobility and user traffic are characterized.

In [38], the authors went a step further, and not only analyzed the WLAN traces but also used them to design a mobility model that can generate new traces similar to the real traces. Figure 10.8 shows the framework of the model used in [38]: WLAN traces recording the session length of each user at each access point (AP) are processed, and relevant statistics (for session length and number of APs visited by a user) are extracted from the traces. The mobility model uses the same distributions to create a "realistic" number of nodes at each AP, with a "realistic" session length. According to the real traces most users use only a single AP; however, some use more than one (some as many as 20); therefore, for the corresponding percentage of users, a roaming mobility model where users move to nearby APs is implemented.

The approach in [38] certainly represents a first step toward creating realistic models, while allowing for significant diversification. For example, the number of users or the total simulation time can be changed by the user of the model, and the model is likely to produce reasonably realistic traces. However, the model is limited to modeling users in a WLAN environment, where the users are assumed to stop at an AP, open their laptop, work, close their laptop, and move again. Thus, this model is not generalized to other MANET scenarios.

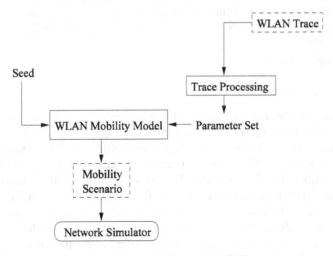

**Fig. 10.8** The framework of the WLAN mobility model in [38]

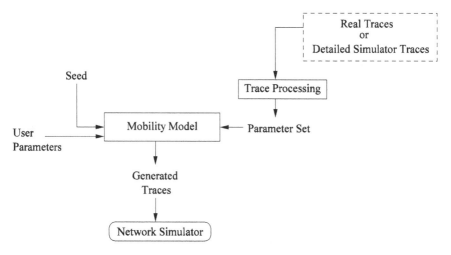

**Fig. 10.9** The generalized framework for trace-based mobility models

The framework employed in [38] can be generalized as shown in Fig. 10.9. A general trace-based mobility model could not only use WLAN traces, but any traces of users either real traces (e.g., collected by GPS) or generated by highly realistic traces detailed mobility models. These real(istic) traces can then be analyzed and relevant statistics can be extracted. In [38], session length and number of APs visited were considered. However, in a more general setting, a larger number of parameters are likely important:

- The inter-arrival and inter-departure times of the nodes (and thus, the resulting number of nodes).
- The existence of hotspots in the scenario. For example, red wolfs, which have dens, have significantly different mobility patterns from zebras.
- The existence of groups (time and size), perhaps quantified by the average number of neighbors.
- Inter-meeting times.
- The distribution of pause times.
- The distribution of node speeds.

The list of parameters to be extracted and matched by the mobility model can, of course, be arbitrary long. However, an interesting question is "Which of those parameters are actually important and have to be reproduced? ". For example, is an accurate reproduction of inter-arrival times important, or using an average (and constant) number of nodes is sufficient? There is no answer to this question yet.

The statistics of the parameters extracted from the real traces can then be used in the mobility model that can also take additional inputs from the users of the model, e.g., not only the number of users and duration of the simulation but also the size of the scenario, and perhaps override some of the parameters extracted from the trace, like the size of the hotspot, speed, and pause distribution, etc.

A mobility model built on the framework in Fig. 10.9 would allow the generation of diversified traces similar to the ones in the real traces. For example, given traces of students in a small campus, similar traces can be generated for a larger campus with far more students.

## 10.4 Impact of Mobility Models on the Performance of Wireless Networks

Previous works [4, 7, 19, 39] repeatedly demonstrated that using different mobility models, the performance of a protocol can vary significantly. Three different components of the problem have been exposed [7]:

- On one hand, the *absolute value* of a performance measure (e.g., throughput, overhead, or delay) of a protocol can vary widely with the mobility model. This creates problems when a network simulation with an inappropriate mobility model is used to predict the performance of a protocol that will later be used in the real world. The simulation may, for example, show a very low overhead (e.g., 5% of the total bandwidth) and, when deployed, a signifi-cantly larger percentage (e.g., 50%) may be actually used.
- On the other hand, the variation of a performance metric with changes in one of the parameters (e.g., transmission range) can change *qualitatively*, for example, the simulations may show that when the transmission range increases, the throughput will also increase, while upon deployment, it may in fact decrease.
- Finally, a side-effect of the first two points is that the relative performance of two protocols is also dependent of the mobility model. For example, it was shown in [7] that at high speed with a RWP mobility model the throughput of AODV [20] is considerably better than that of DSDV [40] (84% vs. 59%). However, when a group mobility model is used (RPGM), the situation is reversed, with DSDV delivering 97% of the packets while AODV remains around 87%.

This variation in performance metrics is especially damaging for system designers that may rely on simulations for evaluating the performance of a protocol that they plan to deploy for a particular application. Similarly, stan-dardization work in the IETF often compares competing draft proposals by using simulations. The suitability of one or the other of the proposals may be very well determined by the choice of mobility model.

In [7] two different types of metrics have been proposed for a mobility model: protocol-independent and protocol-dependent metrics.

Protocol-independent metrics, as the name suggests, are independent of the protocols that are running in the network (as well as of traffic loads, etc.). They can be extracted directly from the traces resulting from the mobility model and quantify several characteristics of the traces. Several such metrics have been introduced in [7]:

- *Degree of spatial dependence* describing the similarity of the velocities of nearby nodes. In the case of a car or military column, this will be very high.
- *Degree of temporal dependence* describing the velocity of a node at two nearby time instances.
- *Relative speed* of two nodes.
- *Geographic restrictions* describing constraints on the degrees of freedom of the nodes (e.g., obstacles in Section 10.3.3).

In addition to those metrics, if the transmission radius of the transceivers is known, the following topology metrics can be defined [7]:

- *Number of link changes* for a pair of nodes is the number of times a link makes and breaks between these two nodes.
- *Link duration* for a pair of nodes is the average time the link remains active once it becomes available.
- *Path availability* is the fraction of time a path is available between a pair of nodes.
- *Number of neighbors* is another indicator that can reveal the existence of hotspots, or groups in the mobility model.

In turn, these protocol-independent metrics can influence the traditional protocol-dependent metrics, which are usually the object of performance evaluation, such as throughput, delay, packet delivery ratio, overhead. For example, a low path availability will likely result in a low throughput and/or high delay.

Because the MANETs depend on the availability of intermediate nodes for packet forwarding, the mobility model influences the performance of MANET protocols far more than for simpler infrastructure, single-hop networks (e.g., WLAN and cellular).

## 10.5 Directions for Future Research

Since mobility models have been shown to have a significant influence on the results (both qualitative and quantitative) of network simulators, it is expected that in the near future researchers will transition from stochastic mobility models to more realistic models. The main challenge is developing a mobility model that diversifies readily, is easy to use, and is computationally efficient. Given the current state of the art, a considerable amount of research is still necessary to produce such a model.

## 10.6 Conclusion

Mobility models have a significant influence especially on the performance of the protocols of MANETs. A meaningful evaluation of a protocol and comparisons with other similar protocols cannot be done without using a realistic mobility model. Stochastic mobility models, while offering ease of diversification and, potentially, stationarity, result in performance results significantly different from those corresponding to realistic mobility models. In turn, detailed models, while realistic for the scenario they target, cannot diversify to other scenarios. The holy grail of mobility models is one that can produce diversified (both in parameters and scenario), yet realistic, node movements.

## 10.7 Terminology

*Mobility Model* is a method of simulating movement of mobile nodes, usually for the purpose of further using the resulting movement for other simulations.

*Realism* is the degree of accuracy of the mobility model, with respect to the movements of mobile nodes in a real scenario.

*Diversification* is the ability of a model to diversify to a large number of different scenarios, including different types of mobile nodes (e.g., vehicles, pedestrians, zebras) and different types of environments (e.g., campus, conference, city).

*Complexity* is a measure of the computational resources required by the mobility model.

*Stochastic Mobility Models* rely primarily on random movements without imposing any (or very few, e.g., borders) constraints on the movement of the nodes.

*Detailed Mobility Models* are custom-built for a particular scenario and may take into account micro-movements of the nodes in response to each others' movements and external stimuli.

*Hybrid Mobility Models* are a combination of stochastic and detailed mobility models (which are just the two end points in a continuum of hybrid models).

*Real Traces* are collections of trajectories of real users in a particular scenario. GPS or other localization techniques are used to collect those traces.

*Random Waypoint* mobility model is, by far, the most common mobility model used for simulations of ad hoc networks. In this model, each node chooses a random destination point (uniformly distributed in the simulated area) and proceeds to this destination with a speed uniformly chosen in a specified interval. It then pauses for a time also uniformly chosen in a specified interval and repeats the process.

*WLAN Trace* is a mobility trace obtained from wireless LAN (usually 802.11) users associating and de-associating with APs in the scenario considered (e.g., a campus).

## Questions

1. When and why do we need mobility models?
2. What is the difference between a mobility model and real traces?
3. What is the main disadvantage of stochastic mobility models?
4. What is the main disadvantage of the detailed mobility models?
5. Why do realistic mobility models are important?
6. What is the motivation for group mobility models?
7. Which parameters are important for the realism of a mobility model?
8. The performance of what protocols is likely to be affected by the choice of mobility models.
9. Is choice of the mobility model more important for ad hoc networks than for wireless local area networks?
10. What variables influence the values of the protocol-independent and protocol-dependent metrics?

## References

1. D. Hong, and S. S. Rappaport, "Traffic model and performance analysis for cellular mobile radio telephone systems with prioritized and nonprioritized handoff procedures," *IEEE Transaction on Vehicular Technology*, vol. 35, pp. 77–92, Aug. 2003.
2. T. Karagiannis, M. Faloutsos, and M. Molle, "Long-range dependence: Ten years of internet traffic modeling," *IEEE Internet Computing*, Special Issue in "Measuring the Internet", vol. 8, no. 5, Sept. 2004.
3. K. Pawlikowski, H.-D. J. Jeong, and J.-S. R. Lee, "On credibility of simulation studies of telecommunication networks," *Communication Magazine, IEEE*, vol. 40, no. 1, pp. 132–139, Jan. 2002.
4. T. Camp, J. Boleng, and V. Davies, "A survey of mobility models for ad hoc network research," *Wireless Communications & Mobile Computing (WCMC) Special issue on Mobile Ad Hoc Networking: Research, Trends and Applications*, vol. 2, no. 5, pp. 483–502, 2002.
5. A. Jardosh, E. M. Belding-Royer, K. Almeroth, and S. Suri, "Towards realistic mobility models for mobile ad hoc networks," in *Proc. of the 9th Annual ACM/IEEE International Conference on Mobile Computing and Networking (MobiCom'03)*, (San Diego, California), Sept. 2003.
6. P. Johansson, T. Larsson, N. Hedman, B. Mielczarek, and M. Degermark, "Scenario-based performance analysis of routing protocols for ad hoc networks," in *Proc. of the 5th Annual ACM/IEEE International Conference on Mobile Computing and Networking (MobiCom'99)*, (Seattle, Washington), 1999.
7. F. Bai, N. Sadagopan, and A. Helmy, "Important: A framework to systematically analyze the impact of mobility on performance of routing protocols for ad hoc networks," in *Proc. of IEEE INFOCOM 2003*, (San Francisco, CA), March/April 2003.
8. J. Yeo, D. Kotz, and T. Henderson, "CRAWDAD: A community resource for archiving wireless data at Dartmouth," *ACM SIGCOMM Computer Communication Review*, vol. 36, no. 2, Apr. 2006.
9. D. B. Johnson and D. A. Maltz, *Dynamic Source Routing in Ad Hoc Wireless Networks*, vol. 353. Kluwer Academic Publishers, Dordrecht, 1996.
10. W. Navidi and T. Camp, "Stationary distributions for the random waypoint mobility model," *IEEE Trans. on Mobile Computing*, vol. 3, no. 1, pp. 99–108, 2004.

11. C. Bettstetter, G. Resta, and P. Santi, "The node distribution of the random waypoint mobility model for wireless ad hoc networks," *IEEE Transaction on Mobile Computing*, vol. 3, July 2003.

12. J. Y. L. Boudec and M. Vojnovic, "Perfect simulation and stationarity of a class of mobility models," in *Proc. of IEEE Infocom 2005*, (Miami, FL), 2005.

13. J.Yoon, M. Liu, and B. Noble, "Random waypoint considered harmful," in *IEEE INFOCOM*, Apr. 2003.

14. H. Cai and D. Y. Eun, "Crossing over the bounded domain: From exponential to power-law inter-meeting time in MANET," in *Proc. of ACM MobiCom*, (Montreal, Canada), Sept. 2007.

15. Z. Haas, "A new routing protocol for the reconfigurable wireless networks," in *Proc. of the IEEE Int. Conf. on Universal Personal Communications*, Oct. 1997.

16. X. Hong, M. Gerla, G. Pei, and C. C. Chiang, "A group mobility model for ad hoc wireless networks," in *Proc. of the 2nd ACM International Workshop on Modeling, Analysis, Simulation of Wireless and Mobile Systems*, (Seattle), 1999.

17. K. Hermann, "Modeling the sociological aspect of mobility in ad hoc networks," in *Proc. of MSWiM'03*, (San Diego, CA), pp. 128–129, Sept. 2003.

18. M. Musoles, S. Hailes, and C. Mascolo, "An ad hoc mobility model founded on social network theory," in *the 7th ACM International Symposium on Modeling, Analysis and Simulation of Wireless and Mobile Systems*, (Venice, Italy), pp. 20–24, 2004.

19. A. Jardosh, E. M. Belding-Royer, K. C. Almeroth, and S. Suri, "Real world environment models for mobile ad hoc networks," *IEEE Journal on Special Areas in Communications – Special Issue on Wireless Ad hoc Networks*, 2005.

20. C. Perkins, "Ad-hoc on-demand distance vector routing," in *Proc. of MILCOM*, Nov. 1997.

21. D. Choffnes and F. Bustamante, "An integrated mobility and traffic model for vehicular wireless networks," in *Proc. of the Second ACM International Workshop on Vehicular Ad Hoc Networks*, (Cologne, Germany), 2005.

22. C. Lochert, M. Caliskan, B. Scheuermann, and M. Mauve, "Multiple simulator inter-linking environment for inter vehicle communication," in *Proc. of the Second ACM International Workshop on Vehicular Ad Hoc Networks*, (Cologne, Germany), 2005.

23. B. Raney, A. Voellmy, M. Vrtic, K. Axhausen, and K. Nagel, "An agent-based micro-simulation model of Swiss travel," *Networks and Spatial Economics*, no. 3, pp. 23–41, 2003.

24. B. Raney, A. Voellmy, M. Vrtic, and K. Nagel, "Towards a microscopic traffic simulation of all of Switzerland," in *Proc. of ICCS'02*, 2002.

25. R. Mangharam, D. Weller, R. Rajkumar, D. Stancil, and J. S. Parikh, "GrooveSim: A topography-accurate simulator for geographic routing in vehicular networks," in *Proc. of the Second ACM International Workshop on Vehicular Ad Hoc Networks*, (Cologne, Germany), 2005.

26. A. Saha and D. Johnson, "Modeling mobility for vehicular ad hoc networks," in *Proc. of the first ACM workshop on Vehicular Ad Hoc Networks*, pp. 91–92, 2004.

27. C. Rokitansky, "SIMCON2: Simulator for performance evaluation of vehicle-beacon and inter-vehicle communication protocols (media access / knowledge-based routing)," in *Proc. of the 41st IEEE Vehicular Technology Conference*, pp. 893–899, 1991.

28. M. Antoniotti and A. Göllü, "SHIFT and SmartAHS: A language for hybrid systems engineering, modeling, and simulation," in *Proc. of the USENIX Conference of Domain Specific Languages*, (Santa Barbara, CA, U.S.A.), Oct. 1997.

29. "CORSIM user manual, version 1.01." Federal Highway Administration, Office of Safety and Traffic Operations R&D, McLean, VA, Aug. 1996.

30. J. Yin, T. ElBatt, G. Yeung, B. Ryu, S. Habermas, H. Krishnan, and T. Talty, "Performance evaluation of safety applications over DSRC vehicular ad hoc networks," in *Proc. of the First ACM Workshop on Vehicular Ad Hoc Networks*, pp. 1–9, 2004.

31. H. Wu, R. Fujimoto, R. Guensler, and M. Hunter, "MDDV: A mobility-centric data dissemination algorithm for vehicular networks," in *Proc. of the First ACM Workshop on Vehicular Ad Hoc Networks*, pp. 47–56, 2004.

32. J. Blum, A. Eskandarian, and L. Hoffman, "Challenges of intervehicle ad hoc networks," *IEEE Transaction on Intelligent Trasnportation Systems*, vol. 5, no. 4, pp. 347–351, 2004.

33. J. Blum, A. Eskandarian, and L. Hoffman, "Mobility management in IVC networks," in *Proc. of the IEEE Intelligent Vehicles Symposium*, pp. 150–155, 2003.

34. Z. Chen, H. Kung, and D. Vlah, "Ad hoc relay wireless networks over moving vehicles on highways," in *Proc. of the 2nd ACM International Symposium on Mobile Ad Hoc Networking & Computing*, pp. 247–250, 2001.

35. "TRANSIMS home page." http://www.ccs.lanl.gov/transims/index.shtml.

36. W. Hsu and A. Helmy, "Impact: Investigation of mobile-user patterns across university campuses using WLAN trace analysis," tech. rep., USC, Los Angeles, CA, July 2005.

37. M. Balazinska and P. Castro, "Characterizing mobility and network usage in a corporate wireless local-area network," in *1st International Conference on Mobile Systems, Applications, and Services (MobiSys)*, (San Francisco, CA), May 2003.

38. C. Tuduce and T. Gross, "A mobility model based on WLAN traces and its validation," in *IEEE INFOCOM*, (Miami, FL), Mar. 2005.

39. D. D. Perkins, H. D. Hughes, and C. B. Owen, "Factors affecting the performance of ad hoc networks," in *Proc. of Int. Conf. Comm. (ICC'02)*, vol. 4, p. 2048â"2052, Apr. 2002.

40. C. Bhagwat, "Highly dynamic destination-sequenced distance vector routing (DSDV) for mobile computers," in *Proc. of ACM SIGCOMM*, pp. 234–244, Sept. 1994.

# Chapter 11
# Models for Realistic Mobility and Radio Wave Propagation for Ad Hoc Network Simulations

Mesut Güneş and Martin Wenig

**Abstract** An ad hoc network is realized by mobile devices which communicate over radio. Since, experiments with real devices are very difficult, simulation is used very often. Among many other important properties that have to be defined for simulative experiments, the *mobility model* and the *radio propagation model* have to be selected carefully. Both have strong impact on the performance of mobile ad hoc networks, e.g., the performance of routing protocols changes with these models. There are many mobility and radio propagation models proposed in literature. Each of them was developed with different intentions and is not suited for every scenario. In this chapter we introduce well-known models for mobility and radio propagation, and discuss their advantages, drawbacks, and limitations in respect to the simulation of mobile ad hoc networks.

## 11.1 Introduction

A mobile ad hoc network [1] is created by a collection of nodes (computers), which communicate using radio interfaces and do not rely on any pre-installed infrastructure. Due to the lack of infrastructure, the nodes have to provide the functionality of the infrastructure by themselves. Since the nodes are assumed to be mobile, the network topology changes with the movement of the nodes. Furthermore, it is supposed that mobile ad hoc networks are inherently adaptive and auto-configured, thus providing high flexibility in means of setting up a network.

In recent years the interest in the deployment of ad hoc networks for real-world scenarios grew. Still the number of real-world ad hoc networks is quite low and most of the testbeds [2] consist only of a small number of nodes. At the

M. Güneş (✉)
Institute of Computer Science, Computer Systems and Telematics (CST), Distributed, embedded Systems (DeS), Freie Universität Berlin, Takustr. 9, 14195 Berlin, Germany
e-mail: guenes@inf.fu-berlin.de

S. Misra et al. (eds.), *Guide to Wireless Ad Hoc Networks*,
Computer Communications and Networks, DOI 10.1007/978-1-84800-328-6_11,
© Springer-Verlag London Limited 2009

same time the experimentation with real networks requires, on the technical side as well as on the non-technical side, high efforts and repetition/replication of experiments are very difficult. Therefore, the development and testing of algorithms and methods for mobile ad hoc networks nowadays relies heavily on network simulations. Simulating wireless networks, and especially mobile ad hoc networks, is not a trivial task and, consequently, there have been discussions about the validity of the presented simulation results [3, 4]. This chapter does not deal with the methodological background used to analyze the output of the simulation; instead, it deals with the question how accurate the simulation output is. For a sound presentation of output analysis, the interested reader is referred to [5, 6].

We argue in this text that the mobility and the radio wave propagation models have an important impact on simulation results accuracy. Therefore, it is of great importance for researchers to better understand the implications and influences of all parts of the simulation environment. This chapter is focused on two basic components for MANET simulations: the *mobility model*, which describes the movement pattern of nodes (on the simulation area), and the *radio wave propagation model*, which defines how the radio transmission takes place between nodes. In the past, very simplistic mobility models were used to generate node movements. It has been shown before that these models yield unrealistic behavior [7, 8]. One reason for this is that movement patterns of humans are much more complex and cannot be modeled by only one of these models. Yet, we state here, that well-known and analyzed models can be used to model smaller parts of the simulation setup, e.g., the random direction model can be useful to generate movements for pedestrians on inner-city places, but it is not suited well to model cars on a street. The second component with which we are dealing is the radio wave propagation model. The impact of the radio wave propagation model on the simulation results is obvious. In most available network simulation packages, nodes are assumed to have a circular transmission range with fixed radius, independent of the current location of the communicating nodes. This might be realistic in open spaces, but it is certainly not true in buildings or in a city. Simplified propagation models will yield much better simulation results than achievable in reality.

To get an idea about the awareness of the research community to this topic, we have done a survey on the publications of one famous conference. Taking the publications of the MobiHoc conferences of the past two years as an example, it is obvious that there is a need for better tool support for simulation designers. Out of 52 papers, 35 presented simulation results (around 67%). Six papers did not give any information about the used mobility model, 10 used *random waypoint* to model mobility, and 14 considered static scenarios. Only two papers showed the results obtained from considering more than one mobility model. Examining the used radio wave propagation model, the findings are even more surprising: only two papers mention the used radio wave propagation model, 10 papers gave no indication about the used model, and 22 used a fixed radius. Assuming that all papers which did not specify their propagation model used a fixed range, it can be concluded that all papers used circular, bidirectional links. None of the presented papers used a small-scale (*fading*) model.

From the previous paragraph it is clear that results and conclusions drawn from network simulations have to be considered with some skepticism. The preparation of network simulation experiments requires thorough planning in which the used models have to be selected carefully with a thought of the application scenario.

In the remainder of this text we will present and discuss related models with their advantages, restrictions, and limitations. Furthermore, we give an overview of popular network simulation packages and discuss the support provided by these packages in respect to the topic of this text. Finally, we present an approach to realize more realistic network simulations.

## 11.2  Background: Mobility Models

A *mobility model* describes the movement of an entity on a given space. The space may be one-, two-, or three-dimensional. Very often only two-dimensional spaces are considered; however, this does not impose any general restrictions on the mobility model. Furthermore, it is assumed to have an empty space without any obstacles, like buildings, or structuring elements, like streets or bridges. This results in the assumption that all coordinates of the space can be occupied by entities. Although this assumption is not very realistic and does not represent the real world, it is very useful for general considerations of mobility models. It simplifies the implementation, testing, and analysis of mobility models. At the other side, this assumption simplifies the real world too much and the resulting mobility patterns do not match with general human movement. Thus, the generated mobility patterns are not suitable for real-world scenarios.

The mobility models proposed in literature can be distinguished in various ways. One way of classification is to distinguish the mobility models based on the number of considered individuals.

- Entity mobility models: Entity mobility models consider only the movement of one individual, e.g., a human or an animal. Typically, each entity is considered independently from others and the model does not take into consideration the number of entities existing on the movement area.
- Group mobility models: In contrast to entity mobility models, group mobility models consider a set of individuals as a group, and the movement of the entities in the group is somehow related to each other. Very often there is a group leader which specifies the movement of the group and the others are somehow gathered around this entity.

Another classification of mobility models can be done based on the original source of the mobility, e.g., copying the movement of animals or natural phenomenon.

- Human mobility: Mobility models which try to mimic the mobility of human beings in their appropriate habitation, e.g., in buildings, in a city, or during work.
- Vehicle mobility: The mobility of vehicles is typically restricted on streets, roads, and highways. The environment imposes particular restrictions on the mobility patterns realized by vehicles.
- Natural or physical phenomena: These mobility models are based on physical phenomena, e.g., the movement of molecules and fluids.

A third classification of mobility models can be done based on scenarios:

- Normal situations: Mobility models which assume the normal existing or living scenarios, e.g., pedestrians walking in a city and vehicles moving on streets.
- Special situations: Mobility models which try to mimic the mobility pattern in a special situation, like an emergency, fire, and earthquake. However, mobility models which mimic the behavior of soldiers, a search troop, and the police may also belong to this class.

It is obvious that the discussed classes are not disjoint and mobility models may belong to several classes. The goal of a general mobility framework may be the construction of mobility patterns by specifying characterizations from the discussed classes.

## 11.2.1 Entity Mobility Models

The most simple random mobility model is called *Random Walk*, also known as *Brownian motion* [9]. In this model, a mobile node selects randomly a direction and speed from predefined ranges [$\varphi_{min}$:$\varphi_{max}$] and [$v_{min}$:$v_{max}$], respectively. Each movement is bound either by travel time or by travel distance. There are many variants of this model. This model is proper to model somebody who walks around without a target, but restricted on an area. The disadvantage of this model is that the generated mobility pattern contains sharp changes in direction and velocity. Given that the only parameter of the model is the velocity interval [$v_{min}$:$v_{max}$]. In the long-term the average of a node with this model will be ($v_{max}$-$v_{min}$)/2. However, network simulations are very often considered over a relatively short time, e.g., many MANET simulations were run over 900 s or 3,000 s. During these short experiment periods, it may happen that nodes move slow and therefore does not match the intended scenario. Therefore, it is important to spend a look on the parameters of the model and also whether it fits with the simulation time. For example, if the speed interval is [0, 1] m/s, the long-term average speed will be ½ m/s. During a simulation time of 900 s, a node will travel in average only 450 m. With a radio range of 250 m the node may leave the communication range of another node only once, i.e., a link break may occur only once.

The *Random Waypoint* mobility model [10] is an extension of Random Walk and integrates a pause time between two consecutive moves. A mobile node stays after a movement a particular time period $t_{pause}$ at the destination location. This model is proper to model somebody on an exhibition or market without a target. The person is traveling directly from one booth (shop) to another and spending sometime there. A disadvantage of this model is the concentration of nodes in the center of the simulation area. An informal explanation of this phenomenon is that the nodes have to cross the center to reach other parts.

To overcome the problem of the accumulation of nodes in the center of the simulation area, the *Random Direction* [11] model can be used. It enforces the nodes to move until they reach the border of the simulation area. This model is particularly useful to model the behavior of people with a target in mind, e.g., people leaving a building and walking directly to another building or people leaving an office crossing the floor to get to another office.

Unfortunately, in all these mobility models nodes have sharp direction changes which does not match the movement behavior of humans. The main reason for this discrepancy from reality is that these mobility models are memoryless, i.e., a node does not consider the visited locations when selecting the next one. However, there are other mobility models which take the visited positions into consideration when selecting new positions. The *Gauss-Markov* [12] model prevents the problem of sharp direction changes by considering the most recent moves into the calculation of the next destination. Therefore, the resulting movement pattern is smoother.

### 11.2.1.1 Vehicle Movement

Beside these general entity models, there are also some models which try to map the characteristics of vehicle movements. In the *Freeway* [13] model, there is at least one lane in each direction of a street. The mobile nodes move on the lanes. The speed of a mobile node depends on other nodes on the same lane. This model is suitable to model individual streets like freeways or highways, but it is not proper to model a city with many crossings and roads. A more suitable mobility model for such a scenario is the *Manhattan* [13] model. As the name reveals, the model is based on Manhattan, New York. The lanes are organized around blocks of buildings and have a direction. The mobile nodes move on the lanes and can change the direction only at intersections.

## *11.2.2 Group Mobility Models*

The mobility models discussed in the previous section describe the movement of only one individual independently of others. In reality, humans and animals show a group behavior, i.e., the movement of one entity is based on the movement of another entity or depends on the movement of another entity. The class of group mobility models specifies how a set of mobile nodes move in respect to each other.

In the *Column* [14, 15] mobility model, a group of nodes build a line and move uniformly to a destination. This mobility model is proper to model the behavior of soldiers or the behavior of a search troop.

In the *Nomadic Community* [14, 15] mobility model, all mobile nodes move to the same location in the same order, but by using different entity mobility models. This allows the modeling of different classes of mobile nodes in a group, which travel to the same destination. Hence, this model is suitable to model mixed groups, e.g., groups of tourists which travel from one object of interest to another during sightseeing. The disadvantage of this model is that the group entities are not bound together, i.e., there is not a given restriction on the distance between the members of the group.

In the *Pursue* [14, 15] mobility model, the movement of a group is determined by a target which the group is tracking or chasing. This mobility model is proper for scenarios in which a special entity has to be tracked in the sense of military or police, i.e., that entity is not a member of the group.

The *Reference Point Group* mobility model [16] specifies the movement of the group, where in contrast to the Pursue mobility model the entity which determines the movement of the group members, the reference point, belongs to the group. In comparison to the *Nomadic Community* mobility model, there is a restriction on the distance of the group members to the reference point. In the example above, this model is suitable to model the movement of one group of tourists, which follow tightly the tourist guide.

### 11.2.2.1 Mobility Based on Social Networks

The group mobility models discussed in the previous section realize a binary relationship among the group and the group members, i.e., either a node belongs to a group or it does not. These models do not allow the building of scenarios where the movement of a node is influenced by more than one other node.

The *Social Relationship* mobility model is based on social networks [17]. The model is divided in two levels. At the first level, artificial social relationships among mobile entities are defined, and at the second level the social organization is mapped onto a topographical space. The artificial social relationship is defined as a matrix $M$

$$M = \begin{pmatrix} 1 & \cdots & \cdot \\ \vdots & \ddots & \vdots \\ \cdot & \cdots & 1 \end{pmatrix},$$

in which the entry $m_{i,j} \in [0,1]$ represents the interaction between the individuals $i$ and $j$; 0 stands for no interaction and 1 for strong interaction. The diagonal elements $m_{i,i}$ represent the interaction of entities with themselves and are set to 1.

Entities belong either to a group or stay alone. The movement of an entity which belongs to a group is given by the group movement and its own movement within the group. Entities which do not belong to a group move merely by themselves. The group mobility as well as the entity mobility of nodes are however defined by entity mobility models.

## 11.2.3 Obstacle Mobility Model

All mobility models discussed so far share the assumption that there are no obstacles, i.e., each point on the simulation area can be occupied by a mobile node. This is especially disturbing if you want to map the movements to real locations like a city or university campus. In the real-world movement, paths are restricted on certain ways, streets, and floors.

One way of incorporating obstacles into the discussed mobility models is to represent them as prohibited sub-areas of the simulation area. Mobile nodes are not allowed to enter the areas specified as obstacles. In [18] the obstacles represent buildings. Upon the definition of buildings, paths between them are calculated. The mobile nodes are randomly distributed on the paths and the destinations of the nodes are selected randomly among the buildings. The nodes move on the defined paths from building to building. Additionally, the radio propagation is affected by the obstacles. It is assumed that radio signals are completely blocked by obstacles. Hence a mobile node inside a building cannot communicate with a mobile node outside the building. Thus, the used radio propagation model is very simplified.

## 11.3 Background: Radio Wave Propagation Models

Radio channels are more complicated to model than wired channels. Their characteristics may change rapidly and randomly and they are dependent on their surrounding (buildings, terrain, etc.). Figure 11.1 shows the three most important propagation effects. When a radio wave hits an object whose dimensions are large compared to the wavelength, reflection occurs. This phenomenon is quite common inside office buildings. The long hallways act as *wave guides*. If the wave hits a rough surface, the energy is diffused in all directions. Diffraction occurs, for example, on edges. The radio wave is *bent* around the obstacle. The useful signal decreases rapidly if the node moves into the region shadowed by the obstacle. However, it might still be possible to maintain connectivity.

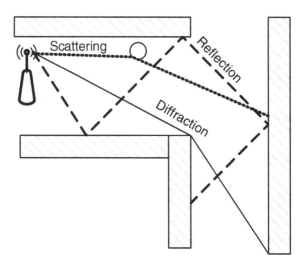

**Fig. 11.1** Typical radio wave propagation phenomena

Since many of these signals take paths that are of different length, the received signal varies heavily in quality. These *multipath waves* lead to fading. In real systems one can observe rapid changes in signal quality over short travel distances or time frames. This severely impacts the communication session. Figure 11.2 shows the effect of Ricean fading on the signal strength. In this example, the black line shows the prediction of the two-ray ground model of ns-2. The gray area shows the effect of Ricean fading. We used the implementation of [19] for our simulation.

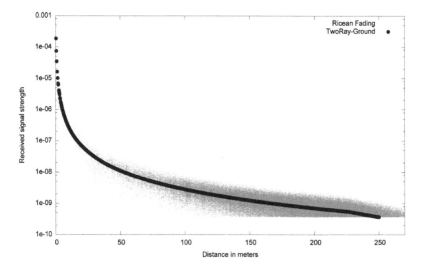

**Fig. 11.2** Comparison of the large- and small-scale propagation

Nevertheless, most wireless network simulators use very simplified propagation models. In general, propagation models can be characterized into two groups: *large-scale* and *small-scale* propagation models. Large-scale models characterize how the transmission power between two nodes changes over long distances and over a long time. Small-scale models account for the fact that small movements (in the order of the wavelength) may have large influence on the transmission quality. Also, due to the previously mentioned multipath propagation, the signal varies heavily even if the nodes do not move. Small-scale propagation models are often called *fading* models.

### 11.3.1 Free Space Model

This is still one of the most common models used during simulations because of its inherent simplicity. The received power only depends on the transmitted power $P_t$, the antenna's gains ($G_s$ and $G_r$), and on the distance between the sender and the receiver. It accounts mainly for the fact that a radio wave which moves away from the sender has to cover a larger area. So the received power decreases with the square of the distance. In this formula $L$ is an additional loss factor independent of the propagation.

$$P_r(d) = \frac{P_t G_t G_r \lambda^2}{(4\pi)^2 d^2 L}.$$

### 11.3.2 Two-Ray Ground

The two-ray ground model assumes that the received energy is the sum of the direct line of sight path and the path including one reflection on the ground between the sender and the receiver. The received power becomes independent of the frequency of the transmitted signal but depends on the height of the transmitter ($h_t$) and receiver ($h_r$).

$$P_r(d) = \frac{P_t G_t G_r h_t h_r}{d^4}.$$

A limitation in ns-2 is that sender and receiver have to be on the same height. ns-2 also uses a combination of these two presented models. If the distance between two nodes is smaller than a threshold $t_m$, then the Free Space model is used, the two-ray ground model otherwise.

$$t_m = \frac{4\pi h_t h_r}{\lambda}$$

As an example, if WLAN and a typical height of 1.3 m for handheld devices is considered, the crossover distance is $t_m \approx 170$ m.

### 11.3.3 Shadowing

The log-normal shadowing model assumes that the average received signal power decreases logarithmically with distance. Measurements, however, have shown that the real value is randomly distributed normally around the predicted value. The path loss is calculated as

$$\overline{PL}(d) = \overline{PL}(d_0) + 10n \log\left(\frac{d}{d_0}\right),$$

where $\overline{PL}(d_0)$ is a reference value measured at distance $d_0$. The value of the path loss exponent $n$ depends on the environment that should be modeled. As a rule of thumb, the more obstructed the environment, the higher should $n$ be. For example, for buildings, values up to $n=6$ are reasonable. To account for the random distribution, a Gaussian distributed random variable $X_\sigma$ with standard deviation $\sigma$ is added to the path loss formula. The value of $\sigma$

$$\overline{PL}(d) = \overline{PL}(d_0) + 10n \log\left(\frac{d}{d_0}\right) + X_\sigma.$$

### 11.3.4 Fading

The previously mentioned models are large-scale models. For a realistic simulation environment, small-scale propagation effects have to be taken into account. These two models are fading models, meaning that they describe the time correlation of the received signal power. Fading is mostly caused by multipath propagation of the radio waves. If there are multiple indirect paths between the sender and the receiver, Rayleigh fading occurs. If there is one dominant (line of sight) path and multiple indirect signals, Ricean fading occurs.

### 11.3.5 Site-Specific Modeling

All these models share the common property that their transmission range is roughly circular and that the transmission is not dependent on the current location. None of these models is able to correctly model complex scenarios with obstacles. One possibility to overcome this limitation is the use of ray-tracing technologies, known mainly from computer graphics. In [20] an approach using this technique is described. It allows the definition of obstacles in a graphical editor, and this scenario description is used in the simulation to feed a ray-tracing algorithm. The algorithm is started once for every new position the node takes up. The authors state that this approach slows down the simulation by a factor of up to 100. Also, no movement information is generated by this tool.

## 11.4 The CosMos Framework

In this section we introduce CosMos – *The Communication Scenario and Mobility Scenario Framework*. CosMos is a general framework for the design of realistic simulation scenarios for mobile ad hoc networks by integrating mobility models, propagation models, and communication models.

CosMos is implemented in C + + /Qt and provides some convenient features, namely creating scenarios using a graphical editor, generating movement data, and calculating the radio wave propagation using a ray-tracing algorithm. The goal is to show the impact of mobility and radio wave propagation on the result of MANET studies and to give researchers an easy-to-use tool to create own scenarios.

### 11.4.1 The World of CosMos

The main building block of CosMos is the zone: *movement zones* and *obstacle zones*. A mobility model is assigned to each movement zone and radio wave propagation parameters to every obstacle zone. Among the movement zones a neighborhood relationship is defined, which is set explicitly by the user. CosMos builds a directed and weighted graph $G(V, E)$ with zones as nodes and the neighborhood relationship as weighted and directed edges. The weight $w_{i,j}$ of a directed edge $e_{i,j} \in E$ from zone $i$ to zone $j$ specifies the rate with which nodes move from zone $i$ to zone $j$. The higher the weight $w_{i,j}$, the higher the probability that zone $j$ is chosen as destination for a node.

Furthermore, our approach allows the calculation of the spatial distribution of nodes on the simulation area as well as the distribution of the nodes on the defined zones. This allows us to figure out the time when the stationary state is reached. Hence, trustworthy MANET simulations should begin when the stationary state is reached.

Figure 11.3 shows a schematic view of the world of CosMos. The directed and weighted graph for the example in Fig. 11.3 are depicted in Fig. 11.4. In this specific case it is assumed that the zones connecting the places are assigned the Freeway model with one lane in each direction. The different zones are depicted

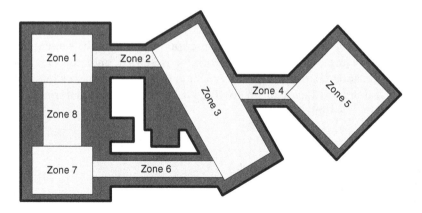

**Fig. 11.3** Environment setup for CosMos

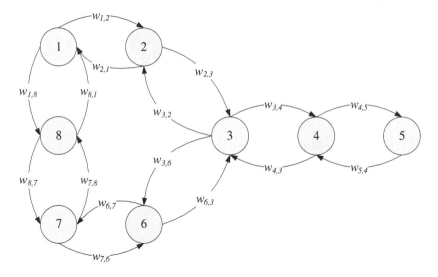

**Fig. 11.4** Graph view of the environment

there. The light and medium gray ones are movement zones with different mobility models, the black ones are obstacles.

## 11.4.2 Scenario Creation

The workflow for CosMos is depicted in Fig. 11.5. First you need to create a scenario definition file. This is done using the graphical frontend of CosMos.

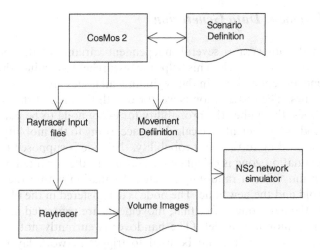

**Fig. 11.5** Workflow using CosMos

A scenario consists of several zones. Zones are either movement zones or obstacles. A mobility model is assigned to each movement zone. The model is given its own set of parameters (e.g., maximum speed) and a certain exit probability. The obstacles are set up with alpha and reflection values and with a height. A scenario also needs *starting points* for the ray-tracer.

The idea behind decomposing the simulation area into a number of smaller areas with their respective mobility models is that the existing mobility models model only parts of the reality. Assuming a man travels from his home to his work place. First, he will walk to his car, then drive towards the freeway, follow the freeway to his destination exit, then drive through an inner-city area, and finally walk to his working place. His movements on each of the mentioned parts of the journey can be characterized by a specific mobility model, but his complete journey cannot be modeled with only one model.

Similar, he crosses several environments which have different properties for the radio wave propagation model. The area around his home may be sparsely covered with buildings, whereas the inner-city area will be dominated by large concrete buildings. Such environments are modeled very badly with common radio wave propagation models.

The creation of complex scenarios is a difficult and time-consuming task. That is one reason why most of the presented simulation studies still use the random waypoint mobility model and the Two-Ray Ground propagation model. To ease the creation of more complex scenarios, CosMos offers a graphical user interface, which allows fast building of the scenario. Movement zones and obstacles can be created by drawing them with the mouse and assigning them the needed properties. Currently, we are in the process of creating a database of existing scenarios. The scenarios will be available for download on our website [21].

### 11.4.3 Movement Data Generation

CosMos per default creates several independent variants of the movements according to the given models. This helps the researcher to conduct simulations with independent replications. In the beginning, all nodes are placed randomly within the zones. The distribution is weighted with the area of the zones: The larger the zones, the higher the probability for every node to be placed there. Then each node's movements are calculated according to the mobility model in its zone. If, according to the exit probability, the node is supposed to leave its zone, the destination zone is calculated according to the weighted graph. The node is then moved to a randomly selected destination within the handover zone of the old and the new zone. The node is deregistered in the old zone and registered in the new zone. After that, movements are generated according to the new zone's mobility model. (The handover is currently not used in the simulation process, but it could be used to trigger network layer handoffs during the simulation.) This method of handing over a mobile node from one zone to another zone requires the slight modification of the used random mobility models. The last target location of the mobile node in zone $i$ has to be on the handover area, and the first location on zone $j$ has to be exactly this location on the handover area. The slight modification of the used random mobility models in our implementation is not a big issue and does not change the general behavior of these mobility models. Accordingly, movements in CosMos can be characterized into *intra-zone* mobility and *inter-zone* mobility. The intra-zone mobility depends on the zone's mobility model and the inter-zone movement depends on the exit probability and the weighted graph. CosMos offers a special kind of mobility model used only to connect zones. If a node enters a zone using this mobility model, one of the neighbors of the zone is selected and the node moves directly to the selected one.

#### 11.4.3.1 Example

An example scenario for the generation of movement files is presented here. Two places of size $500 \times 500 \, \text{m}^2$ with a distance of 1,000 m are defined. Figure 11.6 shows the map of the scenario, and Fig. 11.7 shows the directed and weighted graph of this scenario for a particular setting. The nodes 0 and 1 depict both the places and node 2 depicts the connecting zone. Both of the two places are assigned the Random Waypoint model, and the connecting street is assigned the Freeway model.

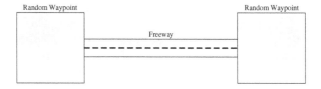

**Fig. 11.6** Setup of the example scenario

**Fig. 11.7** Weighted and directed graph of the example scenario for a particular setting

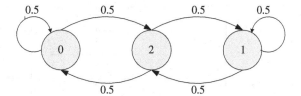

When talking about mobility models, two characteristics are of particular importance: these are the spatial distribution of the nodes on the simulation area and the distribution of the nodes on the individual mobility zones. Figure 11.8 depicts the spatial distribution of the nodes, and Fig. 11.9 depicts the distribution of the nodes on the mobility zones for the presented example. These results are based on the setting where all probabilities are set to 0.5, like in Fig. 11.7. The typical characteristic of the Random Waypoint mobility, i.e., the accumulation of the nodes in the center of the simulation area, is obvious for both of the places in Fig. 11.8. The probability to meet a node is higher on both the places than on the connecting street. The reason for this is that all nodes entering the street leave the street either to one of the places. Figure 11.9 shows the fraction of nodes as a function of the simulation time for each of the three zones. We have run the mobility model for 10,000 s. The transient phase of the simulation lasts around 2,000 s. During this time the number of nodes in each of the zones varies. After this time the mobility enters a stationary state. The network simulation should start after the stationary state is reached.

To get a better idea about the spatial distribution, the distribution of the nodes, and the transient and stationary state, we present here a variant of the example. We changed the exit probability from 0.5 to 0.2 for the place on the right side. Thus the probability that a node leaves the place on the right side is smaller and we expect that more nodes will be on this zone. Figure 11.10 depicts the spatial distribution for this case, and Fig. 11.11 shows the distribution of the nodes on all zones as a function of the simulation time. From Fig. 11.10, it is obvious that the characteristic of the Random Waypoint model is still kept, but the intensity is changed. The

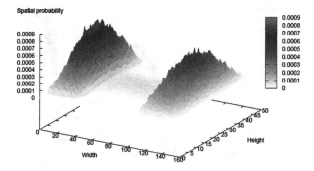

**Fig. 11.8** Spatial distribution of the nodes

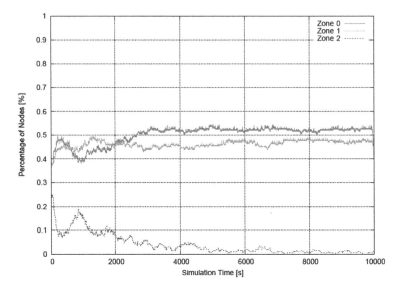

**Fig. 11.9** Distribution of the nodes on the zones

probability to meet a node is different on both of the places; it is higher on the place
on the right side and lower on the place on the left side. Thus, the number of nodes
on the right side must be higher. This assumption is confirmed by Fig. 11.11. During
the transient phase, which lasts again around 2000 s, of the mobility model the
fraction of nodes in each of the zones varies. After reaching the stationary state, the
node distribution is stable. Nearly 85% of the nodes are on the place on the right
side and the rest of the nodes are on both the other zones.

## 11.4.4 Radio Wave Propagation

To use the generated energy distribution maps during the simulation, we
modified the ns-2 network simulator [22]. We added a propagation model

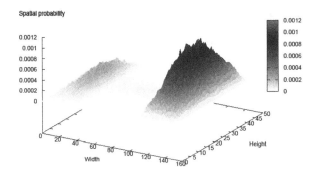

**Fig. 11.10** Spatial distribution of the nodes

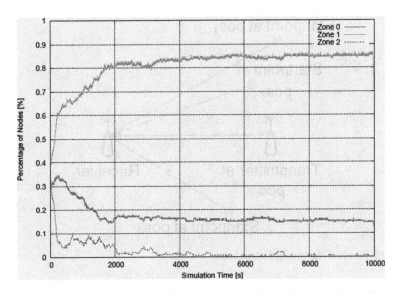

**Fig.11.11** Distribution of the nodes on the zones

that reads in a given set of maps and the corresponding starting points. During the simulation, whenever a node $n_t$ wants to transmit a packet, a $k$-nearest neighbor search is started. Our experiments showed $k = 3$ gave good results. This search finds the $k$ nearest starting points and their corresponding energy distribution maps to the sender's position. For each node inside the maximum interference range of an unobstructed radio wave, the transmission power is calculated. The formula used for the weighted interpolation is given below:

$$s_{t,r} = \frac{\displaystyle\sum_{i=0}^{k-1} \frac{s_i}{\|\mathrm{pos}_i - \mathrm{pos}_t\|^p}}{\displaystyle\sum_{i=0}^{k-1} \frac{1}{\|pos_i - pos_t\|^p}}$$

where $s_{t,r}$ is the signal strength between the transmitter node $n_t$ and the receiver node $n_r$. The position of the transmitter is given as $\mathrm{pos}_t$, $\mathrm{pos}_i$ denotes the position of the starting point of the $i$th closest map. Note that $s_i$ is the predicted signal strength of map $i$ at the position of the receiver $\mathrm{pos}_r$. The exponent $p$ controls how much influence is given to further away maps. In our experiments $p$ was set to 3. Figure 11.12 depicts the interpolation process.

The benefits of our approach are that it is not necessary to rerun the ray-tracing algorithm during simulation time, it is not necessary to divide the simulation area into evenly sized squares, and the accuracy can be increased in areas with a lot of obstacles, simply by adding more points. A real-time evaluation tool has been developed to show the result of the interpolation. Our approach increases the

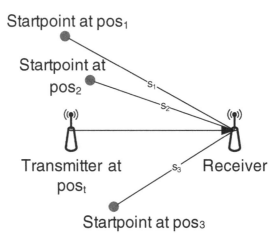

**Fig. 11.12** Distance-dependent interpolation

simulation speed and allows the designer to choose between high accuracy and reduced memory needs [23].

## 11.5 Thoughts for Practitioners

In this section we put the focus on the practical side of research, namely performing experiments by using a network simulation package. Due to space limitations, we cannot give a thorough guide and thus concentrate on the support provided by popular network simulation packages to the topic of this chapter.

### *11.5.1 Preparation, Documentation, and Evaluation of Simulations*

First of all, we would like to encounter some very important issues that should be considered when simulation is used to perform "experiments" to yield performance measurements.

- All used parameter settings should be documented.
- The changes to the "standard" version of the simulator should be described and possible interferences should be discussed.
- A "patch" to the current version of the simulation package should be made available to the research community, so that results can be reproduced independently by other researchers.
- The selection of input data, the range of input values, and the output data should be thoroughly discussed.
- Possible dependency of performance results with input data should be considered.

- The number of simulation replications and the method how replications have been done should be described.
- The type of simulation with respect to output should be described, e.g., *terminating* vs. *steady-state* simulation.
- The seeds of the pseudo random number generator should be documented. Otherwise, an exact reproduction of the simulation cannot be made.
- For the discussion of the simulations results, very often only average values are considered, which do not allow reliable conclusions. At least the standard deviation or better confidence intervals should be used to discuss performance measurements.

Often the space limitations of scientific publications will not allow to put all the mentioned information. In that case the details of the study can be documented in a separate document, e.g., in a technical report, which can be referenced in the publication and made available over the WWW. For a thorough discussion of these topics please consult [5] and [6].

## 11.5.2 Network Simulation Packages

There are some network simulation packages popular within the research community. Most of the network simulation packages are open source and developed by the research community. Sometimes a special simulator is developed if the existing ones do not fit the requirements of the study. There are also some popular commercial packages or commercially available versions of the free available network simulation packages.

The most famous network simulation package in this research area is *ns-2* (Network Simulator 2, http://www.isi.edu/nsnam/ns/). Actually, a successor is being developed, which is called *ns*-3 (Network Simulator 3, http://www.nsnam. org/). The advantage of ns-2 is that it is widely used and results of many papers are obtained by experiments performed with it. Additionally, there is a lot experience around ns-2. The disadvantage of ns-2 is that it lacks modularity, which complicates the development of new simulation models. Furthermore, the documentation of ns-2 is not up to date. We expect that ns-3 will improve many weaknesses of ns-2. In recent years, *OmNeT++* (http://www.omnetpp. org/) got intensive attention from the research community. It has a slim and clear modular architecture that supports the development of new simulation models. However, it is hardly known out of Europe. GloMoSim (Global Mobile Information Systems Simulation Library, http://pcl.cs.ucla.edu/projects/ glomosim/) is based on the *Parsec* compiler and was mainly developed for wireless simulations. There is also a commercial version *Qualnet*. JiST/SWANS (Java in Simulation Time / Scalable Wireless Ad hoc Network Simulator, http://jist.ece. cornell.edu/), as the name reveals is written in Java and uses the mechanisms of the Java virtual machine to accelerate simulation runs. OPNET Modeler (http://www.op net.com/solutions/network_rd/modeler.html) is one of the most famous commercial simulation packages with the support of many kinds of networks.

## 11.5.3 Support for Mobility Models

Table 11.1 shows an overview of the supported mobility models by the popular network simulation packages. In this overview, we only considered the models that are directly realized in the simulation package either as an internal part or as an individual application. We did not consider additional applications and tools, which are available and support various network simulation packages.

It is obvious that only a couple of the mobility models are supported directly by the simulation packages. Typically, only the simplest random mobility models are realized. The more complicated mobility models are not directly supported. Researchers built their experiment scenarios based on these available mobility models, which may lead to incorrect experiment set-ups, resulting in biased results.

One way to use mobility models which are not directly supported by the used network simulation package is to use *trace based* mobility patterns. For this, the simulation package specifies a particular format, which describes the movement of mobile nodes over the simulation time. During simulation, the movement pattern is read in from a file containing the movement pattern. Thus, a researcher can provide his own mobility patterns generated with other tools. We will discuss particular tools to generate movement patterns in the next subsection.

**Table 11.1** Support of mobility models by the established network simulation packages

| Model | ns-2 | ns-3 | OmNeT++ | OPNET | Jist/ SWANS | GlomoSim/ Qualnet |
|---|---|---|---|---|---|---|
| Random Walk | | ● | | | ● | |
| Random Waypoint | ● | ● | | ● | ● | ● |
| Random Direction | | ● | | | | |
| Gauss Markov | | | | | | |
| Freeway | | | | | | |
| Manhattan | | | | | | |
| Column | | | | | | |
| Nomadic Community | | | | | | |
| Pursue | | | | | | |
| Reference Point Group | | | | | | |
| Social based | | | | | | |
| Obstacle | | | | | | |
| Trace based | ● | ● | ● | | ● | ● |

## 11.5.4 Mobility Generation Tools

In this subsection we give an overview of specially designed tools to generate mobility traces and discuss their advantages and drawbacks (in alphabetical order). ANSim (Ad-Hoc Network Simulation, http://www.ansim.info/) is implemented in Java and provides the most famous mobility models. It can generate the mobility patterns based on a rectangular area or on circular area. BonnMotion (http://web.informatik.uni-bonn.de/IV/Mitar beiter/dewaal/BonnMotion/) is also implemented in Java and supports besides the famous entity mobility models also the Manhattan model and the Reference Point Group mobility model.

CosMos (Communication Scenario and Mobility Scenario Framework, http://www.adhoc-nets.de) is implemented in C++/Qt and provides entity mobility models and a graph-based simulation area. Its uniqueness is that it allows the combination of several mobility models in one scenario.

Mobility Generator Program for ns-2 (MGP, http://externe.emt.inrs.ca/ users/nuevo/NSmobgenerator.htm) is implemented in C and supports entity mobility models.

The Obstacle Mobility Model Project (OMM, http://moment.cs.ucsb.edu/ mobility/index.html) provides an implementation for the obstacle mobility model.

Table 11.2 depicts an overview of the discussed mobility generation tools, realized mobility models, and the supported network simulation packages. It is obvious that all the tools support only ns-2/ns-3 and GlomoSim. The other network simulation packages are not directly supported. However, the authors

**Table 11.2** Overview of mobility trace generators and supported mobility models and network simulation packages

| Model | ANSim | BonnMotion | CosMos | MGP | OMM |
|---|---|---|---|---|---|
| Random Walk | | | | • | |
| Random Waypoint | • | • | • | • | |
| Random Direction | • | | • | • | |
| Gauss Markov | • | • | | | |
| Freeway | | | • | | |
| Manhattan | | • | • | | |
| Column | | | | | |
| Nomadic Community | | | | | |
| Pursue | | | | | |
| Reference Point Group | | • | | | |
| Social-based | | | | | |
| Obstacle | | | | | • |
| ns-2/ns-3 | • | • | • | • | • |
| GlomoSim/Qualnet | • | • | • | | • |
| OmNeT++ | | | | | |
| OPNET | | | | | |
| JiSt/SWANS | | | | | |

of the tools state that it is simple to generate mobility traces also for other network simulation packages; however, due to lack of demand, they did not implement. We assume that authors typically support the simulation package which they are using. Besides these mobility generation tools, there are many others that implement only one mobility model for one particular network simulation package. Due to space limitations, we could not consider all of these tools; hence our list of tools is not exhaustive.

### 11.5.5 Support for Radio Propagation

Table 11.3 shows an overview of the supported radio propagation models by the popular network simulation packages. Again, we only considered the models which are directly realized in the simulation package either as an internal part or as an individual application.

All current network simulation packages offer the FreeSpace model. Also, the two-ray ground model is supported by almost all simulators; only ns-3 and the OmNet + + mobility framework do not offer it. Real site-specific modeling is not natively supported by any simulation package. The only option is to define a path loss matrix when using GlomoSim/Qualnet.

Other models not mentioned here because of space-constraints include Walfish-Ikegami and Hata [24] models, the Terrain Integrated Rough Earth Model (TIREM), or the Irregular Terrain Model (ITM).

**Table 11.3** Support of radio propagation models by established network simulation packages

| Model | ns-2 | ns-3 | OmNeT + + | OPNET | Jist/ SWANS | GlomoSim/ Qualnet |
|---|---|---|---|---|---|---|
| FreeSpace | • | • | • | • | • | • |
| Two-ray ground | • | | | • | • | • |
| Shadowing | • | | | | | • |
| Ricean/ Rayleigh Fading | | | | • | • | • |
| Site-specific modeling | | | | | | • (via path loss matrix) |

## 11.6 Directions for Future Research

As discussed throughout the chapter, the realistic simulation of wireless networks contains many subtle issues, and sometimes publications do not give a thorough description of the used simulation environment and the used settings. There are many open questions which have to be addressed by the research community to further improve the quality of simulations and hence the obtained results.

As mobility models influence the topology of the network, it is important that it is suitable for the considered scenario. Unfortunately, most network simulation packages as well as the tools which were designed to generate mobility patterns provide only a limited support for various mobility models. Tables 11.1 and 11.2 provide basically the state-of-the-art support of mobility models by network simulation packages as well as by the mobility generation tools. Even the supported mobility models are only useful for simple and generalized scenarios. Further research should focus on mobility patterns, which provide realistic scenarios. These could be achieved by GIS-based mobility traces. Currently, we are implementing a mobility pattern generator based on publicly available GIS systems.

The accuracy of radio wave propagation influences mainly the quality of wireless links. The simple models are cheap in terms of computations, but the provided accuracy is not useful and especially not appropriate for in-building scenarios. The high-accuracy ray-tracing models at the other side are too expensive in terms of computation. Furthermore, they demand for local terrain and building information to be able to compute useful results. Research is required to develop models with good tradeoff between accuracy and computation expense.

## 11.7 Conclusions

The mobility model and the radio wave propagation model are very important components of mobile ad hoc network simulations. Each component on its own has strong influence on the network topology and therefore a strong influence on the overall network performance. However, there is lack of awareness in the research community in respect to the components, which may lead to biased experiment results or wrong results that in turn allow invalid conclusions.

In this chapter we introduced the problematic of the mobility model and radio wave propagation model, discussed popular models with respect to their potentials and restrictions, gave an overview of popular network simulation packages used in the research community, and investigated the support in respect to the mobility and radio wave propagation models. Subsequently, we presented an integrated approach to realize more realistic simulation setups.

However, there are many open questions regarding the realistic simulation of mobile and wireless networks, which have to be answered to draw reliable conclusions based on the performed simulation experiments. Besides the discussed models, the traffic model is an elementary component, since the communicating peers define the load in the network. Based on the communicating peers, the mobility model, and the radio wave propagation model, the final network topology is created. A random uniform selection of the communicating peers does not seem very realistic.

# Terminologies

*Mobile ad hoc network (MANET)*: A wireless network that does not require any fixed infrastructure

*Node*: An entity in the network, e.g., Laptop, PC, PDA.

*Model*: An abstract representation of something that only contains the details which are important for the consideration. A model typically is application-dependent. It may happen that the same thing is modeled differently for different considerations.

*Mobility model*: A model that describes the movement of mobile nodes

*Topology, network topology*: The interpretation of the structure of a network at a given time. The nodes of the network are interpreted as the nodes of the graph; the edges of the graph are given by the wireless links between nodes in the network.

*Reflection*: When a radio wave hits an object whose dimensions are large compared to the wavelength, reflection occurs.

*Diffraction*: When a radio wave hits a rough surface, the energy is diffused in all directions; diffraction occurs, for example, on edges.

*Multipath propagation*: Many signals take paths that are of different length, thus the received signal varies heavily in quality.

*Large-scale models*: These models characterize how the transmission power between two nodes changes over long distances and over a long time.

*Small-scale models*: These models account for the fact that small movements (in the order of the wavelength) may have large influence on the transmission quality.

# Questions

1. What does the network topology represent?
2. For what purpose was the Brownian motion developed? Describe the Brownian motion model?
3. Which parameters determine the random waypoint mobility model?
4. Describe the mobility pattern of a traveling salesman from his office to a client. The traveling salesman has to walk, ride on a bus, take the metro, and walk again. Can you model such a travel with one mobility model?
5. What has to be considered when planning a network simulation with mobile nodes?
6. Calculate the coverage area of an access point if it transmits with 0.1 W in the 2.4 GHz frequency range. Assume unidirectional antennas without regarding antenna gain. Assume receivers can decode signals down to 2.47e–10 W.
7. Repeat this calculation with the two-ray ground model. Assume that the antennas are at 1.7 m height.
8. Which effects influence the signal propagation in indoor environments?

9. How does the shadowing model take these phenomenon into account?
10. Using the given formula for nearest neighbor interpolation, calculate the following: A transmitter T at position (2,3) wants to send to a receiver at (20,32). Evaluating the propagation map at position $S\_1 = (1,4)$ results in 3, at position $S\_2 = (4,3)$ results in 5. Which result would the interpolation give, if we assume $p = 1$?

# References

1. IETF Working group MANET. Mobile Ad-hoc Networks (MANET) Charter, 2002.
2. M. Günes, and I. Bouazizi. From Biology to Technology: Demonstration Environment for the Ant Routing Algorithm for Mobile Ad-hoc Networks. In Tenth Annual International Conference on Mobile Computing and Networking (ACM MobiCom 2004), http://www.sigmobile.org/mobicom/2004/demos.html, Philadelphia, USA, September 2004.
3. D. Kotz, C. Newport, R. S. Gray, J. Liu, Y. Yuan, and C. Elliott. Experimental Evaluation of Wireless Simulation Assumptions. Technical Report TR2004-507, Dept. of Computer Science, Dartmouth College, June 2004.
4. K. Pawlikowski, H.-D. J. Jeong, and J.-S. R. Lee. On Credibility of Simulation Studies of Telecommunication Networks. IEEE Communications, 40(1):132–139, January 2002.
5. J. Banks, J. S. Carson II, B. L. Nelson, and D. M. Nicol. Discrete-Event System Simulation. Prentice Hall, Upper Saddle River, NJ, 4th edition, 2005.
6. A. M. Law and W. D. Kelton. Simulation Modeling and Analysis. McGraw Hill, New York, NY, 3rd edition, 2000.
7. C. Bettstetter, H. Hartenstein, and X. Perez-Costa. Stochastic Properties of the Random Waypoint Mobility Model: Epoch Length, Direction Distribution, and Cell Change Rate. In MSWiM '02: Proceedings of the 5th ACM International Workshop on Modeling Analysis and Simulation of Wireless and Mobile Systems, Pages 7–14, ACM Press, New York, NY, USA, 2002.
8. C. Bettstetter, G. Resta, and P. Santi. The Node Distribution of the Random Waypoint Mobility Model for Wireless Ad hoc Networks. IEEE Transactions On Mobile Computing, 2(3):257–269, July–September 2003.
9. A. Einstein. Investigations on the Theory of Brownian Movement. Dover Publications, New York, NY, 1956.
10. D. B. Johnson, and D. A. Maltz. Dynamic Source Routing in Ad hoc Wireless Networks. In T. Imielinski and H. Korth, editors, Mobile Computing, Vol. 353. Kluwer, Dordercht, 1996.
11. E. Royer, P. M. Melliar-Smith, and L. Moser. An Analysis of the Optimum Node Density for Ad hoc Mobile Networks. In Proceedings of the IEEE International Conference on Communications (ICC), 2001.
12. B. Liang and Z. J. Haas. Predictive Distance-based Mobility Management for PCS Networks. In INFOCOM (3), pages 1377–1384, 1999.
13. F. Bai, N. Sadagopan, and A. Helmy. The IMPORTANT Framework for Analyzing the Impact of Mobility on Performance of Routing for Ad hoc Networks. AdHoc Networks Journal – Elsevier Science, 1(4):383–403, November 2003.
14. M. Sanchez and P. Manzoni. A Java-based Ad-hoc Networks Simulator. In SCS Western Multiconference. San Francisco, California, January 1999.
15. M. Sanchez. Mobility Models. http://www.disca.upv.es/misan/mobmodel.htm, 2005.

16. X. Hong, M. Gerla, G. Pei, and C.-C. Chiang. A Group Mobility Model for Ad hoc Wireless Networks. In Proceedings of the ACM International Workshop on Modeling and Simulation of Wireless and Mobile Systems (MSWiM), pages 53–60, August 1999.

17. M. Musolesi, S. Hailes, and C. Mascolo. An Ad hoc Mobility Model Founded on Social Network Theory. In Proceedings of the 7th ACM/IEEE International Symposium on Modeling, Analysis and Simulation of Wireless and Mobile Systems (MSWiM 2004), Venezia, Italy, October 2004. ACM.

18. A. Jardosh, E. M. Belding-Royer, K. C. Almeroth, and S. Suri. Towards Realistic Mobility Models for Mobile Ad hoc Networks. In the Ninth Annual International Conference on Mobile Computing and Networking (ACM MobiCom 2003), San Diego, California, USA, September 14–19 2003. ACM.

19. R. J. Punnoose, P. V. Nikitin, and D. D. Stancil. Efficient Simulation of Ricean Fading Within a Packet Simulator. In Vehicular Technology Conference, September 2000.

20. J.-M. Dricot and Ph. De Doncker. High-accuracy Physical Layer Model for Wireless Network Simulations in ns-2. In Proceedings of the International Workshop on Wireless Ad-hoc Networks, 2004.

21. M. Günes, and M. Wenig. Website of the Mobile Communication Group (MCG), http://www-i4.informatik.rwth-aachen.de/mcg/.

22. K. Fall, and K. Varadhan. The ns-2 manual. Technical report, The VINT Project, UC Berkeley, LBL and Xerox PARC, 2003.

23. A. Schmitz and M. Wenig. The Effect of the Radio Wave Propagation Model in Mobile Ad hoc Networks. In MSWiM '06: Proceedings of the 9th ACM international workshop on Modeling, Analysis and Simulation of Wireless and Mobile Systems, New York, NY, USA, 2006. ACM Press.

24. WG2 ÜHF Propagation. Urban Transmission Loss Models for Mobile Radios in the 900- and 1800-MHz Bands. Technical report, COST 231, 1990.

# Chapter 12
# Quality of Service Support in Wireless Ad Hoc Networks

Musfiq Rahman and Ashfaqur Rahman

**Abstract** Quality of Service is the performance level of a service offered by the network to the user. Provision of quality of service guarantees in wireless ad hoc networks is very challenging due to some inherent difficulties like node mobility, multi-hop communications, contention for channel access and lack of central coordination. In this chapter we summarise a comprehensive review of the state-of-the-art research works related to quality of service support in wireless ad hoc network and provide coordination among the research works.

## 12.1 Introduction

A Wireless Ad hoc NETwork (WANET) is a collection of wireless mobile nodes dynamically forming a temporary network. These networks can be formed on the fly, without requiring any fixed infrastructure. This type of dynamic network is useful for battlefield tactical operation or emergency search-and-rescue operations where an infrastructure is beyond imagination. The formation simplicity of WANET also inspired its deployment in civilian forums such as ah hoc conferences, campus recreation and electronic classrooms. On top of that, the rapid development of multimedia applications and potential commercial usage of WANET makes Quality of Service (QoS) inevitable in WANET.

QoS is the performance level of a service offered by the network to the user. The goal of QoS provisioning is to achieve more deterministic network behaviour by proper utilisation of the network resources. A network or a service provider can offer different kinds of services to the users by a set of service requirements such as minimum bandwidth, maximum delay, maximum delay variance and maximum packet loss rate. After accepting a service request from the user, the network is expected to ensure the committed service requirements

M. Rahman (✉)
Department of Computer Science, American International University Bangladesh, Bangladesh
e-mail: musfiq@aiub.edu

S. Misra et al. (eds.), *Guide to Wireless Ad Hoc Networks*, Computer Communications and Networks, DOI 10.1007/978-1-84800-328-6_12,

of the users throughout the communication. Characteristics of WANET, such as lack of central coordination, mobility of hosts and limited availability of resources, make QoS provisioning very challenging.

A careful scrutiny of the existing literature reveals that the major research areas on QoS support in WANET comprises QoS models, QoS resource reservation signalling, QoS routing and QoS Medium Access Control (MAC). A QoS model defines the service architecture of the total QoS framework. QoS signalling is used to reserve the resources required for a QoS session. QoS routing searches a path with enough resources to meet the service requirements. QoS MAC provides a mechanism for resolving medium contention and supporting reliable unicast communication, etc.

The research on QoS support in WANET in the existing literature focuses on certain aspects of QoS in the abovementioned areas. The seemingly different areas are, however, highly correlated to provide a complete QoS support. For example, QoS routing searches for a path with sufficient resources, but the reservation of the resources has to be taken care of by QoS signalling protocol. In a similar fashion, a change in network dynamics (e.g. addition/deletion of a node), as detected by the MAC layer, can provide significant information to the QoS routing protocols to decide on an up-to-date path selection. Altogether, to get an overall picture of the complete QoS support system, it is necessary to understand and evaluate the particular methods as well as their relationships. In this chapter we present a comprehensive review of the state-of-the-art research works on QoS support in WANET. We also provide the coordination among the research works to deliver QoS support in WANET.

## 12.2 Background

This section mainly concentrates on detailing the building blocks of any QoS architectural framework. The following subsections explain a set of modules from the perspective of their role in supporting QoS.

### 12.2.1 Admission Control

This is used to throttle the traffic in such a way that newly admitted traffic does not lead to network overload or service degradation to existing traffic. Admission control in general is based on some policies or sets of rules that can be specific to an Internet Service Provider (ISP) or a Service Level Agreement (SLA) between a subscriber and the ISP. An admission control decision can also depend on the availability of adequate network resources in an attempt to meet the performance objectives of a particular service request. It is often realised either through parameter-based approach or through measurement-based approach. While the parameter-based is more applicable for hard QoS guarantees, the measurement-based approach is more suitable to provide soft or relative service guarantees.

## 12.2.2  Resource Reservation

Through this mechanism, the network sets aside the required resources on demand for delivering the desired network performance. This is in general closely associated with admission control. Since charges are normally based on the use of reserved resources, resource reservation necessitates the support of authentication, authorisation and accounting and settlement between different ISPs. Resource reservation is typically performed with a signalling mechanism such as RSVP or INSIGNIA.

## 12.2.3  Buffer Management

Consider one of the UDP segments generated by an IP phone application. The UDP segment is encapsulated in an IP datagram. As the datagram wanders through the network, it passes through buffers (i.e. queues) in the routers in order to access outbound links. It is possible that one or more buffers in the route from the sender to receiver is full and cannot admit the IP datagram. In this case, the IP datagram is discarded, never to arrive at the receiving application. Therefore, we need a mechanism to deal with the *packet loss*. The buffer management deals with the task of either storing or dropping a packet awaiting transmission. The key mechanisms of buffer management are the backlog controller and the dropper. The backlog controller specifies the time instances when traffic should be dropped, and the dropper specifies the traffic to be dropped. Buffer management is often associated with congestion control.

## 12.2.4  Classifying and Scheduling

It is the mechanism that selects a packet for transmission from the packets waiting in the transmission queue. Hence, the desired service guarantees are realised independently at each router by scheduling. Scheduling is based on a service rate allocation to classes of traffic that share a common buffer. Packet scheduling thus controls bandwidth allocation to different nodes or classes or applications.

## 12.2.5  End-to-End Delay

*End-to-End delay* is the accumulation of transmission, processing and queuing delays in routers; propagation delay in the links; and end-system processing delays. For highly interactive application such as IP phone, end-to-end delays smaller than 150 ms are not perceived by human listeners. The less the end-to-end delay means better performance.

### 12.2.6 Packet Jitter

A crucial component of end-to-end delay is the random queuing delays in the routers. Because of these varying delays within the network, the time from when a packet is generated at the source until it is received at the receiver can fluctuate from packet to packet. This phenomenon is called *jitter*.

### 12.2.7 QoS – Hard vs Soft State

Maintaining the QOS of adaptive flows in WANET is one of the most challenging aspects of the QOS framework. Typically, wireline networks have little quality of service or state management where the route and the reservation between source–destination pairs remain fixed for the duration of a session. This style of *hard-state* connection-oriented communications (e.g. virtual circuit) guarantees QoS for the duration of the session holding time. However, these techniques are not flexible enough in WANET where the paths and reservations need to dynamically respond to topology changes in a timely manner. Therefore, *soft-state* approach to state management at intermediate routing nodes is a suitable approach for the management of reservations in WANET. Soft state relies on the fact that a source sends data packets along an existing path. If a data packet arrives at a mobile router and no reservation exists, then admission control and resource reservations attempt to establish soft state. Subsequent reception of data packets (associated with the reservation) at that router are used to refresh the existing soft-state reservation. This is called a *soft-connection* when considered on an end-to-end basis and in relation to the virtual circuit hard-state model. When an intermediate node receives a data packet that has an existing reservation, it reconfirms the reservation over the next interval. Therefore, the holding time for a *soft-connection* is based on the soft-state timer interval and not based on the session duration holding time. If a new packet is not received within the soft-state timer interval, then resources are released and flow states removed in a fully decentralised manner.

## 12.3 Challenges of QoS Provisioning in WANET

Enormous research is conduced on providing QoS support in conventional wireless networks. This kind of wireless networks often require a fixed wireline backbone where mobile hosts can reach the wireline base stations in one hop radio transmission. In WANET no such fixed infrastructure can be presumed. Thus providing QoS support in WANET is more challenging than conventional wireless networks. The following is a summary of the major challenges [26, 48] to provide QoS support in WANET.

- *Dynamic network topology*: In WANET there is no restriction on mobility. Thus the network topology changes dynamically causing hosts to have imprecise knowledge of the current status. A QoS session may suffer due to frequent path breaks, thereby requiring re-establishment of new paths. The delay incurred in re-establishing a QoS session may cause some of the packets belonging to that session to miss their delay targets/deadlines, which is not acceptable for applications that have stringent QoS requirements.
- *Error-prone wireless channel*: The wireless radio channel by nature is a broadcast medium. The radio waves suffer from several impairments such as attenuation, thermal noise, interference, shadowing and multi-path fading effects [50] during propagation through the wireless medium. This makes it difficult to ensure QoS commitments like hard packet delivery ratio or link longevity guarantees.
- *Lack of central coordination*: Like wireless LAN and cellular network, a WANET does not have central controllers to coordinate the activity of the nodes. A WANET may be set up spontaneously without planning and its members can change dynamically, thus making it difficult to provide any form of centralised control. As a result communications protocols in WANET utilise only locally available state and operate in a distributed manner [43]. This generally increases the overhead and complexity of an algorithm as QoS state information must be disseminated efficiently.
- *Imprecise state information*: The nodes in a WANET mostly maintain link-specific as well as flow-specific state information. The link-specific state information comprises bandwidth, delay, delay jitter, loss rate, error rate, stability, cost and distance values for each link. The flow-specific information includes session ID, source address, destination address and QoS requirements of the flow. Due to dynamic changes in network topology and channel characteristics, these state information are inherently imprecise. This may result in inaccurate routing decisions resulting in some packets missing their deadlines, leading to violation of real-time QoS commitment.
- *Limited availability of resources*: Although mobile devices are becoming increasingly powerful and capable, it still holds true that such devices generally have less computational power, less memory and a limited (battery) power supply, compared to devices such as desktop computers typically employed in wired networks. This factor has a major impact on the provision of QoS assurances, since low memory capacity limits the amount of QoS state that can be stored, necessitating more frequent updates incurring greater overhead. Moreover, QoS routing generally incurs a greater overhead than best-effort routing due to the extra information being disseminated. These factors lead to higher consumption of mobile nodes' limited

battery power. Finally, among the QoS routing problems, many are NP-complete [12], and thus complicated heuristics are required for solving them, which may place an undue strain on mobile nodes' less-powerful processors.

- *Hidden terminal problem*: The hidden terminal problem is inherent in WANET. This problem occurs when packets originating from two or more sender nodes, which are not within the direct transmission range of each other, collide at a common receiver node. It necessitates retransmission of packets, which may not be acceptable for flows that have strict QoS requirements. Some control packet exchange mechanisms [30, 32] reduce the hidden terminal problem only to a certain extent. Some important solution to this problem is proposed in [58, 20].

## 12.4 Factors Affecting QoS Protocol Performance

Even after overcoming the challenges of WANET, a number of factors [26] have major impacts while evaluating the performance of QoS protocols. Some of these parameters are of particular interest considering the characteristics of the WANET environment. They can be summarised as follows:

- *Node mobility*: This factor generally encompasses several parameters: the nodes' maximum and minimum speed, speed pattern and pause time. The node's speed pattern determines whether the node moves at uniform speed at all times or whether it is constantly varying, and also how it accelerates, for example, uniformly or exponentially with time. The pause time determines the length of time nodes remain stationary between each period of movement. Together with maximum and minimum speed, this parameter determines how often the network topology changes and thus how often network state information must be updated. This parameter has been the focus of many studies, e.g. [46, 11];
- *Network size*: Since QoS state has to be gathered or disseminated in some way for routing decisions to be made, the larger the network, the more difficult this becomes in terms of update latency and message overhead. This is the same as with all network state information, such as that used in best-effort protocols [43];
- *Number, type and data rate of traffic sources*: Intuitively, a smaller number of traffic sources results in fewer routes being required and vice versa. Traffic sources can be constant bit rate (CBR) or may generate bits or packets at a rate that varies with time according to the Poisson distribution, or any other mathematical model. The maximum data rate affects the number of packets in the network and hence the network load. All of these factors affect performance significantly [46];
- *Node transmission power*: Some nodes may have the ability to vary their transmission power. This is important, since at a higher power, nodes have more direct neighbours and hence connectivity increases, but the interference

between nodes does as well. Transmission power control can also result in unidirectional 'links' between nodes, which can affect the performance of routing protocols. This factor has also been studied extensively, e.g. [65, 13, 22];

- *Channel characteristics*: As detailed earlier, there are many reasons for the wireless channel being unreliable, i.e. many reasons why bits, and hence data packets, may not be delivered correctly. These all affect the network's ability to provide QoS.

## 12.5  QoS Models

QoS model defines the service architecture of the total QoS framework. That is, it defines the type of services a QoS model should provide and the classifications of the services. It also sets the system level goal that all other components like QoS signalling, QoS Routing and QoS MAC layer of QoS framework should implement.

A lot of work is done to support QoS in the Internet. A QoS model for WANET, however, should be able to overcome the challenges of WANET, e.g. dynamic topology and time-varying link capacity. The QoS model for WANET has extended the traditional Internet models to make them suitable for WANET. We are thus motivated to elaborate the existing QoS models for the Internet such as IntServ [9] and DiffServ [7] as background and highlight their incompatibility for WANET. We then describe FQMW and SWAN – the QoS models for WANET.

### 12.5.1  Integrated Service (IntServ) Model

The IntServ model [9] merges the advantages of two different paradigms: datagram networks and circuit switched networks. It can provide a circuit-switched service in packet-switched networks. IntServ is a framework developed within the IETF to provide individualised QoS guarantees to individual application session. The basic idea of the IntServ [9] model is that the flow-specific states are kept in every IntServ-enabled router [64] . A flow is an application session between a pair of end users.

IntServ proposes two basic service classes: Guaranteed Service [53] and Controlled Load Service [62]. The Guaranteed Service is provided for applications requiring fixed delay bound. The Controlled Load Service is for applications requiring reliable and enhanced best effort service. Because every router keeps the flow state information, the quantitative QoS provided by IntServ is for every individual flow.

In an IntServ-enabled router, IntServ is implemented with four main components [64]: the signalling protocol, the admission control routine, the classifier and the packet scheduler. Other components, such as the routing agent and management agent, are the original mechanisms of the routers and can be kept unchanged. The Resource ReSerVation Protocol (RSVP) [10] is used as the signalling protocol to reserve resources in IntServ. Applications with Guaranteed Service or Controlled-Load Service requirements use RSVP to reserve resources before transmission.

Admission control is used to decide whether to accept the resource requirement. It is invoked at each router to make a local accept/reject decision at the time that a host requests a real-time service along some paths through the Internet. Admission control notifies the application through RSVP if the QoS requirement can be granted or not. The application can transmit its data packets only after the QoS requirement is accepted. When a router receives a data packet, the classifier will perform a Multi-Field (MF) classification [25], which classifies a packet based on multiple fields such as source and destination addresses, source and destination port numbers, Type Of Service (TOS) bits and protocol ID in the IP header (Fig. 12.1). Then the classified packet will be put into a corresponding queue according to the classification result. Finally, the packet scheduler reorders the output queue to meet different QoS requirements.

Unfortunately, IntServ/RSVP model is not suitable for WANETs due to the resource limitation in WANETs: (1) the amount of state information increases proportionally with the number of flows (the scalability problem, which is also a problem for current Internet). Keeping flow-state information will cost a huge storage and processing overhead for the mobile host whose storage and computing resources are scarce. Although the scalability problem may not be likely to happen in current WANET due to its limited bandwidth and relatively small number of flows compared with wired networks, it may occur in near future due to the development of fast radio technology and potential large number of users; (2) the RSVP signalling packets will contend for bandwidth with the data packets and consume a substantial percentage of bandwidth in WANETs; (3) every mobile host must perform the processing of admission control, classification and scheduling. This is a heavy burden for the resource-limited mobile hosts.

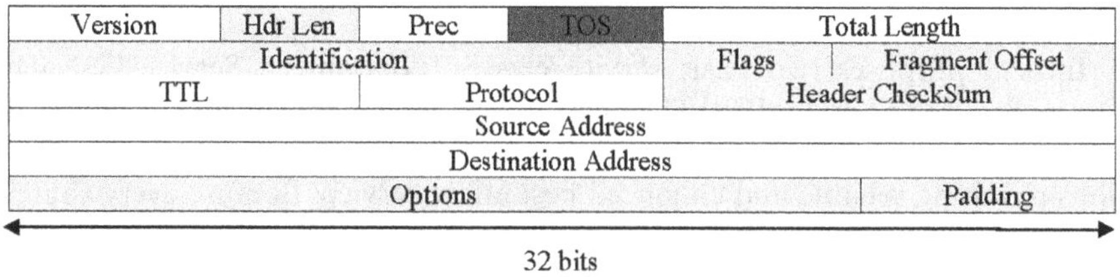

| Version | Hdr Len | Prec | TOS | Total Length | |
|---------|---------|------|-----|--------------|---|
| Identification | | | | Flags | Fragment Offset |
| TTL | | Protocol | | Header CheckSum | |
| Source Address | | | | | |
| Destination Address | | | | | |
| Options | | | | Padding | |

32 bits

**Fig. 12.1** Fields in a Standard IPv4 Packet Header. *Shaded* fields are absent from IPv6 header. TOS represents the Type Of Service bits.

## 12.5.2 *Differentiated Service (DiffServ) Model*

The DiffServ architecture, on the other hand, is designed to overcome the implementing and deploying difficulty of IntServ and RSVP in the Internet backbone [64]. DiffServ also addresses the scalability issues of IntServ by providing a limited number of aggregated classes.

In DiffServ architecture, an edge router controls the traffic entering the network by classifying, marking, policing and shaping mechanism. The policy manager, in the router, assures that no one will violate the type of service it is pre-assigned. DiiffServ also defines the layout of the Type of Service (TOS) bits in the IP header, called the DS field, and a base set of packet forwarding rules, called Per-Hop-Behavior (PHB). When a data packet enters a DiffServ-enabled domain, an edge router marks the packet"s DS field, and the interior routes along the forwarding path forward the packet based on its DS field. Since the DS field only codes very limited service classes, the processing of the core routers is very simple and fast.

Unlike in IntServ, core routers in DiffServ do not need to keep per-flow state information (Fig. 12.2). Many services, such as Premium Service [41], Assured Service [18, 28] and Olympic Service [27] (Nichols et al. 1999), can be supported in the DiffServ model. Premium Service is supposed to provide low loss, low delay, low jitter, and end-to-end assured bandwidth service. Assured Service is for applications requiring better reliability than Best Effort Service. Its purpose is to provide guaranteed or at least expected throughput for applications. Furthermore, it is more qualitative-oriented than quantitative-oriented and thus easy to implement. Olympic Service provides three tiers of services: Gold, Silver and Bronze, with decreasing quality [64].

Because of easy deployment and lightweight core node requirement, Diff-Serv may be a possible solution to the WANET QoS model. In addition, it provides Assured Service, which is a feasible service context in WANET.

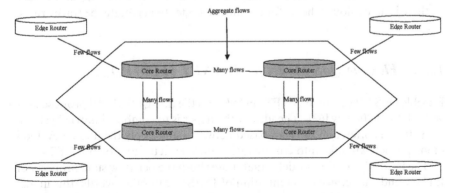

**Fig. 12.2** Flow management in DiffServ. Interior flows are aggregated in the core router and the core routers do not keep per flow information

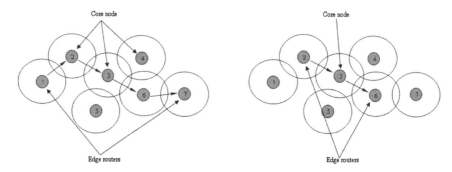

**Fig. 12.3** Ambiguity on edge routers in DiffServ. in the first scenario node 2, 3, 6 act as core nodes and 1, 7 as edge routers; on the other hand, in the second scenario only node 3 act as a core node and 2, 6 as edge routers

However, since DiffServ is designed for fixed wire networks, we still face some challenges to implement DiffServ in WANETs:

- First, it is ambiguous as to what the edge routers in WANETs are (Fig. 12.3). Intuitively, the source nodes play the role of edge routers. Other nodes along the forwarding paths from sources to destinations are core nodes. But every node should have the functionality as both boundary router and interior router because the source nodes cannot be predefined. This arouses a heavy storage cost in every host.
- Second, the concept of Service SLA in the Internet does not exist in WANETs. The SLA is a kind of contract between a customer and its ISP, which specifies the forwarding services the customer should receive. In the Internet, a customer must have a SLA with its ISP in order to receive Differentiated Services. The SLA is indispensable because it includes the whole or partial traffic conditioning rules, which are used to re-mark traffic streams, discard or shape packets according to the traffic characteristics such as rate and burst size. How to make a SLA in WANETs is difficult because there is no obvious scheme for the mobile nodes to negotiate the traffic rules.

### 12.5.3 Flexible QoS Model for WANET (FQMW)

Flexible QoS Model for WANET (FQMW) is the first QoS Model proposed for WANETs in 2000 by [63]. The idea of the paper is to combine knowledge from the solutions offered in the wire-based networks and apply them to a new QoS model, which will take into consideration the characteristics of WANETs.

The basic idea of this model is that it uses both the per-flow state property of IntServ and the service differentiation of DiffServ. In other words, this model proposes to assign highest priority to per-flow provisioning and other priority classes are given per-class provisioning. This model is based on the assumption that not all packets in our network are actually seeking for highest priority because then this model would result in a similar model with IntServ where we have per-flow provisioning for all packets.

As in DiffServ, three kinds of nodes (ingress, core and egress nodes) are defined in FQMW. An ingress node is a mobile node that sends data. Core nodes are the nodes that forward data for other nodes. An egress node is a destination node (Fig. 12.4). The difference though is that in FQMW the type of a node has nothing to do with its physical location in the network, since this would not make any sense in a dynamic network topology. A traffic conditioner is placed at the ingress nodes where the traffic originates. It is responsible for re-marking the traffic streams, discarding or shaping packets according to the traffic profile, which describes the temporal properties of a traffic stream such as rate and burst size.

FQMW is the first attempt at proposing a QoS model for WANETs. However, some problems still need be solved. First, how many sessions could be served by per-flow granularity? Without an explicit control on the number of services with per-flow granularity, the scalability problem still exists. Second, just as in DiffServ, the core nodes forward packets according to a certain PHB that is labelled in the DS field. We argue that it is difficult to code the PHB in the DS field if the PHB includes per-flow granularity, considering the DS field is at most 8 bits without extension. Finally, making a dynamically negotiated traffic profile is very difficult.

Very recently, the FQMW model has been extended to Relative Bandwidth Service Differentiation (RBSD) scheme in order to realise relative service differentiation in WANET [63, 17]. It is termed in [17] as the relative bandwidth service differentiation scheme, where a service profile $\gamma$ for a traffic session is defined as a relative target rate, which is in fact a fraction of the effective capacity of a link and ranges between 0 and 1. The relative target rate of a session is normalised over time according to the traffic distribution in the WANET. Let session $i$ be assigned

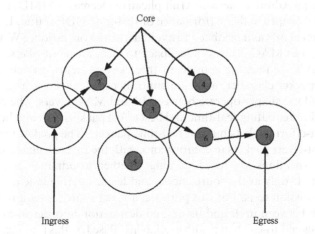

**Fig. 12.4** Ingress, core, and egress nodes in FQMW

a relative target rate $\gamma_i$. At a certain time instance, packets of session $i$ arrive at node M, which altogether handles $m$ number of sessions. The relative target rate of session $i$ at node M is then normalised as:

$$\gamma_i^M = \frac{\gamma_i}{\sum\limits_{k=1}^{m} \gamma_k}. \tag{12.1}$$

In Equation (12.1), $\gamma_k$ is the relative target rate of session $k$. Each node uses a token bucket profile meter whose token rate $\rho$ and bucket length $\beta$ are adaptively adjusted according to the instantaneous value of the effective link capacity. In order to estimate the effective link capacity, RSBD proposes two methods – parameter-based and measurement-based. Like its predecessor (i.e., the FQMW), the present model also present the flaws like how many sessions could be served by per-flow granularity and how to distinguish between core and border node.

## 12.5.4 Service Differentiation in Wireless Ad Hoc Network (SWAN)

SWAN is a stateless approach dealing with service differentiation in mobile ad hoc networks [3]. SWAN assumes the use of best-effort MAC (for example IEEE 802.11 DCF) and uses feedback-based control mechanisms to support soft real-time services and service differentiation in ad hoc networks. In order to ensure that the bandwidth and delay requirements of real-time UDP traffic are met, distributed rate control of TCP and UDP best-effort traffic is performed at every node. Rate control is designed to restrict best-effort traffic – thus yielding the necessary bandwidth required to support real-time traffic. In addition, SWAN uses an Additive Increase Multiplicative Decrease (AIMD) rate control mechanism to improve the performance of real-time UDP traffic. Unlike TCP that uses packet loss as a feedback to avoid network congestion, SWAN, on the other hand, uses MAC delay as a feedback to local rate controllers.

The SWAN architecture consists of three key elements, namely admission controller, packet classifier, and rate controller as depicted in Fig. 12.5. The classifier and the shaper operate between IP and MAC layers. The classifier is capable of differentiating real-time and best effort packets, forcing the shaper to process best-effort packets but not real-time packets. The goal of the shaper is to delay best-effort packets in conformance with the rate calculated by the rate controller. An admission test regarding whether to admit a new real-time session is made only at the source node, and hence intermediate nodes do not perform admission tests. For this purpose, a given source is required to probe the network between itself and its desired destination in order to measure the instantaneous end-to-end bandwidth availability. Based on this probing, the source makes the sole decision. In case of false source-based admission control or traffic

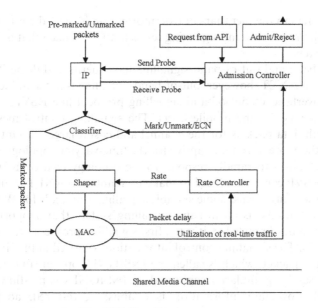

**Fig. 12.5** Key components of the SWAN architecture

violations brought on by the re-routing of real-time sessions, Explicit Congestion notification (ECN) is used to control and regulate UDP real-time traffic.

The weak side of SWAN approach is that it can only provide weak service guarantees and, although it is claimed to be stateless, intermediate nodes may be required to remember whether the flows that traverse them are new or old in order to regulate traffic [3]. In addition, source-based admission control using probing packets is again unrealistic and ineffective in a dynamic environment such as a WANET, as conditions and network topology tend to change fairly frequently. Furthermore, bandwidth calculations in SWAN do not take best-effort traffic into consideration, and hence may lead to a false estimation of the available bandwidth.

## 12.6  QoS Signalling – INSIGNIA

QoS signalling is used to reserve the resources required for a QoS session. A signalling protocol should handle the resource reservation, release, call setup, call tear down and re-negotiation of flows in the network. A good QoS signalling should have two distinct mechanisms. First, the QoS signalling information must be reliably carried between the routers. Second, the QoS signalling information must be correctly interpreted and the relative processing should be activated. Based on the first mechanism, the QoS signalling system can be divided into in-band signalling and out-of-band signalling. The in-band signalling refers to the fact that control information is carried along with data packets [34]; the out-of-band signalling refers to the approach that uses explicit control packets.

Both in-band and out-of-band signalling have merits and demerits in their favour. For WANET, however, out-of-band signalling is not suitable since the signalling overhead of out-ofband signalling protocol like RSVP, as used in IntServ, is heavy for the mobile hosts. The signalling control message will contend with data packets for the channel and cost a large amount of bandwidth. Furthermore, it is not adaptive for the time-varying topology because it has no mechanism to rapidly respond to the topology change in WANETs. However, RSVP can be modified to make to adapt to WANET challenges. In this section we thus discuss some suitable signalling protocols for WANET.

INSIGNIA [2, 34] is an in-band signalling system that supports QoS in WANETs. To our knowledge, it is the first signalling protocol designed solely for WANETs. The signalling control information is carried in the IP option of every IP data packet, which is called the INSIGNIA option (Fig. 12.6). Like RSVP, the service granularity supported by INSIGNIA is per-flow management. Each flow-state information is established, restored, adapted and removed over an end-to-end session in response to topology change and end-to-end quality of service condition.

Figure 12.7 shows the position and the role of INSIGNIA in wireless flow management [2] at a mobile host. The packet-forwarding module classifies the

| Reservation Mode | Service Type | Payload Indicator | Bandwidth Indicator | Bandwidth Request | |
|---|---|---|---|---|---|
| REQ/RES | RT/BE | RT/BE | MAX/MIN | MAX | MIN |
| 1 bit | 1 bit | 1 bit | 1 bit | 16 bits | |

**Fig. 12.6** INSIGNIA IP option field

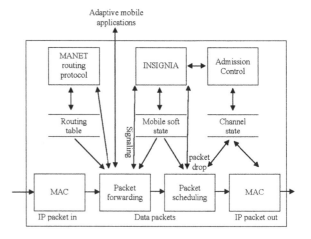

**Fig. 12.7** Wireless flow management model at a mobile host

incoming packets and forwards them to the appropriate modules (routing, INSIGNIA, local applications and packet scheduling modules) [34]. If a received IP packet includes an INSIGNIA option, the control information is forwarded to and processed by the INSIGNIA module. In the meantime, the received packet is delivered to a local application or forwarded to the packet-scheduling module according to the destination address in the IP head. If the mobile host is the destination of the packet, the packet is processed by a local application. Otherwise, the mobile host will forward the packet to the next hop determined by the WANET routing protocol. Before the packets are sent through the MAC component, a packet-scheduling module is used to schedule the output of the flows in order to fairly allocate the resource to different flows. In INSIGNIA, a Weighted Round-Robin (WRR) discipline that takes location-dependent channel conditions into account [38] is implemented. Note that a wide variety of scheduling disciplines could be used to realise the packet scheduling.

The INSIGNIA module is responsible for establishing, restoring, adapting and tearing down real-time flows. It includes fast flow reservation, restoration and adaptation algorithms that are specifically designed to deliver adaptive real-time service in WANET [34]. The flow-state information is managed in soft-state method, that is, the flow-state information is periodically refreshed by the received signalling information. Coordinating with the admission control module, INSIGNIA allocates bandwidth to the flow if the resource requirement can be satisfied. Otherwise, if the required resource is unavailable, the flow will be degraded to best-effort service. To keep the processing simple and light-weight, INSIGNIA does not send rejection and error messages if the resource request is not satisfied. For fast responding to the changes in network topology and end-to-end quality of service conditions, INSIGNIA uses QoS reports to inform the source node of the status of the real-time flows. The destination node actively monitors the received flows and calculates QoS statistical results such as loss rate, delay, and throughput. The QoS reports are periodically sent to the source node. Through this kind of feedback information, the source node can take corresponding actions to adapt the flows to observed network conditions.

As a whole, INSIGNIA is an effective signalling protocol for WANETs. Coordinating with other network components (viz. routing protocol, scheduling and admission control), INSIGNIA can efficiently deliver adaptive real-time flows in WANETs. However, since the flow-state information should be kept in the mobile hosts, the scalability problem may hinder its deployment in the future.

## 12.7  QoS MAC Protocol

QoS supporting components at upper layers, such as QoS signalling and QoS routing, assume the existence of a MAC protocol, which solves the problems of medium contention, supports reliable unicast communication and provides resource reservation for real-time traffic in a distributed wireless environment. A lot of MAC protocols [58, 32, 57, 6] have been proposed for wireless networks. Unfortunately, their design goals are usually to solve medium contention and hidden/exposed terminal problems and improve throughput. Most of them do not provide resource reservation and QoS guarantees to real-time traffic.

The first problem that a MAC protocol in wireless networks should solve is the hidden/exposed terminal problem. For convenience in later discussion, we simply describe the problem and the Request-To-Send (RTS) – Clear-To-Send (CTS) dialogue as its basic solution. As shown in Fig. 12.8, host A and host C cannot hear each other. When A is transmitting a packet to B, C cannot sense the transmission from A. Thus C may transmit a packet to B and cause a collision at B. This is referred as the *hidden terminal* problem since A is hidden from C. Similarly, when B is transmitting a packet to C, A cannot initiate a transmission to D, since this can potentially cause collisions of the control packets at both B and A, thereby disrupting both transmissions. This is called the *exposed terminal* problem since A is exposed to B. An RTS-CTS dialogue can be used to solve the hidden/exposed terminal problem. In Fig. 12.8, when C wants to send a data packet to B, it first sends a RTS message to B. When B receives the RTS, it broadcasts a CTS message to C and A. When C receives the CTS, it begins to transmit the data packet. Upon receiving the CTS, A will defer its data transmission because it knows B will receive data from C. This method avoids the possible collisions at host B and thus solves the hidden terminal (A is hidden from C) and exposed terminal (A is exposed to B) problems.

Another dialogue frequently used in MAC protocols is the PKT-ACK dialogue, which means the sender sends a data packet (PKT) to the receiver and the receiver immediately responds with an acknowledgement packet (ACK) to the sender if the data packet is correctly received. Failure to receive the ACK will prompt a retransmission after a short timeout.

Besides dealing with the hidden/exposed terminal problems, a QoS MAC protocol must provide resource reservation and QoS guarantees to real-time traffic. There are some proposed protocols such as the Multiple Access Collision

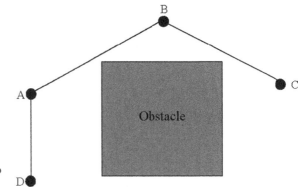

**Fig. 12.8** Node A is hidden from node C and exposed to node B

Avoidance with Piggyback Reservation (GAMA/PR) protocol [40] and the newly proposed Black-Burst (BB) contention mechanism [55], which can provide QoS guarantees to real-time traffic in a distributed wireless environment. However, they are supposed to work in a wireless LAN in which every host can sense each other's transmission, or in a wireless network without hidden hosts.

A MAC layer protocol for QoS support in WANET was proposed by C.R. Lin and M. Gerla in [36]. They proposed Multiple Access Collision Avoidance with Piggyback Reservation (MACA/PR) for multi-hop wireless networks. MACA/PR provides rapid and reliable transmission of non-real-time datagrams as well as guaranteed bandwidth support to real-time traffic.

On the other hand, for the transmission of non-real-time datagrams in MACA/PR, a host with a packet to send must first wait for a free *window* in the Reservation Table (RT), which records all reserved send and receive windows of any station within the transmission range. It then waits for an additional random time on the order of a single hop round trip delay. If it senses that the channel is free, it proceeds with RTS-CTS-PKT-ACK dialogue for a successful packet transmission. If the channel is busy, it waits until the channel becomes idle and repeats the above procedure.

For the transmission of real-time packets, the behaviour of MACA/PR is different. In order to transmit the first data packet of a real-time connection, the sender S initiates an RTS-CTS dialogue and then proceeds with PKT-ACK dialogues if the CTS is received. For subsequent data packets (not the first one) of a real-time connection, only PKT-ACK dialogues are needed. Note that if the sender fails to receive several ACKs, it restarts the connection with the RTS-CTS dialogue again. MACA/PR does not retransmit the real-time packets after collision.

In order to reserve bandwidth for real-time traffic, the real-time scheduling information is carried in the headers of PKTs and ACKs. The sender S piggybacks the reservation information for its next data packet transmission on the current data packet (PKT). The intended receiver D inserts the reservation in its Reservation Table (RT) and confirms it with the ACK to the sender. The neighbours of the receiver D will defer their transmission once receiving the ACK. In addition, from the ACK, they also know the next scheduled receiving time of D and avoid transmission at the time when D is scheduled to receive the next data packet from S. The real-time packets are protected from hidden hosts by the propagation and maintenance of reservation tables among neighbours, not by the RTS-CTS dialogues. Thus, through the piggybacked reservation information and the maintenance of the reservation tables, the bandwidth is reserved and guaranteed for the real-time traffic.

## 12.8 QoS Routing Mechanism

Due to node mobility and limited wireless communication range of nodes in a multi-hop WANET, communication with other node must depend on the neighbour nodes to forward the data packet to the destination node. Hence, a routing protocol for WANET is a protocol that will execute on every node and therefore subject to the limit of the resources at each mobile node. The challenges of WANET, as discussed in Section 12.3, also make the routing protocol more difficult of implement.

Existing literature provides plenty of solutions to the routing problem of WANET. However, not all of them have the capability to assure QoS. A QoS routing protocol can be defined as: Given a source node *s*, a destination node *d*, a set of QoS constraints *C* and a possible optimisation goal, a QoS routing algorithm finds the best feasible path *s* to *d* which satisfies *C*. For example, consider Fig. 12.9 where the numbers next to the radio links represent their respective bandwidth (e.g. megabits per second). To minimise delay and better use of network resources, minimising the number of intermediate hops is one of the principal objectives in determining suitable routes. However, suppose that the packet flow from A to E requires a bandwidth guarantee of 3 Mb/s. QoS routing will then select A-B-C-E over route A-D-E, although the latter has fewer hops. In this chapter we concentrate on WANET QoS routing protocols only.

QoS routing is difficult in WANET due to several reasons. First, the overhead of QoS routing is too high for the bandwidth-limited WANETs because the mobile host should have some mechanisms to store and update link state information. We have to balance the benefit of QoS routing against the bandwidth consumption in WANETs. Second, because of the dynamic nature of WANETs, maintaining the precise link state information is very difficult. Third, the traditional meaning that the required QoS should be ensured once a feasible path is established is no longer true. The reserved resource may not be guaranteed because of the mobility-caused path breakage or power depletion of the mobile hosts. QoS routing should rapidly find a feasible new route to recover the service.

Addressing the abovementioned difficulties, there exist a number of QoS routing protocols for WANET. There also exist different classifications of these protocols based on – (i) interaction with MAC layer, (ii) interaction between

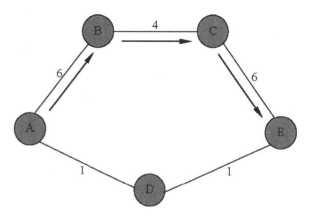

**Fig. 12.9** QoS routing of an example network where the numbers next to the radio links represent their respective bandwidth (e.g. Mbps). QoS routing will select A-B-C-E over route A-D-E under bandwidth guarantee of 3 Mbps, although the other route has fewer hops

route discovery and QoS provisioning mechanism and (iii) approach to route discovery. The basic principles of different routing protocols are summarised in the following sections under the light of different classification schemes.

## 12.8.1  Classification Based on MAC Layer Interaction

Routing protocols were classified in [26] based on the reliance of routing protocols on MAC layer. Three classes of QoS routing solutions are presented in [26]:

- Routing protocols that rely on contention-free MAC protocol
- Routing protocols that rely on contention-based MAC protocol
- Routing protocols that do not require any MAC layer interaction at all.

These major classes (Fig. 12.10) are briefly elaborated in the following sections.

### 12.8.1.1  Protocols Relying on Contention-Free MAC

Routing protocols that rely on accurately quantified resource (commonly channel capacity) availability and resource reservation, and therefore require a contention-free MAC solution such as TDMA, belong to this group. Such protocols are able to provide, what we term, pseudo-hard QoS. Hard QoS guarantees can only be provided in a wired network, where there are no unpredictable channel conditions and node movements. In the solutions that employ a contention-free MAC, the QoS guarantees provided are essentially hard, except for when channel fluctuations or node failures or movements occur, and hence the term 'pseudo-hard'. Due to these unpredictable conditions, a WANET is not a suitable environment for providing truly hard QoS guarantees. Some examples of routing protocols that belong to this group include (i) QoS Routing in a CDMA over TDMA network [15, 37, 29], (ii) ticket-based multi-path routing (Chen and Nahrstedt 1998) [44], (iii) on-demand SIR and bandwidth-guaranteed routing [33], and Node state routing [56].

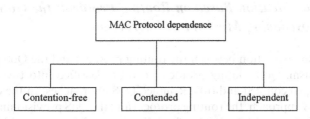

**Fig. 12.10** Classification of QoS routing protocols based on MAC layer dependence. There are three categories: (1) the protocol's operation depends on an underlying contention-free MAC protocol, (2) it can operate with a contended MAC protocol, and (3) it is completely independent of the MAC protocol

#### 12.8.1.2 Protocols Based on Contended MAC

Routing protocols that rely only on a contended MAC protocol and therefore only on the available resources or achievable performance to be statistically estimated belong to this group. Such protocols typically use these estimations to provide statistical or soft guarantees. Implicit resource reservation may still be performed, by not admitting data sessions that are likely to degrade the QoS of previously admitted ones. However, all guarantees are based on contended and unpredictable channel access or are given only with a certain probability and are thus inherently soft. Routing protocols that belong to this group includes (i) core extraction distributed ad hoc routing [54], (ii) interference-aware QoS routing [23], (iii) cross-layer multi-constraint QoS routing [23], (iv) on-demand delay-constrained unicast routing protocol [66], and (v) QoS greedy perimeter stateless routing for ultra wideband WANET [1].

#### 12.8.1.3 Protocols Independent of the Type of MAC

This group consists of those routing protocols that do not require any MAC layer interaction at all and are thus independent from the MAC protocol. Such protocols cannot offer any type of QoS guarantees that rely on a certain level of channel access. They typically estimate node or link states and attempt to route using those nodes and links for which more favourable conditions exist. However, the achievable level of performance is usually not quantified or is only relative, and therefore no promises can be made to applications. The aim of such protocols is typically to foster a better average QoS for all packets according to one or more metrics. This comes often at the cost of trade-offs with other aspects of performance, increased complexity, extra message overhead or limited applicability. The routing protocols of the group includes (i) QoS optimised link state routing [31, 4], (ii) link stability-based routing [49], (iii) hybrid ad hoc routing protocol [42], delay-sensitive adaptive routing protocol [52], (iv) application-aware QoS routing [61], (v) genetic algorithm-based QoS routing [5] and (vi) energy- and reliability-aware routing [39].

### 12.8.2 Classification Based on Routing Protocol: the QoS Provisioning Mechanism Interaction

Based on the interaction between the routing protocol and the QoS provisioning mechanism, QoS routing protocols can be classified into two categories (Fig. 12.11), (i) coupled and (ii) decoupled QoS approaches. In the case of the coupled QoS approach, the routing protocol and the QoS provisioning mechanism closely interact with each other for delivering QoS guarantees. If the routing protocol changes, it may fail to ensure QoS guarantees in coupled category. Some well-known QoS routing protocols that belong to this category include: (i) Ticket-Based QoS routing protocol (TDR) (Chen and Nahrstedt 1998), (ii)

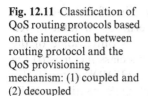

**Fig. 12.11** Classification of QoS routing protocols based on the interaction between routing protocol and the QoS provisioning mechanism: (1) coupled and (2) decoupled

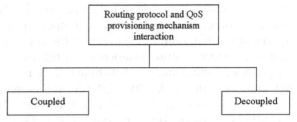

Predictive Location-Based QoS Routing protocol (PLBQR) [51], (iii) trigger-based (on-demand) distributed QoS routing protocol [19] (iv) Bandwidth Routing (BR) protocol [37], (v) On-demand QoS routing (OQR) protocol [35], (vi) On-demand link-state multi-path QoS routing (OLMQR) protocol [16], (vii) Asynchronous QoS Routing (AQR) scheme [59] and (viii) Core Extraction Distributed Ad Hoc Routing (CEDAR) [54],.

In the case of decoupled approach, the QoS provisioning mechanism does not depend on any specific routing protocol to ensure QoS guarantees. Routing protocols that belong to this group include INSIGNIA [2], SWAN [3] and proactive real-time MAC [60].

### 12.8.3 Classification Based on the Routing Information Update Mechanism Employed

Based on the routing information update mechanism employed, QoS approaches can be classified into three categories namely, (i) table-driven, (ii) on-demand and (ii) hybrid QoS approaches, shown in Fig. 12.12.

In the *table-driven approach*, each node maintains routing information to every other node (or nodes located in a specific part) in the network. The routing information is usually kept in a number of different tables. These tables are periodically updated if the network topology changes. Due to the overhead of periodic route update message, this approach is seldom directly used in practice and the predictive location-based QoS routing protocol (PLBQR) [51] is an example of this kind. In the *on-demand approach*, no such tables are maintained at the nodes, and hence

**Fig. 12.12** Classification of QoS routing protocols based on routing information update message employed. There are three categories: (1) table-driven, (2) on-demand and (3) hybrid protocols

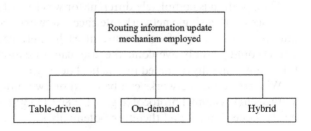

the source node has to discover the route on the fly. Therefore, on-demand routing protocols were designed to reduce the overheads in table-driven approach by maintaining information for active routes only. Some of the well-studied QoS routing protocols that belong to this class includes (i) the trigger-based (on-demand) distributed QoS routing (TDR) protocol [19] (ii) QoS version of AODV [45] (iii) On-Demand QoS routing (OQR) protocol [35] (iv) the On-Demand Link-State Multi-Path QoS routing (OLMQR) protocol [16], (v) Asynchronous QoS Routing (AQR) [59] and (vi) INORA [21]. On the other hand, *hybrid approach* is a new generation of protocol, which is both table-driven and on-demand in nature. This approach is designed to increase scalability by allowing nodes with close proximity to work together to form some sort of a backbone to reduce the route discovery overheads. This is mostly achieved by proactively maintaining routes to nearby nodes and determining routes to far away nodes using a route discovery strategy. Broadband Routing (BR) [37] and Core Extraction Distributed Ad hoc Routing (CEDAR) [54], are examples of this approach.

## 12.9 Thoughts for Practitioners

The support of QoS in WANET combines several routing concepts with the mechanisms for QoS support, at the same time making assumptions as have been made for the QoS support in wired networks. However, the practical applicability of these approaches suffers with several inherent difficulties as stated below.

### *12.9.1 Bandwidth Reservation in WANET*

Guaranteeing a certain amount of bandwidth for a certain flow or service class requires the station providing that guarantee is in control of that bandwidth. In wired network with full-duplex point-to-point links this can be easily assumed. It is also possible to agree on a determined share of bandwidth in a shared wired medium. Since a wired network is comprised of well-defined sub-networks, bandwidth guarantees for flows or service classes can be met by enforcing them in every involved sub-network.

The situation is completely different for wireless ad hoc networks consisting of devices with a single network interface. Networks consisting of devices with multiple network interfaces (such that each interface handles one link exclusively) could possibly overcome the drawbacks we discuss in the following, but this is not generally assumed in practical deployment.

Wireless ad hoc networks can be based on two different MAC technologies. With a single-channel protocol (e.g. IEEE 802.11 [30]), all stations communicate on the same channel and therefore potentially interfere with each other. With a multi-channel protocol in contrast (e.g. Bluetooth [8] or CDMA [47]), stations can communicate on several channels ('piconets' in Bluetooth terminology) simultaneously. Note that this theoretical assumption holds in our definition of multi-channel networks despite the inter-piconet interference in Bluetooth. A station can only be active in a single channel at any given point in time.

In WANET, for a multi-channel MAC, a closed collision domain could easily be formed if any two wireless nodes are in same transmission range. To separate the transmission from different domains, a different channel is assigned for each of the domains. However, a station does not have a separate interface for each sub-network it participates in, which is the case in wired network. This means that the devices have to switch channels regularly, as a result a well-defined, fixed sub-network structure is absent. On the other hand, in the single-channel case, the attempt to identify collision domains fails altogether. These domains would span entire connected components of WANET, since any two neighbours belong to the same collision domain.

However, it depends exactly on the sender–receiver pair, which devices are potential interferers. Both of the discussed cases have in common that a bandwidth reservation mechanism requires a transmission schedule defining time slots, which take their turns periodically [24]. For each slot, its duration and a set of possible simultaneous transmissions must be defined. This need for a transmission schedule is the fundamental limitation in contrast to wired networks. The problem of finding an optimal schedule is even NP-complete [67].

## 12.9.2  Service Differentiation

Apart from the general difficulties related to the bandwidth reservation outlined earlier, there is no chance to integrate classical QoS (based on bandwidth reservations) in WANET, which are to be deployed with off-the-shelf hardware in the near future [24]. And in the face of the complications elaborated above, it can be seriously doubted that it will ever be worthwhile to implement bandwidth reservation mechanisms in WANET as long as a single network interface is used to communicate on several links. As a consequence, IntServ based approaches and naive DiffServ approaches are not applicable in WANET. However, implementing SWAN [3] or FQMW [63] for resource reservation also have some practical difficulties discussed below.

## 12.9.3  Queuing

In service differentiation, the queuing techniques play the most important factor. Implementing traditional priority queuing strategy in WANET is significantly difficult. For example, as proposed in FQMW [63], a simple priority queue ensures that high-priority packets are given unconditional preference over low-priority packets. Secondly, they consider a FIFO queue which they enhance with a mechanism called random early discard with IN/OUT buffer management. Similarly, in SWAN [3] also conceptually utilises a priority queue, but limits the amount of real-time traffic in order to protect the lower-priority traffic from starvation.

### 12.9.4 Dealing with Congestion

FQMW [63] tries to limit network congestion by policing the traffic at the traffic sources. The sources are the equivalent of ingress routers in DiffServ networks. To regulate the traffic, a source node implements a token bucket that determines whether a packet is in-profile or out-out-profile. The source stations have to take great care in regulating their traffic. If the rate of in-profile packets is not chosen properly in areas with little activity, it might accumulate and cause congestion in bottleneck areas.

In contrast to classical DiffServ, FQMW [63] actually lacks service level agreements. For the token bucket metering suggested by the FQMW [63] authors, it remains an open question how to calculate the dynamic parameter that expresses the share of the effective link bandwidth at any given time, given that a single topology change can alter the available bandwidth significantly.

SWAN [3] uses a strict admission control scheme for real-time packets. Bulk-traffic is unconditionally admitted, but only receives the remaining resources. Real-time traffic is admitted by the source node depending on the outcome of probing the network for resources. If the probe packet passes a link on which the total amount of real-time traffic exceeds a certain threshold, the session will not be admitted. This way, it is prevented that the real-time traffic crowds out the bulk traffic. Furthermore, it prevents that real-time traffic suffers too large delays.

It results in a quality degradation of high-priority flows as their volume increases. In effect, all, and not just single users suffer from an excess of high-priority traffic. It is however likely that a regulation of the number of active sessions will result from users that terminate their high-priority connections because they are not willing to bother with the bad transmission quality. Furthermore, it is likely that multimedia applications change their coding scheme to a higher compression when they experience many late or lost packets. This can be seen from a fairness aspect especially if one assumes that, in this way, unimportant connections will be terminated before the important ones. Both approaches, protecting existing sessions or admitting every session request, have their advantages and disadvantages and which one fits to the requirements better depends on the specific scenario.

## 12.9.5 *Dealing with Excessive Delay*

Certain applications have stringent delay bounds for their traffic. This means that packets arriving too late are useless. From the application's point of view, there is no difference between late and lost packets. This implies that it is actually useless to forward real-time packets that stay in a router for more than a threshold amount of time, because they will be discarded at the destination anyway. Dropping those packets instead has the discarded of reducing the load in the network [24]. It would also be possible to accumulate the delay a packet has experienced in each router by adding this information in the packet header, but this would make the approach more complicated. Nevertheless, it could be verified in future work if this is worth the effort. For now, only the delay in each intermediate station is considered separately.

## 12.10  Conclusion and Future Directions

In this chapter we have summarised a comprehensive review of the state-of-the-art research works on QoS support in WANET. We have presented the issues and challenges involved in providing QoS in WANET. The research works on QoS models, QoS resource reservation signalling, QoS routing and QoS MAC, which are required to assure QoS in WANET are described.

Although the importance of QoS is really felt in the WANET research community, research in this area is still in its infancy. Many areas of research in this field of wireless communication provide considerable challenge to enhance the growth and proliferation of WANET and their applications. These areas include power consumption, resource availability, location management, interlayer integration of QoS services, support for heterogeneous WANET, as well as robustness and security. Continued growth is expected in this area of research in order to develop, test and implement the critical building blocks to provide efficient and seamless communications in wireless mobile ad hoc networks.

## Terminologies

*FQMW*: Flexible QoS model for WANET (FQMW) is the first QoS Model proposed for WANETs. The idea is to combine knowledge from the solutions offered in the wire-based networks and apply them to a new QoS model, which will take into consideration the characteristics of WANETs.

*IntServ*: IntServ is an existing QoS model for the Internet, which defines the service architecture of the total QoS framework. The IntServ model merges the advantages of two different paradigms: datagram networks and circuit-switched networks. It can provide a circuit-switched service in packet-switched networks.

*RSVP*: The Resource ReSerVation Protocol (RSVP) is used as the signalling protocol to reserve resources in IntServ. Applications with Guaranteed Service or Controlled-Load Service requirements use RSVP to reserve resources before transmission.

*SWAN*: SWAN is a stateless approach dealing with service differentiation in mobile ad hoc networks. SWAN assumes the use of best-effort MAC (for example IEEE 802.11 DCF) and uses feedback-based control mechanisms to support soft real-time services and service differentiation in ad hoc networks.

*WANET*: A Wireless Ad hoc NETwork (WANET) is a collection of wireless mobile nodes dynamically forming a temporary network. These networks can be formed on the fly, without requiring any fixed infrastructure.

## Questions

1. What are the potential challenges of QoS provisioning in WANET?
2. Briefly describe the factors that may affect the QoS in WANET?
3. Describe why the existing Internet QoS models (that is, IntServ and DiffServ) are not suitable in WANET?
4. Why we need signalling in QoS mechanism? What do you understand by in-band and out-of-band signalling?
5. Why INSIGNIA is said to be a feasible signalling protocol for WANET?
6. Briefly describe the hidden terminal problem?

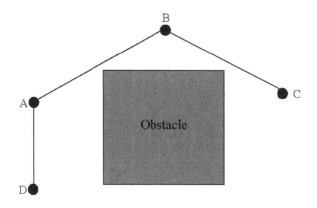

7. Define QoS routing? What are the main classifications of QoS routing protocols?
8. In the figure below, which path the QoS routing should select from A to E and why?

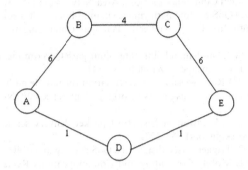

9. What makes QoS routing difficult?
10. Why soft-state approach for flow-state management is preferred in WANET over hard-state approach?

## References

1. Abdrabou A, Zhuang W (2006) A position-based QoS routing scheme for UWB mobile ad hoc networks. IEEE J. Select. Areas Commun. 24:850–856.
2. Ahn G S, Campbell A T, Lee S B, Zhang X (1999) INSIGNIA. Internet Draft. comet. columbia.edu/insignia/draft-ietf-manet-insignia-01.txt Accessed 18 March 2008.
3. Ahn G S, Campbell A T, Veres A, Sun L H (2002) Supporting service differentiation for real-time and best effort traffic in Stateless Wireless Ad Hoc Networks (SWAN), IEEE Transactions on Mobile Computing 1(3):192–207.
4. Badis H, Agha K A (2005) QOLSR: QoS routing for ad hoc wireless networks using OLSR. Wiley European Transactions on Telecommunications 15(4):427–442.
5. Barolli L, Koyama A, Shiratori N (2003) A QoS routing method for ad-hoc networks based on genetic algorithm. Proc. 14th Int. Wksp. Database and Expert Systems Applications 175–179.
6. Bharghavan V, Demers A, Shenker S, Zhang L (1994) MACAW: A media access protocol for wireless LANs. Proc. ACM SIGCOMM 212–225.
7. Blake S, Black D, Carlson M, Davies E, Wang Z, Weiss W (1998) An architecture for differentiated services. IETF RFC2475. www.ietf.org/rfc/rfc2475.txt Accessed 18 March 2008.
8. Bluetooth SIG (2001) Specification of the Bluetooth System – Version 1.1 B.
9. Braden R, Clark D, Shenker S (1994) Integrated services in the internet architecture – an Overview. IETF RFC1633. www.ietf.org/rfc/rfc1633.txt Accessed 18 March 2008.
10. Braden R, Zhang L, Berson S, Herzog S, Jamin S (1997) Resource reSerVation Protocol (RSVP) – Version 1 Functional Specification. RFC 2205. www.ietf.org/rfc/rfc2205.txt Accessed 18 March 2008.
11. Broch J, Maltz D A, Johnson D B, Hu Y C, Jetcheva J (1998) A performance comparison of multi-hop wireless ad hoc network routing protocols. Proc. 4th Annual ACM/IEEE International Conference on Mobile Computing and Networking 85–97.

12. Chakrabarti S, Mishra A (2001) QoS issues in ad hoc wireless networks. IEEE Commu-
    nications Magazine 39:142–148.
13. Chang J H, Tassiulas L (2000) Energy-conserving routing in wireless ad-hoc networks.
    Proc. IEEE INFOCOM 1:22–31.
14. Chen S, Nahrstedt K (1998) Distributed quality-of-service routing in high-speed networks
    based on selective probing. 23rd Annual Conference on Local Area Networks (LCN) 80.
15. Chen T W, Tsai J T, Gerta M (1997) QoS routing performance in multihop, multimedia,
    wireless networks. Proc. IEEE 6th Int. Conf. Universal Personal Communications
    2:557–561.
16. Chen Y, Tseng Y, Sheu J, Kuo P (2002) On-demand, linkstate, multi-path QoS routing in
    a wireless mobile ad-hoc network. Proc. European Wireless 135–141.
17. Chua K C, Xiao H, Seah K G (2003) Relative service differentiation for mobile ad hoc
    networks. Proc. IEEE Wireless Communications and Networking Conference (WCNC)
    2:1379–1384.
18. Clark D D, Fang W (1998) Explicit allocation of best-effort packet delivery service.
    IEEE/ACM Transactions on Networking. 6(4):362–373.
19. De S, Das S K, Wu H, Qiao C (2002) Trigger-based distributed QoS routing in mobile ad
    hoc networks. ACM SIGMOBILE Mobile Computing and Communications Review
    6(3):22–35.
20. Deng J, Haas Z J (1998) Dual busy tone multiple access (DBTMA): a new medium access
    control for packet radio networks. Proc. IEEE ICUPC 1:973–977.
21. Dharmaraju D, Chowdhury A R, Hovareshti P, Baras J S (2002) INORA – A unified
    signalling and routing mechanism for QoS support in mobile ad hoc networks. Proc.
    ICPPW 86–93.
22. Doshi S, Bhandare S, Brown T (2002) An on-demand minimum energy routing protocol
    for a wireless ad-hoc network. Mobile Computing and Communications Review
    6(2):50–66.
23. Fan Z (2004) QoS routing using lower layer information in ad hoc networks. Proc.
    Personal, Indoor and Mobile Radio Communications Conf., 135–139.
24. Gerharz M, de Waal C, Frank M, James P (2003) A practical view on quality-of-service
    support in wireless ad hoc networks. Proc. IEEE Workshop on Applications and Services
    in Wireless Networks (ASWN), citeseer.ist.psu.edu/gerharz03practical.html Accessed 18
    March 2008.
25. Gupta P, McKeown N (1999) Packet classification on multiple fields. Proc. ACM
    SIGCOMM Conference on Applications, Technologies, Architectures, and Protocols
    for Computer Communications 147–160.
26. Hanzo L, Tafazolli R (2007) A survey of QoS routing solutions for mobile ad hoc
    networks. Communications Surveys & Tutorials, IEEE 9(2):50–70.
27. Heinanen J, Baker F, Weiss W, Wroclawski J (1998) Assured Forwarding PHB Group.
    Internet Draft (Work in progress), draft-ietf-diffserv-af-03.txt.
28. Ibanez J, Nichols K (1998) Preliminary simulation evaluation of an assured service.
    Internet Draft (Work in progress) www3.tools.ietf.org/html/draft-ibanez-diffserv-
    assured-eval-00 Accessed 18 March 2008.
29. IEEE Computer Society (2006), Wireless Medium Access Control (MAC) and Physical
    Layer (PHY) Specifications for High-Rate Wireless. Amendment 1: MAC Sublayer IEEE
    Std 802.15.3b-2005 (Amendment to IEEE Std 802.15.3-2003) 1–146.
30. IEEE Standards Board (1999) Part 11: Wireless LAN Medium Access Control (MAC) and
    Physical Layer (PHY) Specifications. The Institute of Electrical and Electronics Engineers
    Inc. www.csse.uwa.edu.au/adhocnets/802.11-1999.pdf Accessed 18 March 2008.
31. Jacquet P, Muhlethaler P, Clausen T, Laouiti A, Qayyum A, Viennot L (2001) Optimized
    link state routing protocol for ad hoc networking. Proc. IEEE Multi Topic Conf. 62–68.
32. Karn P (1990) MACA – a new channel access method for packet radio. Proc. ARRL/
    CRRL Amateur Radio Ninth Computer Networking Conf. 134–140.

33. Kim D, Min C H, Kim S (2004) On-demand SIR and bandwidth-guaranteed routing with transmit power assignment in ad hoc mobile networks. IEEE Transactions on Vehicular Technology 53:1215–1223.
34. Lee S B, Campbell A T (1998) INSIGNIA: in-band signaling support for QOS in mobile ad hoc networks. Proc 5th International Workshop on Mobile Multimedia Communications (MoMuC).
35. Lin C R (2002) On-demand QoS routing in multihop mobile networks. Proc. IEEE INFOCOM 3:1735–1744.
36. Lin C R, Gerla M (1997) MACA/PR: an asynchronous multimedia multihop wireless network. Proc. IEEE INFOCOM. 1:118–125.
37. Lin C R, Liu J (1999) QoS routing in ad hoc wireless networks. IEEE Journal on Selected Areas in Communications 17(8):1426–1438.
38. Lu S, Bharghavan V, Srikant R (1997) Fair scheduling in wireless packet networks. Proc. ACM SIGCOMM. 27(4):63–74.
39. Misra A, Banerjee S (2002) MRPC: maximising network lifetime for reliable routing in wireless environments. Proc. IEEE Wireless Communications and Networking Conf. pages.cs.wisc.edu/~suman/pubs/wcnc02.pdf Accessed 18 March 2008
40. Muir A, Garcia-Luna-Aceves J J (1998) An efficient packet-sensing MAC protocol for wireless networks. ACM Journal on Mobile Networks and Applications 3(2):221–234.
41. Nichols K, Jacobson V, Zhang L (1999) A two-bit differentiated services architecture for the internet. IETF RFC2638. www.ietf.org/rfc/rfc2638.txt Accessed 18 March 2008.
42. Nikaein N, Bonnet C, Nikaein N (2001) Hybrid ad hoc routing protocol – HARP. Proc. Int. Symp. Telecommunications.
43. Perkins C E (2001) Ad Hoc Networking. Ch. 3, Addison Wesley, Reading, MA
44. Perkins C E, Bragwat P (1994) Highly dynamic destination-sequenced distance-vector routing (DSDV) for mobile computers. Proc. ACM SIGCOMM 234–244.
45. Perkins C E, Royer E M, Das S R (2000) Quality of service for ad hoc on-demand distance vector routing. IETF Internet Draft (Work in progress). draft-ietf-manet-aodvqos- 00. txt.
46. Perkins C E, Royer E M, Das S R, Marina M K (2001) Performance comparison of two on-demand routing protocols for ad hoc networks. IEEE Personal Communications Magazine 8:16–28.
47. Rappaport T S (1996) Wireless Communications – Principles & Practice. Prentice Hall Communications Engineering and Emerging Technologies Series, 2nd Edition, ISBN-10: 0130422320, Prentice Hall, Upper Saddle River, NJ
48. Reddy T B, Karthigeyan I, Manoj B S, Murthy C S R (2006) Quality of service provisioning in ad hoc wireless networks: a survey of issues and solutions. Ad Hoc Networks 4:83–124.
49. Rubin I, Liu Y C (2003) Link stability models for QoS ad hoc routing algorithms. Proc. 58th IEEE Vehicular Technology Conf. 5:3084–3088.
50. Saunders S (1999) Antennas and Propagation for Wireless Communication Systems Concept and Design. John Wiley and Sons, New York, NY, USA.
51. Shah S H, Nahrstedt K (2002) Predictive location-based QoS routing in mobile ad hoc networks. Proc. IEEE ICC 2002 2:1022–1027.
52. Sheng M, Li J, Shi Y (2003) Routing protocol with QoS guarantees for ad-hoc network. Electronics Letters 39:143–145.
53. Shenker S, Partridge C, Guerin R (1997) Specification of guaranteed quality of service. RFC 2212.
54. Sivakumar R, Sinha P, Bharghavan V (1999) CEDAR: a coreextraction distributed ad hoc routing algorithm. IEEE Journal on Selected Areas in Communications 17:1454–1465.
55. Sobrinho J L, Krishnakumar A S (1999) Quality-of-service in ad hoc carrier sense multiple access wireless networks. IEEE Journal on Special Areas in Communications 17(8):1353–1368.

56. Stine J, de Veciana G (2004) A paradigm for quality of service in wireless ad hoc networks using synchronous signalling and node states. IEEE Journal on Selected Areas in Communications 22:1301–1321.
57. Talucci F, Gerla M (1997) MACA-BI (MACA By Invitation): A wireless MAC protocol for high speed ad hoc networking. Proc. IEEE ICUPC. 2:913–917.
58. Tobagi F A, Kleinrock L (1975) Packet switching in radio channels: Part II – The hidden terminal problem in carrier sense multiple-access and the busy-tone solution. IEEE Transactions on Communications 23(12):1417–1433.
59. Vidhyashankar V, Manoj B S, Murthy C S R (2003) Slot allocation schemes for delay sensitive traffic support in asynchronous wireless mesh networks. Proc. The International Journal of Computer and Telecommunications Networking 50(15):2595–2613.
60. Vivek V, Sandeep T, Manoj B S, Murthy C S R (2004) A novel out-of-band signaling mechanism for enhanced realtime support in tactical ad hoc wireless networks. Proc. IEEE RTAS 56–63.
61. Wang M, Kuo G S (2005) An application-aware QoS routing scheme with improved stability for multimedia applications in mobile ad hoc networks. Proc. IEEE Vehicular Technology Conf. 1901–1905.
62. Wroclawski J (1997) Specification of the controlled-load network element service. RFC 2211. www.ietf.org/rfc/rfc2211.txt Accessed 18 March 2008
63. Xiao H, Seah W K G, Chua K C (2000) A flexible quality of service model for mobile ad-hoc networks. Proc. IEEE Vehicular Technology Conference (VTC) 1:445–449.
64. Xipeng X, Lionel M N (1999) Internet QoS: a big picture. IEEE Network Magazine 13(2):8–18.
65. Yu C, Lee B, Youn H Y (2003) Energy-efficient routing protocols for mobile ad-hoc networks. Wiley J. Wireless Communications and Mobile Computing Journal 3(8): 959–973.
66. Zhang B, Mouftah H T (2005) QoS routing for wireless ad hoc networks: problems, algorithms and protocols. IEEE Communations Magazine 43:110–117.
67. Zhu C, Corson M S (2002) QoS routing for mobile ad hoc networks. Proc. IEEE INFOCOM, 958–967.

# Chapter 13
# Delay Management in Wireless Ad Hoc Networks

**Wenbo He, Yuan Xue and Klara Nahrstedt**

**Abstract** A good amount of research has been developed to support QoS issues in IEEE 802.11 ad hoc networks. In this chapter, we address QoS management issues through multiple layers for real-time multimedia applications. We investigate the adaptation mechanisms at middleware layer and MAC layer to dynamically adjust the service classes for applications by feedback control theory. Based on our investigation, we propose a cross-layer end-to-end delay management framework for multimedia traffic over multi-hop wireless networks.

## 13.1 Introduction

A key vision of next-generation wireless systems is to support new emerging distributed multimedia services such as voice-over-IP and video-on-demand anytime anywhere. Some applications (e.g. multimedia applications) are very sensitive to end-to-end delay provided by the communication environment. The control of end-to-end delay over 802.11 wireless networks presents a number of technical challenges. First, wireless networks are typically bandwidth constrained in comparison with their traditional wireline peers. Second, the time-varying error characteristics and time-varying channel capacity at the physical layer makes it difficult, if not impossible to provide hard end-to-end delay guarantees. Third, user mobility may induce signal fading, which in turn can trigger rapid degradation in the delivered service quality.

A real-time multimedia application usually has QoS requirements on end-to-end latency. For example, in telephony, one-way delay requirement is from 25 to 400 ms, depending on the requirement of voice quality and existence of echo canceler. It is challenging to manage end-to-end latency to meet the QoS requirement in wireless multi-hop networks.

W. He (✉)
Department of Computer Science, Siebel Center, Urbana, IL, USA
e-mail: wenbohe@uiuc.edu

S. Misra et al. (eds.), *Guide to Wireless Ad Hoc Networks*,
Computer Communications and Networks, DOI 10.1007/978-1-84800-328-6_13,
© Springer-Verlag London Limited 2009

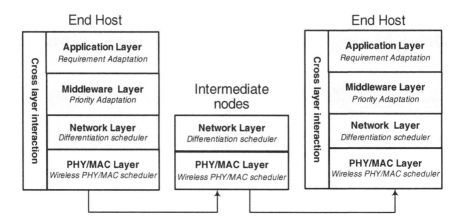

**Fig. 13.1** Communication over multi-hop wireless ad hoc networks

In the following sections, we will present a cross-layer delay management framework based on the findings in [1, 2]. This framework provides comprehensive and flexible control of end-to-end latency over wireless ad hoc networks as shown in Fig. 13.1.

## 13.2 Background

Current efforts on developing delay management and QoS support for wireless networks are mostly focused at the MAC layer. For example, the work of [3, 4] focused on differentiated MAC layer scheduling under IEEE 802.11 DCF. The work of [5–7] studied the fair scheduling in wireless networks. These MAC layer solutions support delay management within one hop and lack the architectural flexibility to accommodate end-to-end application-specific QoS requirements in time-varying wireless networking environments.

Some other efforts provide an end-to-end delay support by adapting traditional QoS models for wireless network. For example, based on the IntServ model, INSIGNIA [8] utilizes resource signaling protocol to reserve per-flow resources. Alternatively, SWAN [9] follows the absolute DiffServ QoS model by defining two service classes: *real-time* and *best-effort* traffic. A measurement-based admission control is performed to support real-time traffic while the ECN bit is marked to reflect the network support for best-effort traffic. Both these approaches use per-application admission control, thus they have poor scalability and may suffer false admission due to the imprecise resource estimation in the highly dynamic wireless environments.

In light of the limitations of existing approaches, it is clear that a complete and efficient end-to-end delay management architecture for wireless network would require (1) lightweight and scalable QoS support with minimum or no negotiation overheads; (2) agile and responsive delay management tools to monitor network resources and make appropriate adaptation decisions for applications so that their individual delay requirements can be satisfied.

## 13.3 Cross-Layer Delay Management Architecture

The cross-layer delay management framework is presented in Fig. 13.2. The architecture operates from the MAC layer up to the middleware layer. At the MAC and network level, packets from different service classes are processed differently via per-hop forwarding mechanisms (e.g., packet scheduling and queue management) to achieve proportional delay differentiation (PDD). At the middleware layer, a monitor component monitors the performance of applications. Based on the monitoring results, it performs appropriate service class adaptation so that different applications are able to meet their required end-to-end latency specifications.

### 13.3.1  At the Middleware Layer in the Wireless end Hosts

(1) Based on the observed end-to-end delay and the delay specifications of the applications, the *Priority Adaptor* decides the appropriate priority for each

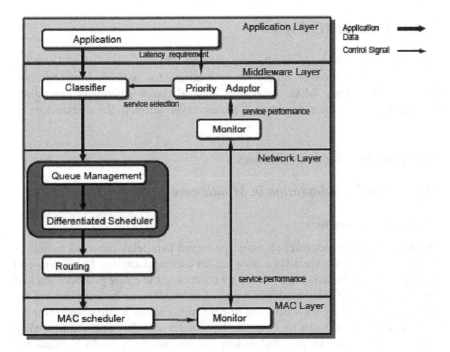

**Fig. 13.2** End-to-end delay management architecture

application and notifies the *Classifier*. The *Classifier* in the middleware determines the service classes for packets according to their priorities. Though the priorities can take a large range of values, the number of service classes can be small. Usually, we choose a small number of service classes for easy deployment and efficient network management.

(2) The packets from applications are delivered through the middleware layer, where the *Classifier* marks the packets with their corresponding service classes by mapping the priorities to service classes. Each service class is represented by the class parameter, which is a number obtained by rounding up the priority in a certain range.

(3) The *Monitor* monitors the performance of each service class and notifies the *Priority Adaptor* of the observed violations of end-to-end delay.

### 13.3.2 At the Network Layer in the Routing Nodes

(1) The *Queue Management* component allocates buffer spaces and marks or drops packets. It deals with packet loss rate differentiation.

(2) The *Differentiated Scheduler* selects a packet to transmit. It performs packet-level QoS enforcement, allocates bandwidth for different flows and provides delay differentiation.

### 13.3.3 At the MAC Layer

The MAC layer scheduler coordinates with the differentiation scheduler at the network layer to provide timely channel access for packets from different wireless nodes. Since MAC layer and network layer schedulers are tightly coupled with each other, we propose a cross layer design in the architecture.

## 13.4 Thoughts for Practitioners

### 13.4.1 Priority Adaptation in Middleware

#### 13.4.1.1 Delay Monitor

The delay monitor measures the average round trip delay incurred to deliver multimedia packets in the ad hoc network for each application. The end-to-end latency contains the delay introduced by traversing the entire protocol stack at the end nodes. The delay manager is placed in middleware because the end-to-end latency caused by traversing the upper layers (network and above) is small at the end nodes and can be ignored. The sender attaches time-stamps when sending the packets to the destination. As the sender gets ACK from the destination, it retrieves the sending time-stamp, and compares it with the current time-stamp, so that a sender can obtain the round-trip delay $d_i$ for packet $i$. We take an average of $N$ round-trip delay measurements $(d_1, d_2, \ldots, d_N)$, and have $d_{avg} = \frac{1}{2N} \sum_{i=1}^{N} d_i$ as the measured value to estimate end-to-end delay, which is used by the *Priority Adaptor* to update the service priority appropriately.

### 13.4.1.2  Classifier

The classifier in the middleware determines the service classes for packets according to their priorities. The goal of the *Classifier* is to map the priorities to service classes in order to keep a small number of service classes. Each service class is represented by the class parameter, which is a number obtained by rounding up the priority in a certain range. The possible priorities generated by *Priority Adaptor* can be divided into $L$ ranges, $R_1, R_2, \ldots, R_L$. If the priority attached to a packet by the *Priority Adaptor* is in the range $R_i$, the service parameter assigned to the packet will be represented by the largest priority in the range $R_i$. Though the priorities can take a large range of values, the number of service classes can be small. Usually, we choose a small number of service classes for easy deployment and efficient network management.

### 13.4.1.3  Priority Adaptor

To design the *Priority Adaptor*, we need first to establish the open-loop dynamic model for the mapping between priority and end-to-end delay. Using system identification techniques [10], we determine the model that captures the relationship between priority and end-to-end delay. After we obtain the control model, we design controller to adjust the priorities of the application to control the end-to-end delay around the required delay level.

The goal of middleware adaptation is to dynamically adjust the priority of a multimedia application, thus make the multimedia application to meet the end-to-end delay requirement. In this section, we focus on the design of middleware *Priority Adaptor*. PID controller is a widely used controller for dynamic systems to maintain the output level at the pre-determined value (reference value), where PID stands for Proportional, Integral, Derivative. A proportional (P) controller has the effect of reducing the rise time and decreasing the steady-state error. An integral (I) controller has the effect of eliminating the steady-state error, but it may make the transient response worse. A derivative (D) controller has the effect of increasing the stability of the system, reducing the overshoot, and improving the transient response. However, if the system noise is large, the derivative controller will decrease the stability of the system. For the system under investigation (multimedia over wireless ad hoc networks), both system noise and the measurement noise are large. The system noise is due to the random workload in the ad hoc network, and the nature of randomized algorithms in MAC layer protocols. The measurement noise comes from the

**Table 13.1** The desired properties and the controller design criteria

| Desired system properties | Controller design criteria |
| --- | --- |
| Stability | Place poles within the unit-circle |
| Accuracy of control (zero steady-state error) | Adopt integral (I) control |
| Fast response and less oscillations | Select reasonable parameters of PI controller |

measured data we get from the delay monitor. The measurement noise of the system is caused by a large range of round trip delay in the wireless ad hoc networks. Due to the large noise in the system, the D controller will introduce the undesired oscillation to the system. So we choose the combination of PI controller and do not consider D controller in the middleware adaptation.

To give a good end-to-end delay performance of the system, the design goals of the closed-loop system are shown in Table 13.1.

Figure 13.3 shows the whole control loop including *Priority Adaptor* (*C*), the open loop wireless multi-hop system (*G*), and the *Delay Monitor* (*M*). To calculate all parameters of the *Priority Adaptor C* as a PI controller, we need to determine the wireless end-to-end system as an open loop system model *G*. The *Delay Monitor* detects the end-to-end delay, $d(k)$ of the system. Then the *Priority Adaptor* compares it with the delay requirement $ref_{delay}$ given by the application, and adjusts the priority of the packets to be sent of the multimedia flow. Let $e(k) = d(k) - ref_{delay}$, where $ref_{delay}$ is the desired level of end-to-end delay. If $e(k)$ can be kept around zero, the end-to-end delay of the system will be stabilized around the desired level.

We obtain the parameters of the *Priority Adaptor* in three steps: (1) model identification of *G* for a fixed load condition; (2) PI Controller design of the *Priority Adaptor C* based on off-line model identification; (3) design of adaptive controller, which dynamically changes parameters of the controller so that the controller is able to adapt with different network load conditions.

## Off-Line Model Identification

It is difficult to obtain the model of *G* using first-principles due to the complexity of the multi-hop ad hoc wireless network system. We treat the wireless network system as a black-box and then infer the model from externally observable metrics. The process to infer the open-loop system model is model identification. We use 802.11 wireless ad hoc test-bed to gather data for the

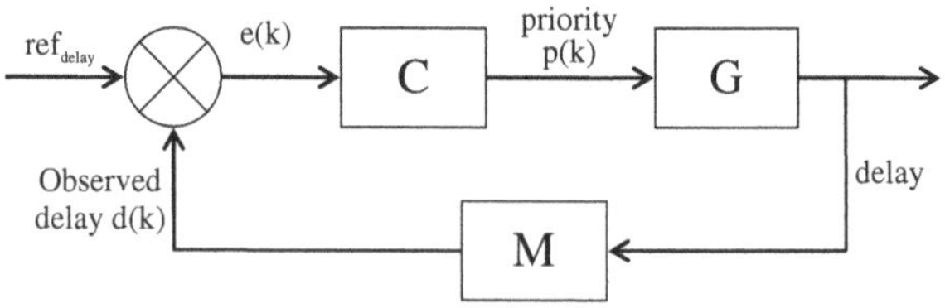

**Fig. 13.3** The block diagram of the closed-loop control system for end-to-end delay

model identification. Note that in model identification, we look at the priority as the system input and the end-to-end delay as output. On the other hand, in the priority adaptor design, the priority is the output of the controller $C$ and the end-to-end delay is the input.

In order to develop an adaptive priority adjustment algorithm to control the end-to-end delay, we need first to understand how priority affects the end-to-end delay under a certain traffic load. In this section, we will show how to get the system model $G$ by model identification.

In the model identification, we use difference equation with unknown parameters as the dynamic model between the input (priority) and the output (end-to-end delay). Such a model estimates the mathematical relationship between the input and the output of the system. We then use a pseudo-random digital white noise generator to stimulate the system by assigning random priorities to multimedia packets and observe the end-to-end delay during a certain time period. We choose a sampling interval of 0.5 s. So we get a priority and delay pair every 0.5 s. With the data we obtained in the experiments, we can apply the auto-regressive (AR) model to get the mathematical relationship from priority to end-to-end delay. We notice that the first-order AR model provides reasonably good prediction for end-to-end delays with different priorities. The first order model is written as:

$$d(k-1) = b_0 p(k) - a_1 d(k) \qquad (13.1)$$

where $d(k)$ and $p(k)$ are the end-to-end delay and priority at time $k$, respectively. The relation of $d(k)$ and $p(k)$ in equation (1) is independent of the number of nodes on the route as long as the route between two end nodes are fixed. Note that the units of parameters in the model are casted to make both sides of equation have consistent units, e.g. the unit of $b_0$ is second, $a_1$ is a constant.

The z-transform is used to take discrete time domain signals into a complex-variable frequency domain to simplify the calculation. It plays a similar role to what the Laplace Transform does in the continuous time domain. Like the Laplace, the z-transform gives a simpler way to solve problems and design discrete time applications. The corresponding z-transform of the transfer function from $p(k)$ to $d(k)$ in equation (13.1) is

$$G(z) = \frac{d(z)}{p(z)} = \frac{b_0}{z - a_1}, \qquad (13.2)$$

where $b_0$ and $a_1$ are to be determined in the model identification. We use 50 input–output pairs of $(p(k), d(k))$ to identify the model (i.e. coefficients $b_0$, $a_1$). The first 25 samples are used for identification, and the remaining 25 samples for validation. We determine the parameters that $b_0 = 0.02425$ and $a_1 = 0.2514$.

PI Controller Design

The form of PI controller is given below. The parameters $k_p$ and $k_i$ are to be determined.

$$p(k+1) = k_p e(k) + k_i \sum_{t=0}^{k} e(t) \tag{13.3}$$

Then,

$$p(k) = k_p e(k-1) + k_i \sum_{t=0}^{k-1} e(t) \tag{13.4}$$

If we subtract (13.4) from (13.3), we have

$$p(k+1) - p(k) = k_p(e(k) - e(k-1)) + k_i e(k) \tag{13.5}$$

Then we get

$$p(k+1) = p(k) + (k_p + k_i)e(k) - k_p e(k-1) \tag{13.6}$$

The z-transform of the priority controller (13.6) is given as

$$C(z) = \frac{(k_p + k_i)z - k_p}{z(z-1)}, \tag{13.7}$$

From the model identification, we know the open loop model $G(z)$ is given as

$$G(z) = \frac{d(z)}{p(z)} = \frac{0.02425}{z - 0.2514}. \tag{13.8}$$

According to discrete control theory, the performance of a system is implied by the poles of its closed loop transfer function $\frac{C(Z)G(z)}{1+C(Z)G(z)}$. We take the advantage of Root Locus, which is a graphical technique that plots the traces of poles of a closed-loop system on the z-plane as the controller parameters change. So with the Root Locus diagram, we can determine parameters $k_p$ and $k_i$, in order to achieve the design goal listed in Table 13.1. We choose $k_i = 1.51$ and $k_p = 1.85$, and the resulting controller is given as

$$p(k-1) = p(k) - 3.36e(k) - 1.85e(k-1) \tag{13.9}$$

Adaptive Controller Design

We have described the design of the PI controller for priority adaptor, which is determined off-line based on the dynamic input–output pairs via system identification. Data collected in the system identification experiments are obtained with fixed load condition. In a real network environment, it is difficult to find a single linear model characterizing the system's behavior under all load conditions. Hence, we need an adaptive scheme which is able to adapt its behavior according to changing load and network condition. In control theory, this goal is achieved through adaptive controller. We implemented a simple self-tuning regulator on our test-bed, where we assume the system model has a fixed structure, but the parameters are time-varying depending on load and network condition. Figure 13.4 shows a block diagram for the self-tuning adaptor to control priorities of multimedia applications. Comparing Fig. 13.4 with Fig. 13.3, we add two blocks: an on-line estimation of identified parameter and self-tuning regulator. In Fig. 13.4, we have

$$C = \frac{(k_\mathrm{p} + k_i)z - k_\mathrm{p}}{z(z - 1)} \text{ and } G = \frac{b_0}{z - a_1}.$$

## 13.4.2 Cross-Layer Proportional Delay Differentiation Scheduler

The middleware layer priority adaptation requires the support of service differentiation from the lower level of the network stack. To enforce the service differentiation in wireless network, a coordination between network layer and MAC layer is needed. In this section, we present a cross-layer delay differentiation scheduling scheme. Here we adopt a proportional delay differentiation model.

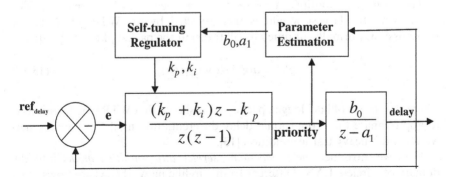

**Fig. 13.4** Block diagram of a self-tuning adaptor

The model of the proportional service differentiation was first introduced as a per-hop-behavior (PHB) for DiffServ in wireline networks [11]. It states that certain class performance metrics should be proportional to the differentiation parameters. In particular, if we consider the case of delay differentiation, in a network with $C$ service classes the proportional delay differentiation model imposes the following constraints for all pairs of classes.

$$\frac{\bar{d}_i(t, t+\tau)}{\bar{d}_j(t, t+\tau)} = \frac{\delta_j}{\delta_i}, \text{for all } i \neq j \text{ and } i, j \in \{1, 2, ..., C\} \qquad (13.10)$$

where $\delta_i$ is the service differentiation parameter for class $i$, and $\bar{d}_i(t, t+\tau)$ is the average delay for class $i$, $(i = 1, 2, ..., C)$ in the time interval $(t, t - \tau)$, where $\tau$ is the monitoring time-scale.

The basic idea of proportional differentiation is that, even though the actual quality level of each class may vary with traffic loads, the quality ratio between classes should remain constant in various time-scales. In addition, such a quality ratio can be controlled by setting the service differentiation parameters, which provides flexible class provisioning and management. Under certain conditions (i.e. the network is well provisioned), applications with absolute delay requirements can select appropriate service classes to meet their requirements [12], even though the network offers only relative differentiation.

One of the packet scheduling algorithms that can realize the proportional delay differentiation model in a short time-scale is the *waiting time priority* (WTP) scheduler [13]. In this algorithm, a packet is assigned with a weight, which increases proportionally to the packet's waiting time. Service classes with higher differentiation parameters have larger weight-increase factors. The packet with the largest weight is served first in non-preemptive order. Formally, let $\text{wt}_{\text{pkt}}(t)$ be the waiting time of a packet pkt of class $i$ at time $t$, we define its *normalized waiting time* $\hat{\text{wt}}_{\text{pkt}}(t, i)$ at time $t$ as follows.

$$\hat{\text{wt}}_{\text{pkt}}(t, i) = \text{wt}_{\text{pkt}}(t) \cdot \delta_i. \qquad (13.11)$$

The normalized waiting time is then used as the weight for scheduling. The packet with the largest weight is then selected by the WTP scheduler for transmission. Formally, at time $t$ it will transmit the packet pkt, which satisfies

$$\text{pkt} = \arg\max_{\text{pkt} \in \mathcal{P}} \hat{\text{wt}}_{\text{pkt}}(t, i), \qquad (13.12)$$

where $P$ is the set of backlogged packets. It is shown that WTP scheduler is able to approximate the proportional delay differentiation model in wireline networks under heavy traffic condition [13].

Here we introduce the *proportional service differentiation model* into the domain of wireless LAN. Different from wireline networks, where flows from the same router contend for the same wireline link, in wireless LANs not only do the flows originating from the same node contend with each other but also the flows from different nodes contend for the same wireless channel. To extend the concept of proportional service differentiation to the wireless LAN, the flows among different pairs of nodes are considered. Specifically, our proportional delay differentiation model for wireless LANs states that the relation equation (13.10) holds for all flows within the wireless LAN no matter whether they originate from the same node or not.

As a result of the distributed medium sharing, packet scheduling needs the cooperation among all the nodes. This is in contrast to wireline networks where packets that need to be scheduled originate from the same router, and hence the packet scheduling decision, can be made by the route itself only considering its own packets. We argue that delay differentiation in wireless LANs can only be achieved through a *joint packet scheduling* at the network layer and distributed coordination at the MAC layer. Therefore, we present a *cross-layer waiting time priority scheduling (CWTP) algorithm*, which is able to achieve proportional delay differentiation in wireless LANs.

*The cross-layer waiting time priority scheduling algorithm (CWTP)* divides the scheduling task into two parts, which are performed at two layers in the network stack. At the network layer, *intra-node* scheduling at node $n$ selects a packet $\text{pkt}_n^*$ with the longest normalized waiting time, i.e., a packet $pkt_n^*$, which satisfies

$$\text{pkt}_n^* = \arg \max_{\text{pkt} \in \mathcal{P}_n} \hat{w}t_{\text{pkt}}(t, i) \qquad (13.13)$$

where $\mathcal{P}_n$ is the set of all backlogged packets at node $n$. At the MAC layer, *inter-node* scheduling selects packet $\text{pkt}^*$ among $\text{pkt}_n^*$, which satisfies

$$\text{pkt}^* = \arg \max_{\text{pkt}_n^*, n \in \mathcal{N}} \hat{w}t_{\text{pkt}_n^*}(t, i), \qquad (13.14)$$

where $\mathcal{N}$ is the set of wireless nodes.

Such an intra- and inter-node scheduling algorithm can fit well the environment of wireless LANs. In particular, the intra-node scheduling can be implemented via network layer packet scheduling at each individual node and the inter-node scheduling can be implemented via media access control (MAC), which coordinates packet transmissions among nodes. Figure 13.5 illustrates such a cross-layer scheduling architecture. In this architecture, the packet scheduler at the network layer and the distributed coordination function at the MAC layer are coordinated using normalized packet waiting time $\hat{w}t$ as a cross-layer signal.

At MAC layer, in order to transmit the packet with larger normalized waiting time before the ones with smaller normalized waiting time, we map the normalized waiting time $\hat{w}t$ to the backoff time $b$ via function $b = \Phi(\hat{w}t)$. In [14], we present two mapping schemes, namely linear mapping and piece-wise linear mapping schemes to implement the function $\Phi(\hat{w}t)$.

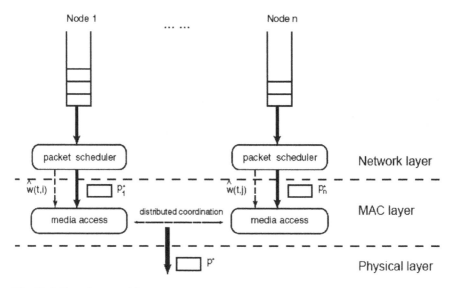

**Fig. 13.5** Cross-layer architecture

In linear mapping scheme, the normalized waiting time of a packet is mapped to its MAC layer backoff time via a linear function. Formally, let us consider a linear function $\phi(x) : \Re^+ \to \Re$,

$$\phi(x) = \beta - \alpha \cdot x \tag{13.15}$$

where $\alpha$, $\beta > 0$ are parameters of this linear function. To ensure it to be a non-negative integer, the backoff time $b$ (in number of time slot) of a packet with normalized waiting time $\hat{w}t$ is chosen as follows,

$$b = \Phi(\hat{w}t) = \lceil [\varphi(\hat{w}t)]^+ \rceil \tag{13.16}$$

where $[x]^+ = \max(0, x)$ and $\lceil \cdot \rceil$ is the ceiling operation. These two operations round up the value of $\phi(\hat{w}t)$ to a non-negative integer. It is obvious that $\alpha$ and $\beta$ determines the effectiveness of the mapping function, and thus the performance of the cross-layer scheduling algorithm. We present a dynamic tuning algorithm of $\alpha$ and $\beta$. Let $\overline{cw}$ be the expected value of contention window under IEEE 802.11 DCF without differentiation. The backoff time $b$ is uniformly chosen from $[0, \overline{cw})$. Let $\hat{w}t_{\max}$ and $\hat{w}t_{\min}$ be the maximum and minimum normalized waiting time, respectively. Preferably, the maximum normalized waiting time $\hat{w}t_{\max}$ can be mapped to the smallest backoff time (0) for efficient channel utilization; and $\hat{w}t_{\min}$ can be mapped to $\overline{cw}$ for similar contention behavior as IEEE 802.11 without differentiation.

Linear mapping scheme neglects the fact that the distribution of the normalized waiting time can be non-uniform. If there is a higher density over a certain

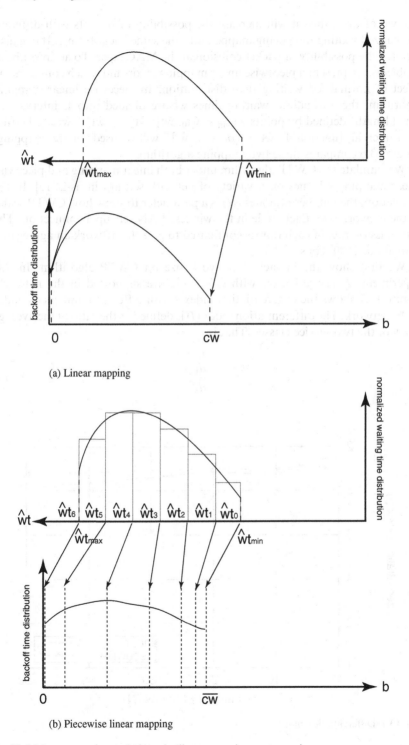

(a) Linear mapping

(b) Piecewise linear mapping

**Fig. 13.6** Linear mapping and piecewise linear mapping: a comparison

interval of time, then it will increase the possibility of packets with different normalized waiting time being mapped into the same backoff time. It can also increase the possibility of packet collision at the MAC layer. To address above problems, we present a piecewise linear mapping algorithm which considers the effect of normalized waiting time distribution. In piecewise linear mapping algorithm, the normalized waiting times $\hat{w}t$ are divided into L intervals of equal lengths defined by points $\hat{w}t_{min} = \hat{w}t_0, \hat{w}t_1, \hat{w}t_2, ..., \hat{w}t_L = \hat{w}t_{max}$. During each interval, function $\Phi_i(\hat{w}t) = [\beta_i - \alpha_i \cdot \hat{w}t]^+$ will be used for the mapping. Figure 13.6 compares these two mapping algorithm.

We simulate the CWTP algorithm under both linear mapping and piecewise linear mapping schemes on a variety of network settings in ns-2 [15]. In the simulation, the number of nodes ($N$) is a parameter to show how CWTP scales to the network size. Each node in the wireless LAN sets up a connection. The transmission rate of each flow is configured to give the network an aggregated load about 1500 Kbps.

We first show the impact of network size on CWTP algorithm. In this experiment, 2 service classes with $\delta_2/\delta_1 = 2$ are supported in the network. Figure 13.7 shows the differentiation index $I$ with different numbers of nodes in the network. The differentiation index ($I$) is defined as the ratio of the average delay of the two service classes. That is

$$I = \frac{\bar{d}_1}{\bar{d}_2},\qquad(13.17)$$

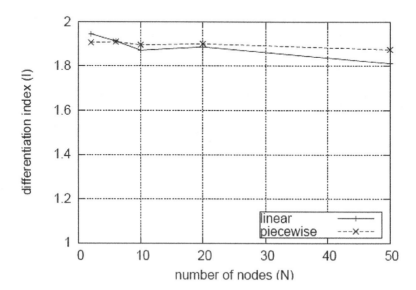

**Fig. 13.7** Differentiation index

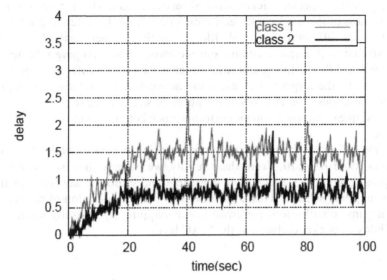

(a) Linear mapping

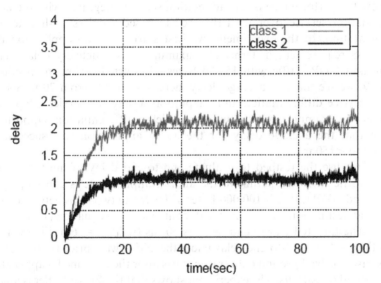

(b) Piecewise linear mapping

**Fig. 13.8** Instantaneous delay behavior

where $\bar{d}_i$ is the expected packet delay of service class $i$. This metric shows the effectiveness of the service differentiation – how close the differentiation result matches the differentiation goal. Ideally, in these experiments $I = 2$. We observe that both linear mapping and piecewise linear mapping schemes can lead the CWTP scheduling algorithm to achieve a delay differentiation index very close to the target value, when the network size is relatively small (the number of nodes $N < 20$). When the network size is large (e.g. $N = 50$), the piecewise linear mapping scheme performs much better than the linear mapping scheme.

In Fig. 13.8, we show the instantaneous delay behaviors under these two schemes when $N = 10$. From these results, we observe that piecewise linear mapping scheme gives much more consistent and smooth delay behavior than linear mapping scheme. This is because with the consideration of normalized waiting time distribution, piecewise linear mapping significantly reduces the possibility of packet collision at the MAC layer.

## 13.5 Evaluation and Implementation

We show the performance of the adaptation service integrated with the delay differentiation service over an IEEE 802.11-based wireless ad-hoc testbed implementation. In the experiment, we first start an audio application that has a QoS requirement in terms of maximum packet delivery delay. Then background UDP traffic with 15,000 Bytes/s is started. From the results in Fig. 13.9, we see that the average delay increases quickly from 70 to 800 ms without service differentiation and adaptation. Using the service adaptation policy in the example and the underlying delay differentiation support, we observe that the average delay for the audio application was successfully bounded to <150 ms.

Next, we use the testbed to evaluate end-to-end delay of multi-hop QoS (audio) traffic. In the experiment, the audio sampling rate is 256 Kbps, and background UDP traffic is 100,000 Bytes/s. There are two hops between audio source and destination. We enable the PI controller as in equation (13.9) on our testbed. The targeted end-to-end delay is 60 ms ($\text{ref}_{\text{delay}} = 60$). Figure 13.10 shows the resulting end-to-end delay under the middleware priority adaptation. In this case, the background data traffic starts after the multimedia application sends around 60 packets. The experiment shows that the PI controller is able to quickly converge the end-to-end delay of a multimedia application to a desired level, when there exist background traffic, which compete with the application with QoS requirement for the network resources in wireless ad hoc environment. We also notice that the delay of most of the audio traffic is in the range from 40 to 80 ms. Therefore, the middleware adaptation method also achieves the latency and playback jitter control, which is very important for multimedia applications.

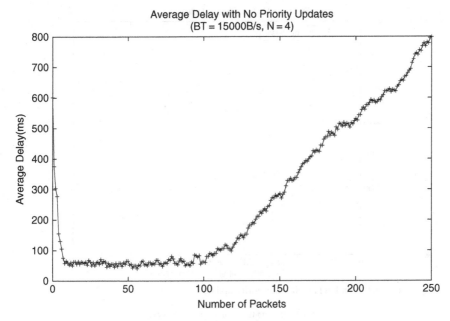

(a) Without service adaptation and differentiation

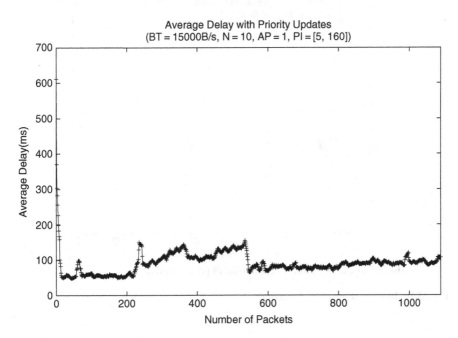

(b) With service adaptation and differentiation.

**Fig. 13.9** Performance comparison

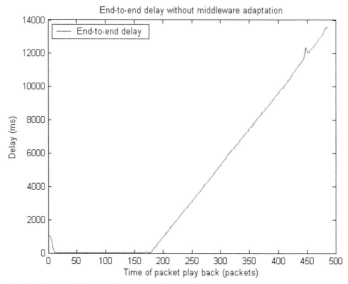

(a) End-to-end delay without priority update

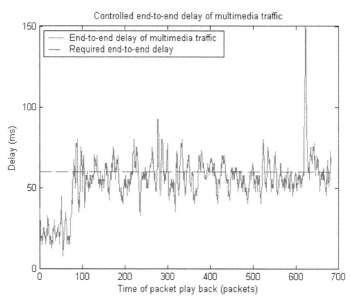

(b) End-to-end delay under PI controller of priority update

**Fig. 13.10** Instantaneous delay behavior

## 13.6  Directions for Future Research

In the cross-layer end-to-end delay control framework, the middleware-layer adaptation in Section 13.4 and network/MAC layer scheduling algorithm co-design in Section 13.5 are localized methods with small overhead. The middleware layer adaptation has the advantage that the adaptation mechanism is hardware-independent, so users do not have to modify the hardware to achieve the QoS requirement. Further, for wireless network with location-dependent contention, the cross-layer support at the network and MAC layer should be adopted.

As the heterogenous networking environment where wireless communication and wired links coexist becomes popular, we will consider the QoS support in such a heterogeneous environment as a future research direction.

## 13.7  Conclusions

In this chapter, we have shown a multi-layer adaptation scheme for end-to-end delay control in multi-hop wireless networks. Based on this cross-layer design, we can control the end-to-end delay given sufficient network bandwidth.

## Terminologies

*Cross-layer design*: Cross-Layer Design has recently become very important in wireless communication systems. Cross-layer protocol interactions, when used appropriately, can lead to increased network efficiency and better QoS support.

*Proportional delay differentiation (PDD)*: In a network with $C$ service classes. the proportional delay differentiation model imposes the following constraints for all pairs of classes.

$$\frac{\bar{d}_i(t,t+\tau)}{\bar{d}_j(t,t+\tau)} = \frac{\delta_j}{\delta_i}, \text{for all } i \neq j \text{ and } i,j \in \{1,2,...,C\}$$

where $\delta_i$ is the service differentiation parameter for class $i$, and $\bar{d}_i(t, t + \tau)$ is the average delay for class $i,(i = 1,2,...,C)$ in the time interval $(t,t - \tau)$, where $\tau$ is the monitoring time scale.

*Delay monitor*: A component in cross-layer architecture to monitoring round-trip end-to-end delay.

*Service classifier*: A component in cross-layer design architecture to determine service class of packets from priority of the packets.

*Priority adaptor*: A component in cross-layer design architecture to adjust priority of packets of a certain flow based on observed end-to-end delay.

*Model identification*: This is the procedure we determine the model which captures the relationship between priority and end-to-end delay.

*PID controller*: A feedback controller which contains proportional, integral and differentiation components.

*Waiting time priority (WTP)*: In WTP scheduling algorithm, a packet is assigned with a weight, which increases proportionally to the packet's waiting time. Service classes with higher differentiation parameters have larger weight-increase factors. The packet with the largest weight is served first in non-preemptive order. Formally, let $wt_{pkt}(t)$ be the waiting time of a packet pkt of class $i$ at time $t$, we define its *normalized waiting time* $\hat{wt}_{pkt}(t, i)$ at time $t$ as $\hat{wt}_{pkt}(t, i) = wt_{pkt}(t) \cdot \delta_i$

*Linear mapping*: A mapping scheme that maps the normalized waiting time of a packet to its MAC layer backoff time via a linear function.

*Piece-wise linear mapping*: A mapping scheme that maps the normalized waiting time of a packet to its MAC layer backoff time via a piece-wise linear function.

## Questions

1. What is a waiting time priority scheduler?
2. Why joint packet scheduling at the network layer and distributed coordination at the MAC layer is needed?
3. Describe the function of each component in the cross-layer delay management architecture.
4. Why piecewise linear mapping algorithm performs better than linear mapping algorithm.
5. Why Priority Adaptor is in the form of PI controller?
6. What is the major advantage of middleware adaptation and what is its disadvantage?
7. What is the purpose of model identification in middleware Priority Adaptor design?
8. Why the number of service classes cannot be very large?
9. Why we need cross-layer design for end-to-end delay control?
10. How normalized waiting time is calculated?

## References

1. W. He and K. Nahrstedt. Impact of upper layer adaptation on end-to-end delay management in wireless ad hoc networks. *12th IEEE Real-Time and Embedded Technology and Applications Symposium (RTAS06)*, 2006.
2. Y. Xue, K. Chen and K. Nahrstedt. Distributed end-to-end proportional delay differentiation in wireless LAN. *IEEE International Conference on Communications (ICC)*, 2004.
3. I. Aad and C. Castelluccia. Differentiation mechanisms for IEEE 802.11. *Proc. of IEEE INFOCOM*, 2001.

4. D. Gu and J. Zhang. QoS enhancement in IEEE 802.11 wireless local area networks. *IEEE Communications Magazine*, 41(6):120–124, 2003.

5. H. Luo, S. Lu, V. Bharghavan. A new model for packet scheduling in multihop wireless networks. *Proc. of ACM Mobicom*, pages 76–86, 2000.

6. N. H. Vaidya and P. Bahl and S. Gupta. Distributed fair scheduling in a wireless LAN. *Proceedings of the 6th Annual International Conference on Mobile computing and Networking*, pages 167–178, 2000.

7. D. Qiao and K.G. Shin. Achieving efficient channel utilization and weighted fairness for data communications in IEEE 802.11 WLAN under the DCF. *Proc. of The Tenth International Workshop on Quality of Service (IWQoS)*, 2002.

8. S.B. Lee, A. Gahng-Seop, X. Zhang and A.T. Campbell. INSIGNIA: An IP-based quality of service framework for mobile ad hoc networks. *Journal of Parallel and Distributed Computing, special issue on Wireless and Mobile Computing and Communications*, 6(4):374–406, 2000.

9. G.-S. Ahn, A.T. Campbell, A. Veres, L. Sun. Supporting service differentiation for real-time and best effort traffic in stateless wireless ad hoc networks (SWAN). *IEEE Transactions on Mobile Computing*, 2002.

10. L. Ljung. System Identification: Theory for the User (2nd Edition), Prentice Hall, Upper Saddle River, NJ 1999.

11. C. Dovrolis and P. Ramanathan. A case for relative differentiated services and the proportional differentiation model. *IEEE Network*, 13(5):26–34, 1999.

12. C. Dovrolis and P. Ramanathan. Dynamic class selection: from relative differentiation to absolute QoS. *IEEE International Conference on Network Protocols*, 2001.

13. C. Dovrolis, P. Ramanathan and D. Stiliadis. Proportional differentiated services: delay differentiation and packet scheduling. *IEEE/ACM Transactions on Networking*, 10:12–26, 2002.

14. Y. Xue, K. Chen and K. Nahrstedt. Achieving proportional delay differentiation in wireless LAN via cross-layer scheduling. *Journal of Wireless Communications and Mobile Computing, special issue on Emerging WLAN Technologies and Applications*, 4(8):849–866, 2004.

15. The Network Simulator - ns-2.

# Chapter 14
# Address Allocation Mechanisms for Mobile Ad Hoc Networks

Xiaowen Chu, Jiangchuan Liu, and Yi Sun

**Abstract** A Mobile Ad hoc Network (MANET) is an independent self-organizing network, in which each node functions as an end host and a wireless relay. This form of wireless network is created by mobile nodes without any existing or fixed infrastructure. The nodes in the MANET need mutually exclusive identities before participating in any form of communication; in particular, each end host in the MANET needs to be uniquely addressed so that the packets can be relayed hop by hop and delivered ultimately to the destination. The mobility of the nodes, however, makes address allocation a challenging task in MANETs.

In the past decade, many address allocation schemes have been proposed to address these challenges. In this chapter, we will present a comprehensive survey on the state-of-the-art of address allocation in MANETs. We will also demonstrate a new address allocation protocol for MANET based on the concept of quadratic residue, which can be applied to large-scale MANETs with low communication overhead, even distribution, and low latency.

## 14.1 Introduction

A Mobile Ad hoc Network (MANET) is an independent self-organizing network, in which each node functions as an end host and a wireless relay. This form of wireless network is created by mobile nodes without any existing or fixed infrastructure. Since the mobile hosts usually have limited transmission range, bandwidth and battery power, multiple hops are generally required in MANETs to exchange data between nodes. In addition, as a community network that relies on the willingness of mobile hosts to forward and relay packets toward the destination, a MANET can be formed and deformed on the fly without the need of any system administrator.

X. Chu (✉)
Department of Computer Science, Hong Kong Baptist University, Hong Kong
e-mail: chxw@comp.hkbu.edu.hk

S. Misra et al. (eds.), *Guide to Wireless Ad Hoc Networks*,
Computer Communications and Networks, DOI 10.1007/978-1-84800-328-6_14,
© Springer-Verlag London Limited 2009

The nodes in the MANET need mutually exclusive identities before participating in any form of communication; in particular, each end host in the MANET needs to be uniquely addressed so that the packets can be relayed hop by hop and delivered ultimately to the destination. Existing routing protocols in MANETs have all assumed a priori that mobile nodes are configured with a valid (conflict free) network address. Note that, because of the multi-hop routing, the MAC address at the link layer level cannot serve for this purpose. On the other hand, address configuration in wired networks, such as the Dynamic Host Configuration Protocol (DHCP) [1], requires the presence of a centralized DHCP server. It does not work well for MANETs due to the mobility of the nodes and the lack of a central authority.

Given these uniqueness, address allocation in MANETs has attracted a significant amount of research. The purpose of address allocation in MANETs is to manage the address space efficiently and effectively. An unconfigured node should be able to allocate a unique network address in a timely manner, without costing excessive network traffic overhead. Once a node leaves the network, its address should be reclaimed for future usage. All these need to well adapt to the distributed and dynamic nature of MANETs. In particular, we have to address the network partitions and mergers. Due to node mobility, MANETs can be split into several disjoint partitions with no communications. These network partitions may or may not merge back later. And such partitioning or merging are often invisible to individual mobile hosts.

In the past decade, many address allocation schemes have been proposed to address these challenges. In this chapter, we will present a comprehensive survey on the state-of-the-art of address allocation in MANETs. We will also demonstrate a new address allocation protocol for MANET based on the concept of quadratic residue, which can be applied to large-scale MANETs with low communication overhead, even distribution, and low latency.

The rest of the chapter is organized as follows. Section 14.2 presents the background and introduces traditional address allocation schemes for IP-based networks. Section 14.3 introduces existing address allocation schemes specifically designed for MANETs. Section 14.4 discusses some performance metrics that should be considered. Section 14.5 presents a quadratic residue–based address allocation scheme. Section 14.6 summarizes the chapter.

## 14.2 Background

In this section, we review the traditional address allocation schemes and explain why they cannot be directly applied in MANETs. These address allocation schemes can be classified into stateful schemes or stateless schemes. The stateful schemes keep state information in a database that keeps track of which addresses have been assigned to which computers; while the stateless schemes let the computers select an address by themselves and perform a procedure, called Duplicate Address Detection (DAD).

## 14.2.1  Traditional Stateful Schemes

### 14.2.1.1  Reverse Address Resolution Protocol (RARP)

In TCP/IP protocol family, RARP allows a computer to obtain its IP address from a RARP server in the bootstrap procedure [2]. Before obtaining an IP address, a computer has to use its MAC address to communicate with others. It first broadcasts a RARP request that specifies itself as a target. The RARP server on the same network keeps the database of IP addresses. Upon receiving a RARP request message, the RARP server looks up the IP address based on the requester's physical address and replies to the requester. Notice that RARP works at a lower layer than IP, and it is generally used in local area networks that support broadcast.

RARP has some limitations, however. First, the reply from the server contains only the 4-octet IP address; second, it cannot be used on networks that dynamically assign physical addresses.

### 14.2.1.2  BOOTstrap Protocol (BOOTP)

BOOTP is developed to overcome some of the drawbacks of RARP [3]. BOOTP uses UDP to carry messages and hence it can be implemented with an application program. Before obtaining an IP address, a computer can broadcast an IP datagram on the local network by using the limited broadcast IP address 255.255.255.255. The BOOTP server then simply broadcasts the reply message on the local network, which contains the requester's IP address, the router's IP address, etc.

BOOTP is designed for a relatively static environment, and it provides only a static mapping from the physical address to the corresponding network parameters. It is not suitable for a dynamic environment.

### 14.2.1.3  Dynamic Host Configuration Protocol (DHCP)

DHCP is developed as a successor to BOOTP [1]. A designated DHCP server allocates network addresses and delivers configuration parameters to dynamically configured computers. The most attractive aspect of DHCP is its dynamic address assignment, in which the DHCP server does not need to know the identity of the client a priori. Auto-configuration becomes possible if the DHCP has been provided with a set of available IP addresses. At present, DHCP is widely used in Ethernets and Wireless LANs.

Both BOOTP and DHCP can use the Relay Agent to permit a computer to contact a server on a non-local network. The Relay Agent must be located at the same local network of the requester, however. When the Relay Agent receives a broadcast request from a client, it forwards the request to the BOOTP server or DHCP server, receives the reply from the server, and forwards the reply to the client.

## 14.2.2 Traditional Stateless Schemes

### 14.2.2.1 IPv6 Stateless Address Autoconfiguration

IPv6 stateless address autoconfiguration is performed only on multicast-capable links [4]. A node begins the auto-configuration process by generating a link-local address for its interface. A link-local address is formed by appending the interface's identifier to the well-known link-local prefix. Before assigning the link-local address to its interface, a node must attempt to verify that this link-local address is not used by another node on the same network. This is done by the Duplicate Address Detection (DAD) procedure. Specifically, it sends a *Neighbor Solicitation* message that contains the tentative address as the target address. Notice that this message uses the well-known unspecified address as source IP address, and the solicited-node multicast address as the destination IP address. If another node is also using that address, it will return a *Neighbor Advertisement* message using the all-nodes multicast address as the destination IP address. If a node finds that its tentative link-local address is not unique in the network, auto-configuration process stops and manual configuration of the interface is required. On the contrary, if a node determines that its tentative link-local address is unique in the network, it assigns the address to itself and starts to communicate with all other nodes using this address.

### 14.2.2.2 Zero Configuration Networking (Zeroconf)

Address configuration without a dedicated server has been investigated by the Zero Configuration Networking (Zeroconf) working group of the Internet Engineering Task Force (IETF). The goal of the Zeroconf Working Group is to enable networking in the absence of configuration and administration.

The Internet draft [5] describes a method for dynamic configuration of IPv4 link-local addresses used for local communications. When a node wishes to configure a link-local address, it selects an address pseudo-randomly, uniformly distributed in the range 169.254.1.0 to 169.254.254.255. Then it tests whether or not this address is already in use by broadcasting an ARP request for the desired address. If no conflicting ARP reply has been received after a predefined time limit, then it can successfully claim the desired link-local address. Otherwise, it needs to select a new pseudo-random address and repeat the process.

### 14.2.3 Issues of Traditional Address Allocation Schemes

The traditional stateful address allocation schemes for IP-based networks require a centralized server to assign addresses to new nodes. They cannot be directly applied to MANETs, however, because MANETs can have a highly dynamic topology and the centralized server may not always be reachable.

The traditional stateless schemes cannot directly apply to MANET either because they require all nodes to be reachable via single-hop broadcast messages, which is generally not the case of MANETs. The Zeroconf solution also performs the DAD based on ARP request/reply messages, which may not be possible for MANETs. IPv6 stateless auto-configuration assumes the 48-bit IEEE-assigned globally unique MAC addresses. This hardware-based addressing scheme has the following limitations: (1) The 48-bit MAC address is too long for an IPv4 address. (2) The 48-bit MAC addresses may not be unique [6]. It is also possible to change the MAC address by reprogramming the EEPROM or by modifying the MAC address in the OS memory. (3) Some devices in MANETs do not use a 48-bit MAC address. (4) The identity of a node can be easily determined from the network address, which raises privacy concerns.

## 14.3 Address Allocation Schemes for MANETs

In the past decade, a lot of research has been done on address allocation schemes for MANETs. Following the previous classification criteria, we classify these address allocation protocols into three groups: stateful schemes, stateless schemes, and hybrid schemes.

### 14.3.1 Stateful Schemes

#### 14.3.1.1 MANETconf

MANETconf prevents concurrent assignment of the same address by maintaining an additional allocation table for pending allocations [7].

A new node $X$ enters the MANET by broadcasting a *Neighbor-Query* message. The neighbors that are already part of the MANET shall respond with a *Neighbor-Reply* message. If no reply is received within a certain period of time, node $X$ assumes that it is the first node in the MANET and assigns an address to itself. Otherwise, it selects one of the responders, say $Y$, as its initiator by sending a *Requester_Request* message to $Y$. Each configured node in the MANET keeps two address tables: allocated table $T_{allocated}$ for the set of all addresses in use in the MANET, and pending table $T_{pending}$ for the set of addresses for which address allocation has been initiated but not yet completed. Upon receiving the *Requester_Request* message from $X$, node $Y$ chooses an address that is not in its address tables, say $ADD_X$. Node $Y$ adds this address to $T_{\text{pending}}$ and then floods an *Initiator_Request* message containing $ADD_X$ to all the configured nodes in the MANET, which aims to seek permission to assign $ADD_X$ to node $X$. A node that receives this message looks up $ADD_X$ in its address tables. If $ADD_X$ is not found, it accepts the allocation by sending an affirmative reply to node $Y$ and also updates its pending table accordingly. If the replies from all nodes are positive, the allocation is regarded as successful, and node $Y$ assigns $ADD_X$ to node $X$, moves $ADD_X$ from $T_{\text{pending}}$ to $T_{\text{allocated}}$, and informs the whole MANET so that all other nodes can also move $ADD_X$ from the pending table to the allocated table. Otherwise, if node $Y$ receives at least one negative response, it repeats the allocation process with another address.

MANETconf also addresses the problem of network partitioning and merger. Each partition is identified by a 2-tuple partition ID, in which the first element is the lowest address in use in this partition and the second element is a universally unique identifier (UUID). If the MANET breaks into two partitions, one partition must contain the node with the lowest address and its partition ID remains the same. All the nodes in this partition need to clean up the addresses that belong to the other partition. In the second partition, the node with the lowest address needs to be discovered first, and then it chooses a new UUID and floods in its partition. Upon receiving this flood, all nodes in the second partition know about the new partition ID. To detect network merger, two previously distant nodes need to exchange their partition IDs if they come within the communication range of each other. A network merger is detected by both nodes if their partition IDs are different. On detecting a merger, both nodes exchange their allocated tables and take the union as the new allocated table. Addresses appeared in both allocated tables are conflicting addresses and, for each conflicting address, one of the two conflicting nodes must give up its old address and acquire a new address. To minimize disruptions in data communication, the conflicting node with fewer and/or short-lived TCP connections are suggested to acquire a new address. The merging is completed when each address conflict has been resolved.

### 14.3.1.2 Buddy Protocol

The buddy protocol is based on the idea of binary split, which splits the address allocation table among all nodes and uses buddy systems for efficient table mergers [8, 9]. Unlike MANETconf, every node has a disjoint set of addresses that it can assign to a new node without asking other nodes for permission. In the beginning, the only node in the network has the whole pool of addresses. The size of the whole address pool is a power of two. When a new node $X$ wants to join the MANET, it broadcasts a request message. Through a handshaking protocol, $X$ selects the configured node, $Y$, who is the first to send an acknowledgement message to $X$, as the initiator. Node $Y$ assigns half of its address pool to $X$. Then node $X$ selects the first address from this pool as its own address and keeps the rest as its available set of addresses. It is possible that node $Y$ has no available addresses to assign. In this case, node $Y$ gets a block of addresses from the node that has the biggest address pool, and then assigns an address block to $X$. If $X$ leaves the network gracefully, it returns its pool of addresses back to any of its neighbors, say $Z$. Node $Z$ then takes the responsibility to handle this set of addresses. It can either keep them or it can find the node who owns the buddy block of $X$'s address block and forward $X$'s address block to the buddy for a merge. If $X$ leaves the network abruptly, $X$'s address block will be lost, which results in the address leak. To resolve this address leak issue, nodes need to synchronize from time to time to keep track of the assigned addresses and detect address leaks. The same idea of buddy system has also been applied in [10].

The main limitation of the Buddy system is the inefficient utilization of address space: if a large number of new nodes join a small area of the MANET, address set can be unevenly distributed. To remedy this issue, the authors in [11] proposed two techniques named remote allocation and leakage collection, where the former is a process to allocate addresses from a remote node when local nodes are short of addresses and the latter is for collecting unoccupied addresses and reusing them for the whole MANET.

### 14.3.1.3 Prophet Allocation Protocol

Prophet allocation scheme makes use of a stateful sequence generating function, say $f(n)$ [12]. The initial state of $f(n)$ is called the seed, and different seeds lead to different sequences. The main desired properties of function $f(n)$ are: (1) the interval between two occurrences of the same number in a sequence is extremely long; (2) the probability of more than one occurrence of the same number in a limited number of different sequences initiated by different seeds during some interval is extremely low. If such sequence generating function $f(n)$ can be designed, then the address auto-configuration algorithm can be as simple as follows:

(1) The first node in the MANET, i.e., the prophet, chooses a random number as its address, and uses a random state value as the seed for its $f(n)$.
(2) A new node $X$ contacts an existing node $Y$ for an address. $Y$ uses its $f(n)$ to obtain an integer, say $m$, and also a state value, say $n$, and provides $(m, n)$ to $X$. Node $Y$ updates its state accordingly.
(3) Node $X$ uses $m$ as its address, and $n$ as the seed for its $f(n)$. Now node $X$ is also able to assign addresses to other nodes.

The main advantage of the above scheme is that only one hop broadcast is required for obtaining the address, which can extensively save communication overheads as compared with multi-hop broadcast communications needed in acquiring permission or conflict detection.

The probability of address conflict depends on the design of function $f(n)$. The authors propose a method based on the canonical factorization theorem in arithmetic that every positive integer can be expressed as a product of primes uniquely. For example, the canonical form of a positive number $n$ is $n = \prod_{i=1}^{k} p_i^{e_i}$, where the primes $p_i$ satisfy $p_i < p_j$ if $i < j$, and the exponents are non-negative integers. The $k$-tuple $(e_1, e_2, \ldots e_k)$ is used as the state. For example, if $k = 4$, the first node has a random address $a$ and an initial state of $(\underline{0}, 0, 0, 0)$. If it assigns an address to a second node, its state will be updated as follows: the underlined element in the 4-tuple is increased by 1. It also needs to assign a state to the second node: its new state will be copied to the second node, but the underline shifts right by 1. So the second node will have the initial state $(1, \underline{0}, 0, 0)$, and its address can be directly calculated based on this state. A much larger value of $k$ is suggested to be used in real applications. Nevertheless, the possibility of duplicate addresses cannot be totally eliminated.

#### 14.3.1.4 Prime DHCP

Prime DHCP follows the design of prophet address allocation [13]. It can allocate addresses without broadcasting over the whole MANET. It configures each node in the MANET as a DHCP proxy, and each node is eligible to assign addresses. Prime DHCP makes use of the Prime Numbering Address Allocation (PNAA) algorithm to guarantee the uniqueness of addresses. PNAA also utilizes the canonical factorization theorem of positive integers. The concept of PNAA can be illustrated by a logical address allocation tree, shown in Fig. 14.1:

The root DHCP proxy of the MANET has an address of 1, and can allocate all prime numbers to new nodes in ascending order. For a non-root DHCP proxy with address $X$, it can assign the set of addresses $\{Y \mid y = X \cdot P, P$ is a prime and $P \geq f(X)$ where $f(X)$ denotes the largest prime factor of $X\}$. For example, in Fig. 14.1, the node with address 9 can assign the set of addresses

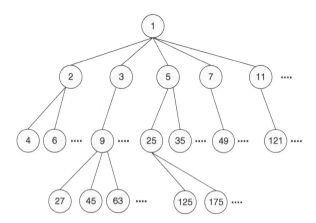

**Fig. 14.1** An example of PNAA logic address allocation tree

{27, 45, 63, ...}. Each node assigns the address in ascending order and it is required to record the last assigned address. It is easy to show that by using PNAA, no two proxies can assign the same address.

## 14.3.2 Stateless Schemes

### 14.3.2.1 Query-Based Duplicate Address Detection

Perkins et al. have proposed a dynamic IP configuration in an Internet draft [14]. This solution performs Duplicate Address Detection (DAD) through multiple rounds of MANET-wide flooding. Following the terminology of [15], we call this solution query-based DAD (QDAD). QDAD is based on the IPv6 stateless address auto-configuration mechanism. If a node, say $X$, wishes to join the network, it first selects a random IP address. Then, it issues an *Address Request* (AREQ) message for that randomly selected address. The purpose of the AREQ message is to find a route to a node with the selected address. If the chosen address is indeed already assigned to another node, then with the routing protocols the route request will result in an *Address Reply* (AREP) message being sent back to node $X$. Absence of an AREP can be used as an indication that no other node has assigned the address chosen by node $X$. If no AREP is returned within a timeout period, node $X$ retires the AREQ up to RREQ_RETRIES times. If after all retires, no AREP is received, node $X$ assumes that the address is not already in use, and assumes that the address can be safely used for its own. Otherwise, node $X$ randomly selects another address and begins the procedures again.

### 14.3.2.2 Weak DAD

A disadvantage of QDAD is that it does not work as intended when unbounded delays due to partitions can occur. Particularly, when partitions merge, the resulting network may contain nodes with duplicate addresses. For correct behavior, this scheme must be augmented with a procedure that detects merging of partitions, and then takes suitable actions to detect duplicate addresses in the merged partitions.

Vaidya identified two version of DAD: strong DAD and weak DAD, in [16]. With strong DAD, if multiple nodes have chosen a common address at a given time, then at least one of these nodes will detect the duplicate within a fixed interval of time. It has been shown that strong DAD cannot be guaranteed if message delays between at least one pair of nodes in the network are unbounded. However, when all message delays are bounded, strong DAD can in fact be achieved. For example, if the MANET can guarantee bounded delay, the QDAD is a strong DAD. However, delays in MANET are not always bounded, particularly in the presence of network partitions and merges. Under such scenario, timeout-based DAD approaches such as QDAD will not work. To this end, Vaidya proposed a mechanism called weak DAD (WDAD), which is functional even for MANETs with unbounded delays. Instead of requiring detection of all duplicate addresses, WDAD only requires that packets meant for one node must not be routed to another node, even if the two nodes have chosen the same address. In WDAD, duplicate addresses can be tolerated so long as packets reach the destination node intended by the sender. To achieve this purpose, the routing protocol has to be changed in concert. WDAD schemes with Link State Routing and Dynamic Source Routing are both given in [16].

Each node generates a unique key at initialization time and distributes this key along with its address in all routing protocol packets. The reason of not using the key as the IP address is that the unique key may be too long to be embedded in the IP address. The key will be used for the purpose of DAD. The keys learned from other nodes are stored in the routing table and distributed further in routing protocol packets. If a node receives a routing protocol packet with an address that is part of its routing table, it compares the corresponding keys. If they are different, a duplicate address is detected and the entry is marked as invalid to prevent misrouting of data packets.

If two nodes happen to select the same address and key, the conflict cannot be detected by WDAD. Although the probability of such a situation decreases with increasing key length, the routing protocol overhead increases, too. Thus, there is a trade-off between reliability and overhead of WDAD.

### 14.3.2.3 Passive DAD

Both query-based DAD and WDAD generate additional traffic in the MANETs during the procedure of conflict detection. To save the communication overhead, Passive DAD (PDAD) is proposed to detect duplicates without disseminating additional traffic in the network [17]. This can be achieved by monitoring the routing protocol traffic. Weniger proposed three approaches of PDAD, all based on the properties of link-state routing protocols. In link-state routing protocols, nodes inform other nodes about their neighborhoods by periodically exchanging link-state packets. Due to limited space, we only briefly introduce the PDAD-SN algorithm, which exploits the fact that link-state packets contain sequence numbers to detect duplicate addresses without generating additional packets other than link-state packets. A properly configured network has the following rules: (1) a node uses each sequence number only once; (2) a node uses sequence numbers incrementally; (3) two nodes do not have the same neighborhood at the same time if they are more than two hops apart from each other. Based on these rules, the authors derived two theorems if no duplicate addresses exist: (1) two messages with the same sequence number and source address are copies of the same message; (2) a node does not receive a link-state packet with its own address as source address and a sequence number that is higher than its own counter value. If one of the above does not apply, an address conflict is present in the MANET. However, the situation of sequence number wrap-arounds has to be carefully handled.

### 14.3.3 Hybrid Scheme

#### 14.3.3.1 PACMAN

The Passive Autoconfiguration for MANETs (PACMAN) protocol exploits ideas from both stateful schemes and stateless schemes, and hence it is classified as a hybrid scheme [18]. A node assigns itself an address using a probabilistic algorithm when joining the MANET. By maintaining an address allocation table (a stateful property), the selected address is unique most of the time. Only in the scenario that many nodes join the network simultaneously (e.g. a network merger) and hence the allocation table is not up-to-date, address conflict may occur.

PACMAN uses PDAD to detect address conflicts. In [18], Weniger proposed a set of PDAD algorithms for proactive link-state routing protocols and on-demand routing algorithms. PDAD algorithms for popular ad hoc routing protocols such as AODV [19], DSR [20], and OLSR [21], are also discussed. To resolve the address conflict, at least one node with conflict address must be notified and get its address updated. PACMAN uses a simple approach to minimize the communication overhead: an *Address Conflict Notification* (ACN) message is unicast in the direction from which the corresponding routing protocol packet was received.

## 14.4 Thoughts for Practitioners

To implement an address allocation scheme for MANETs, the following performance metrics should be considered and evaluated carefully:

- Address allocation latency: the time taken from when a node starts the address autoconfiguration to when it is assigned an address.
- Communication overhead: the number of control packets used for the purpose of address allocation. It usually includes broadcast packets and unicast packets.
- Scalability: If the address allocation scheme requires lot of multi-hop communications, the system scalability will be poor.
- Complexity: mobiles nodes usually have the limited computational power, memory size, transmission speed, and power supply. A too complicated scheme may not be feasible in real applications.

In [22], Kim et al. presented a set of analytical models to evaluate the efficiency of four different address allocation schemes in terms of the address allocation latency and communication overhead: query-based DAD, MANET-conf, token-based scheme [23], and neighbor-based schemes. The models can be used to numerically evaluate the impact of network parameters on the efficiency of address allocation schemes.

## 14.5 Directions for Future Research

As we have seen, the number theory plays an important role in conflict-free address allocation. The future of MANET address allocation relies on the adoption of advanced mechanisms from the number theory. In this section, we present necessary background in this aspect together with a novel address allocation scheme based on quadratic residue.

### 14.5.1 Preliminaries of Number Theory

We first briefly review several concepts in number theory that are fundamental to our address allocation scheme.

**Definition 1:** Suppose $n$ is an integer. Integer $a$ is defined to be a *quadratic residue* (QR) modulo $n$ if $a$ 0 (mod $n$) and the congruence $y^2 \equiv a$ (mod $n$) has a solution $y \in Z_n$. $a$ is defined to be a *quadratic non-residue* (QNR) modulo $n$ if $a$ 0 (mod $n$) and $a$ is not a quadratic residue modulo $n$.

**Definition 2:** Suppose $p$ is an odd prime and $a$ is an integer. Define the *Legendre symbol* $\left(\frac{a}{p}\right)$ as follows:

$$\left(\frac{a}{p}\right) = \begin{cases} 0 & \text{if } a \equiv 0 \text{ (mod } p) \\ 1 & \text{if } a \text{ is a quadratic residue modulo } p \\ -1 & \text{if } a \text{ is a quadratic non } - \text{ residue modulo } p \end{cases}$$

**Theorem 1:** *For prime number $p$, $a$ is a quadratic residue modulo $p$ if and only if* $a^{(p-1)/2} \equiv 1$ (mod $p$).

**Definition 3:** Suppose $n$ is an odd positive integer, and the prime power factorization of $n$ is $n = \prod_{i=1}^{k} p_i^{e_i}$. Let $a$ be an integer. The *Jacobi symbol* $\left(\frac{a}{n}\right)$ is defined to be $\left(\frac{a}{n}\right) = \prod_{i=1}^{k} \left(\frac{a}{p_i}\right)^{e_i}$, where the symbols on the right side are the Legendre symbols.

### 14.5.2 Quadratic Residue Cycle

We use an example to illustrate the idea of quadratic residue cycle. From Table 14.1, it is easy to see that 1, 3, 4, 5, 9 are quadratic residues modulo 11. If we calculate quadratic residue from 1, it gives back 1 repeatedly. If we start at any other point, for example at 4 or 9, we get the next quadratic residue as shown in Fig. 14.2. The sequence of quadratic residues is referred to as a quadratic residue cycle.

### 14.5.3 Address Space and Quadratic Residue Algorithm

Assume we have two odd prime numbers $p$ and $q$, then calculate $N = p \times q$. Denote $\phi$ as Euler's phi-function, i.e., $\phi(n)$ is the number of integers $a$, $1 \leq a \leq n$ and $\gcd(a, n) = 1$. Then the number of quadratic residues modulo $N$ can be shown to be $\phi(n)/4 = (p - 1)(q - 1)/4$.

We want the quadratic residue modulo $N$ to occur in distinct cycles. A quadratic residue lying in one cycle would not be present in any other cycle modulo $N$. Initial quadratic residue of a cycle acts as a seed $S_0$ to generate a sequence of numbers until the seed repeats again. Period or interval is the gap between the first and the second occurrence of a same number in a sequence, i.e., the length of the cycle. We are interested in long intervals. In Fig. 14.3, $S_0$ is a seed, such that it is the first quadratic residue which generates a sequence of quadratic residues. These sequence repeats in cycles and the last quadratic residue before an interval ends is denoted by $S_\pi$.

For explanation purpose, we will give a small example. Let $p = 23$ and $q = 7$. Then $N = 161$ and $\phi(n) = (p-1) \times (q-1) = 132$. The total number of quadratic residues modulo $N$ is 33 and there are five cycles in total as shown in Table 14.2: one cycle of length one, one cycle of length two, and three cycles of length ten. Note that if we use large primes, we could get cycles of long lengths. Quadratic

**Table 14.1** Quadratic residue modulo 11

| a | 1 | 2 | 3 | 4 | 5 | 6 | 7 | 8 | 9 | 10 |
|---|---|---|---|---|---|---|---|---|---|----|
| $a^2 \bmod 11$ | 1 | 4 | 9 | 5 | 3 | 3 | 5 | 9 | 4 | 1 |

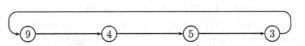

**Fig. 14.2** Quadratic residue cycle

**Fig. 14.3** Quadratic residue sequence

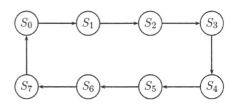

**Table 14.2** Seed generated sequence

| Seed $S_0$ | Sequence $S_i$ |
| --- | --- |
| 1 | 1 |
| 93 | 93, 116 |
| 2 | 2, 4, 16, 95, 9, 81, 121, 151, 100, 18 |
| 8 | 8, 64, 71, 50, 85, 141, 78, 127, 29, 36 |
| 25 | 25, 142, 39, 72, 32, 58, 144, 128, 123, 156 |

residues lying in these long cycles are used to allocate addresses in our protocol. In this way the address space is reclaimed automatically and each node can be assigned a unique address. Note that each of these cycles belongs to a unique seed. The intersection of any two different cycles is always a null set.

Now we present an algorithm to find the quadratic residue cycles for $N = p \times q$ where $p$ and $q$ are two prime numbers. We used Jacobi symbols to find quadratic residues because its computational complexity $O(\log p)^2$ is less than the Euler criterion of $O(\log p)^3$ [24]. To test whether an integer $a$ is a quadratic residue modulo $n$, we can calculate the Jacobi symbol $\left(\frac{a}{n}\right)$ and Legendre symbol $\left(\frac{a}{p}\right)$. If $\left(\frac{a}{n}\right) = -1$, then $a$ is a quadratic non-residue modulo $n$. If $\left(\frac{a}{n}\right) = 1$, since $\left(\frac{a}{n}\right) = \left(\frac{a}{p}\right)\left(\frac{a}{q}\right)$, there are two cases: (1) $\left(\frac{a}{p}\right) = \left(\frac{a}{q}\right) = -1$; (2) $\left(\frac{a}{p}\right) = \left(\frac{a}{q}\right) = 1$. In case (1), $a$ is not a quadratic residue modulo $p$, hence it cannot be a quadratic residue modulo $n$. In case (2), $a$ can be shown, from Chinese remainder theorem, to be a quadratic residue modulo $n$. Based on this observation, our algorithm to find quadratic residue cycles is shown in Fig. 14.4.

## 14.5.4 Address Allocation Protocol

When a mobile node switches on to ad hoc mode, it starts the timer, sends a DISCOVER message, and waits for a reply. If it receives no reply, it repeats the process for a suitable number of times. If all attempts fail, it assumes that it is the first node in the MANET and chooses two prime numbers $p$ and $q$ that are congruent to 3 mod 4. Prime numbers of this form are chosen because their square roots are easy to calculate. Then it computes $N = p \times q$ and $\phi(N) = (p - 1) \times (q - 1)$. This initiator node calculates the number of distinct cycles and length of each long cycle (address block). If the length of the long cycle is small, the initiator

**Fig. 14.4** Quadratic residue
algorithm

Algorithm: QRCycles($p, q$)

$T = \phi$ ;

$n = p \times q$;

$phi = (p-1) \times (q-1)$;

for ($i \leftarrow 1$ to $phi/4$)

{

  if ($i \in T$) break;

  $C = \phi$ ;

  $x = i$;

  if ($\left(\dfrac{x}{n}\right) == 1$ and $\left(\dfrac{x}{p}\right) == 1$)

  {

    while ($x \notin C$) do

      $C = C \cup \{x\}$;

      $x = x^2 \bmod n$ ;

    $T = T \cup C$ ;

    output($C$);

  } // end of if

} // end of for

can repeat the process until it finds long cycles. The initiator (first node) then configures itself with an IP address, keeps the seed $S_0$ value of each distinct long cycle and its corresponding range.

When a new node joins the network, it sends a DISCOVER message. An already configured node provides it an IP address. Along with an IP address a new node also receives a set of seeds $S_0$, and the corresponding range of each seed. Therefore, a new node gets an IP address, state value (seed), and range. Now the new node along with participating in communication is also capable of assigning a unique conflict free IP address without taking permission from any other node in the MANET. Each state represents the sequence of addresses in that cycle. A newly joined node after configuration will calculate the next address by squaring the current quadratic residue modulo $N$. Each cycle of quadratic residue is disjoint; therefore, a node using a cycle $x$ knows that a cycle $y$ would not have any address in common. Therefore, the probability of duplicate address assignment is zero. This greatly reduces the delay associated with address assignment. In other techniques, a node needs to run DAD or has to require permission from remaining nodes in the MANET for address assignment. Our algorithm not only decreases the latency and delay but also reduces the communication overhead and saves bandwidth.

After maximum address range has been reached by a node, it has two options: either to start re-assigning the address as they repeat; or increment its state value by choosing an unused seed. If the re-assignment of first address will lead to duplication, the node will increment its state value to another seed. To confirm that the seed is not in use by another node in the MANET, it floods the network with a NEWSEED message and waits for the reply from all other nodes. If it receives one negative acknowledgment (NACK), it chooses another seed and repeats the process. If it does not receive any NACK, it assumes that the seed is free and will use it to assign addresses to new nodes.

### 14.5.5 Network Partitioning and Network Merger

If a network gets partitioned, node in each partition can still continue assigning unique address. As each QR cycle is disjoint, no address duplication will occur. When partitions merge back later, address uniqueness is guaranteed. This way we handle network partitioning without incurring any extra communication overhead. Consider two independent MANET as in Fig. 14.5. All of the nodes in each MANET have unique IP addresses. Now we will see, what happens when two independent networks merge. Network ID (NID) is piggy backed in HELLO messages. A network merger is detected when a node hears a HELLO message with a different NID. Node D is communicating without any problem with node A having IP address 'a'. Note that in the second MANET there is node K that also has an IP address "a". As these nodes belong to different networks, there is no address duplication or communication problem. Let us consider what happens when these two independent networks come close to each other and get connected to form one big MANET. Misrouting can occur because of duplicate addresses as shown in Fig. 14.6. Nodes A and K have the same IP address and therefore we have duplicate addresses in the new MANET. As a result packets that were meant to be routed to node A could be misrouted to node K. We have to solve this duplicate address problem and/or prevent misrouting after network merger.

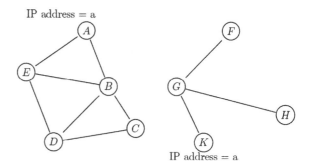

**Fig. 14.5** Two independent MANETs

**Fig. 14.6** Network merger

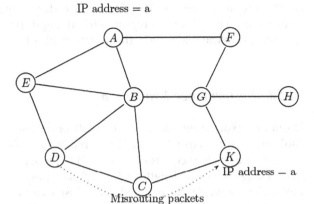

We have two solutions of handling a network merger. One makes use of DAD and the other does not depend on DAD. In the first solution there is no need to invoke an explicit procedure upon merging of the two independent MANETs. We can associate $\phi(N)$ of each MANET to be the key of the respective MANET. The idea is similar to Weak DAD, but we associate a single network key to all nodes of one network instead of using a separate key for each node. The logic is to distinguish between the nodes having the same IP addresses and belonging to two separate networks during network merging. We will show that this does not increase communication overhead and we can still prevent misrouting. When we choose two large primes, $\phi(N)$ of each network would be different with a high probability, as each network selects $p$ and $q$ independently. Even two nodes have the same address, misrouting can still be prevented by associating $\phi(N)$ with IP address to identify each node. In this way, we can easily distinguish between two nodes holding the same IP address but having different key $\phi(N)$. Only if the two nodes have the same address and the two MANETs have the same value of $\phi(N)$, the conflict cannot be detected. This probability can be controlled to be very small by selecting large primes of $p$ and $q$.

In the second approach, we run DAD to remove duplicate addresses. Similarly, network merger is detected when a node hears a HELLO message with a different NID. The smaller NID is chosen to be the new NID of the merged network. In this approach two nodes from the intersection of MANET exchange $N$ and the set of seeds $S_0$ of their individual MANET. Now these nodes will only check for the conflicted QR values by generating sequence of $S_i$ of each cycle. If two values are found to be the same, duplicate addresses are detected. For example, if same QRs are found in both MANETs, then they are in conflict with each other. All such conflicted addresses are checked and, if two nodes hold the same IP address, then one of these nodes has to give up its address and acquire a new one. In our approach, we let the node with fewer TCP connections to change its IP address, in order to break minimum ongoing connections. This process is repeated for all duplicate addresses till there is no remaining duplicate address in the merged network.

### 14.5.6 Performance Evaluation

In our first experiment, we choose two safe primes of 12 bits: $p = 2207$ and $q = 3467$. We then can have $N = 7,651,669$ and $\phi(N) = 7,645,996$. The total number of quadratic residues equals $\phi(N)/4 = 1,911,499$. By running our algorithm, we find 38 long cycles with length 50,228 each, as shown in Table 14.3, which means that we can have 38 clusters and each cluster can configure 50,228 nodes. If some clusters get partitioned from the original MANET, they can still assign unique addresses to newly joining nodes and when the partitions merge back there would be no duplication. As $N$ is 24 bits long, we can fix the 8 bits prefix for network address. When a node hears a message from a node with different prefix, it assumes that it is a network merger and runs a network merging algorithm. Safe primes here were of 12 bit each, but we can also choose bigger safe prime of 16 bit each and get $N$ of 32 bits. In that case, we would be able to configure billions of nodes and we can piggyback the NID in hello messages, generated by the first node in the MANET.

In our second experiment, we chose two doubly safe primes instead of just safe primes to see how big a cycle (address block) we can get. We noticed that the length of the cycles we get is huge. This results in less number of distinct cycles because the length of a cycle is extremely big. The two doubly safe primes we choose were 13 bits each: $p = 4799$ and $q = 4919$. The results are shown in Table 14.4.

In our approach, the network part of the IP address can vary depending on the length of primes we have chosen. For example, consider Table 14.4, we

**Table 14.3** Results of safe prime experiment

| Cycle length | Number of cycles | Number of QRs |
|---|---|---|
| 1 | 1 | 1 |
| 29 | 38 | 1,102 |
| 1,732 | 1 | 1,732 |

**Table 14.4** Results of doubly safe prime experiment

| Cycle length | Number of cycles | Number of QRs |
|---|---|---|
| 1 | 1 | 1 |
| 1,199 | 2 | 2,398 |
| 2,458 | 1 | 2,458 |
| 2,947,142 | 2 | 5,894,284 |

chose two doubly safe primes of 13 bits each and get there product $N$ to be 25 bits. As we are working in modulo, there will be no host part of the IP address generated by our program, which will exceed 25 bits for this particular scenario. Therefore, we support Classless Interdomain routing(CIDR) [25]. With so-called CIDRized network addresses, the network part of an IP address can be any number of bits long, rather than being constrained to 8, 16, or 24 bits.

## 14.6  Conclusions

Address allocation is a critical problem in autonomous MANETs. This chapter has discussed the challenges and potential problems when traditional allocation schemes for infrastructure networks are used MANETs, and surveyed the existing address allocation schemes for MANETs.

We have also presented a novel address allocation protocol using quadratic residues. Our approach assigns unique addresses to new nodes with low latency without relying on periodic flooding, which consumes a considerable amount of bandwidth. It distinguishes between concurrent address request and replies, and handles the mobility scenarios at the time of address assignment. Our protocol also provides reliable state synchronization in presence of packet loss, delays, and network merging. In our approach, network partitioning and subsequent merging does not induce duplicate addresses. Our solution ensures scalable and extremely low overhead solution for network mergers. In the best case scenario we do not have to run the DAD mechanism to check for duplicate address entries.

This approach is capable of configuring at least millions of nodes with even distribution. We have provided support for proactive, reactive, and hybrid routing protocols in MANET. For IPV-6 addressing format we can get a unique IP address by embedding 48 bit MAC address plus a Quadratic Residue or a randomly generated large number. In the future, we would also like to enhance our security measure, to prevent a malicious node from holding or exhausting the address space.

**Acknowledgments**  Xiaowen Chu's work was supported by Research Grant Council, Hong Kong, China, under Grant RGC/HKBU2159/04E and RGC/HKBU210605.

## Terminologies

*Address assignment*: The process of configuring a generated address on an interface.
*Address conflict*: A situation when two or more nodes in a network are assigned the same address.

*Address discovery*: The process by which a node in an ad hoc network discovers whether an address is already claimed within an ad hoc network.

*Address generation*: The process of selecting a tentative address in view to configure an interface.

*Duplicate address detection*: The process by which a node, which lacks an address, determines whether a candidate address it has selected is available. A node already equipped with a network address participates in DAD in order to protect its address from being accidentally misappropriated for use by another node.

*Link-local address*: An address having link-only scope that can be used to reach neighboring nodes attached to the same link.

*Neighbor*: Two nodes are neighbor if one can send/receive packets to the other without passing through any intermediaries.

*Network merger*: The process by which two or more previously disjoint ad hoc networks get connected.

*Network partitioning*: The process by which an ad hoc network splits into two or more disconnected ad hoc networks.

*Standalone MANET*: An independent ad hoc network, which does not contain a border router through which it is connected to the Internet.

## Questions

1. What are the main problems of hardware-based addressing allocation schemes, such as IPv6 stateless auto-configuration, when applied in MANETs?
2. Consider Perkins' query-based DAD scheme. What is the address allocation delay for the case of address allocation success?
3. Consider MANETconf protocol. If more than one initiator choose to assign the same IP address, there will be a conflict. What do you suggest to resolve this problem?
4. List some advantages and disadvantages of Prophet address allocation scheme.
5. Weak DAD scheme assumes that a unique key is pre-assigned to each node. What is the purpose of this unique key, and why not use it in the IP address directly?
6. What are the main advantage and disadvantage of Passive DAD scheme?
7. Find the square root of 71 mod 77.
8. Prove Theorem 1: For prime number $p$, $a$ is a quadratic residue modulo $p$ if and only if $a^{(p-1)/2} \equiv 1 \pmod{p}$.
9. Consider Legendre symbol $\left(\frac{a}{p}\right)$. Prove that for an odd prime $p$, if $ab \neq 0 \pmod{p}$, then $\left(\frac{ab}{p}\right) = \left(\frac{a}{p}\right)\left(\frac{b}{p}\right)$.
10. Calculate the value of Jacobi symbol $\left(\frac{2}{135}\right)$.

# References

1. R. Droms. Dynamic host configuration protocol. In Internet Engineering Task Force (IETF), RFC 2131, March 1997.
2. R. Finlayson, T. Mann, J. Mogul, and M. Theimer. A reverse address resolution protocol. RFC 903, June 1984.
3. B. Croft, and J. Gilmore. BOOTSTRAP PROTOCOL (BOOTP). RFC 951, September 1985.
4. S. Thomson, and T. Norten. Ipv6 stateless address autoconfiguration. RFC 2462, December 1998.
5. B. Aboba, S. Cheshire, and E. Guttman. Dynamic configuration of ipv4 link-local addresses. In IETF Internet draft, July 2004, Work in Progress, http://files.zeroconf. org/draft-ietf-zeroconfipv4-linklocal.txt, July 2004.
6. Duplicate MAC Addresses on Cisco 3600 Series, http://www.cisco.com/warp/public/770/ 7.html, 1997.
7. S. Nesargi, and R. Prakash. MANETconf: Configuration of hosts in a mobile ad hoc network. In Proc. of IEEE INFOCOM 2002, June 2002.
8. J. L. Peterson, and T. A. Norman. Buddy systems. In Proceedings of Comm. ACM, June 1977.
9. M. Mohsin, and R. Prakash. IP address assignment in mobile ad hoc networks. In Proc. of IEEE Milcom, Anaheim, USA, October 2002.
10. A. P. Tayal, and L. M. Patnaik. An address assignment for the automatic configuration of mobile ad hoc networks. In Proc. of ACM Personal and Ubiquitous Computing, 2004.
11. J. Lee, S. Kim, and I. Yeom. Advanced disjoint address allocation for mobile ad hoc networks. In Proc. of IEEE VTC 2007, Dublin, Ireland, April 2007.
12. M. Mutka, H. Zhou, and L., Ni. Prophet address allocation for large scale MANETs. In Proc. of IEEE INFOCOM, San Francisco, CA, April 2003.
13. Y.Y. Hsu, and C.C. Tseng. Prime DHCP: a prime numbering address allocation mechanism for MANETs. In IEEE Communications Letters, Vol. 9, No. 8, pages 712–714, August 2005.
14. C. E. Perkins, E. M. Royer, and S. R. Das, IP address autoconfiguration for ad hoc networks. Internet Engineering Task Force (IETF), Internet Draft, http://people.nokia. net/ charliep/txt/aodvid/autoconf.txt, November 2001.
15. K. Weniger, and M. Zitterbart. Address autoconfiguration in mobile ad hoc networks: current approaches and future directions. In IEEE Network, pages 6–11, July/August 2004.
16. N. Vaidya. Weak duplicate address detection in mobile ad hoc networks. In Proc. of ACM Mobihoc, June 2002.
17. K. Weniger. Passive duplicate address detection in mobile ad hoc network. In Proc. of IEEE WCNC 2003, New Orleans, USA, March 2003.
18. K. Weniger. PACMAN: Passive autoconfiguration for mobile ad hoc networks. In IEEE JSAC, Special Issue on Wireless Ad Hoc Networks, January 2005.
19. E. Belding-Royer, C. Perkins, and S. Das. Ad hoc on demand distance vector (aodv) routing. In Internet Engineering Task Force (IETF), RFC 3561, July 2003.
20. D. Maltz, D. Johnson, and Y. Hu. The dynamic source routing protocol for mobile ad hoc networks (dsr). In IETF Internet draft, draft-ietf-manet-dsr-10.txt.
21. A. Laouiti, T. Clausen, and P. Jaqcuet. Optimized link state routing protocol (olsr). In Internet Engineering Task Force (IETF), RFC 3626, October 2003.
22. S. Kim, J. Lee, and I. Yeom. Modeling and performance analysis of address allocation schemes for mobile ad hoc networks. In IEEE Transactions on Vehicle Technology, September 2007.

23. S. Kim, J. Lee, and I. Yeom. A token-based dynamic address allocation protocol for mobile ad hoc networks. Computer Science, KAIST, Tech. Rep., 2005.
24. D. Stinson. Cryptography: Theory and Practice, pages 173–178, Chapman and Hall/ CRC-Press Inc, Boca Raton, FL, 2nd edition, 2002.
25. J. Yu, V. Fuller, T. Li, and K. Varadhan. Classless inter-domain routing (cidr): an address assignment and aggregation strategy. In Internet Engineering Task Force (IETF), RFC 1519, September 1993.

# Chapter 15
# Congestion Control in Wireless Ad Hoc Networks

Vinay Kolar, Sameer Tilak, and Nael Abu-Ghazaleh

**Abstract** Multi-hop Wireless Networks (MHWNs) are anticipated to play an important role at the edge of the Internet, enabling a large number of innovative applications. The great demand for capacity from a large number of users and applications, coupled with the sparse bandwidth available on the wireless channel, place particular emphasis on effective congestion management approaches. Effective and well-studied algorithms for congestion control at the transport layer exist in wired networks. However, for a number of reasons, these approaches do not translate directly to wireless environments. In this chapter, we first describe the problem of congestion control in MHWNs, and discuss approaches for solving it. The presentation is organized into two components: (1) a review of the causes of congestion and algorithms for congestion avoidance in MHWNs at different layers of protocol stack; and (2) a review of analytical models for the rate control problem and their use for congestion control.

## 15.1 Introduction

Multi-hop wireless networks (MHWNs) are emerging as a critical technology that will play an important role at the edge of the Internet. Mesh networks [4, 51] provide an extremely cost-effective last mile technology for broadband access (for example, a mesh network providing broadband access to the city of Philadelphia, covering 135 square miles, at less than half the cost of traditional broadband access is currently being built [17]); ad hoc networks have many applications in the military, industry, and everyday life [10]; and sensor networks hold the promise of revolutionizing sensing across a broad range of applications and scientific disciplines – they are forecast to play a critical role as

V. Kolar (✉)
State University of New York, Binghamton, USA; Department of Wireless Networks, RWTH Aachen University, Kackertstrasse 9, 52072 Aachen, Germany
e-mail: vinkolar@cs.binghamton.edu; vko@mobnets.rwth-aachen.de

S. Misra et al. (eds.), *Guide to Wireless Ad Hoc Networks*,
Computer Communications and Networks, DOI 10.1007/978-1-84800-328-6_15,
© Springer-Verlag London Limited 2009

the bridge between the physical and the digital worlds [29]. The bandwidth demand from the anticipated range of heterogeneous applications, coupled with the limited capacity of the wireless network, places a particular emphasis on effective use of the available resources.

Thus, avoiding the inefficiencies resulting from congestion is essential. In addition to the limited wireless bandwidth, a number of factors in MHWNs further reduce the effective capacity and complicate capacity and congestion estimation. The time-varying nature of the channel, the impact of interference, the intricacies of MAC protocol scheduling [26, 68, 69], and the self-interference among packets as they are relayed down multi-hop wireless paths [46] combine to restrict the use of conventional approaches for congestion control [5]. Thus, congestion control in MHWN is a different and challenging problem that requires significantly different solutions that are aware of the characteristics of MHWNs.

Congestion is caused when the traffic in the network exceeds its capacity. Conventional approaches avoid congestion by controlling the rate at which the senders inject the traffic into the network. One of the primary reasons for adopting such an approach is due to an "end-to-end" argument [58]. The end hosts on the wired network estimate the available capacity in the network through parameters like packet drops and end-to-end packet delays and the control messages provided by the routers. Another alternative to reduce the congestion is to re-routing packets through areas with lower congestion in the network. Efficient routing is still an open problem in MHWNs. Source routing protocols allow the end hosts to choose an appropriate routes [20, 37]. Studies from the wired networks have already shown that a congestion-aware routing protocol can greatly increase the network performance [23]. The additional degrees of freedom to avoid congestion at the lower layers, like routing and MAC, have a great impact on the performance of applications. Section 15.4 discusses the causes of congestion at different layers in MHWNs.

Efforts in congestion management in MHWNs can be organized into two groups: (1) design of heuristic protocols to avoid congestion; and (2) modeling resource allocation in general, and congestion control in particular, using formal techniques to characterize the problem and deriving effective protocols that converge toward near-optimal solutions. We review the important works in both groups. Section 15.5 reviews the simulation-based work on avoiding congestion. Using the observed behavior, we categorize the congestion control problem into three sub-problems: (1) adapting packet sending rate at the sources; (2) routing; and (3) scheduling. The causes for congestion for each are discussed. We survey existing protocols for congestion management in MHWNs. The design of the protocols illustrates the need for capturing the cross-layer interactions for avoiding congestion. A majority of the congestion control protocols use the MAC layer and routing layer input to avoid congestion. The practical issues of managing congestion in a network are discussed in Section 15.6. Modeling an optimal solution to congestion control problem aids in formally specifying the required functionality at each layer and provides deep insight into the general MHWN networking problem and the shape of effective solutions. Section 15.7 discusses the research in modeling congestion control in MHWNs. Section 15.8 discusses the limitations and research challenges for congestion control in MHWNs. Finally, Section 15.9 presents the concluding remarks.

## 15.2 Background

In Multi-Hop Wireless Networks (MHWNs), the devices self-configure to cooperatively communicate without the need for access points. The wireless nodes cooperate to route each others' packet to provide connectivity. MHWNs increase the coverage as existing devices can self-configure to provide access to each other. While the overall architecture and layering of MHWN are similar to conventional networks, the functionality is quite different. For example, in such settings, the devices are often mobile, causing the topology of the network to change, and making it difficult to use ideas such as hierarchy and address aggregation that have served us well in wired networks. In this section, we outline the challenges and functionality of important layers that are relevant for congestion control in MHWNs. The background for modeling congestion control is later discussed in Section 15.3.1.

### 15.2.1 Physical Layer

The physical layer is responsible for the transmission of bits over the wireless channel. In wireless transmission, as the signal from a sender propagates over the channel, it attenuates with distance; it also suffers from physical propagation due to interactions with the physical environment (e.g., passing through obstacles). A receiver receives the signal after attenuation and other propagation effects, and attempts to decode it. If the received signal strength is sufficiently higher than the sum of the noise and signal from interfering signals, the signal can be decoded successfully (with low error rate); otherwise, the transmission cannot be received [56]. Thus, interference from concurrently transmitting nodes plays an important effect in determining whether correct reception or a collision occurs. In order to reduce the occurrence of collisions, while maintaining distributed access, carrier sensing is often used. More precisely, senders are able to sense that the channel is busy if the sum of the noise and the signal from interfering transmissions is above a given power threshold, called the Receiver Sensitivity Threshold (*RxThreshold*). If the channel is busy, the transmission can be deferred until a free channel is observed. The Channel State at a given node for a given point of time represents if the channel sensed at the node is busy or idle. A single strong signal or multiple weak signals over the same channel may together add up to create a busy channel state.

## 15.2.2 Medium Access Control (MAC) Layer

In the presence of competition for accessing the medium from interfering senders, the Medium Access Control (MAC) protocol plays the role of moderating access to reduce collisions while facilitating channel reuse and fairness. In the rest of the chapter we use scheduler to refer to the functionality of the MAC layer scheduler. Under an ideal MAC, the throughput between a sender–receiver pair depends on the number of active nodes (senders or receivers) that interfere with the pair. However, practical MAC schedulers are rarely able to archive this ideal behavior. The IEEE 802.11 MAC protocol [63] is the de facto standard for Wireless LAN (WLAN) and MHWNs, and much of the existing research is based on this protocol. IEEE 802.11 uses Carrier Sense Multiple Access with Collision Avoidance (CSMA/CA). In contention-based protocols such as IEEE 802.11, the channel state at the receiver is not known at the sender, which gives to the well-known hidden and exposed terminal problems [6]. To counter these effects, IEEE 802.11 uses an aggressive CSMA (low receiver sensitivity) in order to prevent far-away interfering sources from transmitting together. However, the aggressive CSMA also potentially preventing non-interfering sources from transmitting; hidden terminal is reduced, but exposed terminal increased. Optionally, the standard allows the use of short Request-to-Send (RTS)/Clear-to-Send (CTS) control packets to attempt to reserve the medium before the transmission of a data packet. The successful reception of DATA packet is acknowledged by an ACK packet. The source switches to a conservative transmission of the packets when an unsuccessful handshake is detected. For every unsuccessful transmission, the source exponentially increases the waiting time when it attempts to re-transmit the packet (Exponential Backoff). Also, the MAC layer assumes that the connection to the next-hop is broken after a failure to transmit the packet for a given number of times (termed as Retransmit limit). As a result of these mechanisms, many, but not all, collisions are prevented. Depending on the relative location of contending sources and their receivers to each other (or more accurately, the state of the channels between them), they are either able to effectively handshake using the MAC protocol or continue to collide [26, 57]. The problem is complicated because multiple concurrent transmissions may occur and collectively result in a collision, when individually they do not. Similarly, the aggressive MAC mechanisms can prevent some possible concurrent transmissions from proceeding even though they do not cause a collision. In summary, the MAC protocol and the relative locations of contending sources play an important role in how a set of nearby sources interact and the resulting quality of the link observed by each of them.

## 15.2.3 Routing Layer

Routing in MHWNs is responsible for constructing routes for a network that is self-configuring and potentially mobile. The first generation of routing protocols evolved from conventional routing algorithms, but attempted to reduce their overhead while working in a highly dynamic environment. As such, the quality of a path was measured in terms of the hop count between the source and the destination, as is typical in wired routing protocols [37, 55]. However, the hop-count-based routing ignores the link quality of each hop, which may result in increased packet collisions. Recent research studies [19] have concluded that link-quality-based routing is more effective in MHWNs. The routing protocol forwards packets to the MAC protocol using a queue where packets are inserted by the routing protocol and are picked by the MAC protocol when it is ready to transmit. First-in First-out (FIFO) queuing policy is a standard queue implementation in networks where the packet that is picked up for transmission is the packet that arrived earliest; however, some protocols favor control packets over data packets. We assume that the newly arrived packet is dropped when the queue size is full (Drop Tail). Multiple queues, each for a given priority level, may exist if QoS-aware routing or MAC protocols are used [2]. Unless mentioned, we refer to a single FIFO queue.

### 15.2.4  Transport Layer and Congestion Control

Several transport protocols are used at transport layer. The Transmission Control Protocol (TCP) [22] is the primary transport protocol on the Internet with congestion control support. In order to avoid congestion, TCP maintains congestion window size – the expected number of bytes that can be sent over the network without causing congestion. At a given instance, the number of bytes that are not acknowledged by the receiver indicates the bytes in-transit over the network. A new packet is sent only if the sum of unacknowledged packets and the in-transit packets is lesser than the congestion window size. The congestion window size is estimated initially according to the Slow Start algorithm. Under slow start, the congestion window is initially set to 1 unit of MSS (maximum segment size). The congestion window is increased exponentially until a threshold value called Slow Start Threshold. After this threshold, it is increased conservatively in a linear fashion. When a packet drop is detected, the congestion window size is reset to 1 MSS and the Slow Start Threshold is halved. This process ensures: (1) an aggressive exponential start until slow start threshold; (2) a conservative linear increase; and (3) an aggressive multiplicative decrease in packet sending rate when a packet drop is detected.

### 15.3  Modeling Resource Allocation and Congestion Control – Preliminaries

Resource allocation models describe a network as an optimization problem whose solution provides an optimal spatial (transport and routing) and temporal (MAC) control of packet over the network. Optimality is relative to the objective function for the optimization problem, which can be any performance metric like maximizing the overall throughput of the network or ensuring appropriate fairness levels to all the connections, or combinations of these and other performance metrics. The congestion control problem fixes the routing configuration and attempts to derive the optimal source rates. This problem often has a cross-layered solution where each source transmits the packets at an optimal rate, keeping in mind the scheduling intricacies and the routing configuration in a given MHWN. This cross-layered optimization problem is discussed in detail in Section 15.7. We now overview the modeling components that are often used to formulate a congestion control problem in MHWNs. We discuss the structure and representations of a general optimization problem and then overview the interference models that are used in congestion control models.

### 15.3.1 Optimization Problem

An optimization problem maximizes (or minimizes) a given function based on the set of constraints. Optimization is extensively used in many areas including modeling congestion control in MHWNs. Equations (15.1) and (15.2) express a general optimization problem where $\mathbf{x} = [x1,x2,x3]$ denotes the a vector of $n$ variables corresponding to the different rates allocated to the sources. The possible values that the variables can take (known as feasible values) are restricted by a set of $m$ constraints denoted by Equation (15.2). The problem is to choose an appropriate $\mathbf{x}$ such that the objective function $f(\mathbf{x})$ is maximized.

$$\max_{\mathbf{x}>0} f(\mathbf{x})$$
$$\text{subject to :}$$
(15.1)

$$g_i(\mathbf{x}) \leq 0 \ \forall i = 1, 2,..., m$$
(15.2)

Congestion control models in MHWNs define the variables ($\mathbf{x}$) and the functions $f(\mathbf{x})$, $g_i(\mathbf{x})$ such that fairness/throughput of each connection is maximized. Various constraint functions $g_i(\mathbf{x})$ are defined to restrict the values of $\mathbf{x}$. For example, concurrent transmissions of the interfering links are restricted. We discuss the detailed problem in Section 15.7.

#### 15.3.1.1 Dual Problem and Lagrange Multipliers

The main optimization problem in Equations (15.1) and (15.2) is called the Primal Problem. In optimization theory, the primal problem can be transformed into its dual problem. Sometimes the structure of the dual problem leads to useful insights into the structure of the problem. For example, distributed algorithms are derived by converting the primal congestion control problems into its dual counterpart. In this section, we briefly explain the procedure to arrive at the dual problem from the primal problem. In the primal problem, the constraints (Equation (15.2)) impose a strict all-or-none policy in the choice of the variables $\mathbf{x}$: either $\mathbf{x}$ is feasible or infeasible. The main idea for constructing the dual problem is to relax these hard-constraints of the primal problem and introduce them as soft-constraints in the objective of the dual. Weights are assigned for each constraint such that the interior points of the feasible region ($g_i(\mathbf{x}) \leq 0$) are penalized lesser and the penalty gradually increases for the points that lie toward the infeasible region. This biases the optimizer to choose feasible points instead of the infeasible ones. To enable such soft constraints, each constraint is associated with a non-negative number that denotes the weight of the penalty. These weights are known as Lagrange multipliers and the Lagrange multiplier for $i$thconstraint is denoted by $g_i(\mathbf{x})$. A Lagrangian function is defined as:

$$L(\mathbf{x}, \mathbf{q}) = f(\mathbf{x}) - \sum_{i=1}^{m} q_i g_i(x), \qquad (15.3)$$

and the dual function is defined as:

$$D(\mathbf{q}) = \max_{x \geq 0} L(\mathbf{x}, \mathbf{q}). \qquad (15.4)$$

Equation (15.3) biases the optimization problem to choose feasible points for constant values of $g_i(\mathbf{x})$. However, the dual problem should also maximize (or minimize) the objective of the primal problem in addition to the choice of feasible points to preserve a correct transformation from primal to dual. Equation (15.4) enables this by choosing appropriate values for Lagrangian multipliers. The dual problem is given by Equation (15.5).

$$\min_{q \geq 0} D(\mathbf{q}). \qquad (15.5)$$

It can be shown that the solution to the dual problem solves the primal problem if the duality gap is zero [53]. The construction of the dual problem has several attractive properties. In congestion control models, it will be later seen in Section 15.7.3 that the dual naturally decomposes the problem into sub-problems that can be solved in a distributed fashion.

## 15.3.2 Interference Modeling

Interference in wireless networks is a complex phenomenon that depends upon the time-varying nature of the wireless channel and several environmental factors. Hence, an accurate formulation of interference is infeasible in practice.

**Fig. 15.1** Interference
models

However, tracking interference is vital in modeling and simulation of wireless
networks because of the critical impact it has on performance. In the context of
modeling congestion control, Section 15.7.3.2 shows that an effective solution
schedules links that do not interfere with each other. The following approximate interference models are used while solving this scheduling problem.

- Primary interference model: In this model, only links that do not share a
  common node can be active at a given time. For example, in the Fig. 15.1,
  link (A, B) and (B, C) cannot be simultaneously active while (A, B) and
  (C, D) can be concurrently scheduled. Node-exclusive interference model [9]
  is a similar model where a node cannot transmit and listen at the same time.
- Unit-disk model: The primary interference model does not consider the
  signal strength or distance between two nodes to deduce interference metrics.
  Unit-disk interference model is proposed to account the distance factor.
  A node can receive packet and interfere with all the neighboring nodes within
  a threshold called interference range. Consider the node placement and
  interference range as shown in Fig. 15.1. In a unit-disk model, the links
  (A, B) and (C, D) cannot transmit concurrently since A and C are within the
  interference range of each other. However, a primary interference model,
  which is unaware of the interference range, does not prevent such concurrent
  transmissions.
- Two-disk model: In a two-disk model, a node can receive a packet from all
  the neighbors within a "reception range", but interferes with all the nodes
  that are within "interference range". Generally, interference range is more
  than twice the reception range.

While advanced interference models like Signal-to-Interference Noise Ratio
(SINR) exist, we do not review them in this chapter since most of the scheduling
models use the above two interference models.

## 15.4 Analysis of Congestion

The causes and effects of congestion in conventional wired networks are well
studied [23, 33]. However, congestion control in MHWNs poses new challenges
that restrict the use of conventional approaches to solve the problem. In this
section, we contrast the congestion problem in MHWNs with widely used
Ethernet-based networks [62], henceforth referred as wired networks. We
study the adverse effects of using conventional congestion control mechanisms
in MHWNs and describe alternative mechanisms for congestion avoidance.

Congestion avoidance mechanisms can be broadly divided into detection and control. Congestion can be detected by monitoring the queuing delays and packet drops at the nodes. Congestion control is generally achieved by regulating the packet-sending rate at the source node of the connection. The source node infers the congestion information along the connection route for effective congestion control. To enable the source node to control packet-sending rate, the intermediate routers proactively send the observed congestion levels to the participating sources. In such proactive schemes, the routers detect congestion by monitoring the queuing delays and packet drops at intermediate routers and send the required information to the source nodes. Another alternative is a passive mechanism where the source node infers the congestion level by measuring parameters like end-to-end packet delays. A hybrid of active and passive approach is used in the Internet.

In a wired network, the packet drop information accurately conveys the congestion information since packet drops are primarily due overflow of queues at intermediate nodes. Packet drops due to channel error or ineffective MAC protocol is very rare. The end-to-end packet delays are primarily due to the intermediate queuing delays since the transmission time is low. However, detecting and avoiding congestion is more complex in MHWNs. The packet drops and delays cease to serve as effective indicators of congestion due to the new challenges in MHWN. For example, packet drops can be caused due to a high channel error rate. The end-to-end delays can be skewed due to the wireless MAC layer transmission scheme, which can significantly increase the transmission time of a packet.

### 15.4.1 Conventional Congestion Control

In this section, we study the adverse effects of using conventional congestion control approaches in MHWNs. The characteristics of MHWNs that results in the ineffectiveness of these mechanisms are analyzed. We then focus on alternative mechanisms to control congestion at different layers in MHWNs. Later, Section 10.5 describes effective heuristics to avoid these issues in MHWNs.

#### 15.4.1.1  Effect of Mobility

Unlike the wired networks, end-to-end delays and packet drops does not always signify the lack of available capacity in MHWNs. Hence, the use of conventional congestion control approaches results in false congestion alarms. The primary reason for false congestion alarms is the mobility in MHWNs.

The presence of mobility introduces route disconnections when a node moves out of range with its next hop. After attempting to re-transmit the packet for re-transmit-limit number of times, the IEEE 802.11 MAC protocol reports the route-disconnection to the routing layer. The routing protocol then prunes the disconnected route and searches for other possible routes. If such a route is not present in its cache, certain routing protocols [37, 55] initiate a new Route Discovery process. Even if an alternative route is present in its cache, it has a high probability of being a stale route due to constant mobility. It has been shown that performance of MHWN is better without route caching [30]. Certain routing protocols may also send an explicit route disconnection message to the source [37], thus enabling the information about the packet drop.

The time between the disconnection and the discovery of a new path may introduce substantial delays due to the discovery of the new route. The large delays and/or the packet drop information are interpreted as congestion by conventional protocols. For example, TCP assumes congestion when it does not get an ACK for three data packets or if it receives packet drop information. Such an approach does not always signify congestion level in a MHWN under mobility and hence is inappropriate to assume congestion. Many research studies have addressed the problem of avoiding false congestion alarms, which we discuss in Section 15.5.1.1.

### 15.4.1.2 Bursty Traffic

CSMA/CA-based MAC protocols suffer from short-term unfairness where a node wins unfair channel capacity over short intervals of time [42]. This leads to a scenario where a burst of packets is transmitted in succession, followed by intervals of inactivity. Higher-layer protocols may be affected adversely due to the resulting congestion bursts, especially in the protocols that use Window-based transmission (like TCP). In TCP a burst of TCP-ACK packets increases the number of bytes that can be transmitted by the source, thus creating a burst of TCP-data packets. The sudden injection of data into the network results in an acute congestion build up in the network, thus resulting in packet drops. The TCP source aggressively backs off due to packet drops, which results in ineffective use of the channel. While this effect occurs in wired network too, the effect of CSMA/CA protocols like IEEE 802.11 exacerbates the effect of bursty traffic in MHWNs.

## 15.4.2 Alternative Congestion Avoidance Mechanisms

A majority of the wired networks employ congestion control mechanisms at end hosts with limited router support since it is infeasible to alter the core of the network. However, this limitation is absent in MHWNs where the nodes can support custom lower-layer services and protocols. For example, the routing layer can employ a capacity-aware protocol to enable greater traffic and alleviate the congestion levels in the network. While altering the MAC layer is infeasible in a majority of the wireless network cards, modifying the routing or other higher layer protocols to avoid congestion is relatively easier. In this section, we study the causes ad effects at the lower layer, which can aid in congestion avoidance at higher layers.

### 15.4.2.1  Link Quality

The packet error rate in the wireless channel is significantly higher than the wired counterpart. Unlike the wired links, two wireless links can experience a dramatically different packet loss probability even in the absence of interference from other sources. The exponential backoff of the IEEE 802.11 causes idle times at the source when a packet loss is observed. Thus, the effective available capacity at the higher layers is decreased due to backoffs. Hence, transmitting the packet over stronger links will reduce the congestion level experienced by a connection.

Routing protocols that choose hop-count as the sole metric are more vulnerable to such packet losses. Reducing the number of hops results in longer hops, and the quality of the link decreases as the distance between two nodes increases. Several routing protocols attempt to increase the realized capacity by using link-quality-aware metrics [1, 18, 20]. Specifically, the Expected transmission count (ETX) metric [18], which captures the link-quality, performs much better than the other routing metrics [19]. However, the choice of strong links may lead to shorter hop-length, resulting in longer number of hops and greater end-to-end delay of the connection. In spite of longer hops, ETX is shown to outperform a hop-count-based routing protocol in terms of TCP throughput [19, 70].

### 15.4.2.2  Locality of Congestion Regions

Congestion experienced by nodes on a wired local area network (LAN) is fairly uniform and hence it is appropriate to view congestion at the network level. However, the congestion experienced by the nodes in a given region can be drastically different in MHWNs. This is due to the fact that the available capacity of a node can be significantly different from the neighboring nodes. This effect arises due to CSMA/CA-based MAC protocol effects that cause hidden terminals and unfairness.

Figure 15.2(a) explains the effect of hidden terminal on the available capacity of the link. A CSMA-based MAC protocol (like IEEE 802.11) does not prevent the concurrent transmission of links (A, B) and (C, D). Transmission from node C results in a packet collision at the link (A, B), while handshake at link (C, D) is always successful. The realized available capacity of link (A, B) is extremely less than that of the link (C, D) due to constant backoff by the node A.

Even in the absence of hidden terminals, some links may get unfair channel capacity when compared to the neighboring links. A typical example is "Flow in the middle" problem as demonstrated in Fig. 15.2(b). The links (A, B) and (E, F) can transmit concurrently, while link (C, D) can transmit only when the other two links are idle. If the traffic is saturated at all the links, (C, D) gets an

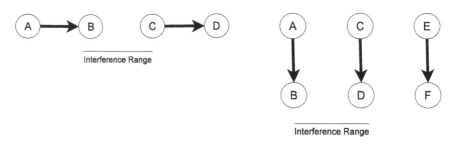

(a) Effect of hidden terminal                    (b) Flow-in-the middle

**Fig. 15.2** Effect of MAC layer on available capacity

unfairly lesser access to channel since either (A, B) or (E, F) or both are active most of the times, thus C not being able to capture the medium. Simulation results indicate that the throughput of (C, D) is 1% of the total throughput [14].

The above examples show that congestion in MHWN is not uniformly distributed around a spatial region, but is a function of the location of the nodes. Slight changes to the node locations can prevent hidden terminals and unfairness. This creates a dramatically greater available capacity at the links, thus reducing the congestion experienced by the links. Choice of effective links by the routing protocols can enhance the performance metrics by reducing the congestion at a connection level.

## 15.5 Protocol Design for Congestion Avoidance

The above discussion shows that congestion in a MHWN is a cross-layer problem, and the causes can be traced at various layers. In this section, we discuss the various approaches to reduce congestion at different layers of a MHWN.

### 15.5.1 Rate-Control

The choice of appropriate packet-sending rates has been an important aspect of controlling congestion in the conventional networks. Transport protocols (like TCP) and application protocols [59] adjust the rate at which the source injects the packet as a function of the estimated congestion level in the network. The dynamics of congestion buildup in the network and the reaction of the sources to such events are described in the rate-control problem.

The rate-control-based congestion avoidance is generally a cross-layered approach that uses the congestion information from the MAC and routing layers. We now discuss the cross-layered rate-control approaches that solve congestion control. The discussion is limited to transport protocols to cover the general set of ideas for controlling congestion. However, general congestion control at the application layer protocol can use similar ideas [59]. Several cross-layered solutions are proposed for a congestion-aware transport protocol. While some approaches provide a backward compatibility with the popularly used TCP protocol [27, 30, 31, 61, 49], others revisit the problem using a clean-slate approach [60]. The primary characteristics of MHWNs that restricts the applicability of the conventional transport protocols to avoid congestion are recognized as: (1) false congestion alarms that are caused due to link failures; and (2) burstiness of traffic in MHWNs. We now discuss each characteristic and the congestion avoidance mechanisms to solve the problems.

### 15.5.1.1   Avoiding False Congestion Alarms

Recall that false congestion alarms are the activation of congestion control algorithms when a packet is dropped due to mobility. Several schemes are proposed to improve the performance of TCP over MHWNs during route disconnections [13, 30, 31].

Holland et al. [30] recognize the problem of false congestion alarms in TCP-Reno during a route disconnection. Packet drop information is inferred at the TCP source when it receives three duplicate ACKs or when the TCP source times out. This triggers the congestion control algorithm at the TCP source. It backs off aggressively by increasing its packet timeouts interval. In addition, after a route reconstruction, it observes a slow-start where the packet-sending rate is increased conservatively. The backoff may lead to TCP idling even when an active route is present, and the slow start may lead to a long delay in estimating the available capacity for a connection. In the meantime, mobility may again cause a route disconnection, causing the entire process to start again.

A cross-layered approach is used where Explicit Link Failure Notification (ELFN) message is transmitted by the node that experienced the disconnection to the source. When the MAC layer detects that a packet retransmission failure for several times (as indicated by Retransmit limit), it informs the routing layer about possible disconnection of the next-hop. The routing layer then sends an ELFN message to the TCP source. The TCP source freezes its congestion variables, including the congestion window size under such cases. Explicit message for route-reconstruction and probing by the TCP source is used to detect a route reconstruction. After the route repair, the source re-activates the congestion variables from the frozen state (thus avoiding a slow start) and resumes transmitting packets at the corresponding rate.

However, the problem with such an approach is the misinterpretation of all the packet re-transmission failures as mobility-based losses. Packet transmission failures may also happen in static networks when the available capacity at a node is less, thus indicating congestion. For example, a hidden-terminal scenario (Fig. 15.2 (a)) indicates a true congestion on the links. ELFN messages under such scenarios cause aggressive TCP behavior, thus sometimes leading to degradation of performance [61].

### 15.5.1.2 Avoiding Burstiness

Sundaresan et al. [60] propose a clean-slate approach for overcoming burstiness in MHWNs called Ad hoc Transport Protocol (ATP). ATP uses a rate-based transmission instead of TCP's Window-based transmission where the rate at which packets needs to be transmitted is informed to the source by the ATP-ACKs instead of indicating the permissible bytes in-transit. This avoids congestion in the network due to bursty traffic. Each packet records the delay at all the nodes along the route. The delay includes the queue waiting time and the transmit time. As the packet traverses the route, the bottleneck delay is recorded. The ATP receiver averages these delays over an epoch and feeds back this information to the sender. The source uses this average delay information to infer the rate at which the packets need to be sent. Congestion is indicated when (1) the source realizes a large average delay; (2) when the receiver feedback information is lost. The source multiplicatively decreases the packet-sending rate upon detecting congestion.

Since a majority of the losses in MHWNs are due to link failures (after being unable to retransmit for a given number of times) rather than queue drops, ATP uses explicit notification mechanisms to avoid false alarms on link-failure-based losses. As in the TCP-ELFN protocol, this mechanism also helps ATP to avoid the slow-start phase.

As we discuss in the Section 15.7, the optimal rate-control approach is a function of queue-lengths at intermediate nodes along the route. Inferring the queue-length information at connection sources using an end-to-end approach is infeasible in wired networks since the policies and protocols at the intermediate router infrastructure cannot be altered. However, routers in MHWNs have the advantage of cooperating with connection sources.

## 15.5.2 Routing

The protocols discussed above solve the rate-control problem using the MAC information. The general trend in such protocols is to capture the scheduling effects of the MAC protocol to regulate the traffic. However, the protocols assume that effective routing is provided by the routing layer. Studies have shown that the choice of the routing protocol has a huge effect on the performance of the congestion control protocols like TCP [8, 19, 21].

Various routing metrics have been proposed to capture the quality of the intermediate links like the expected unsuccessful transmission attempts and queuing delays [1, 12, 18, 20, 54, 72]. It is shown that link-quality-aware routing protocols perform significantly better than the hop-count-based routing protocols [19]. We now review Expected Transmission Count (ETX) [18], one of the primary link-quality-aware metric that outperforms other metrics [19]. The intuition behind ETX is that the cost of a link is proportional to the expected number of attempts required to successfully transmit a packet. The ETX of a link is defined as $1/d_f d_r$ where $d_f$ is the probability that the packet is successfully received over a link in forward direction (for data transmission) and $d_r$ is the success probability of packet transfer in the reverse direction (for ACK transmission). Each node periodically broadcasts a probe packet at a constant rate of W packets per second. The nodes record the number of probe packets received from each neighbor. This count is included in the periodic probe packet sent by the node in order to return the link information to the sources. The packet loss rate is calculated from the data recorded by the node and the data from the probe packet of neighboring nodes. Route selection is then based on choosing the links with least ETX.

However, in addition to link-quality, the routing should be aware of interference regions and dynamics of the network. For example, the available capacity fluctuates significantly after a change in the network traffic. While the above protocols propose heuristics to capture the link-quality, they do not directly answer the question about routing policies upon congestion. Routing packets through low-interference regions avoids congestion in the network and increases the available end-to-end bandwidth for the connection. Studies have discussed routing models that optimizes different performance metrics by choosing paths through the low-interference regions [44, 45]. However, such a routing protocol needs non-local information about the interference regions in the network, thus making it harder for distributed solutions. An intelligent routing protocol that is aware of the estimated interference levels in a MHWN and the link-quality can enable effective congestion control.

### 15.5.3  Scheduling

The above discussion shows that the most commonly used CSMA-based MAC protocols have hidden-terminal and unfairness that causes congestion. Advanced MAC protocols [6, 32, 38, 40] solve the above problems. However, such protocols require additional infrastructure and communication overhead to reserve the channel. These protocols are harder to realize since most of the wireless cards come with an inbuilt IEEE 802.11 protocol. Thus, changing the entire functionality of the MAC protocol is restricted by the embedded implementation.

The optimal MAC scheduling to reduce congestion is analytically shown through Time-Division Multiple Access (TDMA) where nodes transmit in their allocated timeslots. The optimal schedules are determined by coordination between the nodes to determine the interfering links and the non-interfering links are scheduled in each timeslot. Determining such optimal MAC schedules is a hard problem in MHWN with complex interference patterns. We discuss the optimal scheduling problem to avoid congestion in detail in Section 15.7.3.2.

## 15.6 Thoughts for Practitioners

The previous sections analyzed the causes for congestion and surveyed the protocol design for controlling congestion in MHWNs. The effects of lower layers on congestion were studied and the necessity for a cross-layered design was discussed. In this section, we will briefly discuss the practical issues in selecting the protocol stack based on the taxonomy of the network and the providing compatibility for standard congestion control protocol like TCP over a wide variety of routing protocols.

### 15.6.1 Topology and Traffic-Specific Protocol Selection

Practitioners should be aware of the choice of the protocols based on the network topology and the expected traffic characteristics. Networks with mobility, like vehicular networks, are more vulnerable to false congestion alarms due to constant route failures. In such networks, it is imperative that control information like ELFN is necessary to overcome the false congestion alarms. On the other hand, the route failures due to physical disconnection between the nodes are very rare in static networks like mesh networks and sensor networks. While route errors are still possible in such networks, a link-quality-aware routing protocol [18, 54, 72] reduces the chances of such errors. The deterministic nature of the topology in a static network enables deploying a routing protocol that is aware of link-quality and the capacity of the nodes. In such scenarios, a standard higher-layer congestion control protocol like TCP can be used without many modifications.

### 15.6.2 Standardization of Control Messages

Unlike the internet, a wide variety of routing protocols, varying significantly in their functionality, are deployed across different MHWNs. While a subset of routing protocols like Optimized Link State Routing (OLSR) proactively maintain the routes at all times, other routing protocols like Ad-hoc On-demand Distance Vector (AODV) starts the route discovery process only when an application requests for a route. Enabling cross-layer communication between the widely used congestion controlling protocol TCP and the routing protocols needs standardization efforts. Several papers [49, 71] assume control information from routing to physical layer, like Internet Control Message Protocol (ICMP) and Explicit link failure notification (ELFN), for effective congestion management. Standardization effort to promote the flow of control information from routing to TCP enables the widespread use of congestion control mechanisms in diverse MHWNs.

## 15.7  Congestion Control Modeling

In this section, we survey papers that model the congestion control problem. We first briefly discuss the need for congestion models in MHWNs. We later focus on models that attempt to solve the congestion control problem in MHWNs. We conclude the section with a discussion about the necessary modeling efforts for a realistic and feasible characterization of congestion control problem.

### 15.7.1  Motivation

The analysis of congestion and the heuristic solutions for its avoidance were discussed in Section 15.5. The congestion control problem was approached by means of experimental observations and educated guesses. However, the nature of the congestion problem in MHWNs is hard due to several factors like the sparse resources, complex interference patterns, scheduling intricacies, cross-layered communication, and the dynamism of the network. In such a complex system, heuristic solutions that solve one sub-problem often impact the other problems in unintuitive ways. For example, the ELFN-based protocols deduced all packet losses was due to mobility, thus ignoring the possible hidden terminal scenarios. It was also seen that the rate-control solutions ignore the routing policies, even though the performance of the network was greatly dependent upon the routing policy [8, 19, 21]. While considering all the parameters adds to the complexity of distributed protocols, it helps to understand the nature of the problem and quantify the effect of each parameter. Modeling the system solves a problem by formally characterizing the problem and capturing the relationships between the parameters in a methodical approach. It is often used for design of optimal system where efficient operation is guaranteed under minimum resources. It also provides an insightful understanding of the nature of the problem and is an invaluable tool in guiding the development of efficient protocols. For example, in a wired network, FAST TCP [36, 66], which was developed by formally characterizing the functionality of each layer, has shown significant performance improvement. Hence, modeling congestion control in MHWNs has several advantages.

### 15.7.2  Basic Congestion Control Model

In this section, we first describe the notations that are used for modeling the problem of congestion control in MHWNs. We then study the structure and the approaches to solve the problem.

### 15.7.2.1 Notations

The MHWN is modeled as a graph G(**N**,**L**) where **N** denotes the set of wireless nodes (numbered 1,2,...,$N$) and **L** denotes the communication links or edges. An edge $(i, j)$ between two nodes $i$ and j exists if node $i$ is able to communicate directly to node $j$. Edges in the MHWNs can be derived by the distances between the nodes or by signal measurement–based approaches. Let $S$ denote the number of connections. The source of a connection $s$ is represented by $f_s$ and the destination by $d_s$.

Let $x_s$ denote rate of connection $s$(say, in bits/sec) and let $M_s$ be the maximum value for the $x_s$. Let vector $\mathbf{x} = [x_s, s = 1... S]$. The problem of congestion control deals with optimizing these rates at which the packets are sent for connections $(x_s)$ under certain constraints. The problem is generally formulated as a "utility maximization problem". Let $U_\alpha(x_s)$ be a function that denotes the benefit (or the utility) for the connection $s$ when it transmits the packet at the rate $x_s$. The overall problem is to maximize the utilities of all the connections. Equation (15.6) captures this overall utility maximization objective function. The rates of the connection $(x_s)$ should be allocated such that the network is stable[1] under some scheduling policy. The set of rates that is feasible under some scheduling policy is represented by *capacity region* $\Lambda$. Equation (15.7) captures the constraint that the feasible rates should be inside the capacity region.

$$\text{Maximize} \sum_{s=1}^{S} U_\alpha(x_s) \tag{15.6}$$

$$\text{subject to } \mathbf{x} \in \Lambda \tag{15.7}$$

Capturing the feasible rates is explained in detail in Section 15.7.2.4. The time is divided into equally spaced timeslots and the optimization problem selects the optimal rates ($\mathbf{x}$) at each timeslot. The scheduling component is responsible for selecting a set of links that will be activated at each timeslot for achieving the optimal rates. In this survey, we consider a simple scheme where a link $l$ transmits at a constant capacity $(c_l)$ if it is activated. Extensions to this scheme include activating the links at various rates provided by the MAC protocol [15, 47]. The optimization problem defined in Equations (15.6) and (15.7) has two sub-problems: (1) the congestion control component and (2) the scheduling component.

### 15.7.2.2 Rate Control Component

The rate-control problem deals with finding the optimal rates such that the total utility function is maximized (Equation (15.6)). The crux of the problem is to

---

[1] A queuing system is stable when the queue lengths at every node is bounded.

define the utility functions ($U_\alpha(\cdot)$ that have certain properties. Specifically, if the utility functions are strictly concave, non-decreasing, and continuously differentiable on$[0, M_s]$, then efficient algorithms can be developed for solving the resulting optimization problem.

Various utility functions have been defined in the context of wired networks to formulate the objective function [52]. Mo et al. [52] show that utility functions can be defined for capturing the various notions of fairness among the competing connections. Equation (15.8) is one of the widely used utility function that is capable of representing various fairness notions by varying the value of $\alpha$. The utility function defines proportional fairness [39] when $\alpha = 1$, and max–min fairness [34] property is exhibited as the value of $\alpha$ approaches infinity. These definitions of utility functions can be borrowed directly for the congestion control problem in MHWNs.

$$U_\alpha(x_s) = \begin{cases} \log(x_s) & \text{if} \alpha = 1 \\ (1 - \alpha)^{-1} x_s^{1-\alpha} & \text{otherwise} \end{cases} \tag{15.8}$$

### 15.7.2.3 Scheduling Component

The scheduling problem is hidden inside the optimization problem, generally as a set of constraints. The scheduling problem restricts the activation of links such that they do not interfere. They are also responsible for maintaining the stability of the system by controlling the packet-sending rates at the links (such that the queue-lengths are bounded). The former problem is relatively easier in the wired networks where a set of scheduling constraints limit the flow of traffic on any link to its capacity. However, due to the shared nature of the wireless channel, the scheduling constraints are more complex (for example, two links that are within interference range cannot be scheduled in MHWNs). This scheduling problem in MHWNs is known to be NP-hard [35] even under simple interference models. The congestion control models that are surveyed below primarily assume a *Node-exclusive interference model* [9, 41]. Recall that under this model of interference, a node interferes with all its one-hop neighbors. Hence, if a link $(i, j)$ is scheduled to transmit a packet, none of the other links that originate or terminate at $i$ or $j$ can be scheduled. For example, in Fig. 15.3, if the link (B, E) is

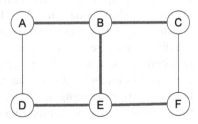

**Fig. 15.3** Node exclusive
interference model

scheduled, then the links (A, B), (B,C), (D,E), and (E, F) cannot be simultaneously scheduled.

Let $r_s^l(t)$ be the rate at which the link $l$ is transmitting for the connection $s$ at
timeslot $t$. Then, the overall rate of link $l$ at timeslot $t$ is $r_s(t) = \sum_s r_s^l(t)$. Let $\mathbf{r}(t)$
be a vector that denotes the rates for all the links, i.e., $[r_{ij},(i,j) \in \mathbf{L}]$. For
simplicity, denote $r_l(t)$ [and $\mathbf{r}(t)$] as $r_l$ (and $\mathbf{r}$).

Let $\mathfrak{R}$ denote the region of feasible rates. The feasible $\mathbf{r}$ can be obtained by
interference models. For example, consider all links to have the unit capacity.
Two links $i$ and $j$ that interfere with each other cannot be scheduled together.
Hence, the rate vector $\mathbf{r}$ is chosen such that for any two interfering links, $r_i - r_j \leq 1$.
Considering only two interfering links $i$ and $j$ in the network, the feasible set of
rates $r_i$ and $r_j$ is given by the shaded triangle OAB in Fig. 15.4.

The combination of such non-interfering set of links can be mapped to a
well-known graph theory problem of *matching*. Then, $\mathfrak{R}$ is the set of all match-
ings (feasible rate vectors). The *convex hull* [53] of the set $\mathfrak{R}$, denoted by $Co(\mathfrak{R})$,
defines the boundary for the feasible rates. In the example shown in Fig. 15.4, all
the points inside the triangle OAB denotes $\mathfrak{R}$ and the boundary of the triangle
denotes $Co(\mathfrak{R})$.

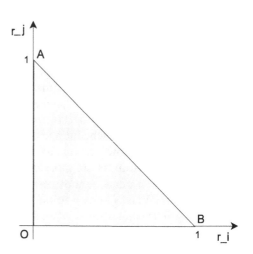

**Fig. 15.4** Feasible rate
region ($\mathfrak{R}$)

### 15.7.2.4 Capturing the Capacity Region

In this paragraph, we expand on the definition of the capacity region $\Lambda$ that is
needed to define Equation (15.7). It is to be noted that $\mathbf{r}$ is a $|\mathbf{L}|$ dimensional
vector, with each element corresponding to the rate at a given link. However, $\mathbf{x}$ is
a $S$ dimensional vector, where each element shows the rate at which the source
of a connection transmits the packet. While a feasible $r$ should lie in the $\mathfrak{R}$, a
feasible $x$ should lie in the $\Lambda$. The scheduling problem assigns the rates at each
links $\mathbf{r}$ (say, by examining the "matching" of non-interfering links). We now map
the constraints of feasible link rates ($\mathfrak{R}$) to feasible source rates ($\Lambda$). We consider

two cases of this problem: (1) the route-dependent case and (2) the route-independent case. The former models a problem where the set of routes for each connection is pre-specified and the latter solves the problem when the routes are unknown.

## Route-Dependent Case

In the route-independent case [9, 47], we introduce another parameter to define the set of routes that are active. Let $H_l^s$ denote a indicator variable, which is set to 1 if link $l$ is used in the route of connection $s$(otherwise, it is set to 0)[2][2]. Then the total rate at which the link $l$ transmits is given by $r_l = \sum_s H_l^s x_s$. The rates at all the links should lie within $Co(\Re)$. Hence, the capacity region can be defined by Equation (15.9) and the rate constraints by Equation (15.10).

$$\Lambda = \{x : [\sum_s H_l^s x_s, l \in \mathbf{L}] \in Co(R)\}$$

$$\sum_s H_l^s x_s \leq r_l, \forall l \in \mathbf{L} \tag{15.9}$$

$$\mathbf{r} \in Co(R) \tag{15.10}$$

## Route-Independent Case

In the route-independent case [47, 48], an additional problem of routing needs to be solved with the problem of congestion control. A well-known approach for solving routing problems in network flow problems is by using mass balance constraints [3]. These constraints restrict the inflow at each node and outflow from each node such that it creates a route from source to destination. Equation (15.11) denotes the complete optimization problems with the rate constraints.

$$\max_{x \geq 0} \sum_s U_\alpha(x_s)$$

subjectto :

$$r_{ij}^s \geq 0, (i,j) \in L, \forall s$$

$$\sum_{j:(i,j)\forall L} r_{ij}^s - \sum_{j:(j,i)\forall L} r_{ji}^s - \sum_{s:f(s)=i} x_s \geq 0, \forall s, \forall i \neq d(s) \tag{15.11}$$

$$[\sum_s r_{ij}^s] \in Co(R)$$

---

[2] This formulation enables a single route for each connection. It can be easily extended to support multiple routes per connection [47].

### 15.7.3 Decomposition

At this point, we have defined the complete optimization problem of congestion control in MHWN. We now discuss a formal approach called "decomposition" that breaks this large optimization problem into sub-problems. The decomposition approach has attractive features of formally arriving at distributed solutions to achieve the optimal performance [16]. Such approach has been proven to yield efficient solutions in wired network (e.g., Design of FAST TCP [36, 66]). Moreover, decomposition reduces the computation time since solving several smaller sub-problems is much faster than solving a single large problem. In the context of congestion control, we survey various approaches that use decomposition to arrive at distributed algorithms.

In this survey, we review decomposition of a route-independent congestion control model, which is given by the Equations (15.11) [47]. Henceforth, we refer to this optimization problem as the Primal Problem.

#### 15.7.3.1 Decomposing the Congestion Control Problem

Lin and Shroff [47] formulate a dual problem for Equation (15.11) and have shown that the duality gap is zero under route-independent congestion control model. Similar approach has been followed to formulate the dual problems for other congestion control models [9, 15]. It can be shown that the dual problem for Equation (15.11) is given by Equation (15.12).

$$\min_{q \geq 0} D(q)$$

$$\text{where}$$

$$\begin{aligned}
D(q) &= \sum_s B_s(q_{f_s}^s) + V(\mathbf{q}) \\
B_s(q) &= \max_{x_s \geq 0, r_{ij}^s \geq 0} U_\alpha(x_s) - x_s q \\
V(\mathbf{q}) &= \max_{\mathbf{r}} \sum_{(i,j) \in L} r_{ij} \max_d (q_i^d - q_j^d)
\end{aligned} \tag{15.12}$$

For simplicity, we refer to the Lagrange multiplier for connection $s$ with destination $d = d_s$ at node $i$ as $q_i^d$. It has been shown that this Lagrange multiplier can be directly mapped to the queue-length at node $i$ for connection $s$. The value of $q_i^d$ belongs to a non-negative real numbers ($q_i^d \in R^+$) and $q_i^d = 0$ when $i = d$. The vector of Lagrange multipliers is denoted by $\mathbf{q}$. The above dual problem can be readily decomposed into two sub-problems, the rate-control problem and the scheduling problem. The rate-control problem (Equation (15.13)) finds the optimal packet-sending rates ($\mathbf{x}$) at the sources for a given queue-length information ($\mathbf{q}$) at time $t$. The rate-control problem can be viewed as selecting the source rates such that the utility is maximized (first part of Equation (15.13)) and the penalty due to queue buildup is reduced (the second part). If the queue-length is zero at the source ($q_{f_s}^{d_s} = 0$), then a rate that corresponds to the maximal utility is selected. During non-empty queues, there is a balance between the maximum utility and the queue-length.

$$x_s(t) = \arg\max_{0 \leq x_s \leq M_s} [\, U_\alpha(x_s) - x_s q_{f_s}^{d_s} \,] \tag{15.13}$$

The scheduling problem, given in Equation (15.14), finds the optimal schedule of the links **r** for the given queue-length information (**q**).

$$\mathbf{r}(t) = \arg\max_{\mathbf{r}} [\, \sum_{(i,j) \in L} r_{ij} \max_d (q_i^d - q_j^d) \,] \tag{15.14}$$

Equation (15.14) can be interpreted as the optimal choice of link activation and it considers the feasible rate vectors (that may be based on "matching" of non-interfering links) and the difference in the queue-lengths at the nodes of a link ($q_i^d - q_j^d = 0$). Higher priority is given to the links that have the maximum differential queue backlog.

At each iteration of $t$ the multipliers [$\mathbf{q}(t-1)$] is updated according to the following expression for the computed values of $\mathbf{x}(t)$ and $\mathbf{r}(t)$:

$$q_i^d(t+1) = \{q_i^d(t) - h_t [\, \sum_{j:(i,j) \forall L} r_{ij}^d(t) - \sum_{j:(j,i) \forall L} r_{ji}^d(t) - \sum_{s:f(s)=i,d_s=d} x_s(t)]\}^+ \tag{15.15}$$

In the above expression, $h_t$ is a non-negative integer step-size parameter, which is chosen as a constant in a majority of the studies [9, 15, 47, 48]. The value of the expression $\{p\}^+$ maps the value $p$ to a non-negative real number, i.e., if $p$ is non-negative, then $\{p\}^+ = p$, else $\{p\}^+ = 0$. The physical interpretation of this update can be explained as follows. The expression in the square brackets denotes the difference between outflow and inflow at time $t$ observed at a node $i$. If this difference is negative (i.e., the inflow is greater than the outflow), then there is an increase in queue-length, thus increasing the queue-length penalty at time $(t+1)$.

The above decomposition has several attractive features while still solving the complete optimization problem of Equation (15.11). The parameters **x** and the **r** have been nicely decoupled in the rate-control and scheduling problems. This means that the source for the connection (**x**) does not need to know the sending rate of each link (**r**), which enables low-overhead distributed solutions. Also, the coupling Lagrange multiplier (**q**) has been nicely mapped to physical queue-length. This queue-length information coordinates between the rate-control and the scheduling problems to arrive at the optimal solution. Such a decomposition procedure has also aided to isolate the most complex component in congestion control – the scheduling problem.

### 15.7.3.2 The Scheduling Problem

In the previous sections, we explained the decomposition of the congestion control optimization problem. We now survey the scheduling component, its complexity, the approximation mechanisms and their trade-offs in this section. We describe the main problem following the argument by Chen et al. [15]. We later discuss the other approaches of solving the scheduling problems [9, 48].

The rate-control component of the problem can be easily solved by well-known convex programming approaches. Distributed solutions can be derived for controlling the packet-sending rate at the source based on the queue-length information. However, an optimal solution to scheduling problem needs global information about the possible non-interfering links that can be concurrently scheduled. Equation (15.14) has two parts: to find the set of feasible rate vectors ($r \in Co(R)$) and the choice of the optimal rate region. The former part deals with finding the set of non-interfering links and the latter part in selecting the set of the edges that have a maximum weighted sum. The differential queue-lengths ($\max_d(q_i^d - q_j^d)$) of Equation (15.14) act as constant weights for the scheduling problem [15]. We describe this problem and analyze the complexity of the solution under a more general *unit-disk interference model* and later for *node-exclusive interference model*.

Recall that under a unit-disk interference model, nodes within the interference range will interfere with each other. The set of interferers for a link will thus be the union of the set of interferers for the source or the destination of the link. The conflict graph for a given set of nodes and links is a transformed graph where each link is represented as a vertex and there is an edge between two vertices if the links interfere with each other. The set of vertices in the conflict graph that are not connected by an edge denotes the links which do not interfere with each other. This set of edges (or wireless links) is known as "independent set", and such wireless links can be scheduled concurrently without interfering with each other. Each independent set can be represented by a rate vector **r** in which $r_{ij}$ is equal to the capacity of the link ($c_{ij}$) if the link is present in the independent set; otherwise $r_{ij} = 0$. The set of all such independent sets gives all the combination of links that can be concurrently activated, i.e., the set of feasible schedules $\mathfrak{R}$. The scheduling problem transforms to the problem of "weighted maximal independent set problem" [15], which is NP-hard.

However, the scheduling problem can be solved in a polynomial time algorithm under the node-exclusive model of interference. Under such a model, the problem of finding the set of non-interfering links reduces to the "matching problem" [53]. The matching problem in a graph selects the set of edges that do not have common nodes, which is the set of non-interfering links in the node-exclusive model. The complete scheduling problem is shown to be an instance of "weighted maximal matching" [15], which is solvable in polynomial time [53].

Enabling Distributed Solutions

While a low-complexity algorithm exists to solve the scheduling problem, it still requires a centralized implementation where a central agent selects the optimal schedule by gathering the queue-length information from all the nodes in the network. Such information dissemination from all the nodes and at each time-slot has a large overhead of communication. Studies have proposed approximation algorithms to enable distributed solutions. Several studies [15, 67] propose distributed algorithm that guarantees at least half the performance of a centralized solution. While this is the worst-case performance, it has been observed that some solutions [15] typically provide 4/5 the optimal performance. The main approach of such algorithms is:

1. *Passing the Queue-length information:* Collect the queue-length information to compute the weight for each link in the maximal matching; this needs collecting the values of **q** for all the neighboring nodes.
2. *Computing the maximum weight:* Communicate the maximum weight $(\max(q_i^d - q_j^d))$ observed at each node to neighboring nodes.
3. *Coordinating for matching:* Each node $i$ then sends a match-request message to the best neighbor $j$ (based on the weight) and sends them a message about the willingness to match. If the neighbor is free, then it accepts the match request and the link $(i, j)$ is scheduled. If the neighbor finds a better match, it transmits a match-request to this neighbor and sends a drop-match message to the originally matched neighbor. The dropped neighbor then searches for next best matching neighbor (Step 2). The process continues till a node is matched with one of its neighbors or all its neighbors are matched with other nodes.

## 15.8  Directions for Future Research

In this section, we sketch the plan for further research in congestion control; both in the area of heuristic protocol development and modeling.

### 15.8.1  Heuristic Protocol Development

As discussed in previous sections, the link-layer and routing layer play an important role in deciding the end-to-end congestion observed by a connection. Supporting the information flow from a lower layer to the higher layers enables effective cross-layered mechanisms for congestion control. Collection of relevant information at the lower layers like MAC layer provides valuable information about the performance of the network. For example, the exponential backoff and unfairness of the IEEE 802.11 MAC protocol affects the critical performance parameters like packet transmission delay and jitter. If the information is provided to the routing protocol, then packets can be re-routed through less congested paths with lesser delay and jitter or congestion can be avoided by adjusting the sending rate. One primary research challenge is to decide the architecture for the cross-layer communication. We now discuss the two main issues for designing a cross-layered architecture: (1) measurement challenges and (2) providing the right abstractions.

### 15.8.1.1 Measurement Challenges

Sections 15.4.2 and 15.5.2 have discussed the use of detailed MAC level information at the routing layer to avoid congestion at the higher layers. However, measurement of the link-level information is hard due to two reasons. Firstly, the link experiences varying interference levels due to the dynamic traffic in the network. Hence, the quality of the link fluctuates with time. Constant measurement of the link-quality under such a dynamical system is a major research challenge. Routing protocols have evolved to capture the essential MAC level information to optimize the performance of the protocol [18, 19, 54, 72]. However, the overhead involved to capture and propagate accurate link-quality details in a dynamic, and possibly in an already congested, network poses new challenges. Secondly, the MAC protocol that provides the link-level details is generally embedded into the network cards. Hence, very few parameters are exposed to the networking stack in the operating system. While some of the device drivers for the cards are proprietary, open source device drivers for certain chipsets are available [50], which provide a subset of the MAC parameters.

### 15.8.1.2 Providing the Right Abstractions

This problem deals with abstracting the measurements at the lower layer information at the right level such that a vast majority of the common protocols at the lower layer are able to gather the required information. For example, it was seen in Section 15.5.1.1 that ELFN message, which is communicated from routing layer to TCP, was very effective in controlling false congestion alarms. Cross-layered API that passes such information from any routing protocol to the TCP will reduce the strong coupling between a specific routing protocol and the TCP, further increasing the portability of TCP over MHWN routing protocols. Furthermore, the cross-layered architecture should also support the backward communication, where the higher layers can request manipulation of the important parameters of the lower layers. With proper authentication mechanisms, such a backward communication provides custom quality of service for the higher layers. For example, the real-time applications like Voice-over-IP (VoIP) can request the packet to be dropped at the network queue if the packet is delayed beyond its deadline. This avoids further congestion in the network by avoiding the transmission of packets that will be dropped at the end destination.

## 15.8.2  Toward Realizing Congestion Control Models

The models studied in Section 15.7 characterize the optimal congestion control mechanism and support near-optimal distributed protocols. It was seen that the complexity of the problem resulted in several assumptions, like simplified interference models, which abstracts the realistic wireless network in an ineffective way. In this section, we discuss the challenges of using congestion control models in realistic networks. Supplementary models that solve the subproblems required for effective congestion control are also briefly studied.

### 15.8.2.1  Timeslots

The congestion control models studied above assume a slotted time in which the optimal rates are computed for both the sources ($x$) and individual links ($r$). At each timeslot, there is a set of message exchanges (e.g., for matching the non-interfering links) and the packet delivery is started after these control messages. While efficient algorithms have been found, computing the congestion control problem at each timeslot is unrealistic for several online applications of the model. The numerical results [15] indicate that optimal source rates evaluated at each timeslot converges to a global average optimum rate. These results are important since they answer the question of whether such an average near-optimal rate exists or not. A desirable feature in such online applications is to evaluate the converged near-optimal source packet rates. The model can be re-evaluated whenever the connection dynamics change (e.g., during an initiation or termination of a connection).

### 15.8.2.2  Realistic Interference Models

A majority of the congestion control models assume a node-exclusive interference model for simplicity. But such an interference model does not account for the signal strength or the distance between the sender and the receiver. Unit-disk model and two-disk model of interference are extended models that characterize interference as a function of distance.

Under both the models, the scheduling problem of finding all the non-interfering links maps to the "maximal independent set" problem of graph theory, which is NP-hard. However, approximation algorithms have been proposed to overcome the high complexity [35]. Based on such an algorithm, Jain et al. [35] propose a routing model to calculate the efficient routes.

Kolar et al. [44] study the problem of computing interference-separated routes for a set of connections under two-disk model of interference. In order to avoid the NP-hard problem of computing maximal independent sets, they approximate interference observed at a node as the sum of all the traffic within the interference region of a node. Based on this approximation, they formulate a multi-commodity flow [3] problem that provides maximum throughput routes. Since the throughput is dictated by the nodes that observe maximum interference (bottleneck nodes), the problem minimizes the bottleneck nodes in each connection for computing the *throughput optimal routes*. The objective function for such an approach is proposed and evaluated. However, minimizing the bottleneck node for every connection becomes a Mixed-Integer Linear Program (MILP), which is NP-hard too.

While the above routing formulations do not solve the rate-control sub-problem, it helps controlling congestion in a network by providing routes for the connections that observe less interference. The framework of such models can also be extended for including the rate-control problem. However, a low-complexity approximation of such formulations is necessary for realistic use of such models in MHWNs. Another drawback of such formulations is the assumption ideal scheduling that discounts for the inconsistencies of a realistic CSMA-based scheduling protocol like IEEE 802.11.

### Effect of Contention-Based Schedulers

In the previous discussions, it was seen that the models compute or assume the "optimal schedules". However, the majority of the existing MHWNs operate under a contention-based CSMA schedulers like IEEE 802.11. As discussed in the Section 15.2, many strange interactions arise in the operation of realistic schedulers like the IEEE 802.11 MAC protocol [26]. The heuristic protocol development [13, 30, 31, 61] that solves the congestion control problem has showed the huge impact of CSMA scheduling intricacies on the performance of the networks. Most of the congestion models in MHWNs either use very simplistic scheduling models or derive the optimal schedules, which cannot be attained by realistic CSMA schedulers. Hence, the direct application of these optimization models under CSMA-based schedulers can be prone to several side effects. Capturing the scheduling behavior is critical for accurate modeling of congestion in MHWNs.

Precise characterization of the effects of CSMA schedulers on throughput and fairness has been studied extensively and several stochastic models have been proposed [7, 11, 24, 25, 64, 65]. However, such models are complex and efforts to approximate these effects into the optimization models have not been extensively studied. Also, the scheduling interactions are determined only when the traffic on each link is specified. Under a CSMA protocol, even slight alterations of the active links can lead to significant changes in the scheduling behavior. For example, a newly added link can cause a hidden terminal to an already existing link, thus changing the dynamics of the network. Hence, in a naive approach, the CSMA-aware routing model has to evaluate the CSMA effectiveness for each combination of the routes to pick the best set of routes. The problem of enumerating all the possible combination of routes and evaluating CSMA effectiveness has a high complexity. Since, the scheduling model is often used as a sub-component while modeling the complete problem of congestion control. Hence, low-complexity models that approximate the scheduling issues are preferred. Intelligent methods have to be developed to model routing using the scheduling component in a realistic network. We survey one such approach in the next paragraph.

Kolar et al. [43] propose such a low-complexity scheduling model and integrate it with a routing model. The approach is to first evaluate the interference-separated routes using a model similar to [44]. A low-complexity scheduling model that determines the CSMA effectiveness, like hidden terminals, is evaluated on the set of routes. Conflicting links are mutually excluded and the routing model is re-evaluated. The process is repeated until an acceptable set of routes that is interference-separated and effective under CSMA is determined. However, proving the convergence properties of the approach is a hard problem due to the sensitivity of CSMA to small alterations in the routes. Nevertheless, such a model aids in determining the near-optimal routing in real MHWN like mesh networks. According to the best of our knowledge, congestion control models under such realistic constraints are not yet solved. However, the models implicitly control the congestion by the choice of superior routes, thus enabling greater performance. The framework of such models can also be extended for solving the rate-control problem.

## 15.9  Conclusion

In this chapter, we reviewed the problem of congestion control in MHWNs. The chapter analyzes the causes of congestion at different layers. The manifestation of congestion at different layers was analyzed and the need for cross-layered design was motivated. Heuristic protocols utilizing the cross-layered ideas to avoid congestion were surveyed. Modeling congestion control in MHWN is advantageous due to a formal characterization and futuristic protocol design of the problem. We reviewed the modeling efforts for optimal congestion control in MHWNs. The various sub-problems, their complexities, and approximations were studied. However, practical utility of such congestion control models in real networks still remains an active research area. The chapter discussed the remaining work for realizing congestion control models.

## Terminologies

*Capacity region*: The set of feasible transmission rates from the connection
   sources such that the queue-lengths at all the nodes are bounded.
*Carrier sense multiple access* (CSMA): CSMA MAC protocols transmits the
   packet only if the channel is sensed idle. If the channel is sensed busy, then
   the node tries transmitting the packet after the channel becomes idle.
*Carrier sense multiple access / collision avoidance* (CSMA/CA): A modified
   version of the CSMA protocol where different mechanisms to avoid

collisions are employed. The mechanisms may like exponential backoff and RTS-CTS are common examples of collision avoidance mechanisms in multi-hop wireless network.

*Dual optimization problem* (dual problem): An alternative formulation of the primary optimization problem (primal problem).

*Explicit link failure notification messages*: The message transmitted from the routing layer to the TCP when a *route error* occurs. This is used for controlling false congestion alarms in MHWNs.

*False congestion alarm*: False congestion alarms are the misinterpretation of certain messages from the lower layers as congestion. In MHWN, false congestion alarms are triggered when routing layer loses a packet due to route disconnection and the TCP layer assumes such an event as network congestion.

*Feasible rate region*: The transmission rate of each link during a timeslot such that there exists some schedule where none of the links interfere with each other.

*Hidden terminal scenario*: A scenario where packet collision is observed at a receiving node due to interference from an interfering node. The actual source node of the packet and the interfering node are unable to listen to each other.

*Lagrange dual problem:* The dual problem that is formed by one of the well-known methodologies. In the primal problem, each constraint imposes a strict yes-or-no value for any given point $x$. Lagrangian dual problem removes the hard constraints and includes the constraints as weighted expressions in the objective function. In order to choose feasible points without the hard-constraints, the points are rated such that the points far away from the feasible region have very high weight, thus making it very costly to choose infeasible points.

*Lagrange multipliers*: The weights chosen for each constraint during the formulation of the *Lagrange dual problem*.

*Primary interference model (node exclusive interference model):* In this interference model, only links that share a common node interfere with each other.

*Rate-based transmission:* A transmission scheme where the node transmits at a given rate instead of using a Window-based scheme.

*Route discovery process:* The process when the routing layer initiates a set of messages to construct a route to a given destination.

*Route error message:* The message sent by the routing layer when it concludes that the link to a neighboring node is broken.

*Two-disk interference model*: The interference is dictated by the same rule as that of Primary Interference model. However, two nodes can receive packets from each other only if the distance between the two nodes is lesser than the reception range. Normally, reception range is approximately half the interference range.

*Unit-disk interference model*: In this interference model, neighboring nodes that are within a certain distance, called interference range, from a given node interferes with the node.

*Window-based transmission:* A transmission scheme where the node limits the number of bytes that are in-transit toward a destination. This is generally maintained by keeping a moving Window that indicates the bytes that can be transmitted. This is usually followed in popular congestion control protocol like TCP.

## Questions

1. In order to avoid false congestion alarms, the routing protocol triggers ELFN messages to inform TCP about the link breakage. In such routing protocols, the route is used as long as the MAC layer is able to transmit the packet to the next-hop and route error message is sent once the link is broken. Route discovery is initiated to search for an alternative route. Instead of waiting for the complete link breakage, certain class of routing protocols [28] observes the signs for link disconnection and will switch to alternative routes if needed. Metrics like signal strength are used to guess the effectiveness of the link. Such routing protocols are called Pre-emptive Routing Protocols. Compare and contrast the effectiveness of ELFN messages in pre-emptive routing protocols in: (1) static networks (where nodes do not have mobility); (2) mobile networks.
2. Apart form the ELFN messages, propose cross-layered techniques that can be used to improve TCP performance which can under a pre-emptive routing protocol.
3. Assume a static mesh network where mesh routers form the backbone for routing traffic from end-nodes. The backbone route between the mesh routers is constructed such that the backbone links do not experience any packet losses (thus, no re-transmissions at MAC layer) and communicate on a single channel. Further assume that the mesh routers can listen to two different channels: (1) the backbone channel through which they connect to other mesh routers; and (2) end host channel, which is used to connect to the end hosts. Sketch the challenges of controlling congestion in such networks and comment about the false congestion alarms.
4. Assume a MHWN with homogeneous nodes and Software Defined Radio (SDR) running on the IEEE 802.11 MAC protocol. While MAC protocol is embedded in regular wireless cards, SDR has a software-based-MAC that enables easy reprogramming of the MAC functionality. Consider a real-time application like VoIP that is running on the network equipped with SDR radios. If end-to-end delay of the packets is greater than a certain threshold, then the packet is useless to the application and discards the packet. Create cross-layered APIs between (1) the routing layer and the MAC layer; and (2) application layer and routing layer to enable the delivery of packets across least congested path. Avoid altering the functionality of IEEE 802.11 protocol completely. You can solve the question by altering the key IEEE 802.11 variables.
5. Figure 15.4 shows the feasible rate region for two interfering links. Sketch the feasible rate region for the Flow-in-the-middle scenario (Fig. 15.2 (b)).

Also, list the vertices of the convex hull of the feasible rate region Co(R). Consider a channel of capacity 2 Mbps.

6. Consider a simple chain topology as shown in Fig. 15.1. Consider a primary-interference model (Link A-B interferes with B-C and B-C interferes with C-D) instead of unit-disk interference model. There is one connection between node A to node D. Formulate the complete primal optimization problem as given in Equation (15.11). Use the utility function $U_\in x_s = \log(x_s)$. Assume that the channel capacity is 2 Mbps for all the links and there are links that are unidirectional (A to B, B to C and C to D). Compare the Co(R) of the scenario with that of Question 5.

7. Optimization problems can be run within MATLAB. Make sure that Optimization Toolbox is installed. Using a function called as fmincon, the user can provide objective functions and the constraints of an optimization problem (type help fmincon within the MATLAB console for usage). Write a MATLAB program for the chain scenario in Question 6 and compute the optimal rates at which the node A should transmit.

8. Form the Lagrange dual problem for the scenario described in Question 6.

9. Write the MATLAB code, which iterates for each timeslot and solves the dual optimization problem in Question 8. Update the Lagrange multipliers using Equation (15.15). Assume a reasonable value of ht. After how many iterations does the source rate converge? Comment on relaxing the primary-interference model and including a two-disk model of interference in congestion control models. Specifically, how would you calculate the feasible rate region? Extend the algorithm provided in Section 15.7.3.2 to provide an approximate computation for links that can be active in a given timeslot using the above interference model.

# References

1. A. Adya, P. Bahl, J. Padhye, A.Wolman, and L. Zhou. A multi-radio unification protocol for IEEE 802.11 wireless networks. In BroadNets, 2004.
2. G.-S. Ahn, L.-H. Sun, A. Veres, and A. Campbell. SWAN: service differentiation in stateless wireless ad hoc networks. In Proc. of Infocom 2002, 2002.
3. R. K. Ahuja, T. L. Magnanti, and J. B. Orlin. Network Flows: Theory, Algorithms, and Applications. Prentice-Hall, Inc., Englewood Cliffs, NJ, 1993.
4. I. F. Akyildiz, X. Wang, and W. Wang. Wireless mesh networks: a survey. In Computer Networks, 47(4):445–487, 2005.
5. M. Allman, V. Paxson, and W. Stevens. TCP congestion control, 1999.
6. V. Bharghavan, A. Demers, S. Shenker, and L. Zhang. MACAW: A media access protocol for wireless LANs. In SIGCOMM, 1994.
7. R. R. Boorstyn, A. Kershenbaum, B. Maglaris, and V. Sahin. Throughput analysis in multihop CSMA packet radio networks. IEEE Trans. on Communication, 1987.
8. J. Broch, D. A. Maltz, D. B. Johnson, Y-C. Hu, and J. Jetcheva. A performance comparison of multi-hop wireless ad hoc network routing protocols. In MobiCom '98: Proceedings of the 4th Annual ACM/IEEE International Conference on Mobile Computing and Networking, pages 85–97, ACM Press, New York, NY, USA, 1998, ISBN: 1-58113-035-X

9. L. X. Bui, A. Eryilmaz, R. Srikant, and X. Wu. Joint asynchronous congestion control and distributed scheduling for multi-hop wireless networks. In INFOCOM, 2006.

10. Building the business case for implementation of wireless mesh networks, 2004.

11. M. M. Carvalho, and J. J. Garcia-Luna-Aceves. A scalable model for channel access protocols in multihop ad hoc networks. In MobiCom, 2004.

12. A. Cerpa, J. L. Wong, M. Potkonjak, and D. Estrin. Temporal properties of low power wireless links: modeling and implications on multi-hop routing. In MobiHoc '05, 2005.

13. K. Chandran, S. Raghunathan, S. Venkatesan, and R. Prakash. A feedback based scheme for improving TCP performance in ad-hoc wireless networks. In ICDCS '98: Proceedings of the 18th International Conference on Distributed Computing Systems, page 472, IEEE Computer Society, Washington, DC, USA, 1998.

14. C. Chaudet, I. G. Lassous, E. Thierry, and B. Gaujal. Study of the impact of asymmetry and carrier sense mechanism in IEEE 802.11 multi-hops networks through a basic case. In PE-WASUN '04: Proceedings of the 1st ACM International Workshop on Performance Evaluation of Wireless Ad Hoc, Sensor, and Ubiquitous Networks, pages 1–7, New York, NY, USA, 2004. ACM Press, New York, NY.

15. L. Chen, S. H. Low, M. Chiang, and J. C. Doyle. Cross-layer congestion control, routing and scheduling design in ad hoc wireless networks. In INFOCOM, 2006.

16. M. Chiang, S. H. Low, A. R. Calderbank, and J. C. Doyle. Layering as optimization decomposition: A mathematical theory of network architectures. In Proceedings of IEEE, 2007.

17. J. Cox, Network World. Philadelphia wireless win launches earthlink new strategy, October 2005. http://www.networkworld.com/news/2005/100605-earthlink-wireless. html

18. D. S. J. De Couto, D. Aguayo, J. Bicket, and R. Morris. A high-throughput path metric for multi-hop wireless routing. In MobiCom '03, 2003.

19. R. Draves, J. Padhye, and B. Zill. Comparison of routing metrics for static multi-hop wireless networks. In SIGCOMM, 2004.

20. R. Draves, J. Padhye, and B. Zill. Routing in multi-radio, multi-hop wireless mesh networks. In MobiCom '04: Proceedings of the 10th Annual International Conference on Mobile Computing and Networking, pages 114–128, New York, NY, USA, 2004. ACM Press, New York, NY

21. T. D. Dyer, and R. V. Boppana. A comparison of TCP performance over three routing protocols for mobile ad hoc networks. In MobiHoc '01: Proceedings of the 2nd ACM International Symposium on Mobile Ad Hoc Networking & Computing, pages 56–66, New York, NY, USA, 2001. ACM Press, New York, NY.

22. S. Floyd. TCP and explicit congestion notification. SIGCOMM Comput. Commun. Rev., 24(5):8–23, 1994.

23. S. Floyd, and K. Fall. Promoting the use of end-to-end congestion control in the internet. IEEE/ACM Trans. Netw., 7(4):458–472, 1999.

24. Y. Gao, D-M. Chiu, and J. C.S. Lui. Determining the end-to-end throughput capacity in multi-hop networks: methodology and applications. SIGMETRICS Perform. Eval. Rev., 34(1):39–50, 2006.

25. M. Garetto, T. Salonidis, and E. W. Knightly. Modeling per-flow throughput and capturing starvation in CSMA multi-hop wireless networks. In IEEE INFOCOM, 2006.

26. M. Garetto, J. Shi, and E. W. Knightly. Modeling media access in embedded two-flow topologies of multi-hop wireless networks. In MobiCom '05, pages 200–214, 2005.

27. M. Gerla, K. Tang, and R. Bagrodia. TCP performance in wireless multi-hop networks. In WMCSA '99: Proceedings of the Second IEEE Workshop on Mobile Computer Systems and Applications, page 41, IEEE Computer Society, Washington, DC, USA, 1999.

28. T. Goff, N. Abu-Ghazaleh, D. Phatak, and R. Kahvecioglu. Preemptive routing in ad hoc networks. J. Parallel Distrib. Comput., 63(2):123–140, 2003, ISSN: 0743-7315.

29. J. Hill, M. Horton, R. Kling, and L. Krishnamurthy. The platforms enabling wireless sensor networks. In Communications of the ACM, 47(6): 41–46, 2004, ISSN: 0001-0782.
30. G. Holland, and N. Vaidya. Analysis of TCP performance over mobile ad hoc networks. Wirel. Netw., 8(2/3):275–288, 2002.
31. G. Holland and N. H. Vaidya. Impact of routing and link layers on TCP performance in mobile ad hoc networks. In IEEE Wireless Communications and Networking Conference (WCNC), volume 3, pages 1323–1327, 1999.
32. IEEE 802.11e: Wireless medium access control (MAC) and physical layer (PHY) specifications: Medium access control (MAC) enhancements for quality of service (QoS).
33. V. Jacobson. Congestion avoidance and control. SIGCOMM Comput. Commun. Rev., 25(1):157–187, 1995.
34. J. Jaffe. Bottleneck flow control. IEEE Transactions on Communications, 29:954–962, 1981.
35. K. Jain, J. Padhye, V. N. Padmanabhan, and L. Qiu. Impact of interference on multi-hop wireless network performance. In MobiCom, 2003.
36. C. Jin, D. Wei, S. H. Low, J. Bunn, H. D. Choe, J. C. Doyle, H. Newman, S. Ravot, S. Singh, F. Paganini, G. Buhrmaster, L. Cottrell, O. Martin, and W. C. Feng. Fast TCP: from theory to experiments. In IEEE Network, volume 19, pages 4–11, 2005.
37. D. B. Johnson, D. A. Maltz, Y-C. Hu, and J. G. Jetcheva. The dynamic source routing protocol for mobile ad hoc networks (DSR), 2002.
38. S-S. Kang, and M. W. Mutka. Provisioning service differentiation in ad hoc networks by modification of the backoff algorithm. In Computer Communications and Networks, 2001.
39. F. P. Kelly, A. K. Maulloo, and D. K. H. Tan. Rate control for communication networks: shadow prices, proportional fairness and stability. Journal of the Operational Research Society, 49(3):237–252, 1998.
40. L. Kleinrock, and F. Tobagi. Packet switching in radio channels. In IEEE Transactions on Communications, 23(12):1400–1433, 1975 (Part I and Part II), ISSN: 0096-2244.
41. M. Kodialam, and T. Nandagopal. Characterizing achievable rates in multi-hop wireless networks: the joint routing and scheduling problem. In MobiCom, 42–54, 2003, ISBN: 58113-753-2.
42. C. E. Koksal, H. Kassab, and H. Balakrishnan. An analysis of short-term fairness in wireless media access protocols (poster session). In SIGMETRICS '00: Proceedings of the 2000 ACM SIGMETRICS International Conference on Measurement and Modeling of Computer Systems, pages 118–119, ACM, New York, NY, USA, 2000.
43. V. Kolar, and N. Abu-Ghazaleh. Scheduling aware network flow models for multi-hop wireless networks. In IEEE International Symposium on a World of Wireless, Mobile and Multimedia Networks (WOWMOM), 2008.
44. V. Kolar, and N. B. Abu-Ghazaleh. A multi-commodity flow approach to globally aware routing in multi-hop wireless networks. In IEEE PerCom, 2006.
45. V. Kolar and N. B. Abu-Ghazaleh. Towards interference-aware routing for real-time traffic in multi-hop wireless networks. In IEEE Distributed Simulation and Real Time Applications, 2007.
46. J. Li, C. Blake, D. S. J. De Couto, H. I. Lee, and R. Morris. Capacity of ad hoc wireless networks. In MobiCom, 2001.
47. X. Lin, and N. B. Shroff. Joint rate control and scheduling in multihop wireless networks. 43rd IEEE Conference on Decision and Control, 2:1484–1489, 2004.
48. X. Lin, and N. B. Shroff. The impact of imperfect scheduling on cross-layer congestion control in wireless networks. IEEE/ACM Trans. Netw., 14(2):302–315, 2006.
49. J. Liu and S. Singh. ATCP: TCP for mobile ad hoc networks. In IEEE Journal on Selected Areas in Communications, 19: 1300–1315, July 2001.
50. Madwifi. http://madwifi.org/
51. Meshdynamics inc. http://www.meshdynamics.com/

52. J. Mo, and J. Walrand. Fair end-to-end window-based congestion control. IEEE/ACM Trans. Netw., 8(5):556–567, 2000.
53. C. H. Papadimitriou, and K. Steiglitz. Combinatorial Optimization: Algorithms and Complexity. Prentice-Hall, Inc., Upper Saddle River, NJ, USA, 1982.
54. J. C. Park and S. K. Kasera. Expected data rate: an accurate high-throughput path metric for multi-hop wireless routing. In SECON: Second Annual IEEE Communications Society Conference on Sensor and Ad Hoc Communications and Networks, 2005.
55. C. E. Perkins, E. M. Belding-Royer, and S. Das. Ad hoc On-Demand Distance Vector (AODV) Routing, 2003.
56. T. Rappaport. Wireless Communications: Principles and Practice. Prentice Hall PTR, Upper Saddle River, NJ, USA, 2001.
57. S. Razak, V. Kolar, and N. Abu-Ghazaleh. Modeling and analysis of two-flow interactions in wireless networks. In Fifth Annual Conference on Wireless On Demand Network Systems and Services(WONS), 2008.
58. J. H. Saltzer, D. P. Reed, and D. D. Clark. End-to-end arguments in system design. ACM Trans. Comput. Syst., 2(4):277–288, 1984.
59. E. Setton, X. Zhu, and B. Girod. Congestion-optimized scheduling of video over wireless ad hoc networks. In IEEE International Symposium on Circuits and Systems (ISCAS), volume 4, pages 3531–3534, 2005.
60. K. Sundaresan, V. Anantharaman, H-Y. Hsieh, and R. Sivakumar. ATP: a reliable transport protocol for ad-hoc networks. In MobiHoc '03: Proceedings of the 4th ACM International Symposium on Mobile Ad Hoc Networking & Computing, pages 64–75, ACM Press, New York, NY, USA, 2003.
61. P. Sinha, J. P. Monks, and V. Bharghavan. Limitations of TCP-ELFN for ad hoc networks. In MOMUC, 2000.
62. A.S. Tanenbaum. Computer Networks, 3rd ed. Prentice-Hall, Inc., Englewood Cliffs, NJ, 1996.
63. The IEEE Working Group for WLAN Standards. IEEE 802.11 Wireless Local Area Networks. http://grouper.ieee.org/groups/802/11/, 2002.
64. F. A. Tobagi, and J. M. Brazio. Throughput analysis of multihop packet radio network under various channel access schemes. IEEE INFOCOM, 1983.
65. X. Wang, and K. Kar. Throughput modelling and fairness issues in CSMA/CA based ad-hoc networks. In INFOCOM, 2005.
66. D. X. Wei, C. Jin, S. H. Low, and S. Hegde. Fast TCP: motivation, architecture, algorithms, performance. IEEE/ACM Trans. Netw., 14(6):1246–1259, 2006.
67. X. Wu and R. Srikant. Regulated maximal matching: a distributed scheduling algorithm for multi-hop wireless networks with node-exclusive spectrum sharing. 44th IEEE Conference on Decision and Control, 2005 and 2005 European Control Conference. CDC-ECC '05, pages 5342– 5347, 2005.
68. K. Xu, M. Gerla, and S. Bae. How effective is the IEEE 802.11 RTS/CTS Handshake in Ad Hoc Networks? In Globecom, 2002.
69. S. Xu and T. Saadawi. Revealing the problems with 802.11 medium access control protocol in multi-hop wireless ad hoc networks. Computer Networks, 38(4):531–548, 2002.
70. Yaling Yang, Jun Wang, and Robin Kravets. Designing routing metrics for mesh networks. In WiMesh: First IEEE Workshop on Wireless Mesh Networks, 2005.
71. Xin Yu. Improving TCP performance over mobile ad hoc networks by exploiting cross-layer information awareness. In MobiCom '04: Proceedings of the 10th Annual International Conference on Mobile Computing and Networking, pages 231–244, ACM, New York, NY, USA, 2004
72. H. Zhang, A. Arora, and P. Sinha. Learn on the fly: data-driven link estimation and routing in sensor network backbones. In 25th IEEE International Conference on Computer Communications (INFOCOM), 2006.

# Chapter 16
# Security in Wireless Ad Hoc Networks

Klara Nahrstedt, Wenbo He, and Ying Huang

**Abstract** Operating in open and shared media, wireless communication is inherently less secure than wired communication. Even worse, mobile wireless devices usually have limited resources, such as bandwidth, storage space, processing capability, and energy, which makes security enforcement hard. Compared with infrastructure-based wireless networks, security management for wireless ad hoc networks is more challenging due to unreliable communication, intermittent connection, node mobility, and dynamic topology. A complete security solution should include three components of prevention, detection, and reaction, and provides security properties of authentication, confidentiality, non-repudiation, integrity, and availability. It should be adaptive in order to trade-off service performance and security performance under resource limitation. In this chapter, we will focus on the preventive mechanism for key management and broadcast authentication with resource constraints.

## 16.1 Introduction

In wireless networks, signals are transmitted via open and shared media. Without protection, anyone in the transmission range of the sender can intercept the sender's signal. Therefore, wireless communications are inherently less secure than their wired counterparts. Furthermore, wireless (mobile) devices usually have limited bandwidth, storage space, and processing capacities. It is harder to reinforce security in wireless networks than in wired networks.

There are two types of wireless networks: wireless LAN (WLAN) and wireless ad hoc network. The former requires the use of one or more access points (or base stations). Those access points connect wireless users which are one hop away and centrally control their access to Internet and the other WLANs. The

K. Nahrstedt (✉)
Department of Computer Science, University of Illinois at Urbana-Champaign, Siebel Center, 201 N. Goodwin Ave., Urbana, IL 61801
e-mail: klara@cs.uiuc.edu

S. Misra et al. (eds.), *Guide to Wireless Ad Hoc Networks*,
Computer Communications and Networks, DOI 10.1007/978-1-84800-328-6_16,
© Springer-Verlag London Limited 2009

ad hoc form of communications is based on radio to radio multi-hopping. Wireless ad hoc networks have been evolving to serve a growing number of applications, including military communications, emergency rescue operations, and disaster recoveries efforts. Benefiting from the ease of deployment, wireless ad hoc networks show great potential. Compared with WLANs, the security management in wireless ad hoc networks is much tougher due to the following characteristics.

1. Resource Constraints: The wireless devices usually have limited bandwidth, memory and processing power. This means costly security solutions may not be affordable in wireless ad hoc networks.
2. Unreliable Communications: The shared-medium nature and unstable channel quality of wireless links may result in high packet-loss rate and re-routing instability, which is a common phenomenon that leads to throughput drops in multi-hop networks. This implies that the security solution in wireless ad hoc networks cannot rely on reliable communication.
3. Node mobility and dynamic topology: The network topology of wireless ad hoc network may change rapidly and unpredictably over time, since the connectivity among the nodes may vary with time due to node departures, node arrivals, and the mobility of nodes. This emphasizes the need for secure solutions to be adaptive to dynamic topology.
4. Scalability: Due to the limited memory and processing power on mobile devices, the scalability is a key problem when we consider a large network size. Networks of 10,000 or even 100,000 nodes are envisioned, and scalability is one of the major design concerns.

Performance in wireless ad hoc networks is strongly related to the strength of security. However, without satisfactory network performance, security is meaningless. Therefore, in this chapter, we address network performance perspectives in security protocol design rather than cryptanalysis or formal verification of security protocols. The following requirements need to be considered for secure real-time communications.

- **Authentication:** Authentication is the process to verify the identity of the sender of a communication. Without authentication, malicious attackers can access resource, gain-sensitive information, and interfere with the operation of other nodes very easily.
- **Confidentiality:** Confidentiality means certain information is only accessible to authorized recipients. Participating parties to handle an emergency event need to cooperate with each other, while keeping the confidentiality of the traffic traversing the network.
- **Non-repudiation:** Non-repudiation ensures that the origin of a message cannot deny having sent the message. It is useful for detection and isolation of compromised nodes.
- **Integrity:** The integrity of a message is the property that the message cannot be modified without detection. Without integrity, attackers can easily

corrupt and modify the data and therefore cause mobile devices to make wrong decisions based on the corrupted data.

- **Availability:** Availability ensures the survivability of network services despite denial of service attacks. In unreliable wireless communications with highly dynamic topology, availability affects network performance greatly.

From security point of view, multiple lines of defense against attacks are desired. A complete security solution for wireless ad hoc networks should contain three components: prevention, detection, and reaction. In this chapter, we focus on the preventive protection in mobile ad hoc wireless networks. The topics described include *key management* and *broadcast authentication* issues.

## 16.2 Background

### 16.2.1 Key Management in Wireless Networks

Security solutions in wireless ad hoc networks rely on key management mechanisms. In this section, we briefly introduce symmetric key management and asymmetric (public) key management.

#### 16.2.1.1 Symmetric Key Management

Symmetric key systems, like DES, AES and keyed hash functions, are based on shared key information between two parties in communications. In this case, if the sender uses the secret key to encrypt a message, the receiver uses the same secret key to decrypt the message. Symmetric key techniques are attractive due to their energy efficiency. Therefore, a number of techniques have been developed for a specific type of ad hoc networks—wireless sensors networks, since sensors are inexpensive and low-power devices.

In symmetric key cryptography, a sender and a receiver must establish a shared key before communication. In the context of sensor networks, shared keys are distributed to sensors before their deployment. It is very challenging to design key distribution schemes with the following two concerns in a large-scale sensor network under limited memory resources:

- Connectivity: High percentage of the neighboring sensor nodes should share at least one secret key.
- Resilience: When some nodes are compromised by an adversary, other sensors are still able to maintain secure communications.

Random Key Distribution

In [10], key distribution consists of three phases: (1) key pre-distribution, (2) shared-key discovery, and (3) path-key establishment. In the pre-distribution phase, a large *key-pool* of $K$ keys and their corresponding identities are generated. For each sensor within the sensor network, $k$ keys are randomly drawn from the *key-pool*. These $k$ keys form a *key ring* for a sensor node. During the key-discovery phase, each sensor node finds out which neighbors share a common key with itself by exchanging discovery messages. If two neighboring nodes share a common key, then there is a secure link between two nodes. In the path-key establishment phase, a path-key is negotiated for each pair of neighboring sensor nodes who do not share a common key but can be connected by two or more multi-hop secure links at the end of the shared-key discovery phase. In the random key distribution mechanism mentioned above, the probability that any pair of nodes possesses at least one common key is:

$$p = 1 - \frac{((K-k)!)^2}{(K-2k)!K!} = 1 - \frac{(1-\frac{k}{K})^{2(K-k+\frac{1}{2})}}{(1-\frac{2k}{K})^{(K-2k+\frac{1}{2})}}. \tag{16.1}$$

A modification of the basic random key distribution scheme has been made in [4], where multiple common keys are needed to establish a secure link in the key-setup phase, instead of one common key. Such a modification increases the network resilience against node compromise. Random key distribution schemes can be further improved if the location of a sensor after deployment is predictable. Deployment information is able to reduce the memory requirements and increase the resilience against node compromise [7, 22].

## Combinatorial Design on Key Distribution

The combinatorial design [2] supports $q^2 + q + 1$ nodes in the network. The size of key-pool is $q^2 + q + 1$, and each node has $q + 1$ keys. The scheme is based on finite projective plane of order $q$, where $q$ is a prime number, to generate a symmetric design with parameters $(q^2 + q + 1, q + 1, 1)$. Therefore, every pair of nodes has exactly one key in common, and every key is owned by exactly $q + 1$ nodes. Thus, probability of key sharing among a pair of sensor node is 1. When a sensor node is captured by an adversary, the probability that a link is compromised is about $1/q$.

The disadvantage of the combinatorial design is that the parameter $q$ is a prime number, hence not all network sizes can be supported. To support arbitrary network sizes, the combinatorial design techniques and random key distribution approaches can be used together. Assuming the target network size is $n$, we can use the combinatorial design to determine the key distribution on $M$ nodes, where $M < n$. Then we can employ the random key distribution approach to assign keys to the remaining $N - M$ nodes. Such a hybrid design improves scalability and resilience of combinatorial design solution, at the cost of degraded connectivity (key sharing probability between neighboring nodes).

## Schemes Based on Blom's $\lambda$-secure Key Pre-distribution

Blom proposed a key pre-distribution method with $\lambda$-secure property [1]. It means that if no more than $\lambda$ nodes are compromised, the communications in the network are secure. Blom's scheme guarantees that any given pair of nodes in the network has a shared secret key. If we draw an edge between every two nodes sharing a secret key, the resulting graph will be a complete graph, and we get full connectivity. To achieve better resilience against node capture, we can sacrifice the connectivity moderately and let each sensor node carry less keys. Along this direction, multiple key-space Blom's schemes have been proposed in [8].

There are many other efforts addressing symmetric key distribution in recent literature, including [5, 21, 29, 33]. We observe that these schemes trade off conflicting design requirements among memory usage, connectivity, scalability, and resilience. A brief summary of performance issues in symmetric-key algorithms is given below:

1. Speed: Symmetric-key algorithms are generally much less computationally intensive than asymmetric key algorithms. Therefore, symmetric key algorithms are very popular in resource-limited wireless sensor networks.
2. Scalability: In symmetric-key algorithms, to ensure secure communications between everyone in a network of size $n$, a total of $n \times (n-1)/2$ keys are needed. However, current design of key management in wireless sensor networks does not require that each pair of nodes share a unique secret key.
3. Management: Symmetric-key algorithms require a shared secret key known at both sides during communication. To prevent adversaries from discovering the cryptographic keys, keys should be changed regularly. It is difficult to keep shared keys secure during key distribution.

### 16.2.1.2 Public Key Management

Unlike symmetric-key algorithms, asymmetric (or public) key algorithms (e.g., RSA, ECC) use two different keys, namely private and public keys, for encryption, decryption, authentication, and verification. For instance, a user knowing the public key of an asymmetric algorithm can encrypt messages destined for a receiver. The nodes other than the receiver do not know the receiver's private key and thus cannot decrypt encrypted messages. Compared to symmetric-key algorithms, public-key cryptography increases security and convenience, because private keys never need to be transmitted or revealed to anyone. Another advantage is that public-key cryptography can provide digital signatures which cannot be repudiated.

Because both asymmetric and symmetric key algorithms have their advantages and disadvantages, we wish to use virtue of both. Therefore, asymmetric keys are used to negotiate symmetric keys, and then symmetric keys are used to secure communications in the wireless ad hoc networks.

Public-key approaches were originally targeted at the Internet [19]. Recently, Elliptic Curve Cryptography (ECC) has been emerging as an attractive public key cryptography scheme for mobile/wireless environments [12, 17, 25]. In public key cryptography, any two nodes can establish a secure channel between them without necessarily carrying pre-distributed keys. However, if nodes do not carry pre-distributed public keys, one or more trusted certificate authorities (CAs) are needed. In wireless ad hoc networks, authentication process by CAs is very costly in terms of wireless communication overhead. In order to tailor public-key approaches to ad hoc networks, Zhou and Haas proposed a distributed public-key management scheme for ad hoc networks [32], where multiple distributed certificate authorities are used. To sign a certificate, each authority generates a partial signature for the certificate and submits the partial signature to a coordinator that calculates the signature from the partial signatures. Kong et al. described a fully distributed scheme [20], where every node carries a share of the private key of the service. This scheme increases availability of authentication, but it increases communication overhead for authentication. Capkun, Buttyan, and Hubaux proposed a self-organized public key management system [3], where users issue certificates based on their personal acquaintances. Each user maintains a local certificate repository. When two users want to verify the public keys of each other, they merge their local certificate repositories and try to find (within the merged repository) appropriate certificate chains that make the verification possible. We notice that in these certificate-based schemes, communication overhead to transfer certificates around is large in bandwidth-restricted wireless ad hoc networks. Next, we will present a public key management scheme for resource-limited environments in details.

## 16.2.2 Broadcast Packet Authentication

In large-scale wireless environment, data packets are often forwarded multiple hops until arrival at intended receivers. Operating in an open or hostile environment, it is easy for the packets to be modified or impersonated. False injection not only relays misleading information, which results in malfunction or catastrophe, but also consumes excessive communication, computational resources, and energy if they fail to be dropped at the very beginning. Therefore, it is critical to prevent packet manipulation, carefully grant authorized access and consistently assure the resource availability. A pessimistic way is to authenticate every packet in a hop-by-hop fashion and only packets from a legitimate device are forwarded. Hop-by-hop authentication is acceptable for unicast traffic, thanks to efficient symmetric key scheme and instantaneous one-way hash function scheme [34]. It is still acceptable for routing control packets, which could be either unicast or broadcast, because of their low rate and criticality. Example works can be found in secure routing protocols [14, 15]. However, hop-by-hop authentication scheme imposes remarkable penalty on end-to-end delay for legitimate broadcast data due to authentication delay at each intermediate hop. The accumulated delay postpones packet delivery to nodes far away from the sources, and the maximal delay is proportional to network diameter in hops.

Hence, the following requirements are desired for broadcast authentication schemes.

- **Containment:** Faked messages are dropped near the initiators so that unnecessary bandwidth and CPU consumption is avoid. The ideal case is to stop the false packets in one-hop range of their initiators.
- **Timeliness:** Authentic messages are delivered to majority of nodes quickly.
- **No Single Point of Failure:** Security schemes should embody randomness and distribution to avoid a single point of failure or targeted attacks.
- **Load Balancing:** Nodes receive approximately equal workload to avoid battery depletion, channel congestion and resulting network partition. Exceptions are nodes in one-hop range of attackers because they must authenticate every packet in order to properly filter out false packets.

With limited resources at mobile devices, the above requirements cannot be achieved simultaneously, and tradeoff exists. Next, we will summarize common assumptions, authentication primitives, and classification of broadcast authentication protocols.

### 16.2.2.1 Assumptions

Design of authentication protocol highly relies on assumption of network and attackers. If deployed mobile devices are powerful enough, advanced cryptography can be applied without much degradation on service performance, similar as wired network. If attackers are powerful enough to perform physical compromise, authentication protocol must take compromise and sybil attacks into account. We list common assumptions in prior arts below.

### Network Model

It is reasonable to assume that all the devices are dispatched from a single administrative domain, in which case integrity and authenticity of packets are valued most. Before being deployed on field, all devices are loaded with the necessary public and private keys or certificates from trusted authorities. After mission starts, they are able to establish pair-wise trust relationship without help from trust authorities.

Assumptions on network model cover the following aspects: (a) node mobility and the resulting dynamics of network topology; (b) lifetime range from hours, days, to years; (c) hardware and software trustworthy, which correlates with attackers' capability of physical compromise, secret key information exposure and software turnover; (d) traffic pattern, such as the numbers of senders and traffic rate; and (e) clock synchronization, which is required by TESLA [26] and its variations.

Attacker Model

Attackers can target at various resources, like energy, CPU, bandwidth and memory. Assumptions of attackers' capability, resources and typical behaviors largely determine the quality of protection level demanded by authentication protocols.

They could (a) physically compromise legitimate devices and subsequently extract private cryptographic information; (b) overhear over channels, possibly across a large network region; (c) understand protocol details, send semantics-compatible faked messages, and replay overheard packets (They may occasionally forward authentic messages to confuse intrusion detection mechanism.); (d) manipulate control fields if they are not authenticated; and (e) collude with each other.

### 16.2.2.2 Authentication Primitives

There are three cryptographic candidates for broadcast authentication: public key cryptography (PKC), symmetric keys, and one-way hash function.

Public Key Cryptography

In PKC, either private/public key pre-distribution or certificate approach are used. Receivers use sender's public key to validate signed packets immediately after reception. There is no need for online trust maintenance. However, for low-end mobile devices or sensors, public key signature generation and verification processes have heavy computation and memory overhead, thus limiting integrated traffic rate processed per node.

Symmetric Keys

Symmetric keys [9], widely used in low-end sensor networks to secure one-hop transmission, are suitable for pairwise trust establishment. However, it is inappropriate for broadcast authentication. If a unique secret key is assigned for each pair of nodes, broadcast traffic needs to carry separate signatures for all destinations, incurring tremendous bandwidth, computational and memory overhead. On the other hand, if one secret key is shared by a group of nodes, any compromised node can forge packets and impersonate others. Revocation of compromised keys is difficult.

One-way Hash Function

One-way hash function, represented by TESLA [26] and its variations, has low computation overhead (several order smaller than that of PKC) and memory overhead, which makes it a superior alternative to defend against false data injection in a light-weighted way [27, 35].

InTESLA , a sender splits time into even time intervals, called round, and generates a one-way chain using a predetermined one-way hash function $H$ and a seed $S$. A hash value is assigned to the corresponding round and used to authenticate packets generated in that round. Each hash value also has a release time, several time intervals after the assigned interval. After negotiation with the sender about the key disclosure schedule, $S$ and $H$, receivers are ready to validate packets sent by the senders. When sending a message, the sender attaches a message authentication code (MAC), which is generated based on the key associated with the round plus the most recent key which it can disclose. When receiving a message, the receiver checks that the message is not replayed and the disclosed key is legitimate, and then buffers the message. Finally, it removes all the buffered packets whose authentication keys are released and authenticates the packets. In summary, authentication by TESLA is featured with low communication and computation overhead, scalability to a large number of receivers and tolerance to packet loss.

Due to delayed key disclosure, authentication is delayed until the instant authentication keys are delivered to receivers. We call this delay keying delay. Even though immediate key disclosure schemes are proposed, they are vulnerable to online replaying attacks. In addition, TESLA requires trust negotiation and periodical trust maintenance. Therefore, it does not fit for high-mobility scenarios.

### 16.2.2.3 Classification of Broadcast Authentication Protocols

We can divide existing broadcast protocols according to node roles and adaptability.

|          | Flat                               | Hierarchical                 |
| -------- | ---------------------------------- | ---------------------------- |
| Static   | Authenticate first/Forward first   | Trusted base-station [6, 11] |
| Adaptive | Dynamic Window Scheme [30]         | –                            |

When every node has equal responsibility, static scheme adopts the policy of either authenticate-first or forward-first. While the former suffers from the significant end-to-end delay for legitimate traffic, the latter suffers from the wasted computational and communication resources network-wide to authenticate faked messages and low delivery ratio due to buffer overflow and network partition.

Drissi and Gu considered a large sensor network with a few predetermined trusted and better secured nodes, which divides the whole network into subsets [6, 11]. Trusted nodes use TESLA to broadcast messages to its subset and each ordinary node, upon receiving a broadcast message from its subset, rebroadcasts it before validation. Trusted nodes exchange broadcast messages among themselves. A source other than the trusted nodes sends broadcast messages to a nearby trusted node. However, the predetermined set of trusted nodes may attract attacks and too much authentication and coordination overhead consumes their battery and the battery of nodes around them quickly. Besides, how messages are sent from source to trusted nodes, among trusted nodes and from trusted nodes to subnets may introduce additional vulnerability, such as DoS attacks and jamming

Wang, Du, and Ning proposed a dynamic window scheme to contain bogus data by the public key cryptography (PKC) authentication [30], where sensor nodes determine whether to verify a message first or forward the message first by estimating their distance from the malicious attackers and how many hops the incoming message has passed without authentication. Despite the fact that Dynamic Window scheme displays good potential to improve end-to-end delay of authentic traffic, it is not robust against smart attackers who maliciously manipulate the AIMD scheme. Additional, during the decreasing process of AIMD, a large number of false packets are broadcast to the whole network.

## 16.3 Public Key Management with Resource Constraints

### 16.3.1 Introduction

The characteristics of mission-critical ad hoc networks pose the following challenges for the design of public key management schemes that would support secure communication over wireless ad hoc networks:

(1) **Communication Overhead for Authentication:** As we mentioned before, certificate-based schemes incur communication overhead during authentication.
(2) **Unreliable Communications and Network Dynamics:** Due to shared-medium nature of wireless links, flows may frequently interfere with each other. Moreover, a network may be partitioned frequently due to node mobility and poor channel condition. Mobile nodes may leave and join the ad hoc network frequently, and new legitimated nodes may join the network later after some nodes have already been deployed in the field. Mobility increases the complexity for trust management.
(3) **Large Scale:** The number of ad hoc wireless devices deployed at an incident scene depends on specific nature of the incident. In general, the network size can be very large. In addition, an ad hoc network should be able to accommodate more mobile devices if necessary. Therefore, it is necessary to have newly deployed devices and previously deployed devices trust each other without introducing too much overhead.
(4) **Resource Constraints:** The wireless devices usually have limited bandwidth, memory and processing power. Among these constraints, communication bandwidth and memory consumptions are the two primary concerns for key management schemes. Wireless bandwidth is the scarcest resource in wireless network. And memory concern for key storage is more and more evident as the requirement on network scalability (or network size) is increasing.

Given the above challenges (1) and (2), a node in a network may encounter unreliable communication and large communication overhead for authentication. Therefore, we need a self-contained key management scheme. We will present a self-contained public-key management scheme, where all necessary cryptographic keys (certificates) are stored at individual nodes before the deployment of the nodes. As a result, we can expect almost zero communication overhead for authentication. In contrast to traditional certificate-based schemes, our authentication procedure does not require the communication of certificate which binds the node's ID to its public key and is signed by an off-line trusted authority. The required storage space for traditional self-contained public key management schemes is of $O(n)$ order. With challenges (3) and (4), storage space at individual nodes may be too small to accommodate self-contained security service, when network size $n$ is large. Hence, we discuss a Scalable Method Of Cryptographic Key (*SMOCK*) management scheme, which scales logarithmically with network size, $O(log\ n)$, with respect to storage space.

In order for *SMOCK* to use smaller set of cryptographic keys, a sender uses multiple keys to encrypt a message and a receiver needs multiple keys to decrypt the message. We then use the public key cryptography as follows. Each node possesses a unique combination of private keys, and knows all public keys. The private key combination pattern is unambiguously associated with the node ID. It means, if a sender $A$ wants to send a message to receiver $B$, $A$ will first acquire $B$'s ID to infer a set of private keys owned by. $B$ Then $A$ will encrypt the message with the public key set that corresponds to the private keys owned by $B$. We have evaluated *SMOCK* with respect to the communication overhead for key management, memory footprint, and resilience to node break-in by adversaries. Note that it is likely that adversaries may eventually break into a limited number of nodes over a certain period before a network detects the break-in and revokes the compromised keys. However, before the system detects break-ins, a majority of network nodes under the *SMOCK* will operate securely even when a small amount of nodes is compromised.

## 16.3.2  Overview

In SMOCK, network nodes want to exchange correspondence securely in a pair-wise fashion. Symbols and terms used throughout this section are shown as in Table 16.1. The key pool $\mathcal{K}$ of such a group consists of a set of private–public key pairs, and is maintained by an off-line trusted server. Each key pair consists of two mathematically related keys. The $i$th key pair in the key pool is represented by $(k_{priv}^i, k_{pub}^i)$. To support secure communication in the group, each member is loaded with all public keys of the group and assigned a distinct subset of private keys. Let $\mathcal{K}_{Alice}^{priv}$ denote a subset of private keys held by Alice, and $\mathcal{K}_{Alice}^{pub}$ represents Alice's corresponding public key subset. If Bob wants to send a secret message to Alice, he needs to know $\mathcal{K}_{Alice}^{pub}$, where $\mathcal{K}_{Alice}^{pub} \not\subset \mathcal{K}_{anybody\_else}^{priv}$.

**Table 16.1** Notations and symbols

| | |
|---|---|
| $\mathcal{K}$ | A Key pool: a set of public-private key pairs |
| $privateKey_{ij}$ | $j$ th private key hold by user $i$ |
| $privateKey_{ij}$ | $j$ th public key hold by user $i$ |
| $\mathcal{K}_i^{priv}$ | A set of private keys held by user $i$, $\mathcal{K}_i^{priv} = \{privateKey_{ij}\}$ |
| $\mathcal{K}_i^{pub}$ | A set of public keys corresponding to $\mathcal{K}_i^{priv}$ |
| $\mathcal{K}_i$ | A set of public–private key pairs held by user $i$, $$\mathcal{K}_i = \{(k_{priv}, k_{pub}) | k_{priv} \in \mathcal{K}_i^{priv}$$ $$\& \; corresponding \; k_{pub} \in \mathcal{K}_i^{pub}\}$$ |
| M | Memory size for key storage |
| $a$ | Number of distinct key pairs $a = |\mathcal{K}|$ |
| $b$ | Number of private keys held by each user under isometric key allocation, $b = |\mathcal{K}_1| = |\mathcal{K}_2| = \cdots = |\mathcal{K}_n|$ |
| $k_c(x)$ | Expected number of disclosed keys when $x$ nodes are broken in |
| $k_v(x)$ | Maximum number of disclosed keys when $x$ nodes are broken in |
| $V_x(a,b)$ | Vulnerability metrics as $x$ nodes are broken in. |
| $C(a,b)$ | Abbreviation of $a$ choose $b$, $\left(\frac{a}{b}\right)$ |
| $V$ | A set of nodes in the ad hoc wireless network |
| $n$ | Total number of nodes in the network, $n = |V|$ |

Bob is able to pass the secret message to Alice, using the public keys $\mathcal{K}_{Alice}^{pub}$ to encrypt the message. The message can be opened only by Alice, who has the private key set $\mathcal{K}_{Alice}^{pub}$, but others do not.

Consider an example of a small group with 10 users. In *SMOCK*, we need five distinct public–private key pairs to build pair-wise secure communication channels among 10 users. They are $(k_{priv}^1, k_{pub}^1)$, $(k_{priv}^2, k_{pub}^2)$, $(k_{priv}^3, k_{pub}^3)$, $(k_{priv}^4, k_{pub}^4)$, $(k_{priv}^5, k_{pub}^5)$. Each user keeps five public keys and two private keys. The unique private key set allocation for each user is then shown in Table 16.2.

**Table 16.2** An example private key allocation

| User | $\mathcal{K}_i^{priv}$ private-key set held by user $i$ | User | $\mathcal{K}_i^{priv}$ private-key set held by user $i$ |
|---|---|---|---|
| 1 | $\mathcal{K}_1^{priv} = \left\{k_{priv}^1, k_{priv}^2\right\}$ | 6 | $\mathcal{K}_6^{priv} = \left\{k_{priv}^2, k_{priv}^4\right\}$ |
| 2 | $\mathcal{K}_2^{priv} = \left\{k_{priv}^1, k_{priv}^3\right\}$ | 7 | $\mathcal{K}_7^{priv} = \left\{k_{priv}^2, k_{priv}^5\right\}$ |
| 3 | $\mathcal{K}_3^{priv} = \left\{k_{priv}^1, k_{priv}^4\right\}$ | 8 | $\mathcal{K}_8^{priv} = \left\{k_{priv}^3, k_{priv}^4\right\}$ |
| 4 | $\mathcal{K}_4^{priv} = \left\{k_{priv}^1, k_{priv}^5\right\}$ | 9 | $\mathcal{K}_9^{priv} = \left\{k_{priv}^3, k_{priv}^5\right\}$ |
| 5 | $\mathcal{K}_5^{priv} = \left\{k_{priv}^2, k_{priv}^3\right\}$ | 10 | $\mathcal{K}_{10}^{priv} = \left\{k_{priv}^4, k_{priv}^5\right\}$ |

In this scenario, we know that

- Each node keeps a predetermined subset of private keys, and no one else has all the private keys in that subset.
- For a public–private key pair, multiple copies of the private key can be held by different users. In the given scenario, each private key has four copies.
- A message is encrypted by multiple public keys, and it can only be read by a user who has the corresponding private keys. For example, if user 1 encrypts a message $m$ by public keys $k_{pub}^2$ and $k_{pub}^5$ as $Enc(Enc(m, k_{pub}^2), k_{pub}^5)$, then only user 7 can decrypt it with private keys $k_{priv}^2$ and $k_{priv}^5$.

In traditional public management schemes, each user holds one public–private key pair. Therefore, a user should store $n$ public keys and 1 private key to achieve self-contained key management in a network of size $n$. In *SMOCK*, the storage for public and private keys is much smaller. In the above 10-user example, a user only needs to store 7 keys (5 public keys and 2 private keys), which is smaller than 11 keys (10 public keys and 1 private keys) in traditional schemes. We will show that in *SMOCK* the total number of keys held by each user is approximately $O(log(n))$, but it is $O(n)$ under traditional key management schemes.

## 16.3.3  Design Objectives

Before we elaborate *SMOCK*, we introduce several definitions and design objectives.

**Definition 1:** A *key allocation KA*: $2^{\mathcal{K}} \rightarrow V$, maps the key pairs in $\mathcal{K}$ to a set of users in $V$, so that $\nu \in V$ is assigned a subset of key pairs $\mathcal{K}_i$ ($\mathcal{K}_i \subset \mathcal{K}$). To guarantee the secure communication between each pair of nodes $i$ and $j$, we have $\forall i \forall j \mathcal{K}_i \not\subseteq \mathcal{K}_j$ (the same as $\mathcal{K}_i^{priv} \not\subseteq \mathcal{K}_j^{priv}$) and $\mathcal{K}_j \not\subseteq \mathcal{K}_i$ (the same as $\mathcal{K}_j^{priv} \not\subseteq \mathcal{K}_i^{priv}$), *iff* $i \neq j$. If this property holds, the *key allocation* is valid.

**Definition 2:** We say that a key allocation is isometric, if $|\mathcal{K}_1| = |\mathcal{K}_2| = \ldots = |\mathcal{K}_n| = b$; otherwise, the key allocation is non-isometric.

**Definition 3:** We say that the key assignment to user $i$ and $j$ conflicts, if either $\mathcal{K}_i^{priv} \subseteq \mathcal{K}_j^{priv}$ or $\mathcal{K}_j^{priv} \subseteq \mathcal{K}_i^{priv}$. For a valid key allocation, there does not exist conflicting key assignments for any pair of the users.

Generally, we desire the key management to be memory efficient for key storage, computationally efficient during encryption and decryption, and resilient to break-ins. Therefore, we define multiple objectives of the *SMOCK* key allocation mechanism as follows:

**Objective 1:** *Memory Efficiency*: Given a network of size $n$, we need to find a *key pool* $\mathcal{K}$ and a *key allocation KA* to achieve

$$\begin{cases} min \quad |\mathcal{K}| + \max_{i \in V} |\mathcal{K}_i^{priv}| \\ s.t. \quad \mathcal{K}_i \nsubseteq \mathcal{K}_j \text{ and } \mathcal{K}_i \nsupseteq \mathcal{K}_j \quad \forall i \neq j \end{cases} \qquad (16.2)$$

where $|\mathcal{K}_i^{priv}| = |\mathcal{K}_i|$ is the total number of private keys stored at node $i$. $|\mathcal{K}|$ is the total number of public keys stored at each node. Note that each node stores all public keys , but it only stores a small subset of private keys $\mathcal{K}_i^{priv}$ for user $i$. Therefore, $|\mathcal{K}| + |\mathcal{K}_i^{priv}|$ is the number of memory slots at node $i$ to store the public keys and private keys for secure communications.

**Objective 2:** *Computational Complexity*: To simplify security operation, each user wants to use a small number of public keys to encrypt the outgoing messages, and a small number of private keys to decrypt incoming messages. Therefore, we have the following objective

$$\begin{cases} min \quad \max_{i \in V} |\mathcal{K}_i^{priv}| \\ s.t. \quad \mathcal{K}_i \nsubseteq \mathcal{K}_j, \mathcal{K}_i \nsupseteq \mathcal{K}_j (\forall i \neq j) \text{ and } |\mathcal{K}| \leq M \end{cases} \qquad (16.3)$$

where $M$ is the total number of memory slots for key storage at each node.

We have proved that isometric allocation of keys performs better than non-isometric allocation in terms of *Objective 1* and *Objective 2* in [13]. Therefore, **we assume isometric key allocation throughout the section**.

**Objective 3:** *Resilience Requirement*: Under isometric key allocation scheme, we denote $a = |\mathcal{K}|$ and $b = |\mathcal{K}_i| = |\mathcal{K}_i^{priv}|$. Each user needs only to carry $b$ private keys and $a$ public keys under isometric key allocation, wherein $b \ll a \ll (a + b) \ll n$. Clearly, if a node is compromised, all its keys are compromised, regardless of the number of private keys it carries. Therefore, on average $C(k_c(x), b)$ distinct key-sets are compromised when adversaries break into $x$ nodes, wherein $k_c = \lfloor a - (a - b) \left( \frac{(a-b)}{a} \right)^{x-1} \rfloor$ is the number of disclosed keys [13].

We denote a vulnerability metric by $V_x(a, b)$, which is the percentage of communications being compromised when $x$ nodes are broken in. It follows that $V_x(a, b)$ is $\frac{C(k_c(x), b)}{C(a,b)}$ on average. We define the resilience requirement as

$$V_x(a, b) = \frac{C(k_c(x), b)}{C(a, b)} \leq \mathcal{P}, \qquad (16.4)$$

where $\mathcal{P}$ is the resilience bound representing the upper-bound of the compromised communications when $x$ nodes are randomly compromised, each with equal likelihood.

We observe that $C(k_c(x), b)$ and $C(k_c(x), b)$ do not compare favorably with $x$. But by increasing the value of $a$, we can make $C(a, b) \gg n$, therefore making $V_x(a, b)$ compared favorably with $x/n$, which we refer to as *benchmark resilience.* There is a trade-off between memory usage and resilience against break-ins: For a larger number of public–private key pairs, we can get better resilience against break-ins at the cost of larger memory footprint.

## 16.3.4 Key Allocation Algorithm

In this section we show: (1) For a given network, how to determine $a$ and $b$; (2) How to allocate distinct private key sets to users to achieve secure communication between each pair of users. To determine value of $a$ and $b$, we first specify an algorithm to obtain the optimal key allocation solution in terms of both *Objective 1* and *Objective 2*, with the resilience requirement constraint specified in *Objective 3*. Observing the trade-off between memory usage and resilience against break-ins, we then present an algorithm to fully utilize memory space to achieve better resilience by slightly relaxing the optimality of *Objective 1* and *Objective 2*. With the given value of $a$ and $b$, we will discuss key allocation details of *SMOCK*.

### 16.3.4.1 Optimization of Design Objectives

Value $b$ affects the complexity of encryption and decryption. Therefore, we would like to relax $a$ to allow $b$ to be small. The extreme case is that $a = n$ and $b = 1$,[1] where each user keeps a key and every key only has a single copy. The following algorithm helps to determine $a$ and $b$ to achieve the design objectives. Assume the network size is $n$.

*Objective 1* requires $\frac{a}{n}$ to be small for key storage efficiency. Meanwhile *Objective 3* requires $\frac{a}{b}$ to be large for good resilience. Therefore, there are two conflicting objectives. *Algorithm 16.1* trades off between memory efficiency and good resilience.

---

**Algorithm 16.1 Determine value $a$ and $b$**

(1)  Initialize $l = 2$.
     While $(C(l, \lfloor \frac{l}{2} \rfloor) < n)$ do $\{l = l + 1\}$;
     $a = l, \quad b = \lfloor \frac{l}{2} \rfloor$;
(2)  While $(C(a, b - 1) > n)$ do $\{b = b - 1\}$;
(3)  While $(C(a + 1, b - 1) > n)$
     do $\{a = a + 1, \quad b = b - 1\}$
(4)  While (Equation (16.4) is not satisfied)
     do $\{$
        if$(C(a + 1, b - 1) > n)$
          $\{a = a + 1, \quad b = b - 1\}$
        else $\{a = a + 1\}$
     $\}$
(5)  $|\mathcal{K}| = a$ and $|\mathcal{K}_i| = b$.

---

[1] This is exactly the traditional public key management scenario.

Step (1) of Algorithm 16.1 calculates the minimum number of memory slots to store public keys in order to support the secure communication among $n$ nodes. Step (2) minimizes *Objective 1*. Step (3) further optimizes the *Objective 2* while keeping *Objective 1* unchanged. Step (4) ensures that the key allocation meet *Objective 3*. If the resulting $a$ and $b$ do not satisfy the resilience requirement specified by *Objective 3*, we either increase $a$, or simultaneously increase $a$ and decrease $b$. Thus $\frac{a}{b}$ is increased by $\frac{1}{n}$ and $\frac{a}{b}$ is increased by $\frac{1}{b}$ or $\frac{b+1}{b(b-1)}$. For $n \gg b$, it is a reasonable trade-off of memory slots to achieve better resilience.

#### 16.3.4.2 Meeting Key Storage Constraint

Total memory slots for key storage are often limited by $M$, where $M$ is large enough to support $n$ nodes. In this case, we should fully utilize the memory slots to optimize *Objective 2* and achieve the best resilience given by $\dfrac{C(k_c(x), b)}{C(a, b)}$ in Equation 16.4. Thus, we come up with Algorithm 16.2.

---

Algorithm 16.2 Determine value $a$ and $b$ with storage constraint

(1)    Let $a = \lceil \frac{2M}{3} \rceil, b = \lfloor \frac{M}{3} \rfloor$;

(2)    While $(C(a + 1, b - 1) > n)$
       do $\{a = a + 1, b = b - 1\}$;

(3)    Then $|\mathcal{K}| = a$ and $|\mathcal{K}_i| = b$.

---

### 16.3.5 Key Allocation

For a given network size $n$, we have determined $a$ and $b$. The key assignment should satisfy $\mathcal{K}_i \not\subseteq \mathcal{K}_j$ and $\mathcal{K}_i \not\supseteq \mathcal{K}_j$, so that the *key allocation* described above can support the pair-wise secure communication for a network of size $n = C(a, b)$. Assuming a single private key can be assigned to at most $y$ nodes, we have $b \times n = a \times y$ (both sides indicate the total copies of private keys in the system). Therefore, $y = \frac{b}{a}n = \frac{b}{a}C(a, b)$. We randomly assign $b$ private keys to network nodes in the key allocation, where a single key should be assigned to at most $\frac{b}{a}C(a, b)$ nodes. Otherwise, we cannot get a valid key allocation. In the example given in Table 16.2, each key is assigned four times, where $a = 5$, $b = 2$. For a key assignment, we just need to assign a random unused private key combination to a node (totally, there are $C(a, b)$ possible combinations). Algorithm 16.3 illustrates the procedure to assign a subset private keys to a node. Note that very small $a$ and $b$ can support a very large network. For example, if we ignore the resilience requirement, $a = 20, b = 4$, the network size can be as large as 4845.

---

**Algorithm 16.3 Key allocation**

---

(1)  For the *i*th node ($i \leq C(a,b)$), randomly select *b* distinct private keys to generate a subset
of keys, where either of these *b* private keys has been assigned more than $\frac{b}{a} C(a,b)$ times;
(2)  If (the generated key set = an assigned key set) Adjust key by key in the generated key set
to get unassigned key set;
(3)  Assign the generated key set to node *i*.

---

## 16.3.6 Secure Communication Protocols

We have shown how to determine *a*, *b* and how to assign private-key set to
a node if network size *n* is given. In this section we specify detailed protocols
used for initialization, communication, and bootstrapping when new nodes
are deployed. The initialization phase is performed before deployment. Since
communication and bootstrapping are on-line procedures, they have to be
very efficient in terms of communication overhead (using a small number of
messages).

### 16.3.6.1 Initialization

The initialization phase is to assign keys and identifications to each node.
A node's identification (ID) is a good indicator to show what subset of
private keys the node carries. If two nodes want to exchange a secure
message, each needs to know the ID of the other. Node IDs do not have
to form a contiguous range. After key allocation, each node knows the
private keys assigned to it, and all the public keys. We label the keys by
numbers $0, 1, 2 \ldots$. Let $keyID_i^j$ be the *i*-th private key held by node *j*. For
each node *j*, we have $keyID_1^j < keyID_2^j < \cdots < keyID_b^j$. The ID field spans
$b \times \lceil \log_2 a \rceil$ bits as shown in Fig. 16.1. Each $keyID_i^j$ takes $\lceil \log_2 a \rceil$ bits. It
is clear that the node ID is unique as long as each node is assigned a
unique subset of private keys.

In the example shown in Table 16.2, user 7's private key set is
$\mathcal{K}_7^{priv} = \{k_{priv}^2, k_{priv}^5\}$. Correspondingly, the ID of the user 7 is "010|101", where
$a = 5, b = 2$. We can see that a node automatically obtains an ID after it has been
assigned a private-key set. If other peer nodes know user 7's ID, they can infer
that user 7 has private key number 2($k_{priv}^2$) and private key number 5($k_{priv}^5$). If user
7 claims a fake identity, other nodes will use public keys represented by the fake
identity to encrypt the messages. Therefore, the user 7 cannot decrypt the
message. In this way, *SMOCK* scheme is able to resist against the *Sybil* attack.

| $keyID_1^j$ | $keyID_2^j$ | ... | $keyID_b^j$ |
|---|---|---|---|

**Fig. 16.1** ID field of node *j*

### 16.3.6.2 Secure Communication

Figure 16.2 shows a protocol of secure communication between Alice and Bob, where Alice and Bob establish a secure communication channel. If Alice already knows Bob's ID, she can send an encrypted message (EncMsg) directly to Bob. Otherwise, she needs to send a ID request message to Bob, and Bob replies with his ID (possibly in plain text). After Alice receives Bob's ID, she can figure out which private keys Bob is associated with, and she encrypts the message correspondingly before she sends the message.

### 16.3.6.3 Bootstrapping to Accommodate New Nodes

In some cases, we need to deploy new nodes to an existing ad hoc network. In *SMOCK*, it is easy to make previously deployed nodes to trust newly deployed nodes. Let us assume that $n$ nodes are already deployed in a network with $a$ public keys and each node stores $b$ private keys, and $m$ new nodes are being assigned into the network. If $n + m(a, b)$ and resilience requirement (Equation (16.4)) are still satisfied after we deploy $m$ more nodes, then no bootstrapping is necessary, since the newly deployed nodes can be assigned with unused combinations of private keys from the existing key pool owned by off-line trusted server before they are deployed. However, if network size $n + m(a, b)$ or resilience requirement is violated after incremental deployment, then the system needs to generate more key pairs, say $a'$ new key pairs. We can still assign $b$ private keys to the additional nodes before their deployment. Additionally, the newly generated public keys and a' are broadcast to those previous deployed nodes. Since we fix $b$, those previously deployed nodes can adjust the existing ID field to span $b \times \lceil \log_2(a + a') \rceil$ bits.

It can be verified that, given $C(a,b)$, the increment of $a$ by 1 brings $C(a, b-1)$ new valid key sets for new nodes. Therefore, with $a'$ new key pairs, the network is able to accommodate $\sum_{i=0}^{a'-1} C(a + i, b - 1)$ new nodes. Note that keeping $b$ unchanged and increasing $a$ does not violate the resilience bound $\mathcal{P}$ given in *Objective 3*.

## 16.3.7 Evaluations

Next, we will show the performance evaluation of *SMOCK* based on key storage usage, communication overhead and resilience to break-ins.

**Fig. 16.2** Secure
Communication Protocol
Between Alice and Bob

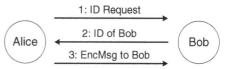

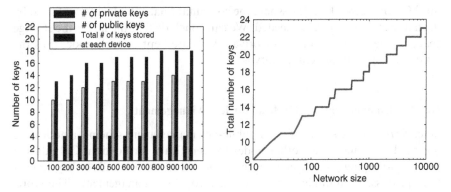

**Fig. 16.3** The minimum number of keys needed

### 16.3.7.1 Small Memory Footprint

In *SMOCK*, a few key pairs can support secure communication of a very large network. According to *Algorithm 16.1*, 18 key pairs in the network can support end-to-end secure communication among up to 1000 nodes without resilience consideration. In Fig. 16.3, we show the minimum number of keys needed at each node for typical mission-critical network sizes. Therefore, we can achieve very small memory footprint under the *SMOCK* scheme.

A total of $a$ public keys can support at most $C(a, \lfloor \frac{a}{2} \rfloor)$ nodes in the network. By Stirling's Approximation, the total number of key pairs required is at a level of $\Theta(\lg n)$, which can be verified by Fig. 16.3. We conclude that the *SMOCK* scheme yields very small memory footprint.

If we relax the storage limitation, the number of private keys needed decreases, and computational complexity is reduced accordingly. Figure 16.4 shows the tradeoff between computational complexity and key storage space

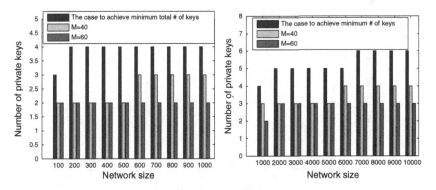

**Fig. 16.4** Tradeoff between storage space and computational complexity ($M$ is total memory slots for key storage)

for different network scales, where the computational complexity is inferred by the number of private keys needed. We can conclude that the larger the storage space is, the smaller number of private keys are kept at each node, thus the smaller computational complexity is.

### 16.3.7.2 Communication Overhead for Key Management

Since $SMOCK$ is a self-contained public-key management scheme, a node does not need to contact/trust other nodes for certificate verification. Only during the bootstrapping phase when new nodes join the network and the key revocation process, communication is needed for key management. Therefore, $SMOCK$ has little communication overhead for key management.

### 16.3.7.3 Resilience to Break-ins

The break-in of any single node by an adversary does not release enough information to the adversary to break secure communication for any pair of nodes. However, break-ins of multiple nodes may compromise a set of other nodes. Assume $x$ nodes are compromised and $k_c(x)$ is the expected number of keys disclosed correspondingly. As shown before, $k_c(x) = \lfloor a - (a - b)\left(\frac{a-b}{a}\right)^{x-1} \rfloor$. Then $\frac{C(k_c(x),b)}{C(a,b)}$ percentage of nodes will be compromised. Let us assume $n = 100$, Fig. 16.5 shows the average case percentage of compromised nodes when a small portion of nodes are controlled by adversaries.

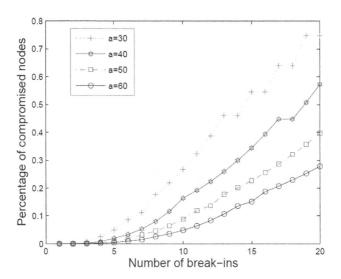

**Fig. 16.5** Percentage of compromised nodes with break-ins ($b = 4$)

## 16.4  Broadcast Authentication with Resource Constraints

Public key scheme and one-way hash function scheme (TESLA) are two common cryptographic primitives for broadcast authentication (Other alternatives are extensively surveyed in [24]). TESLA is an efficient protocol that utilizes one-way hash chains and delayed key disclosure to authenticate broadcast traffic. However, TESLA incurs security vulnerability if used together with the "authenticate-first" policy. This is because at the time the intermediate nodes are able to authenticate a packet *Msg*, the secret hash key *hKey* used to sign *Msg* is already released by the source. Afterwards, no nodes can trust any newly received packets signed by*hKey*. Hence, the intermediate hop has to resign *Msg* under his identity before forwarding. By manipulating resigning process, a compromised node could flood as many packets as it wants and viciously claim that those packets are sent by other innocent sources.

In this section, we present a novel broadcast authentication scheme, called *DREAM*[16]. It effectively limits false data injection via frequently using "authenticate-first" policy based on public-key authentication. It also reduces the end-to-end delay by allowing a small percentage of unverified packets forwarded probabilistically via "forward-first" policy so that remote nodes obtain the broadcast messages quickly. The assumption we make are : (a) Devices are static or with moderate mobility. (b) Free-to-move attackers want to permeate as many false packets throughout the network as possible.

### 16.4.1  Main Idea

*DREAM* addresses the end-to-end delay issue by relaxing the containment requirement. The main idea is to allow a small and controlled number of packets transmitted to remote locations quickly without authentication in a probabilistic manner. A path segment over which packet *P* is not validated before forwarding is referred to as an unverified forwarding path for *P*. Nodes along the unverified forwarding paths forward packets before authenticating; while the other nodes, representing the majority of the network, authenticate packets before forwarding. Those unverified transmissions virtually reduce the network radius from the broadcast sources and thus decrease end-to-end delay.[2]

DREAM is integrated with broadcast protocol (because of the diversified suppression policies of broadcast protocols) and independently runs at each device. The architecture of DREAM contains three modules as shown in Fig. 16.6.

---

[2] This may introduce out-of-order packets with large delay variance. We assume that applications will reorder packets if in-order delivery is required.

**Fig. 16.6** DREAM
architecture

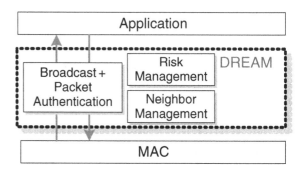

- *Packet Authentication Module* is responsible for signing and validating packets. There are two operating modes yielding different delay and containment trade-offs.
- *Risk Management Module* continuously monitors the contextual threat. When the evidence of false packet injection shows up, the module adjusts the operating mode in Packet Authentication Module to a more defensive and secure mode.
- *Neighbor Management Module* periodically exchanges *hello* messages with one-hop neighbors. *Hello* messages flag both a node's liveness and its one-hop neighborhood size. In order to prevent malicious tampering, hello messages are signed and verified.

Next two sections are devoted to Packet Authentication and Risk Management modules separately.

## *16.4.2 Packet Authentication*

Packet Authentication Module comprises two parts: (a) signing at sources and (b) verification and forwarding at receivers. Whenever a source sends out a broadcast packet, it signs the packet. Upon receiving a broadcast message, a node probabilistically determines to forward it first or to authenticate it first. When a node authenticates the packet first, we call the node an authenticating node. No matter which choice is made, each node will (a) only forward a unique broadcast packet at most once (it does not forward identified false packets); and (b) validate the packet and send the authentic one to applications. The message format is:

$$ID_{src}, Seqno, Msg, PubKeySign_{src}, Ht, Id_{fwder}$$

wherein

$$PubKeySign_{src} = Sign^{src}_{PubKey}(ID_{src}, Seqno, Msg)$$

$ID_{src}$ and $ID_{fwder}$ are IDs of the source and the last forwarder, respectively. When the source signs $Msg$, $ID_{fwder}$ is set to $ID_{src}$ since source is the last forwarder. $PubKeySign_{src}$ is the public key signature signed by the source and is never changed during forwarding process. $HT$ is the number of hops traversed since the last authenticating node. It is reset to 0 at every authenticating node. Sources always set $HT$ to 0. Each forwarder, who forwards the unverified copies, increases $HT$ by 1. Here, we assume that receivers know the public key of the source; otherwise, sources' certificates should be broadcast as well.

Next, we will describe *packet verification and forwarding* algorithm at receiver sides based on flooding.

We reduce average end-to-end delay at the cost of imperfect containment. Therefore, it is crucial to carefully control unverified forwarding. We have the following requirements:

- The decision is independently made to avoid message negotiation.
- The decision is probabilistically made to achieve load balancing and to prevent a single point of failure.
- The number of hops that an unverified message is allowed to travel is controlled so that, on one hand, broadcast packets spread out in space fast and, on the other hand, false injected packets are contained near their originators.
- The number of "forward-first" nodes is kept small to reduce the cost of communication and public key computation wasted on false packets.

Public key signature verification is an expensive operation and usually takes time in seconds or tens of milliseconds in resource-constrained wireless mobile devices. Hence, a *verification queue* is placed in DREAM to buffer the packets waiting for signature verification. We assume for simplicity that, at any moment, only one signature verification is performed so that spare CPU resources are reserved for other tasks. A *verification demon* process, whenever free, continuously monitors the verification queue. Once a packet is found at the head of the queue, the verification process removes the packet from queue, verifies its public key signature, and relays the authentic one to applications. In addition, if the authentic packet has not been forwarded yet, it rebroadcasts the packet not under suppression. When the demon process completes processing a packet, it becomes free.

Upon receiving a broadcast message $m$, a node $\nu$ probabilistically determines whether to forward $m$ first or authenticate $m$ first according to Algorithm 16.4. *Rand* is a random number generated uniformly from [0, 1]. $b$, $c$ and $K$ are system parameters. $b$ and $c$ are the expected numbers of neighbors in the one-hop neighborhood of the source and the last forwarder $m.ID_{fwder}$ (other than the source), respectively, who forwards $m$ first. $K$ is the maximum number of hops that $m$ is allowed to travel without verification.

---

**Algorithm 16.4** Verification and forwarding algorithm

**input:** An overheardpc broadcast message $m$

1  **if** (*overheard messages with the same* $(m.ID_{src}, m.seqno)$ *before*) **then** return;

2  **if** ($m.HT == 0$ & $m.ID_{fwder}$ *is unknown neighbor*) **then** return;

3  **if** (*I am one-hop neighbor of* $m.ID_{src}$) **then**

4      $prob = b/ \mid Nbr(m.ID_{fwder}) \mid$;

5  **else**

6      $prob = 2 * c/ \mid Nbr(m.ID_{fwder}) \mid$;

7  **end**

8  **if** (*Rand* $> prob$ or $m.HT == K$) **then**

9      // *authenticate m first;*

10     $m.HT = 0$;

11     place $m$ into verification queue;

12 **else**

13     // *forward m first;*

14     $m.HT + +$;

15     rebroadcast $m$;

16     place $m$ into verification queue

17 **end**

---

Algorithm 16.4 incorporates 2 steps:

(i) Suppression and Filtering (*Line 1–2*): If $\nu$ has received at least one message with the same sequence number from the source before, message $m$ is dropped. Thereby, a node forwards each broadcast message at most once. If $\nu$ is one-hop away from the last authenticator, which is an unknown neighbor, message $m$ is ignored due to lack of trust.

(ii) Probabilistic Pruning (*Line 3–17*): $\nu$ decides whether to forward $m$ first or authenticate $m$ first probabilistically. $|Nbr(m.ID_{fwder})|$ is the neighborhood size of last forwarder $m.ID_{fwder}$. This value is available via *Neighbor Management module*. $b$ is usually set to an appropriate value so that unverified broadcast messages can cover most directions (4 for instance). $c$ is selected from [1, 2] to make the event that all the one-hop neighbors are pruned unlikely. Both $b$ and $c$ cannot be too large, because we need to control the number of unverified packets. The constant 2 before $c$ in line 6 accounts for the fact that on average, half of the neighbors of the last forwarder have already heard $m$. Considering a case that $m$ is forwarded from node $A$ to $\nu$, via $B$, averagely half neighbors of $B$ have already heard $m$ broadcast by $A$, thus suppressing $m$ according to line 1.

If $m$ has already traversed $K$ hops without verification, it is better to authenticate $m$ first to confine its permeation in case of a false packet. If $\nu$ decides to authenticate $m$ first, it resets $HT$ field and places $m$ in the verification queue. If $\nu$

decides to forward $m$ first, it increases $HT$ field by 1, forwards $m$ and then places $m$ in the verification queue. The verification process is responsible for rebroadcast after signature verification. If $m.HT > 0$, verification process knows that $m$ has already been forwarded once.

Since we use $|Nbr(m.ID_{fwder})|$ to make probabilistic forwarding decision (in Line 4 and 6), network topology should be relative stable so that during the Hello period, the neighbor information is accurate. Otherwise, inaccurate and outdated neighborhood sizes may result in more or less number of nodes forwarding the packet *first* from the last forwarder.

### 16.4.3  Risk Management

Nodes working as in previous section are said to be in *Normal Mode*. Via signature verification, they are able to detect emergence of false packet attacks if the number of packets received with invalid signature in a time interval exceeds a predetermined threshold. They switch to hop-by-hop authentication scheme and totally disable the "forward first" policy to contain false packet injection in a defensive way. They are then said to be in *Alert Mode*. Conversely, when the false packet attack lessens, nodes switch back to Normal Mode to trade for improved end-to-end delay. The transition between two modes is shown in Fig. 16.7.

Each node continuously monitors the number of detected false packets every *riskWindow* interval. The node switches to Alert Mode if this number reaches $\alpha$ and switches back to Normal Mode if this number drops to $\beta$. $\alpha$ and *riskWindow* are related to the tolerable percentage of CPU resource spent in evaluating false packets before switching to Alert Mode. Suppose the processing time to validate one public key signature is $T_{val}$ seconds, a node can tolerate

$$\frac{T_{val}{}^{*}\alpha}{riskWindow}$$

CPU resource wasted on validating false packets. For mission critical networks, this percentage can be set to a small value. $\beta$ is there to avoid frequent switches and instability.

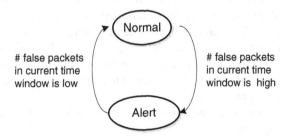

**Fig. 16.7** Adaptation to contextual threat

Switching to Alert Mode adversely increases end-to-end delay. However, we isolate the rest of network from infection and their computation and communication resources are protected. It is expected that, when converging, only the nodes around the attackers enter Alert Mode. Other nodes can still receive the packets through alternative broadcast paths quickly.

### 16.4.4 Evaluation

In this section, we evaluate the performance of DREAM in ns2 network simulator by comparing it with hop-by-hop authentication scheme and Dynamic Window (DW) scheme proposed in [30]. In Dynamic Window scheme, the forwarding decision is made by comparing the size of a locally maintained dynamic window on sensor nodes and the number of hops the incoming message traversing after its last authentication: if window size is the larger, they use forwarding-first; otherwise, they use authentication-first. Additive Increase Multiplicative Decrease (AIMD) technique is used to dynamically manage the window on sensor nodes: if an authentic message is received, the window size increases; otherwise, window size decreases.

The criteria of our evaluation represents two aspects: penalty on authentic messages and containment capability of false messages. We have the following metric for the legitimate packets:

- The average end-to-end delay: End-to-end (authentication) delay of packet $m$ from broadcast source $src$ at node $\nu$ is defined as the interval between the moment $src$ broadcasts $m$ into wireless networks and the moment $\nu$ finishes verification of $m$.

We have the following metrics for false packets:

- The number of nodes forwarding the false packets.
- The number of nodes receiving the false packets.
- The number of public-key signature validation.

The default public-key signature validating time is 0.5 s. At any moment, a device can perform only a single public key validation and all the other validation requests (up to 50) are queued. The public-key signature has size 40 bytes. Each node is equipped with an omni-directional antenna operating on a single channel. The channel model we use in simulation is two-ray path-loss propagation model. The broadcast traffic is sent via CBR (constant bit rate) traffic with packet size 64 bytes in UDP. 802.11 DCF Medium Access Control protocol is used with default configuration. Transmission range is 250 m.

We investigate two network topologies, i.e., grid topology and random topology.

### 16.4.5 Grid Topology

The first scenario we study is grid topology and each data point is averaged over three runs of simulations. Since delay penalty is critical in large-scale networks with long diameter, we choose to place 400 nodes in a grid. Each row (column) contains 20 nodes equally spaced with distance 240 m apart. There is a single broadcast source located at the left top corner. It continuously sends CBR traffic at the rate of a packet every 2 s. The average network radius from the broadcast source is 10 hops based on the setting.

First, we study end-to-end authentication delay for legitimate traffic in the absence of attackers. We vary $K$ from 3 to 5 and $c$ from 1.0 to 2.0. Due to space limitation, we do not evaluate the sensitivity to $b$ and $b$ is set equal to $c$. The length of *riskWindow* is 6 s. $\alpha = 6$ and $\beta = 1$. For Dynamic Window Scheme, the initial window size is set to 64. Additive increase and multiplicative values are 1 and 2 separately, default as in authors' paper.

As shown in Fig. 16.8, the average end-to-end delay in hop-by-hop authentication is the worst, above 4.6 s because the average length of paths from the broadcast source to destinations is 10 hops and the public-key validation delay is 0.5 s. Clearly, the average end-to-end delay in Dynamic Window scheme is optimal at about half a second. Because there are no false packets detected, the window size is at least 64, which is always greater than $HT$, the number of hops traversed since last authenticator. The performance of DREAM is shown by the middle four lines. As $c$ increases, the end-to-end delay decreases since more

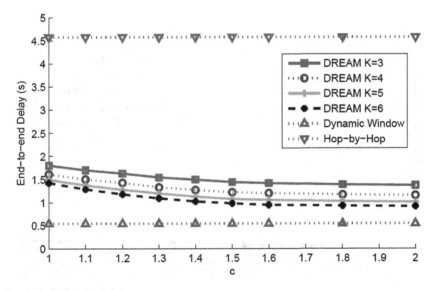

**Fig. 16.8** End-to-End delay

number of nodes are inclined to forward the packets first, thus virtually reducing the end-to-end path length. However, as $c$ increases above 1.5, the degree of improvement levels off. When $K$ increases, the end-to-end delay decreases because the distance between two successive authenticators is lengthened and the entire network can be quickly covered.

Next, we study the containment capability of DREAM and its response to malicious injection from Fig. 16.9 to 16.10. Clearly, if attackers always flood less than $\alpha$ false packets in *riskWindow* interval, DREAM never switches to Alert Mode. However, this is not the best interest for attackers. Therefore, we focus on cases of high-rate false injection. Again, we use the same configuration

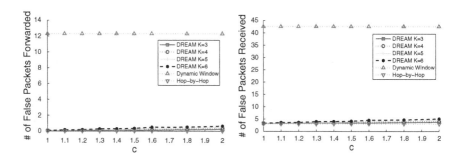

**Fig. 16.9** Resilience against false injection

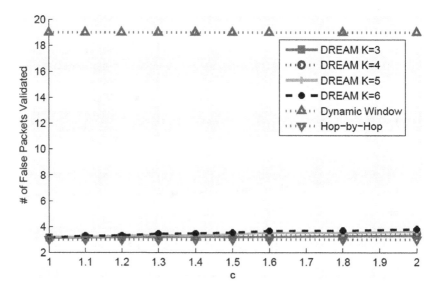

**Fig. 16.10** Normalized number of false packets validated

as before except that an attacker is flooding simultaneously near the source at the left top corner of grid. The attacker floods packets at the rate of 1 packet per second. During the false injection attack, it is out of question that all the neighboring nodes have to receive the false packets and validate them. But our scheme can effectively confine the false packets within a small number of nodes. The number of false packets is normalized by the total number of false packets sent by the attackers. The normalized value is below 45 for both DREAM and Dynamic Window scheme since they both switch to defensive mode. However, before the dynamic window size in DW scheme is adjusted down, many false packets are broadcast to the whole network. Furthermore, the mixing traffics of good and false packets interdict the slowdown of dynamic window. For every packet sent by attacker, Dynamic Window scheme has 12 nodes to forward, 43 nodes to receive, and 19 nodes to verify the message. On the contrary, the containment performance of DREAM is close to hop-by-hop authentication. The infection of false injection increases as $c$ or $K$ increases. However, waste on both transmission and computational resources by the false packets are under control.

The end-to-end authentication delay in presence of the attacker is shown in Fig. 16.11. Delay for both DREAM and Dynamic Window scheme increases, compared with the case in absence of attackers.

From the above measurements, we see clear tradeoff between end-to-end authentication delay and containment capability, i.e. public-key signature computational overhead and forwarding overhead for false packets.

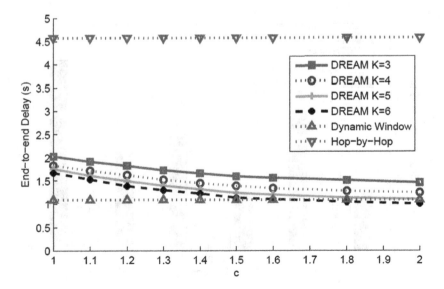

**Fig. 16.11** End-to-End delay

### *16.4.6 Random Topology*

The second scenario we study is random topology with 400 nodes randomly placed in a 4000 × 4000 m network. We make sure that the resulting topology is almost connected. There are 1 broadcast source randomly selected to send CBR traffic at rate of 2 packets per 3 s and two attackers randomly selected to send CBR traffic at rate of 1 packets per 2 s each. Nodes switch to Alert Mode whenever the number of false packets in 6-second interval reaches 3 and falls back to Normal Mode when this number drops to 1. Since the processing time for one public-key signature validation is 0.5 s, a node can tolerate at most $3 \times 0.5/6 = 25\%$ CPU spend in evaluating false signatures.

The end-to-end delay for two example configurations of DREAM is compared with the delay for Dynamic Window and hop-by-hop authentication schemes in Fig. 16.12. $X$-axis shows three random scenarios with different location deployments. As usual, hop-by-hop authentication has the worst end-to-end delay for authentic traffic. DREAM with $c$ set to 1.3 halves the end-to-end delay with further improvement with $c$ set to 1.5. Even though Dynamic Window Scheme has the best performance in terms of end-to-end delay, it is not resilient to false packet injection mixed with the legitimate traffics. As shown in Table 16.3, for the total 330 false packets, Dynamic Window scheme forwards more than 5000 false packets in order to achieve the desired delay. But DREAM only forwards tens of false packets before nodes around attackers switch to Alert Mode. Due to space limit, we do not place the results for the number of false packets received and validated here. Those two results are similar as the number of false packets being forwarded.

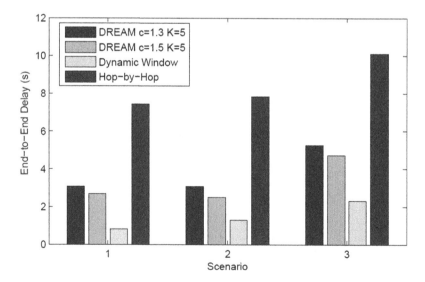

**Fig. 16.12** End-to-End delay

**Table 16.3** Normalized number of false packets forwarded

|               | Scenario I | Scenario II | Scenario III |
| ------------- | ---------- | ----------- | ------------ |
| DREAM $c=1.3$ | 28         | 12          | 12           |
| DREAM $c=1.5$ | 30         | 25          | 32           |
| Dynamic Window | 6992      | 5768        | 14207        |
| Hop-by-Hop    | 0          | 0           | 0            |

## 16.5 Thoughts for Practitioners

We implemented SMOCK under the context of Trustworthy Cyber Infrastructure for Power grid (TCIP) by C language in Linux operating system. In our implementation, all SMOCK public keys are sent via SSL channel from trusted authority before secure communication. We measured the time taken to encrypt and decrypt a message in a standard IBM T60 laptop. For a pair of private keys held per node, the measured encryption and decryption time are 9.82 ms and 4.88 ms for 1024-bit key and 32-byte data packet, respectively. However, if applied to low-end sensors or hand-held devices, encryption and decryption time could be much longer, in the order of seconds. SMOCK can then be used to negotiate light-weighted session keys, such as symmetric keys.

Bearing the similar deficiency as dynamic window scheme, DREAM relies on public key cryptography. For low broadcast traffic rate and infrequent communication, it is fine. However, how to build efficient broadcast authentication using light-weighted cryptographic primitive is still needed.

## 16.6 Direction for Future Research

SMOCK presents our initial effort to make memory overhead for public key pre-distribution scalable to a large number of devices. Future improvement can be done in the following directions: (1) to perform private key combination update upon compromise detection; (2) to further reduce memory consumption by location-aware combination assignment. (For example, a group of nodes, which are likely to communicate with each other, is assigned a set of private-key combinations with many keys shared.)

Idea of DREAM to reduce end-to-end authentication delay is applicable to other broadcast protocols than flooding. Further effort can be made to integrate DREAM with advanced broadcast forms [28, 31] and test their performance. DREAM relies on probabilistic decision to push unverified packets remotely. But the unverified forwarding path may not be optimal for quick coverage. GPSR routing protocol [18] can be used to deterministically forward unverified packets further along the best direction. Additionally, counter measurement for attacks on broadcast suppression is worthy investigation, wherein attackers pretend to be the broadcast source and send packets with higher or equal sequence numbers before the true source.

## 16.7  Conclusions

Security is a challenging and important issue for wireless ad hoc networks. To design better security solution, we need to correctly understand the network model and attacker model. Moreover, security incurs overhead. Resource constraints of mobile devices, such as memory, computation, communication and energy, needs to be carefully considered in the solution. To balance the resource constraint, security, and real-time performance requirement, adaptivity is a promising way for the mission critical wireless network.

## Terminologies

1. Self-contained key management scheme
   A key management scheme where all necessary cryptographic keys (certificates) are stored at individual nodes before deployment.
2. Key Pool
   A set of public–private key pairs generated at trusted server in the initialization phase.
3. Private key set
   A set of private keys held by a user
4. A valid key allocation
   A key allocation satisfying the following property:

$$\forall i \forall j \, \text{not} \, (K_j \subseteq K_j \text{ or } K_j \subseteq K_i), \text{ if}$$

5. Isometric key allocation
   A key allocation is isometric, if $|K_1| = |K_2| = \cdots = |K_n| = b$.
6. Vulnerability metrics
   The percentage of communications being compromised when $x$nodes are broken in.
7. False injection containment
   Faked messages are detected and dropped near the injector.
8. Hop-by-Hop Message Authentication
   An authentic packet is forwarded to next-hop after being validated. False packets are dropped after validation failure.
9. Normal Mode
   Nodes probabilistically forward some packets without authentication so as to achieve small end-to-end delay by trading imperfect false packet containment capability.
10. Alert Mode
    Upon emergence of false packet attacks, nodes switch to hop-by-hop authentication and work in a defensive way

## Questions

(1) Give a formal definition for vulnerability to compromise. For different key management schemes mentioned in Section 16.2 and Section 16.3, please compare their vulnerability metrics.

(2) How to eliminate collision of private key sets assigned to a pair of users?

(3) Which key management schemes do not support full connectivity?

(4) Compare SMOCK and certificate approach?

(5) In random key distribution mechanism, the probability that any pair of nodes possesses at least one common key is

$$1 - \frac{((K-k)!)^2}{(K-2k)!k!}$$

In the modification of basic random key distribution scheme, multiple common keys are needed to establish a secure link in key-setup phase so that resilience against node compromise improves. What is the probability that any pair of nodes possesses at least i common keys? K and k are defined as before.

(6) Suppose, $b=3$, derive the value of for a network size of 10 000 without considering memory and resilience objectives. In this setting, what is the size of ID field? What is the ID for Node 300 if it has been assigned private key set $\{k_{priv}^{30}, k_{priv}^{2}, k_{priv}^{15}\}$.

(7) Explain why hop-by-hop TESLA does not work in broadcast authentication.

(8) In DREAM, why a packet includes HT field?

(9) Design a packet authentication scheme with containment capability for unicast data traffic.

(10) Is DREAM resilient to compromise?

## References

1. R. Blom. An optimal class of symmetric key generation systems. *Lecture Notes in Computer Science, Springer-Verlag*, 1985.

2. S. A. Camtepe and B. Yener. Combinatorial design of key distribution mechanisms for wireless sensor networks. *In Proceedings of 9th European Symposium On Research in Computer Security (ESORICS '04)*, 2004.

3. S. Capkun, L. Buttyan and J. P. Hubaux. Self-organized public-key management for mobile ad hoc networks. *IEEE Transactions on Mobile Computing*, 2003.

4. H. Chan, A. Perrig and D. Song. Random key predistribution schemes for sensor networks. *In IEEE Symposium on Research in Security and Privacy*, 2003.

5. F. Delgosha and F. Fekri. Threshold key-establishment in distributed sensor networks using a multivariate scheme. *INFOCOM*, 2006.

6. J. Drissi and Q. Gu. Localized broadcast authentication in large sensor networks. *icns*, 0:25, 2006.

7. W. Du, J. Deng, Y. S. Han, S. Chen and P. K. Varshney. A key management scheme for wireless sensor networks using deployment knowledge. *IEEE INFOCOM*, 2004.
8. W. Du, J. Deng, Y. Han, P. Varshney. A pairwise key pre-distribution scheme for wireless sensor networks. *In Proceedings of 10th ACM Conference on Computer and Communications Security (CCS)*, 2003.
9. L. Eschenauer and V. Gligor. A key management scheme for distributed sensor networks. *IEEE Symposium on Security and Privacy*, 2002.
10. L. Eschenauer and V. D. Gligor. A key-management scheme for distributed sensor networks. *In Proceedings of the 9th ACM Conference on Computer and Communications Security*, 2002.
11. Q. Gu and J. Drissi. Dominating Set based Overhead Reduction for Broadcast Authentication in Large Sensor Networks. *ICNS '07: Proceedings of the Third International Conference on Networking and Services*, pages 81, Washington, DC, USA, 2007. IEEE Computer Society.
12. V. Gupta, M. Millard, S. Fung, Y. Zhu, N. Gura, H. Eberle and S. Chang. Sizzle: A standards-based end-to-end security architecture for the embedded internet. *In Proceedings of the 3rd IEEE Percom*, 2005.
13. W. He, Y. Huang, K. Nahrstedt, and W.C. Lee. SMOCK: A self-contained public key management scheme for mission-critical wireless Ad Hoc networks. *Percom*, 201–210, 2007.
14. Y.-C. Hu, A. Perrig, and D.B. Johnson. Ariadne: A secure on-demand routing protocol for Ad Hoc networks. *Proceedings Fourth IEEE Workshop on Mobile Computing Systems and Applications*, 2002.
15. Y.-C. Hu, A. Perrig and D.B. Johnson. Ariadne: A secure on-demand routing protocol for Ad Hoc networks. *Wireless Networks*, 2005.
16. Y. Huang, W. He, K. Nahrstedt, and W.C. Lee. DoS-resistant broadcast authentication protocol with low end-to-end delay. *UIUC Technical Report, UIUCDCS-R-2008-2953 March 2008*.
17. G. Gaubatz, J. Kaps and B. Sunar. Public keys cryptography in sensor networks—revisited. *In The Proceedings of the 1st European Workshop on Security in Ad-Hoc and Sensor Networks (ESAS)*, 2004.
18. B. Karp and H.T. Kung. GPSR: greedy perimeter stateless routing for wireless networks. *Mobile Computing and Networking*, 243–254, 2000.
19. S. Kent and T. Polk. Public-key infrastructure (x.509) (pkix) charter. *Available at* http://www.ietf.org/html.charters/pkix-charter.html.
20. J. Kong, and P. Zerfos and H. Luo and S. Lu and L. Zhang. Providing robust and ubiquitous security support for mobile ad-hoc networks. *In Proceedings of the 9th IEEE International Conference on Network Protocols (ICNP)*, 2001.
21. D. Liu and P. Ning. Establishing pairwise keys in distributed sensor networks. *In Proc. of 10th ACM Conference on Computer and Communications Security (CCS)*, 2003.
22. D. Liu and P. Ning. Improving key predistribution with deployment knowledge in static sensor networks. *ACM Transaction Sensor Network,*, 1(2):204–239, 2005.
24. D. Liu, P. Ning, S. Zhu and S. Jajodia. Practical broadcast authentication in sensor networks. *MOBIQUITOUS '05*, pages 118–132, Washington, DC, USA, 2005. IEEE Computer Society.
24. D. Liu, P. Ning, S. Zhu and S. Jajodia. Practical broadcast authentication in sensor networks. *MOBIQUITOUS '05*, pages 118–132, Washington, DC, USA, 2005. IEEE Computer Society.
25. D.J. Malan, M. Welsh and M.D. Smith. A public-key infrastructure for key distribution in TinyOS based on elliptic curve cryptography. *In The First IEEE International Conference on Sensor and Ad Hoc Communications and Networks*, 2004.
26. A. Perrig, R. Canetti, D. Tygar and D. Song. The TESLA broadcast authentication protocol, 2002.

27. A. Perrig, J.D. Tygar, D. Song and R. Canetti. Efficient authentication and signing of multicast streams over lossy channels. *Proceedings of IEEE Symposium on Security and Privacy'00*, pages 56, 2000. IEEE Computer Society.
28. Y. Sasson, D. Cavin and A. Schiper. Probabilistic broadcast for flooding in wireless mobile ad hoc networks, 2003.
29. P. Traynor, H. Choi, G. Cao, S. Zhu and T.L. Porta. Establishing pair-wise keys in heterogeneous sensor networks. *INFOCOM*, 2006.
30. R. Wang, W. Du and P. Nings. Containing denial-of-service attacks in broadcast authentication in sensor networks. *MobiHoc*, 2007.
31. J.E. Wieselthier, G.D. Nguyen and A. Ephremides. On the construction of energy-efficient broadcast and multicast trees in wireless networks. *INFOCOM (2)*, 585–594, 2000.
32. L. Zhou and Z.J. Haas. Securing Ad Hoc Networks. *IEEE Network Magazine*, 1999.
33. S. Zhu, S. Xu, S. Setia and S. Jajodia. Establishing pairwise keys for secure communication in ad hoc networks: a probabilistic approach. *In Proceedings of the 11th IEEE International Conference on Network Protocols* (Nov. 04–07, 2003). *IEEE Computer Society* (Washington, DC), 326.
34. S. Zhu, S. Xu, S. Setia and S. Jajodia. LHAP: A lightweight hop-by-hop authentication protocol for Ad-Hoc networks, 2003.
35. S. Zhu, S. Setia and S. Jajodia. LEAP: Efficient security mechanisms for large-scale distributed sensor networks. *CCS '03*, pages 62–72, New York, USA, 2003. ACM Press.

# Chapter 17
# Intrusion Detection in Mobile Ad Hoc Networks

Sevil Şen and John Andrew Clark

**Abstract** In recent years, mobile ad hoc networks (MANETs) have become a
very popular research topic. By providing communications in the absence of a
fixed infrastructure, MANETs are an attractive technology for many applications such as rescue operations, tactical operations, environmental monitoring,
conferences, and the like. However, this flexibility introduces new security risks.
Since prevention techniques are never enough, intrusion detection systems
(IDSs), which monitor system activities and detect intrusions, are generally
used to complement other security mechanisms.

Intrusion detection for MANETs is a complex and difficult task mainly due
to the dynamic nature of MANETs, their highly constrained nodes, and the
lack of central monitoring points. Conventional IDSs are not easily applied to
them. New approaches need to be developed or else existing approaches need to
be adapted for MANETs. This chapter outlines the issues of intrusion detection
for MANETs and reviews the main solutions proposed in the literature.

## 17.1 Introduction

Wireless networking is now the medium of choice for many applications. In
addition, modern manufacturing techniques allow increasingly sophisticated
functionality to reside in devices that are ever smaller, and so increasingly
mobile. Mobile ad hoc networks (MANETs) combine wireless communication
with a high degree of node mobility. Limited-range wireless communication
and high node mobility means that the nodes must cooperate with each other to
provide essential networking, with the underlying network dynamically changing to ensure needs are continually met. The dynamic nature of the protocols
that enable MANET operation means they are readily suited to deployment in
extreme or volatile circumstances. MANETs have consequently become a very

S. Şen (✉)
Department of Computer Science, University of York, Heslington, York, YO10 5DD, UK
e-mail: ssen@cs.york.ac.uk

S. Misra et al. (eds.), *Guide to Wireless Ad Hoc Networks*,
Computer Communications and Networks, DOI 10.1007/978-1-84800-328-6_17,

popular research topic and have been proposed for use in many areas such as rescue operations, tactical operations, environmental monitoring, conferences, and the like.

MANETs by their very nature are more vulnerable to attack than wired networks. The flexibility provided by the open broadcast medium and the cooperativeness of the mobile devices (which have generally different resource and computational capacities, and run usually on battery power) introduces new security risks. As part of rational risk management, we must be able to identify these risks and take appropriate action. In some cases we may be able to design out particular risks cost-effectively. In other cases, we may have to accept that vulnerabilities exist and seek to take appropriate action when we believe someone is attacking us. As a result, intrusion detection is an indispensable part of security for MANETs.

Many intrusion detection systems (IDS) have been proposed in the literature for wired networks, but MANETs' specific features make direct application of these approaches to MANETs impossible. New approaches need to be developed or else existing approaches need to be adapted for MANETs. In this chapter, we examine special IDS issues of MANETs and proposed IDSs for MANET-specific systems to find out how well-proposed systems address these issues. In the next section, an introduction to intrusion detection systems is given. Then, intrusion detection on MANETs is discussed along with proposed IDSs. In conclusion, thoughts for practitioners and ideas for future research are given.

## 17.2 Intrusion Detection Systems

Intrusion is any set of actions that attempt to compromise the integrity, confidentiality, or availability of a resource [6] and an intrusion detection system (IDS) is a system for the detection of such intrusions. There are three main components of an IDS: data collection, detection and response.

The *data collection component* is responsible for collection and pre-processing data tasks: transferring data to a common format, data storage and sending data to the detection module [14]. IDS can use different data sources as inputs to the system: system logs, network packets, etc. In the *detection component* data are analyzed to detect intrusion attempts, and indications of detected intrusions are sent to the *response component*.

In the literature, three intrusion detection techniques are used. The first technique is *anomaly-based intrusion detection*, which profiles the symptoms of normal behaviours of the system such as usage frequency of commands, CPU usage for programs and the like. It detects intrusions as anomalies, i.e. deviations from the normal behaviours. Various techniques have been applied for anomaly detection, e.g. statistical approaches and artificial intelligence techniques like data mining and neural networks. Defining normal behaviour is a major challenge. Normal behaviour can change over time and intrusion detection systems must be kept up to date. False positives – the normal activities that are detected as anomalies by IDS – can be high in anomaly-based detection. On the other hand, it is capable of detecting previously unknown attacks. This is very important in an environment where new attacks and new vulnerabilities of systems are announced constantly.

*Misuse-based intrusion detection* compares known attack signatures with current system activities. It is generally preferred by commercial IDSs since it is efficient and has a low false positive rate. The drawback of this approach is that it cannot detect new attacks. The system is only as strong as its signature database, and this needs frequent updating for new attacks. Both anomaly-based and misuse-based approaches have their strengths and weaknesses. Therefore, both techniques are generally employed for effective intrusion detection.

The last technique is *specification-based intrusion detection*. In this approach, a set of constraints on a program or a protocol are specified and intrusions are detected as runtime violations of these specifications. It is introduced as a promising alternative that combines the strengths of anomaly-based and misuse-based detection techniques, providing detection of known and unknown attacks with a lower false positive rate [26]. It can detect new attacks that do not follow the system specifications. Moreover, it does not trigger false alarms when the program or protocol has unusual but legitimate behaviour, since it uses the legitimate specifications of the program or protocol [26]. It has been applied to ARP (Address Resolution Protocol), DHCP (Dynamic Host Configuration Protocol) [25] and many MANET routing protocols. Defining detailed specifications for each program/protocol can be a very time-consuming job. New specifications are also needed for each new program/protocol and the approach cannot detect some kind of attacks such as DoS (Denial of Service) attacks since these do not violate program specifications directly [9].

When an intrusion is detected, an appropriate response is triggered according to the response policy. Responses to detected intrusions can be passive or active. Passive responses simply raise alarms and notify the proper authority. Active responses try to mitigate effects of intrusions and are divided into two groups: those that seek control over the attacked system and those that seek control over the attacking system [3]. The former tries to restore the damaged system by killing processes, terminating network connections and the like. The latter tries to prevent attacker's future attempts, which can be necessary for military applications.

## 17.3  Intrusion Detection Issues in MANETs

Different characteristics of MANETs make conventional IDSs ineffective and inefficient for this new environment. Consequently, researchers have been working recently on developing new IDSs for MANETs or changing the current IDSs to be applicable to MANETs. There are new issues which should be taken into account when a new IDS is being designed for MANETs.

*Lack of central points:* MANETs do not have any entry points such as routers, gateways, etc. These are typically present in wired networks and can be used to monitor all network traffic that passes through them. A node of a mobile ad hoc network can see only a portion of a network: the packets it sends or receives together with other packets within its radio range. Since wireless ad hoc networks are distributed and cooperative, the intrusion detection and response systems in MANETs may also need to be distributed and cooperative [28]. This introduces some difficulties. For example, distribution and cooperativeness of IDS agents are difficult in an environment where resources such as bandwidth, processor speed and power are limited. Furthermore, storing attack signatures in a central database and distributing them to IDS agents for misuse-based intrusion detection systems is not suited to this environment.

*Mobility:* MANET nodes can leave and join the network and move independently, so the network topology can change frequently. The highly dynamic operation of a MANET can cause traditional techniques of IDS to be unreliable. For example, it is hard for anomaly-based approaches to distinguish whether a node emitting out-of-date information has been compromised or whether that node has yet to receive update information [7]. Another mobility effect on IDS is that IDS architecture may change with changes to the network topology.

*Wireless links:* Wireless networks have more constrained bandwidth than wired networks and link breakages are common. IDS agents need to communicate with other IDS agents to obtain data or alerts and need to be aware of wireless links. Because heavy IDS traffic could cause congestion and so limit normal traffic, IDS agents need to minimize their data transfers [18]. Bandwidth limitations may cause ineffective IDS operation. For example, an IDS may not be able to respond to an attack in real-time due to communication delay. Furthermore, IDS agents may become disconnected due to link breakages. An IDS must be capable of tolerating lost messages whilst maintaining reasonable detection accuracy [24].

*Limited resources:* Mobile nodes generally use battery power and have different capacities. MANET devices are varied, e.g. laptops, handheld devices like PDAs (personal digital assistants) and mobile phones. The computational and storage capacities vary too. The variety of nodes, generally with scarce resources, affects effectiveness and efficiency of the IDS agents they support. For example, nodes may drop packets to conserve resources (causing difficulties in distinguishing failed or selfish nodes from attacker or compromised nodes) and memory constraints may prevent one IDS agent processing a significant number of alerts coming from others. The detection algorithm must take into account limited resources. For example, misuse-based detection algorithm must take into account memory constraints for signatures and anomaly-based detection algorithm needs to be optimized to reduce resource usage.

*Lack of a clear line of defense and secure communication:* MANETs do not have a clear line of defense; attacks can come from all directions [28]. For instance, there are no central points on MANETs where access control mechanisms can be placed. Unlike wired networks, attackers do not need to gain physical access to the network to exploit some kinds of attacks such as passive eavesdropping and active interference (these require only radio contact) [28]. Furthermore, the critical nodes (servers, etc.) cannot be assumed to be secured in cabinets and nodes with inadequate protection have high risk of compromise and capture. IDS traffic should be encrypted to avoid attackers learning how the IDS works [18]. However, cryptography and authentication are difficult tasks in a mobile wireless environment since they consume significant resources. In many cases IDS agents risk being captured or compromised with drastic consequences in a distributed environment. They can send false alerts and make the IDS ineffective. IDS communication can also be impeded by blocking and jamming communications on the network.

*Cooperativeness* MANET routing protocols are usually highly cooperative. This can make them the target of new attacks. For example, a node can pose as a neighbour to the other nodes and participate in decision mechanisms, possibly affecting significant parts of the network.

## 17.4 Background

### 17.4.1 Proposed IDSs

IDSs on MANETs use a variety of intrusion detection methods. The most commonly proposed intrusion detection method to date is specification-based detection. This can detect attacks against routing protocols with a low rate of false positives. However, it cannot detect some kind of attacks, such as DoS attacks. There are also some anomaly-based detection systems implemented in MANETs. Unfortunately, mobility of MANETs increases the rate of false positives in these systems. There have been few misuse-based IDSs developed for MANETs and little research on signatures of attacks against MANETs. Updating attack signatures is an important problem for this approach. Some systems use promiscuous monitoring of wireless communications in the neighbourhood of nodes.

Since nodes in MANETs have only local data, a distributed and cooperative IDS architecture is generally used to provide a more informed detection approach. In this architecture, every node has its local IDS agent and communicates with other nodes' agents to exchange information, to reach decisions and respond. Other IDS architectures in MANETs are stand-alone and hierarchical IDSs [1]. In stand-alone IDS architectures, every node in the network has an IDS agent and detects attacks on its own without collaborating with other nodes. Because this architecture cannot detect network attacks (network scans, distributed attacks, etc.) with the partial network data on the local node, it is generally not preferred. Hierarchical IDSs are also a kind of distributed and cooperative architecture. In this architecture, the network can be divided into groups such as clusters, zones where some nodes (clusterheads, interzone nodes etc.) have more responsibility (providing communication with other clusters, zones) than other nodes in the same group. Each node in a cluster/zone carries out local detection while clusterheads and inter-zone nodes carry out global detection. It is more suitable for multi-layered networks [1]. Distributed IDS agents (nodes) are generally divided into small groups such as clusters, zones, and one-hop away nodes, enabling them to be managed in a more efficient way. Communication between these IDS agents is provided either by exchanging data directly or by use of mobile agents.

Two different decision-making mechanisms are used in distributed and cooperative IDSs: collaborative decision-making, where each node can take active part in the intrusion detection process, and independent decision-making, where particular nodes are responsible for decision-making [12]. Both decision-making mechanisms have pros and cons. Collaborative decision-making systems are more reliable. If all nodes contribute to a decision, a few malicious nodes cannot easily disrupt the decision-making. However, if any node can trigger a full-force response, it can affect the entire network and be vulnerable to a DoS attack [12]. A collaborative decision-making approach is also more resilient to benign failure of nodes. On the other hand, failing or compromise of particular nodes in independent decision-making systems can have drastic effects. However, these systems are less prone to spoofed intrusion attacks than collaborative decision-making systems [12].

The main proposed IDSs for MANETs in the literature are described below.

### 17.4.1.1 Distributed and Cooperative IDS [28, 29]

The first IDS for MANETs proposed by Zhang and Lee is a distributed and cooperative IDS. In this architecture, every node has an IDS agent, which detects intrusions locally and collaborates with neighbouring nodes (through high-confidence communication channels) for global detection, whenever available evidence is inconclusive and a broader search is needed. When an intrusion is detected, an IDS agent can either trigger a local response (e.g. alerting the local user) or a global response (which coordinates actions among neighbouring nodes).

Since expert rules can detect only known attacks and the rules cannot easily be updated across a wireless ad hoc network, statistical anomaly-based detection is chosen over misuse-based detection. The local data are relied on for statistical anomaly-based detection: the node's movement (distance, direction, velocity) and the change of routing table (PCR: percentage of changed routes, PCH: percentage of changes in the sum of hops all the routes).

A multi-layer integrated intrusion detection and response is proposed allow-ing different attacks to be detected at the most effective layer. It is believed to achieve a higher detection rate with a lower false positive rate.

The RIPPER and SVM-Light classification algorithms are used. In their subsequent research [29], these algorithms are evaluated on three routing pro-tocols: AODV, DSR and DSDV using detection rate and false alarm rate metrics. SVM-Light is shown to have better performance than RIPPER. It is also shown that the protocols with strong correlation among changes of differ-ent types of information (location, routing, etc.) have better performance, so reactive (on-demand) protocols are more appropriate for this system than proactive (table-driven) protocols. Moreover, it is stated that the IDS works better with protocols which include some redundancy (such as path redundancy in DSR). However, the mobility effect is not discussed.

This is one of the few approaches considering mobility by monitoring node movements. This can decrease false positives resulting from the node's mobility. However, it only reflects the local mobility not the network's mobility. Also, every node has to have a built-in GPS (Global Positioning System) to obtain this mobility data. It is emphasized that it can be applied to all routing protocols since it uses the minimal routing information. It also allows addition of new features for a specific protocol. From the security point of view, the system is reliable unless the majority of nodes are compromised [28]. (These can send falsified data.) Furthermore, the collaborative detection mechanism can be prone to denial of service and spoofed intrusion attacks [12].

### 17.4.1.2 Cooperative IDS Using Cross-Feature Analysis in MANETs [7, 8]

Huang et al. use data-mining techniques to automatically construct an anomaly detection model [8]. They use an analysis technique that targets multiple fea-tures and which acknowledges the characteristic patterns of correlation between them. The basic assumption here for anomaly detection is that normal and abnormal events have different feature vectors that can be differentiated.

In cross-feature analysis, they train the following classification model $C_i$ from normal data based on exploring the correlation between each feature and all other features [7]:

$$C_i : \{f_1, f_2, ..., f_{i-1}, f_{i+1}, ..., f_L\} \rightarrow f_i \text{ where } \{f_1, f_2, ..., f_L\} \text{is the feature set.}$$

In practice, each feature $f_i$ is analyzed and compared with the predicted values of $f_i$. Then, the average match count is evaluated by dividing the number of total true matches of all features by $L$ and used to detect anomalies that are below the threshold. Instead of count values, probabilities can also be used. Different classification algorithms C4.5, Ripper and NBC are investigated to calculate the probability function [7]. Since C4.5 shows better performance, it is the chosen method in their subsequent research [8].

Due to resource-constraints in MANETs, they propose a cluster-based IDS architecture. A fair and secure cluster-head assignment is presented. Cluster-heads are selected randomly, which also facilitates security. Equal service time is assigned to all selected cluster-heads.

Simple rules are also introduced to determine attack types and sometimes attackers. The rules are executed after an anomaly is detected. They are based on statistics such as the number of incoming/outgoing packets on the monitored node and are pre-computed for known attacks. For example, unconditional packet dropping of a node m is formulated as follows [8]:

$$FP_m(\text{forward percentage}) = \frac{\text{packets actually forwarded}}{\text{packets to be forwarded}}$$

If the denominator is not zero and $FP_m$ is 0, it means that node m is dropping all packets. The attacker is identified by a neighbour of node m who can promiscuously overhear node m's traffic.

It is implemented on the NS-2 simulator by using traffic related and non-traffic-related features. Traffic-related features are packet type, flow direction, sampling periods and statistics measures (counts and standard deviations of inter-packet intervals). Non-traffic-related features represent a view of network topology and routing operations and comprise information such as the number of routes added by route discovery, total route change and absolute velocity (the physical velocity of a node). The AODV protocol is targeted and the following metrics are used for evaluation: detection rate, false positive rate and attack type detection rate. The results are promising.

It is the first approach that uses feature correlations. They propose to investigate how computational cost can be reduced [7]. Attacker identification and attacks against the IDS (a major issue for a cluster-head architecture) are identified as future research [8].

### 17.4.1.3 Zone-Based Intrusion Detection System [22]

In [22], a non-overlapping zone-based IDS is proposed. In this architecture, the network is divided into zones based on geographic partitioning to save communication bandwidth while improving detection performance by obtaining data from many nodes. The nodes in a zone are called *intrazone* nodes, and the nodes which work as a bridge to other zones are called *interzone* (gateway) nodes. As shown in Fig. 17.1 there can be more than one gateway node in a zone, for instance, the nodes 1, 6, 7 are gateway nodes in zone 5. Each node in the zone is responsible for local detection and sending alerts to the interzone nodes.

Their framework aims to allow the use of different detection techniques in each IDS agent; however, they use only Markov chain anomaly detection in their research. Inputs to IDS agents are the routing table updates (PCR and PCH) as in [28, 29].

**Fig. 17.1** Zone-based IDS architecture in MANETs

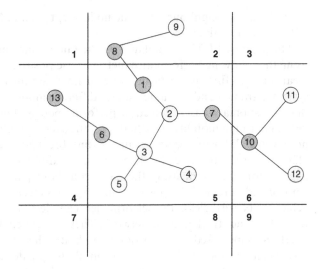

Intrazone nodes carry out local aggregation and correlation, while gateway nodes are responsible for global aggregation and correlation to make final decisions and send alarms. So only gateway nodes participate in intrusion detection. The alerts sent by interzone nodes simply show an assessment of the probability of intrusion; the alarms generated by gateway nodes are based on the combined information received. In their aggregation algorithm, gateway nodes use the following similarities in the alerts to detect intrusions: classification similarity (classification of attacks), time similarity (time of attack happening and time of attack detection) and source similarity (attack sources). Source similarity is the main similarity used, so the detection performance of aggregation algorithm could decrease with increasing number of attackers [22].

One of the contributions in this paper is MIDMEF (MANET Intrusion Detection Message Exchange Format), which defines the format of information exchange between IDS agents. It is consistent with Intrusion Detection Message Exchange Format (IDMEF) proposed by the Internet Engineering Task Force (IETF) [10].

Previous work [21] analyzed how to consider mobility when designing an IDS. Link change rate is proposed to reflect different mobility levels. Suitable normal profiling and proper thresholds can then be adaptively adopted by IDS agents using this measure. Furthermore, it is shown that link change rate reflects the mobility model of the network better than the generally used mobile speed measure. Link change rate of a node is defined as [21]:

$$\frac{|\mathbf{N}_1 - \mathbf{N}_2| + |\mathbf{N}_2 - \mathbf{N}_1|}{|t_2 - t_1|},$$

where $\mathbf{N_1}$ is the neighbour set of the node at $t_1$ time and $\mathbf{N_2}$ is the neighbour set of the node at $t_2$ time.

The proposed IDS is simulated on the GlomoSim simulator and evaluated using the following performance metrics: false positive rate, detection rate and mean time of first alarm (a measure of how fast intrusion is detected). The system is trained and evaluated under different mobility levels and it is shown that the anomaly-based detection performs poorly due to the irregularity of data under high mobility. Furthermore, the presence of partial victims who do not receive all falsified data because of link breakages resulting from mobility [22] is claimed to make the detection more difficult. The advantages of an aggregation algorithm using the data from both partial and full victims are emphasized: lower false positive and higher detection rate than local IDS achieves. Nevertheless, its performance can decrease with the existence of more than one attacker in the network. They also conclude that communication overhead is increased in proportion to mobility where local IDSs generate more false positives and send more intrusion alerts to gateway nodes. In addition, aggregating data and alerts at interzone nodes can result in detection and response latency, when there is sufficient data for intrusion detection even at intrazone nodes. The authors plan to investigate further attack scenarios at the routing and other layers, as well as constructing further security-related features and misuse-based detection approaches.

### 17.4.1.4 General Cooperative Intrusion Detection Architecture [20]

In [20], Sterne et al. present a cooperative and dynamic hierarchical IDS architecture, which uses multiple-layering clustering. Figure 17.2 shows a network with two-level clusters. The nodes annotated with "1" are the first-level cluster-heads, essentially acting as a management focus for IDS activity of immediately surrounding nodes. These level 1 cluster-heads can form a cluster around high-level node "2", second-level cluster-head. This process goes on

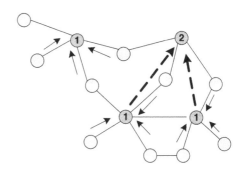

**Fig. 17.2** IDS hierarchy with two-level clusters

──────▶ data flow from 1. level cluster nodes to their cluster head

▬ ▬ ▶ data flow from 2. level cluster nodes to their cluster head until all nodes are assigned to a cluster. To avoid single point of failure, they propose choosing more than one cluster-head for the top-level cluster. The selection of cluster heads is based on topology and other criteria including connectivity, proximity, resistance to compromise, accessibility by network security specialists, processing power, storage capacity, energy remaining, bandwidth capabilities and administratively designated properties [20].

In this dynamic hierarchy, data flow is upward, while the command flow is downward. Data are acquired at leaf nodes and aggregated, reduced and analyzed as it flows upward. The key idea is given as detecting intrusions and correlating with other nodes at the lowest levels for reducing detection latency and supporting data reduction, whilst maintaining data sufficiency. It supports both direct reporting by participants and promiscuous monitoring for correlation purposes.

The proposed intrusion detection architecture for MANETs targets military applications. The authors claim that the dynamic hierarchy feature is highly scalable. It also reduces the communication overhead through the hierarchical architecture. However, the cost of configuration of the architecture in dynamic networks should also be considered.

Neither specific intrusion detection techniques nor the implementation of this architecture is covered. Supporting a broad spectrum of intrusion detection techniques is posed as one of the general requirements of IDS. However, applicability of these techniques to mobile ad hoc networks, which can have resource constrained nodes and no central management points, is not addressed. Examples of usage scenarios, which cover MANET-specific and conventional attacks, are presented by indicating different intrusion detection techniques on the architecture. Some attacks can be drastic in this architecture, for example, the capturing of cluster-heads or a malicious node being selected as a cluster-head by sending false criteria. Ongoing areas of investigation are the comparison of existing clustering algorithms and communication overhead metrics. They identify as future work the development of Byzantine-resistant techniques for clustering and for intrusion detection and correlation.

### 17.4.1.5  Intrusion Detection Using Multiple Sensors [12]

Kachirski and Guha propose an IDS solution based on mobile agent technology, which reduces network load by moving computation to data. This is a significant feature for MANETs that have lower bandwidth than wired networks. A modular IDS structure is proposed, which distributes the functional tasks by using three mobile agent classes: monitoring, decision-making and action-taking. The advantages of this structure are given as increased fault-tolerance, communication cost reduction, improved performance of the entire network and scalability [12].

A hierarchical and distributed IDS architecture is given, which divides the network into clusters. Cluster-heads are chosen by vote, with each node voting for a node based on its connectivity. Each node in the network is responsible for

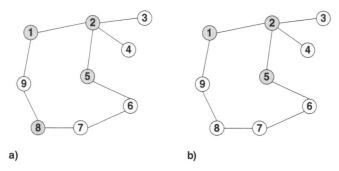

**Fig. 17.3 (a)** one-hop clustered network: nodes 1, 2, 5, 8 are cluster-heads **(b)** two-hop clustered network: nodes 1, 2, 5 are cluster-heads

local detection using system and user-level data. Only cluster-heads are responsible for detection using network-level data and for making decisions. However, depending on the hop attribute of the clusters, network intrusion detection performance can change. For example, every node has direct connection to at least one cluster head in a one-hop clustered network, so each packet in the network can be monitored as shown in Fig. 17.3(a), while three links in Fig. 17.3(b) cannot be monitored by the cluster-heads in a two-hop clustered network. As the degree of monitoring increases the number of cluster heads increases too. So, choosing the hop attribute of the clusters is a trade-off between security and efficiency. However, the nodes not in a cluster-head's communication range can move to the monitoring area of another cluster-head due to mobility. So, having a few links that cannot be monitored by any cluster-head is regarded as acceptable for highly dynamic environments.

Cluster nodes can respond to the intrusions directly if they have strong evidence locally. If the evidence is insufficient, they leave decision-making to cluster heads by sending anomaly reports to them.

In this paper a scalable and bandwidth-efficient IDS is proposed by using mobile agents, but without giving any validation via simulation or implementation. On the other hand, there are urgent security issues for mobile agents that are set to be investigated in the authors' future research. In addition, details of the anomaly-based detection method are not given, with research on more robust and intelligent cooperative detection algorithms left as future research.

### 17.4.1.6 Specification-Based IDS for AODV [25]

The first specification-based IDS in MANETs is proposed by Tseng et al. [25]. They use network monitors (NM), which are assumed to cover all nodes. Nodes moving out of the current network monitoring area are also assumed to move into range of other network monitors. Other assumptions are: (i) network monitors know all nodes' IP and MAC addresses, and MAC addresses cannot be forged. (ii) network monitors and their messages are secure. (iii) if some nodes do not respond to broadcast messages, this will not cause serious problems.

Network monitors employ finite state machines (FSM) as specifications of the operations of AODV, especially for the route discovery process, and maintain a forwarding table for each monitored node. Each route request (RREQ) and route reply (RREP) message in the range of the network monitor are monitored in a request-reply flow. When a network monitor needs information about previous messages or other nodes not in its range, it can ask neighbouring network monitors. In high-mobility conditions, the communication between network monitors increases since monitored nodes or/and packets frequently move out of the range of the monitoring node.

The authors also modify the AODV routing protocol by adding a new field: the *previous node*. Since RREQs are broadcast messages, it is necessary to keep track of the RREQ path. The previous node is needed to detect some kind of attacks such as sending an RREP to a node that is not on the reverse route [25].

Future work includes experimentation via NS-2 network simulation, profiling network QoS (Quality of Service) to reduce false positives by separating packet loss, packet error, and packet generation through defining reasonable thresholds for the current profile, and refining NM architecture using via a P2P (peer-to-peer) approach.

This is a promising approach that can detect both known and unknown attacks against routing protocols, which have clearly defined specifications. It is claimed to detect most of the attacks with minimum overhead in real time. However, some of the assumptions accepted in this paper are not very realistic. For example, assuming the network monitors cover all network nodes and have all nodes' IP and MAC addresses. Scalability is one of the important features on many MANET applications where the nodes can join and leave network independently and move frequently. Assuming MAC addresses cannot be easily forged is unrealistic. Moreover, dropping of some broadcast messages in the network can affect all network services if the node dropping messages is at a critical point. Furthermore, the details of the architecture are not addressed (such as the positions of network monitors in MANETs where the topology changes arbitrarily).

### 17.4.1.7  DEMEM: Distributed Evidence-Driven Message Exchanging ID Model [24]

DEMEM is a distributed and cooperative IDS in which each node is monitored by one-hop neighbour nodes. In addition to one-hop neighbour monitors, 2-hop neighbours can exchange data using intrusion detection (ID) messages [24]. The main contribution of DEMEM as stated by the authors is to introduce these ID messages to help detection, which they term *evidence-driven message exchange*. Evidence is defined as the critical information (specific to a routing protocol) used to validate the correctness of the routing protocol messages, for instance, hop count and node sequence number in AODV [24]. To minimize ID message overhead ID messages are sent only when there is new evidence, so it is called evidence-driven. DEMEM also introduces an ID layer to process these ID messages and detect intrusions between the IP layer and the Routing Layer without modifying the routing protocol, so it can be applied to all routing protocols.

DEMEM uses the specification-based IDS model for OLSR proposed in their previous work [23]. In OLSR [11], there are nodes called Multipoint Relays (MPRs), which serve to reduce the flooding of broadcast packets in the network. These nodes are selected by their neighbouring nodes called MPR selectors. The packets of an MPR node's MPR selectors are only retransmitted by that MPR node. TC (topology control) messages are sent by each node periodically to declare its MPR selectors. The proposed specification-based system uses the following constraints of OLSR to detect intrusions:

C1: neighbours in Hello messages must be reciprocal.
C2: MPRs must reach all two-hop neighbors.
C3: MPR selectors must match corresponding MPRs.
C4: Fidelity of forwarded TC messages must be maintained.

The authors state that the system cannot detect collaborative attacks. For example, two attackers who falsely claim that they are neighbours might not be detected by the above constraints [23].

DEMEM introduces three authenticated ID messages for OLSR. The first one is ID-Evidence, which is designed for two-hop-distant detectors to exchange their evidence concerning one-hop neighbours, MPRs and MPR selectors on OLSR. The second message, ID-Forward, is a request to forward any held ID-Evidence messages to other nodes. This means that a node can request the holder of evidence to forward it directly, rather than sending it itself, so reducing message overhead. The last message, ID-Request, is designed to tolerate message loss of ID-Evidence with low communication overhead. The false positives and delay detection due to message loss are decreased by an ID-Request message. Moreover, they specify a threshold value to decrease false positives due to temporary inconsistencies resulting from mobility. When a detector detects an intrusion, it automatically seeks to correct the falsified data.

DEMEM is simulated on the GlomoSim simulator with the random way-point mobility model and with different speed and pause time sets for mobile nodes. The approach is very effective for mesh networks where nodes in the network do not move: there were no false positives and no false negatives with 0.05% message overhead in a network with 150 nodes and 3% overhead in a network with 10 nodes. Interestingly, the message overheads of DEMEM are decreased as the number of nodes in the network increases, because the number of Hello and TC messages is greater than ID messages in large networks [24]. The message overhead in the simulation varies between 2% and 30% depending on mobility level. They also show how detection accuracy and detection latency of the system vary with the chosen thresholds.

The applicability of DEMEM on other routing protocols, especially on reactive protocols, is addressed. Because reactive protocols produce fewer routing messages with generally smaller size compared to periodic routing messages of proactive protocols, IDS on reactive protocols may have a greater message overhead than proactive protocols [24]. Ongoing research includes implementation of DEMEM on AODV and implementation of a reputation-based cooperative intrusion response model.

### 17.4.1.8 Case-Based Agents for Packet-Level Intrusion Detection [5]

Guha et al. [5] proposed a case-based reasoning system for packet level monitoring based on a hierarchical IDS architecture. In the case-based reasoning approach, known attacks are formulated as cases in the case archive, which stores the features of known problems as well as the actions to solve these problems. The idea is to search for similar cases in the case archive when a problem is detected on the network. The returned similar cases are used either as direct solution to the problem or else as bases on which to formulate the new case. The case concerning the final situation, failure or success, is stored into the archive. In this paper, Snort IDS [19] rules are used as the cases and each node has the database of these rules (which is claimed to be small in size). Since Snort rules need exact matching, this is used instead of searching for similarity in the case archive.

IDS functions (monitoring, decision making and actions) are distributed across several mobile agents. Some of them are presented on all mobile hosts, while others are distributed to only a select of group nodes [5]. All nodes have system-level and user-level monitoring that uses an anomaly-based approach. However, packet-level monitoring, which uses case-based reasoning approach, and decision-making are assigned only to cluster-heads. In their simulation, it is shown that the number of dropped packets by cluster-heads increases as the density of the network increases.

Using both anomaly-based detection for system-level and user-level monitoring, and misuse-based detection for packet-level monitoring increases effectiveness. It is also bandwidth-conscious, since it uses mobile agents. However, the security of the mobile agents still needs research.

### 17.4.1.9 An IDS Architecture with Stationary Secure Database [18]

A distributed architecture consisting of IDS agents and a stationary secure database (SSD) is proposed in [18]. All nodes have IDS agents responsible for local detection and collaborating with other agents in need. IDS agents have five components: local audit trail; local intrusion database (LID); secure communication module; anomaly detection modules (ADMs); and misuse detection modules (MDMs). The local audit trail gathers and stores local audit data – network packets and system audit data. The LID is a database that keeps information for IDS agents such as attack signatures, patterns of normal user behavior, etc. The secure communication module is used only by IDS agents to communicate securely with other IDS agents. ADMs use anomaly-based detection techniques to detect intrusions. There can be more than one ADM module in an IDS agent, for example, using different techniques for different kinds of audit data. There are also MDMs responsible for misuse-based detection to detect known attacks.

The stationary secure database (SSD) maintains the latest attack signatures and latest patterns of normal user behaviours. It is to be held in a secure environment. Mobile agents get the latest information from the SSD and transfer their logs to the SSD for data mining. The SSD has more storage and computation power than mobile nodes, so it is capable of mining rules faster than the nodes in the network and can keep all nodes' logs [18]. Moreover, updating the SSD rather than all nodes in the network is easy. On the other hand, a stationary database is not suited to all kinds of networks. Military tactical environments with control centres are given as examples of the architecture suitable for SSD. However, nodes in hostile environments may not attach to the SSD. Letting the nodes update themselves with the help of other nodes (which can consume significant bandwidth) is proposed as a solution to this problem.

Implementation and evaluation of this architecture are planned for future work. Although it seems to be an effective approach taking advantage of both anomaly-based detection using data mining techniques and misuse-based detection, it has a single point of failure, the SSD. Moreover, a stationary node goes against the nature of MANETs.

### 17.4.1.10  An IDS Model Integrating Different Techniques [9]

Huang and Lee propose an IDS model that uses both specification-based and anomaly-based detection approaches to detect interesting events [9]. A basic (routing) event is defined as the smallest set of casually related routing operations such as receiving/delivering a packet, modifying a routing parameter. An anomalous event is defined as the basic event that does not follow system specifications, such as deleting an entry in the route table, modifying route messages, etc. [9]. A specification-based approach is used to detect anomalous events that directly violate the specifications of AODV. Anomaly-based detection is used to detect events that do not violate specifications of the routing protocol directly and so require statistical measures.

In the specification-based approach extended finite state automata (EFSAs) are used to represent the specifications of AODV. Events which include only local node operations are mapped to the transitions of the automata. In the statistical-based approach, features are determined to detect anomalous events that cannot be detected by the specification-based approach, and then a set of detection rules is generated using RIPPER classifier.

The approach is evaluated using the MobiEmu simulator on some scenarios (not including high a degree of mobility). It is shown that some attacks are not detected effectively by this approach. It is concluded that these attacks cannot be detected locally [9].

The authors propose a taxonomy of attacks, which decomposes an attack into a number of basic events, and also propose a model to detect them. They use only local detection, since the local node is only reliable data source. That is why it cannot detect some kind of attacks which do not trigger anomalous events because of needing data from another layer such as a wormhole attack or needing other nodes such as network scan [9]. The authors plan to investigate multi-layer and global detection. Extracting features for detecting unknown attacks automatically is another issue identified as future research.

**Table 17.1** Outline of the proposed IDSs

| IDS | Contribution | Other MANET IDS issues |
|---|---|---|
| Distributed and Cooperative IDS | First distributed and cooperative IDS<br>Consider mobility | Consider only local mobility |
| IDS Using Cross-Feature Analysis | Use cross-feature analysis<br>Construct anomaly-based detection model automatically<br>Define rules to detect attack(er)s | High computational cost<br>Consider only local mobility<br>Not consider cluster-heads' capabilities |
| Zone-Based IDS | Use zone-based architecture<br>Define MIDMEF<br>Consider mobility based on changes of node's neighbours | Cause detection and response latency even when there is enough evidence on local nodes |
| General Cooperative ID Architecture | Use multiple-layered clustering | High-cost maintenance of the architecture under high mobility |
| IDS Using Multiple Sensors | Use mobile agents for a scalable and bandwidth-efficient system | May not monitor each node on the network due to the hop attribute of clusters |
| Specification-Based IDS for AODV | First application of specification-based detection technique to MANETs | Communication overhead under high mobility |
| DEMEM | Introduce ID messages between IDS agents to help detection | May not detect some kind of distributed and collaborative attacks |
| Case-Based Agents for Packet-Level ID | Use case-based approach and anomaly-based detection technique together | Have difficulties in updating case archives in a distributed environment |
| IDS Architecture with Stationary Database | Have a stationary secure database to keep patterns of normal user behaviours and attack signatures | Have a central point |
| IDS Model Integrating Different Techniques | Use anomaly-based and specification-based detection techniques together | Carry out only local detection, may not detect distributed attacks |

An outline of the proposed IDSs is given in Table 17.1. This shows the contribution/novelty each IDS brings and the MANET issues it does not address. However, security and limited resources issues are not shown in the table for each IDS separately, since all proposed systems usually make assumptions about these issues, or pay no attention to them.

## 17.4.2 Detection of Misbehaving Nodes

Nodes in MANETs rely on other nodes to forward their packets. However, these intermediate nodes can misbehave by dropping or modifying these packets. Several proposed techniques to detect such misbehaviours are given below.

### 17.4.2.1 Watchdog and Pathrater [15]

This is the primary work in detecting misbehaving nodes – nodes that do not carry out what they are assigned to do – and mitigating their effects. Since ad hoc networks maximize total network throughput based on cooperativeness of all nodes for routing and forwarding, misbehaving nodes can be critical for the performance of the network as stated in [15]. In this paper, watchdog and pathrater mechanisms on DSR are proposed to improve throughput of the network in the presence of misbehaving nodes. Nodes can misbehave because they can be overloaded, selfish (wanting to save their own resources), malicious or simply malfunctioning [15].

The watchdog's work is to detect misbehaving nodes by listening to nodes in promiscuous mode. When a node forwards a packet, the watchdog mechanism of that node monitors the next node to confirm that it also forwards the packet properly. It keeps sent packets in a buffer. When the packets are actually forwarded by next nodes, they are removed from the buffer. If the packets remain in the buffer longer than some timeout period, the watchdog increments the failure count of the node implicated. When the failure count of a node exceeds a threshold, the node is identified as a misbehaving node and a notification is sent to the source node. It is stated that watchdog can also detect replay attacks to some extent. However, since it uses promiscuous listening, it is stated that it might not detect misbehaving nodes in the existence of ambiguous collisions, receiver collisions, nodes that control their transmission power to deceive a listener into believing a message has truly been sent, and nodes that falsely report other nodes as misbehaving. It cannot detect partial dropping attacks and collaborative attacks involving at least two consecutive malicious nodes in a route [15].

Pathrater finds the most reliable path by using link reliability data and misbehaving nodes' information from the watchdog. In DSR, there can be many paths from source to destination, but the shortest path is selected. By using pathrater, the most reliable path is selected instead of the shortest path in the presence of misbehaving nodes. The SRR (send extra route request) extension to DSR can be added to find new paths when all paths include misbehaving nodes. Pathrater gives ratings to each node and provides a path metric based on the ratings of the nodes on the path. The authors state that ratings of the nodes should be rearranged to prevent permanently excluding temporary misbehaving nodes from routing and forwarding.

Watchdog and Pathrater with/without SRR is evaluated on the NS simulator with four different mobility levels by using throughput, overhead and false positive rates as metrics. The results show that watchdog and pathrater increase the throughput by 17% in the presence of 40% misbehaving nodes in moderate mobility with 9–17% overhead. Under extreme mobility, they increase throughput by 27% with 12–24% overhead.

The approach detects misbehaving nodes efficiently by using simple techniques without priori trust relationship information. Moreover, it increases the throughput of the network in the existence of misbehaving nodes, and does so with low overhead. On the other hand, it cannot detect collaborative attacks and partial dropping attacks. Additionally, it is applicable only to source routing protocols, because the watchdog needs to know where the packet is going to be forwarded by the next node. Applying the watchdog mechanism to other protocols requires adaptation. DSR needs modification for the SRR extension in the case of existence of misbehaving nodes on all paths. Finally, it rewards and reinforces malicious nodes in their behaviour by forwarding their packets while they do not forward for other nodes [4].

### 17.4.2.2 Nodes Bearing Grudges [4]

This is an interesting approach for detecting and responding to misbehaving nodes, inspired by the biology concept of reciprocal altruism. It detects misbehaving nodes and responds by not forwarding their packets. The aim of this approach is given as increasing fairness, robustness and cooperation in MANETs.

Each node is responsible for monitoring the behaviour of its next-hop neighbors and detecting misbehaving nodes. There is trust architecture and an FSM in each node with four main components: the monitor, the reputation system, the path manager and the trust manager.

The monitor (neighbourhood watch) keeps a copy of recently sent packets. It can compare them with the packets forwarded by the nexthop node and can detect routing and forwarding misbehaviours as deviations from normal expected behaviour. The types of misbehaviour that can be detected by this system are stated to be: no forwarding, unusual traffic attraction, route salvaging, lack of error messages, unusually frequent route updates, and silent route change [4]. When a misbehaving behaviour is detected, a reputation system is called for rating the misbehaving node.

The reputation system (node rating) keeps a local rating list and/or black list, which can be exchanged with friends. The rating of a node can change when there is enough evidence, and is based on the frequency of misbehaviour occurrence [15]. The rate function also uses weights depending on the source detecting misbehaviour. One's own experience has the highest weight, where observations have relatively smaller weights and reported experiences from other nodes have weight based on the trust level of these nodes. The reputation system uses only negative experience; research on positive changes and timeouts still needs attention. A path manager is called to take action when sufficient evidence of misbehaviour is obtained.

The trust level of nodes is managed by the Trust Manager, which is distributed and adaptive. It is also responsible for forwarding alarm messages and filtering incoming messages from other nodes. Trust of a node plays a significant role when exchanging routing information with that node, using it for routing or forwarding, and accepting its forwarding requests.

Path manager may respond to a request from misbehaving nodes in a variety of ways, such as ignoring the request, not replying back to the node, responding to any request for a route that include misbehaving nodes by sending alerts to the source node, re-ranking paths and deleting paths including misbehaving nodes [4].

ALARM messages are an extension to DSR and are used to distribute warning information. An ALARM message contains the type of protocol violation, the number of occurrences observed, whether the message was self-originated by the sender, the address of the reporting node, the address of the observed node and the destination address [4]. When an ALARM received, it is sent to Trust Manager to evaluate its trust level.

Assessment of this approach uses the GlomoSim simulator for evaluation and performance analysis is in progress. Moreover, the use of Game Theory for analytical evaluation is being investigated. One aim of the evaluation is to find the relation between the number of nodes in the network, the number of malicious nodes that can be tolerated, the number of friend nodes that needed for detection. In addition, they are planning to analyze the scalability, the cost/benefit ratio, the increase in the number of bits per unit of time forwarded to the correct destination minus any bits lost or retransmitted and overheads for achieving security (an important consideration for MANETs). The effects of mobility on promiscuous monitoring (which can increase collusions) could be analyzed. Since it uses a threshold mechanism, the effects of different threshold values for different mobility levels could usefully be assessed.

### 17.4.2.3 LiPaD: Lightweight Packet Drop Detection for Ad Hoc Networks [2]

Anjum and Talpade have proposed a practical approach for detecting packet dropping attacks [2]. In this approach every node counts the packets that it receives and forwards and periodically reports these counts to a coordinator node. Promiscuous monitoring is not used since it depends on the link layer characteristics and the link layer encryption approach [2]. That is why every node is responsible for monitoring its packets in LiPaD. The algorithm executed in each node is very simple, which is good for resource-constrained nodes. On the other hand, the network bandwidth consumption can be huge, since every node sends reports of each flow defined by source IP and destination IP to the coordinator node. They suggest compressing and aggregating the reports of multiple flows instead of sending each flow in a packet. However, it still affects network traffic, especially in networks with hundreds nodes. There will be a heavy computation load on the coordinator node (which analyzes all nodes' reports). The coordinator node needs to be a powerful device and must also be secure as it can be the target of the attacks to disable the detection mechanism. For example, it can be target of DoS attacks (by overloading with reports).

Since the coordinator node analyzes the same flow through the reports from all nodes in the route, it can detect liar nodes that pass the wrong information about the statistics of their packets to the coordinator node [2]. If all the nodes on the route are cooperative and malicious, LiPaD cannot detect packet dropping attacks on this route. It is stated that LiPaD detects selective forwarding attacks. It determines a threshold value for permissible packet loss. The coordinator node also implements rewards and punishments depending on the behaviour of the nodes.

It is assumed that IDS messages are encrypted and that nodes use a delivery mechanism for IDS messages to prevent them being dropped.

LiPaD is simulated on a network with 30 nodes using the OP-net simulator. It demonstrates that LiPaD detects malicious packet-dropping nodes even in the presence of non-malicious natural link-loss. On the other hand, the performance of LiPaD needs to be evaluated under high mobility and frequent link-loss. Evaluation of LiPaD performance under increased network traffic and node mobility is needed.

### 17.4.2.4 Intrusion Detection and Response for MANET [17]

Parker et al. extend snooping based methods to detect misbehaviour across routing protocols. A node listens to all nodes in its transmission range, not just the packets forwarded by one of its next nodes (as in watchdog [15]). To detect a malicious node in this approach, it is stated that the node must be in the proximity of a good node and act maliciously. It detects dropping and modification attacks, which exceed the value in the threshold table for the particular attack class. However, a node moving out of range of the monitoring node before it forwards packets can be assumed to be carrying out a dropping attack. This issue will be addressed in future by the authors. Also, this approach cannot detect misrouting attacks, since it does not know the next hop of a packet that it monitors.

The intrusion detection protocol can give either a local or a global response. In a local response, misbehaving nodes in the Bad Node table are isolated. It is emphasized that it is more effective in more dense networks, since more nodes detect intrusive behaviour and prevent malicious nodes from utilizing network resources. In the global response, the maliciousness of node is determined by a vote by all nodes in a cluster. If the majority of the nodes agree that the node is intrusive, an alert will be broadcasted. Voting is initiated by cluster-heads. Cluster-heads can be malicious, but the likelihood of malicious nodes being elected as cluster-heads is relatively small.

The approach is simulated using the GlomoSim simulator. The effect of node density (both malicious and normal nodes) on false positives is stressed. The response mechanism also affects the rate of false positives. It is claimed that global response reduces false positives due to rapid isolation of the intrusive nodes from the network.

## 17.5 Thoughts for Practitioners

Proposed IDSs for MANETs vary significantly, e.g. in terms of their detection technique, architecture, decision-making and response mechanisms. All systems have advantages and disadvantages. On the other hand, every proposed system should be considered in its own context. For example, a system using a misuse-based technique is generally not suited to the very nature of MANETs, since attack databases cannot easily be updated without a central point. On the other hand, it can fit a military network, which has a central location during peace-time.

Mobility, node capabilities and network infrastructure are usually the main features examined for proposed MANET IDSs. For highly mobile networks, IDSs using anomaly-detection techniques may suffer high false positive rates. Furthermore, an IDS architecture that is easy to set up should be preferred for these networks, e.g. IDS agents who collaborate with one-hop away nodes. Besides mobility, node capabilities should also be considered. Simple detection techniques can be more appropriate for nodes with limited resources. Trying to make the techniques simpler can be another approach. For instance, the approach in [27] uses a reduced feature set without significantly decreasing detection rate. Obviously, network infrastructure plays an important role in IDS selection. A hierarchical IDS architecture should be preferred to a multi-layered infrastructure, and distributed and cooperative architecture should be preferred for flat infrastructure [1]. Networks with central points make misuse-based and anomaly-based detection techniques easier to use by maintaining the signature database and user behaviours and analyzing them at these points. There may be an opportunity to use these techniques together in order to increase the effectiveness of the system.

The requirements of the system like high security, low bandwidth should also be satisfied by the IDS. For high-secure networks, the security of IDS and IDS traffic should be considered. For example, use of mobile agents can be avoided. Moreover, IDSs that are able to detect both known and unknown attacks should be preferred. That security requirements of the system can change in different situations (e.g. peace time and war time requirements of a military network may differ) should be borne in mind while designing an IDS. For low-bandwidth networks, communication between IDS agents should be minimized.

None of the proposed systems are necessarily the best solution taking into account different applications. Every organization should choose the appropriate IDS for its network. Moreover, it can change the IDS according to its own requirements and characteristics. For example, it can change the architecture of chosen IDS or put different intrusion detection techniques together. Therefore, defining requirements and determining characteristics of the network are very important factors in determining the most appropriate IDS solution.

## 17.6 Directions for Future Research

MANETs are a new type of distributed network whose properties are complex and ill-understood. Intrusion detection on these complex systems is still an immature research area. There are far fewer proposed IDSs for MANETs than for conventional networks. Researchers can focus on either introducing new IDSs to handle MANET specific features or can adapt existing systems. Hybrid approaches may also prove of significant use.

As stated earlier, IDS in MANETs poses special problems. Table 17.1 shows each proposed IDS reviewed in this chapter, identifying any novel contributions together with an indication of notable specific issues they do not address. In terms of these specific issues, none of the systems are complete. They usually emphasize just a few specific MANET concerns. The range of MANET issues should be considered during design to ensure effective and efficient intrusion detection suited to the environment at hand.

We make the following observations about the proposed IDSs:

- The systems generally cover restricted sets of attacks.
- The systems usually target a specific protocol.
- Some proposed IDS systems do not take into account mobility of the network.
- Inadequate acknowledgement is given to the resource constraints that many nodes are likely to be subject to, and to the likelihood of nodes with different capabilities.
- Several network architectures proposed do not sit well with the dynamic nature of MANETs.
- A more extensive evaluation of many of the systems would seem appropriate.

The proposed systems seek to address the *lack of central points* issue on MANETs by proposing distributed and cooperative IDS architectures. Such architectures raise questions about security, communication and management aspects. Suitability of the architecture to the environment is an important consideration in designing IDS. An architecture should not introduce new weaknesses/overheads to IDS. For instance, some of the proposed architectures like cluster-based approaches are costly to build and maintain for high-mobility networks. Some have critical points of failure.

Appropriate weight should be attached to *mobility*, especially for anomaly-based IDSs. The false positive rate may be greatly affected by mobility level. The system should be aware of its mobility and current network topology. So features having information about mobility should be included to the intrusion detection system being designed. How we get information about the mobility of the network and what features of the nodes or the network are related to mobility should be investigated.

Communication between IDS agents should be minimized due to constrained bandwidth of *wireless links*. This is one of the goals of the approach described in [24]. Other proposed systems usually do not pay attention to this issue. MANET Intrusion Detection Message Exchange Format (MIDMEF) consistent with IDMEF is defined in [22].

Since the nodes are the only data sources on the network, all nodes should contribute to IDS by carrying out local monitoring, detection and providing local data to other nodes when needed. However, nodes can have different computational capabilities. Moreover, some of them cannot be powerful enough for executing complex or large intrusion detection algorithms. There would appear to be insufficient research on the *limited resources* issue. Researchers can consider developing different algorithms for different nodes based on their resources and/or computational capabilities. Besides this, more intense detection algorithms can be applied in order to monitor critical nodes as proposed in [13].

Due to the *lack of clear line of defence* and *cooperativeness* features of MANETs, IDS agents can easily become the target of attackers. The proposed systems usually assume that IDS agents and communication between them are secure. Researchers should address the security of IDS. Detection of malicious IDS agents is an important research goal.

Testing IDS is an open research area for both MANETs and conventional networks. Some of the proposed systems in MANETs have not yet been implemented. Some of them are tested only on very small networks and with few attack scenarios. IDSs should be tested under different mobility levels and with different network topologies. Defining testing criteria for IDSs and preparing test datasets need research.

## 17.7 Conclusions

MANETs are a new technology increasingly used in many applications. These networks are more vulnerable to attacks than wired networks. Since they have different characteristics, conventional security techniques are not directly applicable to them. Researchers currently focus on developing new prevention, detection and response mechanism for MANETs.

In this chapter, we have given a survey of research on IDS for MANETs. Many MANET IDSs have been proposed, with different intrusion detection techniques, architectures, and response mechanisms. We have focused on the contribution/novelty each brings and have identified the specific MANET issues each does not address. Proposed systems generally emphasize few MANET issues. MANETs have most of the problems of wired networks and many more besides. As a consequence, intrusion detection for MANETs remains a complex and challenging topic for security researchers. We recommend the area to the reader for investigation.

## Abbreviations

| | |
|---|---|
| DEMEM | distributed evidence-driven message exchanging ID model |
| DoS | denial of service |
| FSM | finite state machine |
| ID | intrusion detection |
| IDMEF | intrusion detection message exchange format |
| IDS | intrusion detection system |
| IETF | Internet engineering task force |
| LiPaD | lightweight packet drop detection for ad hoc networks |
| MANET | mobile ad hoc network |
| MIDMEF | MANET intrusion detection exchange format |
| PCH | percentage of changes in the sum of hops all the routes |
| PCR | percentage of changed routes |
| RREP | route reply |
| RREQ | route request |

## Terminologies

*Intrusion*: Any set of actions that attempt to compromise the integrity, confidentiality or availability of a resource [6].

*Intrusion Detection System (IDS)*: A system to detect the intrusions against computers and the network, and respond to these detected intrusions.

*False Positives*: Normal activities that are detected as intrusions by IDS.

*Mobile Agent*: A composition of computer software and data, which is able to migrate from one computer to another autonomously and continue its execution on the destination computer [16].

*Intrusion Detection Message Exchange Format (IDMEF)*: The format to define data formats and exchange procedures for sharing information of interest to intrusion detection and response systems, and to the management systems that may need to interact with them [10].

*Denial of Service (DoS) Attacks*: Attacks that aim to make computer/network resources unavailable to the intended users.

*Dropping Attacks*: Attacks where selfish or malicious intermediate nodes drop packets that should be forwarded.

*Eavesdropping Attacks*: Attacks that monitor and find out information about the network. They do not interfere with the network operation.

*Random Waypoint Mobility Model*: A mobility model used for the simulation of MANETs in which a node randomly selects a destination and moves in the direction of the destination with a speed chosen in a defined range.

*Promiscuous monitoring*: The monitoring of all packets in a node's transmission range regardless of their destinations in wireless networks.

## Questions

1. How does node mobility affect IDSs in MANETs? How are wired network IDS techniques affected by mobility?
2. How do computational constraints impact on IDS techniques for MANETs? Are some techniques rendered difficult or infeasible?
3. How does the need to be power-efficient impact on IDS for MANETs? Are some techniques rendered difficult or infeasible?
4. How do memory constraints impact on IDS techniques for MANETs? Are some techniques rendered difficult or infeasible?
5. A variety of architectures are possible for MANET IDS. Why might you wish to have a distributed and collaborative approach?
6. Why might you wish to have a hybrid IDS for a MANET (for example, one that uses both misuse-based detection and also anomaly-based detection)?
7. What attacks on MANETs would be detectable by autonomous systems running on individual nodes (i.e. with no collaboration)?
8. What are the advantages of a collaborative approach to IDS in MANETs? What overheads are incurred by such approaches?
9. How do proposed IDSs for MANETs respond to detected intrusions? Discuss the need for different response mechanisms in different applications of MANETs.
10. What techniques are used for detecting misbehaving nodes in MANETs? How do they respond to these misbehaving nodes?

## References

1. Anantvalee T, Wu J (2006) A Survey on Intrusion Detection in Mobile Ad Hoc Networks (Chapter 7). Edited by Xiao Y, Shen Y, Du D-Z Wirel/Mobil Netw Secur, Springer, 170–196
2. Anjum F, Talpade R (2004) LiPaD: Lightweight Packet Drop Detection for Ad hoc Networks. In Proc of IEEE Veh Technol Conf (VTC) 2: 1233–1237

3. Axelsson S (2000) Intrusion Detection Systems: A Survey and Taxonomy. Technical Report No 99-15, Dept. of Computer Engineering, Chalmers University of Technology
4. Buchegger S, Le Boudec J (2002) Nodes Bearing Grudges: Towards Routing Security, Fairness, and Robustness in Mobile Ad Hoc Network. In Proc of 10th Euromicro Workshop on Parallel, Distrib and Netw-based Process, 403–410
5. Guha R, Kachirski O et al. (2002) Case-Based Agents for Packet-Level Intrusion Detection in Ad Hoc Networks. In Proc of 17th Int Symp on Comput & Inf Sci: 315–230
6. Heady R, Luger G, Maccabe A, Servilla M (1990) The architecture of a network level intrusion detection system. Technical Report, Computer Science Department, University of New Mexico
7. Huang Y, Fan W et al. (2003) Cross-Feature Analysis for Detecting Ad-Hoc Routing Anomalies. In Proc of 23rd IEEE Int Conf on Distrib Comput Syst (ICDCS) 23: 478–487
8. Huang Y, Lee W (2003) A Cooperative Intrusion Detection System for Ad Hoc Networks. In Proc of the 1st ACM Workshop on Secur of Ad Hoc and Sens Netw: 135–147
9. Huang Y, Lee W (2004) Attack Analysis and Detection for Ad Hoc Routing Protocols. In Proc of Recent Adv in Intrusion Detect LNCS 3224: 125–145
10. Intrusion Detection Message Exchange Format (IDMEF), http://www.ietf.org/html. charters/OLD/idwg-charter.html Accessed 30 August 2007
11. Jacquet P, Muhlethaler P et al. (2001) Optimized Link State Routing Protocol for Ad Hoc Networks. In Proc of IEEE INMIC: 62–68
12. Kachirski O, Guha R (2003) Effective Intrusion Detection Using Multiple Sensors in Wireless Ad Hoc Networks. In Proc of the 36th IEEE Int Conf on Syst Sci (HICSS)
13. Karygiannis A, Antonakakis E et al (2006) Detecting Critical Nodes for MANET Intrusion Detection Systems. In Proc. of 2nd Int. Workshop on Secur, Priv and Trust in Pervasive and Ubiquitous Comput (SecPer)
14. Lundin E, Jonsson E. (2002) Survey of Intrusion Detection Research. Technical report 02-04, Dept. of Computer Engineering, Chalmers University of Technology
15. Marti S, Giuli TJ et al. (2000) Mitigating Routing Misbehaviour in Mobile Ad Hoc Networks. In Proc of 6th ACM Int Conf on Mobil Comput and Netw (MobiCom): 255–265
16. Mobile Agent, http://en.wikipedia.org/wiki/Mobile_agent Accessed 30 August 2007
17. Parker J, Undercoffer J et al. (2004) On Intrusion Detection and Response for Mobile Ad Hoc Networks. In Proc of 23rd IEEE Int Perform Comput and Commun Conf
18. Smith AB (2001) An Examination of an Intrusion Detection Architecture for Wireless Ad Hoc Networks. In Proc of 5th Natl Colloq for Inf Syst Secur Educ
19. Snort, http://www.snort.org/ Accessed 30 August 2007
20. Sterne D, Balasubramanyam P et al (2005) A General Cooperative Intrusion Detection Architecture for MANETs. In Proc of the 3rd IEEE IWIA
21. Sun B (2004) Intrusion Detection in Mobile Ad Hoc Networks. PhD Thesis, Computer Science, Texas A&M University
22. Sun B, Wu K et al. (2006) Zone-Based Intrusion Detection System for Mobile Ad Hoc Networks. Int J of Ad Hoc and Sens Wirel Netw 2: 3
23. Tseng CH, Song T et al. (2005) A Specification-Based Intrusion Detection Model for OLSR. In Proc of the 8th Int Symp on Recent Adv in Intrusion Detect LNCS 3858: 330–350
24. Tseng CH, Wang SH (2006) DEMEM: Distributed Evidence Driven Message Exchange Intrusion Detection Model for MANET. In Proc of the 9th Int Symp on Recent Adv in Intrusion Detect LNCS 4219:249–271
25. Tseng C-Y, Balasubramanyam P et al. (2003) A Specification-Based Intrusion Detection System for AODV. In Proc of the ACM Workshop on Secur in Ad Hoc and Sens Netw (SASN)

26. Uppuluri P, Sekar R (2001) Experiences with Specification-based Intrusion Detection. In Proc of the 4th Int Symp on Recent Adv in Intrusion Detect LNCS 2212: 172–189

27. Wang X, Lin T et al. (2005) Feature Selection in Intrusion Detection System over Mobile Ad-hoc Network. Technical Report, Department of Computer Science, Iowa State University

28. Zhang Y, Lee W (2000), Intrusion Detection in Wireless Ad Hoc Networks. In Proc of the 6th Int Conf on Mobil Comput and Netw (MobiCom): 275–283

29. Zhang Y, Lee W (2003) Intrusion Detection Techniques for Mobile Wireless Networks. Wirel Netw 9(5): 545–556

# Chapter 18
# Security Threats in Ad Hoc Routing Protocols

John Felix Charles Joseph, Amitabha Das, and Bu-Sung Lee

**Abstract** The lack of infrastructure and centralized nodes in ad hoc networks have made nodes to depend on the cooperative nature of neighbor nodes for critical services like routing and maintenance. This unique cooperative nature of ad hoc network's routing mechanism has spawned unprecedented security threats in the routing protocol. Attackers have the potential to directly disrupt the routing protocol communications, which is transparent in conventional wired and wireless networks. A routing attack scenario is composed of a sequence of attack behaviors, and an attack behavior is a sequence of elementary attack behaviors, which is executed in particular fashion. The chapter enumerates in detail the routing attack scenarios and their elements. Also, the chapter introduces a security threat analysis that is used to explore different attack scenarios and behaviors in ad hoc routing protocols. The observations made in the threat analysis are useful for understanding the security vulnerabilities and realizing the requisites of a security system design for ad hoc networks.

## 18.1 Introduction

In Mobile Ad hoc Network (MANET) technology, the ad hoc connectivity between nodes in the network makes their network functions self-organized, decentralized, and distributed. For incorporating this ad hoc connectivity, new MANET routing protocols have been designed so that critical services such as routing and forwarding are dependent on willful cooperation of all the nodes in the network. This dependency has given rise to routing insecurity [1, 2] which is a new type of vulnerability not present in conventional networks.

J.F.C. Joseph (✉)
Center for Multimedia and Networks, Division of Computer Communications, School of Computer Science and Engineering, Nanyang Technological University, Nanyang Avenue, Singapore 639798
e-mail: john0007@ntu.edu.sg

S. Misra et al. (eds.), *Guide to Wireless Ad Hoc Networks*,
Computer Communications and Networks, DOI 10.1007/978-1-84800-328-6_18,
© Springer-Verlag London Limited 2009

This chapter examines the various threats and attacks on ad hoc routing protocols, and provides a taxonomy of attack behaviors by organizing them in a systematic hierarchy in terms of the complexity of the attack behaviors and the nature of disruption caused by them. This taxonomy can be considered to be the first step in appreciating the routing threats faced by ad hoc networks.

Once the general nature of the attacks are understood, the next logical step is to be able to analyze various routing protocols to determine the specific mechanism through which such attacks can be mounted on a given protocol. [3]. Conventional security analysis would involve modeling the security vulnerabilities based on the historical information about the protocols, such as known/reported flaws, limitations. In this methodology, the scope of the analysis is confined to the knowledge of the security analyzer(s). Nevertheless, this approach provides a practical insight to factors affecting the security.

In this chapter, we introduce a novel threat analysis methodology to systematically explore the vulnerabilities of an ad hoc routing protocol. The threat analysis will reveal new attack scenarios and behaviors considering the uniqueness of the routing protocol design. This analysis has two major benefits. First of all, it helps to fine-tune protocol design and implementation to minimize vulnerabilities. Secondly, it helps to design efficient security mechanisms by identifying the optimum localities in the data and control path for implanting protection mechanisms.

To demonstrate the threat analysis technique, the Optimized Link State Routing (OLSR) protocol is analyzed. OLSR is chosen mainly for two reasons. First, OLSR is relatively more insecure compared to protocols such as AODV, which has been widely studied elsewhere. This higher level of insecurity of OLSR is caused by its proactive (table-driven) nature, and the fact that it uses selective flooding, which reduces redundancy and thereby reduces the security options for the protocol. The second major reason is that OLSR is a well-established routing protocol for ad hoc networks.

## 18.2 Organization

The following text is organized into five major sections. Section 18.3 lists and explains briefly the existing security analysis methodologies for OLSR and more generally for ad hoc routing protocols. It also discusses their major contributions and their shortcomings. Section 18.4 discusses in detail various attack scenarios, attack behaviors, and elements of attacks behaviors. The next section, introduces the security threat analysis technique. The facts inferred from the observations from the threat analysis are explained in Section 18.5. Finally, the chapter concludes by summarizing the routing threats in ad hoc networks.

## 18.3 Background

Every research paper in literature that addressed security issues of ad hoc routing protocols at least cursorily discussed some security vulnerabilities. Interesting attack scenarios can be found in literature [4, 5]. However, the scenarios explained in these papers are only a few of the myriad attack possibilities. A security system design cannot rely on designing a security system based on partial information. This raises the need for a comprehensive methodology to explore security threats on ad hoc routing protocols.

Huang and Lee [6] systematically analyzed the AODV ad hoc routing protocol. They considered routing protocol internal states to analyze the attack possibilities. Possible malicious states and transitions were identified and routing protocol exploits were analyzed. However, they did not include external factors such as user behavior, mobility and node-to-neighbor interaction. Another interesting work is by Ning and Sun [7], who also extensively analyzed and provided a detailed case study on AODV routing protocol security issues. However, similar to Huang and Lee's approach, they centered more on analyzing AODV rather than providing a generic approach that can be applicable to most ad hoc routing protocols.

Kong et al. [8] discuss various passive attack scenarios against ad hoc systems. Their work details the effect of passive attacks in the perspective of military applications. For a discussion on the various Byzantine attacks possible, refer to Awerbauch et al. [9].

## 18.4 Taxonomy of Ad Hoc Network Routing Attacks

As briefly indicated in the introduction, malicious actions directed at routing protocols can be organized in a hierarchy of behaviors, depending on the complexity of the actions and the extent of damage caused by them. This hierarchy is shown in Fig. 18.1. At the highest level of this hierarchy are the routing attack scenarios, which essentially identify the disruptive effects aimed for by the attacker. A routing attack scenario consists of a set of attack behaviors executed in a particular sequence to achieve a malicious goal. As shown in Fig. 18.1, there are seven major attack scenarios on ad hoc routing protocols.

The attack behaviors identify the broad strategies employed by the malignant party to achieve the ultimate attack goal. They are categorized into three classes, namely, route invasion, route disruption, and route monitoring. These behaviors aim to maliciously add, modify, or delete routes in the network.

An attack behavior can be further decomposed into elements of the attack behavior. The elements of an attack behavior are the simple steps or precise techniques used by the attacker to affect a certain behavior. These include, spoofing, sinking, modifying, rushing, and replaying behavior. In what follows, the three hierarchies of attack behaviors are described in more detail.

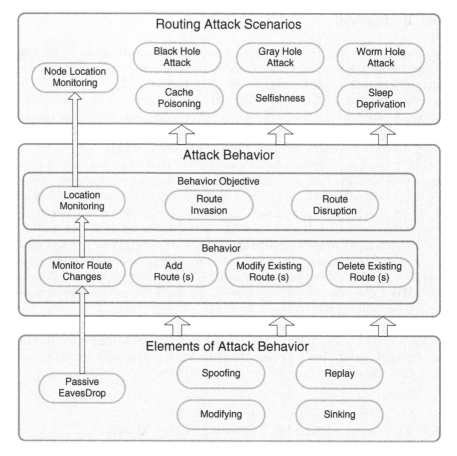

**Fig. 18.1** Routing attacks taxonomy

### 18.4.1  Elements of Attack Behavior

Sinking refers to the intentional dropping of packets by a node instead of forwarding them to the next node along the intended route. The objective of sinking is to either selfishly conserve resources like battery, medium, etc. or drop critical packets like routing messages to disrupt routing functions. While the later can form inconsistent routes, selfish nodes in a sparse network may partition some parts of the network, making them unreachable.

Rushing is a malicious behavior, which is unique to reactive routing protocols. In a reactive routing protocol, when a node wants to establish a path to the destination, it broadcasts a Route request (RREQ) to the neighbor nodes. Neighbor nodes forward this RREQ by flooding it in their respective neighborhood. This process is continued until the path to the destination is found. To control the flooding, a node forwards only the first received RREQ. An attacker can exploit this feature by rushing the RREQ toward the destination, thus getting preference over other nodes. The path established by rushing will certainly contain the malicious node, thereby giving it an advantageous position for further attacks.

The malicious behavior of retransmitting old routing messages to form inconsistent routes is referred to as replay attacks. The objective of replay attacks is to disrupt routing functions and cause Denial of Service. Replay attacks are less severe, but can thwart intrusion prevention mechanisms such as encryption and digital signatures. The scope of this attack is confined to the immediate neighborhood and the extent of damage depends on the transience or mobility of the network nodes. High mobility causes routes to change frequently. In this transient environment, replay attacks can cause serious damage to the routing functions of the network.

Spoofing is a process of impersonating another node in the network. In routing, the attacker can impersonate a node by changing the source address of the routing message. The ability to spoof routing messages gives a malicious node a wide range of possibilities to attack the network routing functions. Spoofing can create non-existent neighbor nodes and can help emulate supporting information to the malicious routing information disseminated by the malicious node. The objective of spoofing include route invasion, route disruption, network partitioning, and Denial of Service (DoS). In proactive routing protocols, spoofing behavior has the potential to corrupt the entire topology.

When an attacker changes other contents in the routing message, like declaration of neighbors, sequence number, etc., instead of the source address, it is usually referred to simply as a modification attack. Modification of routing message contents affects the network functions significantly. The objective of modification is similar to spoofing; however, unlike spoofing, its effect is confined to its neighborhood nodes only.

## 18.4.2  Attack Behavior

As shown in Fig. 18.1, the routing attack behaviors are classified as, route invasion, route disruption, and route monitoring. Basically, these behaviors aim either to *add* new malicious routes or *modify* or *delete* existing routes in the network.

The routing security vulnerability can be exploited in two ways. Either the vulnerability is used indirectly as a platform for more serious attacks on the higher layer protocols or the vulnerability is used directly attacking the routing system itself. For the former kind of threat, a malicious node invades a route and stages a vantage point for higher layer attacks. In the later kind of threat, the attack directly disrupts the network functions by adding inconsistent malicious routes, deleting or modifying existing benign nodes.

Additionally, passive attacks such as route monitoring are also possible, which can reveal the location of the attacked nodes. In many applications, such as military, location information can be sensitive. As stated earlier, passive attacks are hard to detect and prevent.

## 18.4.3 Attack Scenarios

Attack scenarios are complex sequences of attack behavior executed to achieve a malicious objective. In this section, few notorious attack scenarios are discussed in detail.

### 18.4.3.1 Black Hole Attack

Black hole attacks cause packets to disappear from the network without any trace. Most often, black hole attacks are effected by simple sinking behavior, which was discussed in the previous section [10]. Besides sinking, black hole attacks have other variants. In one variant, the data traffic is forwarded to a non-existent or another malicious node, where the data are dropped. This forwarding behavior before the actual sinking will make the detection of sinking behavior hard.

Another variant of black hole is a gray hole attack. In gray hole attacks, only selected data traffic is forwarded to a non-existent or another malicious node for sinking. Gray hole attacks are hard to detect due to the sporadic sinking behavior.

Other than sinking, there are numerous ways by which a malicious node can coerce benign nodes to drop incoming traffic. Essentially, the malicious node achieves this by disrupting the benign route between source and destination. Route disruption can be enforced using spoofing and modification of routing messages.

*Example 1:* In Fig. 18.2, a malicious node M spoofs a benign node V and declares the route availability to the destination node D in the spoofed routing message. Since, node V is not receiving the data or does not have path to node D, node V will not forward the traffic and drop the traffic benignly. Thus the attacker fulfills its goal of sinking traffic. Besides sinking incoming traffic, through the above spoofing behavior the attacker also makes the neighbor nodes to believe that node V is malicious.

### 18.4.3.2 Wormhole

In this attack scenario, two malicious nodes belonging to two different neighborhoods create a tunnel between them [11]. Through the tunnel, routing messages transmitted by malicious nodes' neighbors are exchanged and replayed in the neighborhood, where these nodes are not present. This behavior

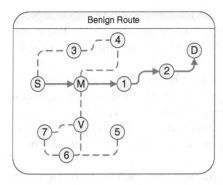 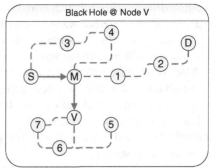

**Fig. 18.2** Black hole attack scenario

makes the neighbors to believe that the two malicious nodes are one-hop neighbors to nodes in the other neighborhood. Thus, all routes established between these nodes in the two neighborhoods will contain at least one of the malicious nodes.

*Example 2:* Consider two malicious nodes M1 and M2 present in two neighborhoods NN1 and NN2, which are connected by more than two hops, as shown in Fig. 18.3. Malicious nodes M1 and M2 create a tunnel and channels all routing message traffic that is received in their corresponding neighborhood through the tunnel to the other malicious node. Both M1 and M2 replay the routing messages received through the channel in their own neighborhood. This will disrupt the perception of topology for the nodes in both the neighborhood. Nodes in NN1 will believe that nodes in NN2 are one-hop away through the node M1. Similarly, nodes in NN2 will have wrong perception that NN1 is at

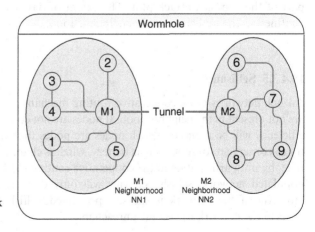

**Fig. 18.3** Wormhole attack scenario

one-hop distance through M2. Now consider the scenario where a node in NN1 and another node NN2 want to establish a route between them. The route established will certainly contain either M1 or M2. Hence, the colluding malicious nodes will control the link.

After the nodes control the link, their next behavior may be a passive or active attack. The malicious nodes can passively monitor the traffic for attacks such as traffic analysis, location monitoring. Passive attacks are grave threat to military applications and tactical communications. If the nodes attack the controlled link actively, such as sinking or manipulating the traffic, this may cause denial of service, etc.

It is interesting to note that security mechanisms, such as encryption and access control, do not prevent this attack. Furthermore, as the traffic is tunneled, detection of wormhole attacks is hard.

### 18.4.3.3 Network Partitioning

A malicious node can delete routes to isolate a part of the network and render the nodes in the isolated network sector unreachable. The above is referred to as network partitioning. Statistically, detection of network partitioning is trivial. However, the malicious node can thwart detection by exhibiting low layer attacks such as channel jamming, MAC flooding.

### 18.4.3.4 Cache Poisoning

Corrupting the routing information (tables) stored locally in the nodes is referred to as cache poisoning. A malicious node can add, delete, or modify routes in a node's routing cache by exhibiting any of the five attack behaviors discussed earlier. The attack scenarios such as black hole, worm hole also corrupt the routing cache. However, in these scenarios cache poisoning is one part of the complex attack plot. The extent of damage by cache poisoning is confined to the malicious node's neighborhood.

### 18.4.3.5 Selfishness

Ad hoc networks use a cooperative routing mechanism for reaching nodes that are not present in the neighborhood. If nodes are not cooperative, then the routing efficiency will be reduced. Selfish nodes are nodes that are non-cooperative and try to conserve resources, such as battery, wireless channel bandwidth, processing time, by not participating in the cooperative routing mechanism. In a reasonably populated network, selfish behavior will only degrade the routing efficiency. However, if the network is sparsely populated, selfish nodes will cause some areas in the network to become unreachable.

### 18.4.3.6 Sleep Deprivation

In ad hoc networks, most nodes are mobile, which have limited resources such as battery power and processing capacity. A malicious node can constantly flood a node with junk routing messages. These junk routing messages will be processed by the victim node, which wastes its processing time and may eventually drain its battery. This attack scenario is referred to as sleep deprivation.

## 18.5  Security Threat Analysis

The security threat analysis of a protocol consists of systematic steps of exploring the security vulnerabilities of the given protocol and the network environment. The threat analysis is carried out in three major stages. In the first stage, the implementation of the protocol is analyzed for understanding the propagation of malicious information and to study the scope of every type of routing message of the protocol. At the second stage, causal relationships between different attack behaviors and the extents of disruption caused are derived for all possible network conditions. Finally, based on the observations of the previous stage, contents of every type of routing message in the protocol are assessed for security risk.

We will use the well-known OLSR protocol to illustrate the security analysis technique.

### 18.5.1  OLSR Fundamentals

OLSR [12] is a table-driven proactive routing protocol. Primarily, OLSR was developed for ad hoc mobile wireless networks. The protocol is very similar to Link-state routing (LSR), where the topology information (link states) is disseminated to all the nodes in the networks. Flooding is used to disseminate the link –states, and every node in the network has the detail information about the current topology of the network. OLSR optimizes LSR by reducing the flooding redundancy and conserving bandwidth. This is achieved through multipoint relaying (MPR).

The fundamental concept of multipoint relaying is to selectively choose the relays/forwarders for information originating from a particular node. These selectively chosen nodes are called multipoint relay set of node. Each node has a set of MPR nodes. The selection of MPR nodes is aimed at reducing the number of nodes in the MPR set. At the same time, the MPR set should have connectivity to all two-hop neighbors.

OLSR has three types of routing control message namely, Hello packet, Topology control packet, and multiple interface declaration (MID) packets. Hello message is used for neighbor sensing and MPR set calculation. TC messages are used to disseminate topology/link states to different nodes in the network through MPR nodes. MID messages are used to declare multiple interfaces for a particular node.

## 18.5.2 Protocol Analysis

The first step is to derive the relationship between the input and the output of the protocol. The objective of this step is to identify the extent to which information received in a routing message is propagated. Additionally, the effects of input on internal/local resources of the protocol is also studied and organized as shown in Fig. 18.4. Events/local resources that generate different information for interacting with the neighbors are also examined. These relationships were derived from the documentation of the protocol, such as RFCs [13]. Since routing control messages are the only possible resources used by the attacker against the routing protocol, the analysis of OLSR starts by understanding the role of each control message and the effect it brings upon the network and the local resources. First, the effect of routing events on the local resources, such as routing tables, are studied. In the second step, the extent of propagation of routing information for various routing messages is examined.

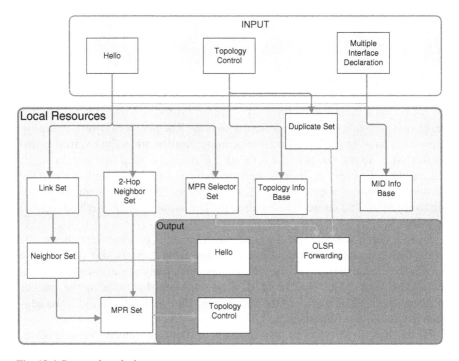

**Fig. 18.4** Protocol analysis

### 18.5.2.1  Local Resources

In OLSR, the local resources in a node consist of routing tables, the MPR set, and a number of information bases as follows:

- Neighbor Nodes Information Base
- Two-Hop Neighbor Nodes Information Base
- Multipoint Relay Selectors (MPRS) Information Base
- Topology Information Base
- Multiple Interface Declaration (MID) Information Base
- MPR Information Base.

Some information bases are influenced by only one type of routing message, whereas some are affected by multiple message types. Foe example, MPR set computation is influenced by the hello message as it affects the one-hop and two-hop neighbor sets. Similarly, MPRS set is controlled by the TC message only. On the other hand, route calculation is an example of a local element that is affected by multiple routing message types, as it is influenced by all three of them. Such resources are critical from the security point of view, as prevention and detection of attacks on such components are practically infeasible.

However, it should be noted that the routing messages cannot directly influence these components. Only the information bases control these components. Thus, it is necessary to prevent any malicious information to propagate from the information bases to the security critical local components. This knowledge gives us insight as to the optimum location of a local security mechanism to prevent or detect attacks.

### 18.5.2.2  Propagation Analysis

The next step in protocol analysis is the examination of routing information propagation to other nodes in the neighborhood. From Fig. 18.4, it can be seen that information in both Hello and TC routing messages are propagated. However, information in hello messages are forwarded only to nodes within two-hop distance, whereas information in TC messages are propagated to nodes in the entire network.

Though only TC messages are directly propagated, hello messages control the information of TC messages generated by one-hop and two-hop neighbor nodes. A TC message contains the current MPRS set of the source node. The neighbor nodes use hello messages of other neighbors to compute the MPR set. Thus, hello messages can influence the MPRS set of two-hop neighbor nodes, thereby influencing the content in TC messages. This shows the potential of hello messages in the propagation of malicious routing information.

The above discussion shows the relationship between hello and TC messages among nodes in neighborhood. The security system should guard the integrity of this relationship and prevent malicious information propagation. It was also revealed that a malicious hello message, if not prevented, will lead to a malicious

TC message generation in the two-hop neighborhood. In most cases, only malicious TC messages are responsible for any kind of routing attack. Malicious hello or MID messages are used to indirectly generate malicious TC messages or modify benign TC messages.

### 18.5.3  Causal Relations – Effects and Behavior

In this stage of threat analysis, the contents of routing message are examined for security vulnerabilities. Every field in all the routing messages is analyzed for security vulnerabilities. The network environment plays a major factor in determining the scope of a malicious behavior. Hence in this step, causal relationships between different attack behaviors and their effects on the network under various network conditions are derived. Figure 18.5 shows graphically the causal relations derived for spoofing behavior. The attack behavior's range of disruption depends on the current network connectivity (network environment) between the attacker and the victim node.

Similar to the causal relations derived for spoofing behavior, we can derive the same for other fields in the routing message. These causal relations between the attack behavior and the malign network effect are used to trace malign network effect to different attack behaviors and vice versa.

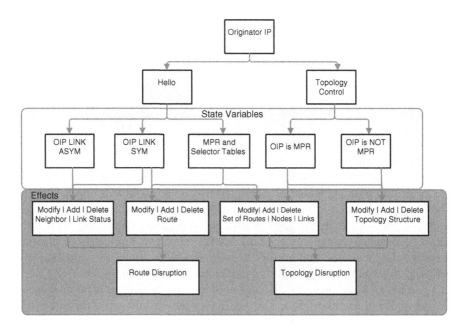

**Fig. 18.5**  Causal relations – spoofing behavior

Consider a spoofing attack scenario, where a node M sends malicious routing message to node B. The forged originator IP address (OIP) used in the spoofed routing message belongs to node C, which may exist in a distant neighborhood or is non-existent. In this scenario, node M can either be malicious or a compromised node.

Given the above scenario, in this stage of threat analysis we explore the possible ways node B will react to the malicious routing information under various network environments. Furthermore, the effect on the node C, whose IP address is used for spoofing, is also examined.

### 18.5.3.1 States of Network Connectivity (Environment)

OLSR consists of three network connectivity states, namely,

- Symmetric channel (SYM)
- Asymmetric channel (ASYM)
- MPR connectivity between node (MPR)

Two nodes having symmetric state of network connectivity means that the communication between them is bidirectional. On the other hand, if the nodes have asymmetric link, then the communication channel is unidirectional. The third state of network connectivity is that one node is an MPR to the other node. For instance, node C or A is a MPR to B. Then, the network state of the link between A → B is MPR and state of link in the reverse direction B → A is MPRS.

An arrival event of a hello or TC routing message on node B may change the current protocol state of the node. In OLSR, since hello message is used for link and neighbor sensing, an arrival of hello message will change the link or neighbor status of the node. Thus, a malicious hello packet that has a forged source IP address from a compromised or malicious node A may affect these states, as shown in Fig. 18.5.

Similarly, a malicious TC routing message affects the network depending upon the link state between nodes A and B. The malicious TC message will affect only the MPR state of the link connecting the nodes A and B.

### 18.5.3.2 Effects of Attacks (Malicious Behavior)

Figure 18.5 shows the effect of spoofing behavior under different network conditions. A spoofed hello routing message can cause disruption by adding/modifying/deleting link status or routes of nodes in its current neighborhood. If the link between A→B or B → A is not MPR, then the scope of the spoofing behavior is limited to the disruption of single route. If link between A and B is MPR, this gives the attacker the potential to cause disruption to many routes.

From Fig. 18.5, it is evident that the malign effect on the network depends not only on the attack behavior but also on the current network connectivity (environment). The level of propagation of malicious information depends on the current link state between nodes A and B. As the scope of the spoofing behavior depends on the extent of malicious information propagation, we assess the bounds of the attack based on the current network conditions.

As stated earlier, the effect of a spoofed TC message depends only on whether link A → B is MPR, or not. Similar to hello spoofing, the MPR link status between A and B gives a wide range of attack possibilities.

It is important to note that the effect of the attack also depends on the IP address which is used to forge the OLSR routing messages. Depending on the IP address used for the spoofing, there are two possible spoofing behaviors. In the first type of spoofing behavior, the attacker uses the IP address of existing node, mostly in a distant neighborhood. In this scenario, the spoofing behavior causes disruption in two neighborhoods, in the attacker neighborhood and also in the victim's (owner of the spoofed address) neighborhood. In the second type of spoofing behavior, the attacker uses the IP address of a non-existing node in the network. In this case, non-existent nodes are created. Usually, the attacker node uses these non-existent nodes to emulate support to its malign cause.

### 18.5.4 Risk Estimation

The final step in security threat analysis is assessing and estimating the security risk of each field in every type of routing message in the routing protocol. The security risk of fields in hello and TC routing messages of OLSR are assessed and shown in Fig. 18.6.

The causal relations derived in the previous step are used to assess the level of vulnerability for each field in OLSR routing messages. For example, in the previous section we illustrated the potential of the attack behavior in which the attacker spoofs the source IP address of the routing message. The extent of

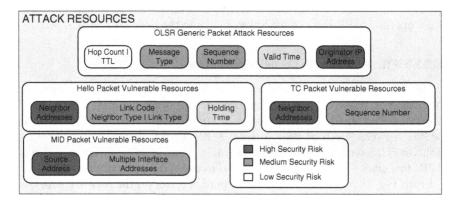

**Fig. 18.6** Risk estimation of routing message contents

potential disruption by this behavior is enormous. Thus, source IP field in routing messages is assessed as a *high security risk field*.

The above security risk estimation of individual elements of routing messages will help a security system design to focus on the critical fields in routing control messages.

## 18.6 Inference

The security threat analysis highlighted many critical aspects that need consideration in the design of an efficient security system for ad hoc routing protocols. In the first stage, the protocol analysis revealed the optimal locality for security and protection mechanisms. The causal relations derived in the second stage of the threat analysis are used to explore new attack scenarios with the consideration of routing protocol design. The causal relations that are derived for the five elementary attack behaviors on various fields in every type of routing message in the protocol will give deep understanding on the routing attack possibilities and limitations. The security risk estimation of every field in routing messages can be used to rank the vulnerability of information contained in the routing message depending on its sensitivity to attacks.

## 18.7 Thought for Practitioners

Research in ad hoc network security is at its infancy and thus many issues remains as an open research challenge. This chapter has listed and explained all the possible elementary attack components. However, the list of attack behavior and scenarios are not complete and exhaustive. Most of these discussed attack scenarios are basically hypothesized. This is primarily because of the fact that ad hoc network has not been implemented in a public network. Thus, we are yet to face the real attacks in ad hoc networks. Future of ad hoc network security will highly depend on the feedback received from a real public ad hoc network.

## 18.8 Directions for Future Research

The next research direction in respect to the security threat analysis is to refine the methodology to by generalizing it to most of the ad hoc routing protocols. This will require identifying the differences and commonalities between different types of routing protocols. For instance, the primary difference between proactive and reactive routing protocols is the way they discover route. Furthermore, they also differ in how routing information is propagated. These characteristic differences and commonalities is used to generalize the security threat analysis methodology.

## 18.9  Conclusion

In this chapter various routing threats in ad hoc networks are identified and discussed in detail. For a better appreciation of the wide variety of threats, a taxonomy of attacks on ad hoc routing has been introduced. Elementary attack behaviors, such as sinking, spoofing, rushing, modifying and replay, are explained. Attacks such as route invasion, route disruption, which comprise-sone or more of the above elementary attack behaviors, are enumerated. Some complex attack scenarios, which are notorious in ad hoc networks, are shown to illustrate the routing protocol vulnerabilities.

Furthermore, a systematic security threat analysis method is introduced using OLSR as a target protocol. The threat analysis can be performed on any ad hoc routing protocol to understand its vulnerabilities. Three major stages of the threat analysis are protocol analysis, causal relation derivation, and risk estimation. The observations of the security threat analysis significantly helps in understanding the vulnerabilities of the routing protocol, and thus helps in understanding the requirements in the design of a security system for the ad hoc network paradigm.

## Terminologies

*Intrusion/Attack*: Intrusion is a behavior of an external or internal node(s) with malign intent, which aims to affect other benign nodes in the network.

*Routing attack*: Attacks targeted at the routing protocol and with an intent to disrupt the routing mechanism of the network.

*Security threat*: A weakness in the network system, which allows a malicious user to exploit that vulnerability to cause disruption in the network.

*Ad hoc network*: A loosely organized network whose nodes rely on the cooperation of neighbor nodes for data forwarding.

*Threat analysis*: A systematic methodology for finding the security vulnerabilities of the network system.

*Security system*: A system designed based on an in-depth understanding of the security vulnerabilities for preventing or detecting malicious behavior.

*Denial of service*: DoS is an attack scenario, in which the attacker attempts to make a network resource unavailable to its intended users. The motives and targets of DoS may vary, but it generally aims to prevent a service from functioning efficiently or, at all, temporarily or indefinitely.

*Cache*: Cache is a temporary storage location in the node, which is used for storing routing information. Cache system is optimized for fast updating and timely purging of old information.

*Routing information propagation:* The property of a routing protocol to distribute neighborhood topology information to nodes in the network. The information is used to help the nodes to calculate optimal routes.

*Network state*: State of connectivity of a node to other neighbor nodes is referred to as network state.

## Problems

1. What are the unique network characteristics that contribute to the unprecedented security vulnerabilities faced by ad hoc networks?
2. Besides routing insecurity, are there any other unique security threats in ad hoc networks?
3. Identify the possible ways of preventing passive attacks. (Please refer to [8] for hints.)
4. Identify the attack objectives for spoofing, sinking, and selfishness.
5. Design a variant of black hole attack that aims to thwart detection.
6. Design a variant of wormhole attack and discuss its potential for disruption.
7. Similar to the protocol implementation analysis performed in Section 18.4, analyze AODV, a reactive routing protocol, and assess its security vulnerabilities.
8. Compare and contrast the security vulnerabilities of proactive versus reactive routing protocols. Which is more secure? Why?
9. Assess and derive the causal relations for modifying behavior on the sequence number field in OLSR routing messages. Consider all possible network conditions to assess the scope of malicious behavior.
10. With the help of inferred information from the chapter, list the prime components/modules necessary for a comprehensive security system for ad hoc networks.

## References

1. Keng Seng, N. and W.K.G. Seah. Routing security and data confidentiality for mobile ad hoc networks. The 57th Semi-annual Vehicular Technology Conference. 2003. 3: 1821–1825.
2. Lidong, Z. and Z.J. Haas. Securing ad hoc networks. Network, IEEE, 1999. 13(6): 24–30.
3. Brutch, P. and C. Ko. Challenges in intrusion detection for wireless ad-hoc networks. in Applications and the Internet Workshops, 2003. Proceedings. 2003 Symposium on. 2003.
4. Hubaux., J.-P., L. Buttyan., and S. Capkun. The Quest for Security in Mobile Ad hoc Networks. in 2nd ACM international Symposium on Mobile Ad hoc Networking & Computing. 2001. Long Beach, CA, USA: ACM Press.
5. Papadimitratos., P. and Z. Haas. Secure Routing for Mobile Ad hoc Networks. in SCS Communication Networks and Distributed Systems Modeling and Simulation Conference (CNDS 2002). 2002. San Antonio, TX.
6. Huang, Y.A. and W. Lee. Attack analysis and detection for ad hoc routing protocols, in Recent Advances in Intrusion Detection, Proceedings. 2004, Springer-Verlag Berlin: Berlin, p. 125–145.

7. Ning, P. and K. Sun. How to misuse AODV: A case study of insider attacks against mobile ad-hoc routing protocols. Ad Hoc Networks, 2005. 3(6): 795–819.

8. Kong, J., X. Hong, and M. Gerla. A new set of passive routing attacks in mobile ad hoc networks. in Military Communications Conference, 2003. MILCOM 2003. IEEE. 2003.

9. Awerbuch., B., et al. Mitigating Byzantine Attacks in Ad Hoc Wireless Networks. March 2004, Department of Computer Science, John Hopkins University.

10. Tseng, H.C. and B.J. Culpepper. Sinkhole intrusion in mobile ad hoc networks: The problem and some detection indicators. Computers & Security, 2005. 24(7): 561–570.

11. Song, N., L. Qian, and X. Li. Wormhole Attacks Detection in Wireless Ad Hoc Networks: A Statistical Analysis Approach. in Parallel and Distributed Processing Symposium, 2005. Proceedings. 19th IEEE International. 2005.

12. Clausen, T. and P. Jacquet. Optimized Link State Routing Protocol. RFC 3626, October 2003.

13. Tonnesen., A. Implementing and Extending Optimized Link State Routing Protocol, in UniK University Graduate Center, University of Oslo. p. 150.

14. Wong, K.D., T.J. Kwon, and V. Varma. Towards commercialization of ad hoc networks. 2004. 2: 1207–1211.

15. Anjum, F. and P. Mouchtaris. Security for wireless Ad hoc networks. 2007. Wiley.

16. Bangnan, X., S. Hischke, and B. Walke. The role of ad hoc networking in future wireless communications. IEEE Network. 2003. 2: 1353–1358.

17. Yang., S. and J.S. Baras. Modeling vulnerabilities of ad hoc routing protocols. in 1st ACM workshop on Security of ad hoc and sensor networks. 2003. Fairfax, Virginia: ACM Press.

# Chapter 19
# Trust Management in Mobile Ad Hoc Networks

Venkat Balakrishnan, Vijay Varadharajan, and Uday Tupakula

**Abstract** In this chapter, we present the state of the art in trust management systems in mobile ad hoc networks (MANET). First, we consider the rationale for trust management systems by demonstrating the shortcomings of secure routing protocols and incentive-based systems. We then establish the context by exploring the concepts and methodology of traditional trust management systems. In sequent, we analyze and compare recently proposed and few well-reviewed trust management based models for MANET. Finally, we present the limitations and shortcomings of these trust models, and then discuss the possible research directions toward the development of efficient trust management systems.

## 19.1 Introduction

The advent of wireless communication between two nodes and the ability of a node to communicate beyond its radio transmission range using intermediate nodes have led to the growth of *Mobile Ad hoc NETworks (MANET)*. These networks can establish themselves on fly without relying on any infrastructure. Further, they can lay platform for the development of similar networks such as wireless sensor, ubiquitous, peer-to-peer, mesh and vehicular ad hoc networks. Nevertheless, features such as wireless communication, mobility, lack of infrastructure, and absence of power cables, which serve as substratum for the growth of MANET, also hinder the successful deployment of the same. In other words, these features translate into the following issues: (a) promiscuous communication channel, (b) dynamically changing topology causing broken links and sporadic connections, (c) difficulty in achieving cooperation among

V. Balakrishnan (✉)
Information Networked System Security (INSS) Research Group, Department of Computing, Division of ICS, Macquarie University, North Ryde, Sydney, NSW, Australia 2109
e-mail: venkat@ics.mq.edu.au

S. Misra et al. (eds.), *Guide to Wireless Ad Hoc Networks*,
Computer Communications and Networks, DOI 10.1007/978-1-84800-328-6_19,
© Springer-Verlag London Limited 2009

nodes, and (d) constrained battery power. Security is paramount in the presence of these issues, since they obstruct the assumption of centralized or distributed online trusted authorities. Perhaps, the fundamental question that needs to be addressed in MANET is – *how to enable a mobile node to enlist trusted intermediate mobile nodes so that they can cooperate in forwarding the information to a target without modifying the information or obstructing the operation of other mobile nodes.* This advocates that the security of MANET heavily relies on the presence of a trustworthy secure communication layer so that services can be delivered at the higher layers.

Initially, several secure routing protocols [1–13] have been developed to deliver secure routes by authenticating intermediate nodes and verifying the integrity of routing messages. Data transmissions can then be protected using the secure routes discovered by these protocols. However, key management [14–17], which is the basis for proper functioning of secure routing, is difficult to achieve, especially in the absence of centralized authority due to dynamically changing topology and resulting broken links and sporadic connections. Since secure routing protocols are only designed to prevent against predefined attacks and assume all available nodes to perform routing and network management, they are prone to overlook the correct execution of critical network functions such as packet forwarding. For this reason, secure routing protocols fail to enforce cooperation among nodes. Henceforth, they are vulnerable to nodes that are liable to maximize their utility in the resource-constrained MANET through non-cooperation. Consequently, incentive-based systems [18–22] have been proposed to stimulate cooperation among nodes. However, these systems assume nodes to be economically rational, and consider tamper-proof hardware or trusted third parties to make unforgeable incentives. Further, incentive-based systems struggle to address the issue of pricing [23], and other related issues such as how to handle deprivation of incentives. In addition, most of the incentive-based systems address only a subset of nodes known as *selfish nodes* that attempt to retain their battery resource by not forwarding packets for other nodes. Also, incentive-based systems fail to defend against malicious nodes that disrupt the transmissions of other nodes by either dropping or flooding packets.

The main reason for the shortcoming of secure routing and incentive-based systems is that they fail to measure the trustworthiness of nodes based on the latter's dynamically changing behavior. This is so required to react to those behaviors. All these have eventually led to the growth of *trust management systems* [24–58], which are synonymously referred as *detection reaction* and *reputation systems* in the literature. Since trust management systems proactively detect and reactively isolate (or select) malicious (or benign) nodes, these systems are also known as *self-policing systems* [32, 51]. Although these systems lack consensus on the definition of trust at a more fundamental level, they adhere to all or some of the following steps. First, these systems tend to collect evidence for the behavior of other nodes through passive monitoring, acknowledgements, or Intrusion Detection Systems (IDS) [59, 60]. They may also consider evidence from other nodes and define suitable approach for distributing evidence, which is often regarded as recommendations. Second, trust management systems subjectively evaluate and rate the evidence, and then utilize those ratings to predict the future behavior of other nodes. Finally, they make context-based decisions based on the policies defined for those contexts and the behavior anticipated for those nodes involved in the context.

In the following, we detail the issues inherent to MANET in order to demonstrate their impact on the secure routing and incentive-based systems, and to establish the rationale for trust management systems. Section 19.3 introduces the concepts and methodology relevant to traditional trust management systems and their emergence in MANET. In Section 19.4, we analyze and compare various trust management systems proposed for MANET. We investigate the limitation and shortcoming of these systems, and then discuss the possible research directions toward the development of efficient trust management systems in Section 19.5. Section 19.6 summarizes the chapter.

## 19.2  Security Issues in MANET

Although features such as wireless communication, mobility, lack of infrastructure, and absence of power cables serve as substratum for the growth of MANET, they also emerge as a hindrance to the deployment of MANET in the form of: (a) promiscuous communication channel, (b) dynamically changing topology, (c) difficulty in attaining cooperation among nodes (which is a prerequisite for the functioning of network), and (d) constrained battery resources. In the following, we reason out the shortcoming of both secure routing and incentive-based systems, and then we establish the rationale for trust management systems.

### 19.2.1  Wireless Medium: Promiscuous

In contrast to wired networks, mobile nodes in MANET can directly communicate with each other or through intermediate mobile nodes depending on whether they are within the proximity of each other or beyond their radio transmission ranges. Although MANET surpass wired network by eliminating the need for communication cables, their promiscuous wireless communication allow malicious mobile nodes to eavesdrop the transmission of other mobile nodes. Also, malicious nodes can take advantage of the wireless medium to modify those transmissions or inject spoofed packets. Such behaviors are further enhanced in the presence of basic routing protocols [61, 62] which are not designed to: (a) authenticate intermediate nodes, (b) verify routing messages, and (c) protect data transmissions. Secure routing protocols [1–13] promptly defend against malicious nodes that eavesdrop, modify, or inject spoofed packets. They rely on key management systems to establish the secret associations between mobile nodes. They then define appropriate rules to authenticate intermediate nodes or (and) to verify the integrity of routing messages using a suitable cryptography mechanism. The secure path discovered by secure routing protocol is utilized to deliver the protected data to destination. It is evident that secure routing protocols efficiently prevent against predefined attacks, while incentive-based systems address a different class of malicious behaviors. However, key management and secure routing can only defend against external attackers since they are vulnerable to internal attackers. This is so true in MANET, where nodes are prone to physical capture and compromise.

## 19.2.2 Mobility: Broken Links and Dynamically Changing Topology

Mobility introduces flexibility in MANET by facilitating nodes to locate various network services. Although mobility has been shown to help security [63] and to increase the capacity of MANET [64], it is irrefutable that mobility leads to broken links and sporadic connections. This opens door for malicious nodes to report genuine links as broken links and in sequence to disrupt communication between nodes. Note that secure routing protocols are incapable of differentiating genuinely reported broken links from maliciously reported broken links due to their static design. This in turn urges for a defense-in-depth strategy since history of security has always shown that a completely intrusion-free system is infeasible [10], no matter how carefully the prevention mechanism is designed.

As mobility yields dynamically changing topology, it becomes hard to transfer key management systems that have been designed for wired networks to MANET. Hence most of the MANET-based key management systems either assume to pre-install the secret associations within mobile nodes prior to deployment or through the certificates generated by a Certificate Authority (CA). Nevertheless, these systems fail to scale with the non-organizational characteristics of MANET, since nodes can enter or leave at any point during the network lifetime. For instance, pre-installing secret associations within nodes would either cause a newly joining node to struggle to establish secret associations with existing nodes or an existing node to struggle to refresh secret associations with other nodes. In certificate-based systems, key revocation becomes challenging due to the non-organizational characteristics of MANET. Further, dynamically changing topology and thereof the susceptibility to single point of failure categorize CA as an unattractive option for key management in MANET. Also, such systems are vulnerable to malicious nodes that aim to disintegrate the operation of secure routing protocols by disrupting either key refreshment or revocation operation in the form of non-cooperation. All these have motivated to the development of threshold cryptography–based key management systems [14–17]. In these systems, the capability of CA is distributed to all or subset of nodes so that a node can obtain a certificate by combining the certificate shares acquired from distributed CAs. However, it is unclear how adequate count of distributed CAs are assumed to be available for all nodes, given distributed CAs may be unavailable for certificate shares due to broken links, sporadic connections, and random network distribution caused by mobility. Alternatively, reducing the proportion of certificate shares to reduce the requirement of higher distribution of CAs consequently reduces the strength of threshold cryptography–based key management system. This may also lead to the compromise of the system.

### 19.2.3 Self-organized and Lack of Infrastructure: Prerequisite Cooperation for Network Operation

Since self-organized multihop communication in MANET eliminates the necessity for an infrastructure, every mobile node is required to cooperate to forward packets for other mobile nodes. This is so necessary to establish the network during the initial stages of deployment and to sustain during the later stages of operation. Such intrinsic requirement allows malicious nodes to launch Denial of Service (DoS) attacks in the form of dropping or (and) flooding packets. Several simulations [65, 66] have shown how such malicious behaviors can degrade the network performance dramatically. Since secure routing protocols are not tailored to defend against DoS attacks, IDS [59, 60] have been developed to detect such malicious behaviors. However, they are restricted only to observable malicious behaviors for which they define poor decision policies and therefore fail to effectively isolate malicious nodes. The main limitation of MANET-based IDS is their inability to differentiate packet drop attacks from accidental packet drops that result from contention and congestion.

### 19.2.4 Battery-Powered Devices: Constrained Resources

Battery-powered nodes in MANET eliminate the usage of power cables to complement mobile communication. However, the heterogeneity that prevails in the network inspire a category of nodes known as selfish nodes to maximize their resources or network utility by either not forwarding packets for other nodes or forwarding only self-generated packets. This has a profound effect in the network performance and, at extreme conditions, it may also lead to an inoperative network. Note that selfish behaviors are analogous to malicious behaviors such as packet drop and flooding attacks. However, selfish nodes differ from malicious nodes since their intention is to maximize the resource or network utility. Selfish nodes powered with low battery resource attempt to save their resources by not forwarding packets for other nodes. Alternatively, selfish nodes powered with high battery resource aim to hijack the wireless medium by boycotting the Medium Access Control (MAC) protocol's contention resolution mechanism. Selfish nodes then maximize the network utility by propagating only self-generated packets. This has led to the development of incentive-based models [18–22] where incentives are used to stimulate cooperation among nodes. These incentive-based systems defend against only selfish nodes and rely on either a tamper-proof hardware or offline centralized authorities for their operation. Further, they struggle not only to address pricing-related issues but also to manage incentives especially when selfish nodes collude to deprive the incentives. They also respond poorly to boundary nodes, for example, even when the boundary nodes are willing to cooperate with other nodes in order to earn incentives, they might be unable to do so due to lack of opportunities. Incentive-based systems fail to address flooding attacks and concentrate only on the packet drop attacks exhibited by selfish nodes. Also, they overlook the packet drop attacks performed by malicious nodes.

## 19.3 Background: Trust Management Systems

In this section, we investigate the concepts: (a) *trust* and (b) *reputation*, and the methodology adopted in trust management systems in order to establish the background for our analysis and their significance in MANET.

### 19.3.1 Concepts: Trust and Reputation

Trust management and trustworthy computing are becoming increasingly significant in a distributed environment, since they assist the systems in making sensible interactions with unknown parties by providing a basis for more detailed and automated decisions [67]. The concepts, *trust* and *reputation*, are closely related in trust management systems [68] and they are firmly rooted in sociology and psychology. Although there is no universal definition for these concepts due to their rich connection with different disciplines, we confine to computing-oriented definition.

In traditional trust management systems, trust enables a *trustor* to reduce uncertainty in its future interactions with a *trustee*, who is beyond the control of trustor but whose actions are of interest to the trustor and affects the state of trustor. In other words, trust is a subjective probability that enables the trustor to take a binary decision by balancing between the known risks and the opinion held for trustee. Here, only known risks are considered for making decisions as it is difficult to prove unknown risks, and the opinion presents the trustor's relationship with the trustee based upon the trustor's experiences. Other factors that influence the decision are time and context, where context accounts for the type of interaction between trustor and trustee, and the nature of application.

In reputation systems, reputation is defined as the opinion held by the trustor towards the trustee depending on its past experiences with the trustee [68]. In other words, reputation generally represents the trustor's direct relationship with the trustee. Also, trustor's relationship with a second trustee based on its direct relationship with a first trustee and the first trustee's direct relationship with the second trustee is known as indirect relationship. This is possible as nodes are allowed to share their opinions in the network.

Although trust and reputation are used interchangeably in MANET, we define them as follows since they are shown to complement each other from the above discussion. Hence, trust can be defined as the prediction of a node's future action in a context such as forwarding routing messages without modification, while reputation then becomes the opinion held for the node based on the node's past actions and the one that influences the prediction [44]. For this reason, we consider the trust definition in [69] to be more appropriate and timely: "Trust is the firm belief in the competence of an entity to act as expected such that this firm belief is not a fixed value associated with the entity but rather subject to the entity's behavior (reputation held for the entity) and applies only within the context and at a given time."

### 19.3.2 Components of Trust Management Systems

Trust management systems are characterized by the following components: (a) evidence manager to collect and classify evidence; (b) mathematical model to formulate evidence into opinion, and then to use those opinions to predict the result of future interactions; and (c) policy manager to define decision policies for making decisions. Traditional trust management systems were monotonic as they were only designed to authenticate users for granting access to services using digital certificates. Here the digital certificate is considered as evidence, and then the certificate is evaluated with the help of cryptosystems to make an opinion regarding to the user identity. Finally, the policy manager defines simple decision policy, such as *grant access to the user whose identity matches with the digital certificate*. Such a simple trust management system may be warranted in closed organizations where there is a centralized authority, but the same fails to hold in a distributed system where there is no hierarchical relationship between entities. Although policy system [70, 71] has undergone significant expansion, decisions are still based on the credentials presented by users.

Meantime, the development of IDS to meet the shortcoming of prevention systems has led the researchers to consider IDS to contribute behavioral evidence to trust management systems [67]. The merge of IDS and trust management systems results from the inflexibility of IDS to take full advantage of behavioral evidence and the inability of trust management systems to adapt to peer-to-peer distributed networks. Accordingly, Lin and Varadharajan [72] refer such behavioral evidence as *soft evidence* and the credentials used for authentication as *hard evidence*.

### 19.3.3 Significance of Trust Management Systems in MANET

Given that the concepts and methodologies have been briefed above, let us now focus on the significance of trust management systems in MANET. Most of MANET-based trust management systems [24–32, 34, 35, 37–39, 42–49, 51–58] are decentralized and deployed at every node in order to make subjective decisions. These systems predominantly consider only behavioral evidence, and collect them through passive monitoring, acknowledgements, or IDS [59, 60]. The behavioral evidences collected by passive monitoring are based upon the assumption that these evidence are observable and classifiable. Alternatively, the feature to employ IDS to collect behavioral evidence facilitates trust management systems to defend against emerging attacks in MANET. This is in contrast to secure routing and payment systems, where the protocol design has to be changed for every new type of attack. Further, few trust management systems [24–30, 51] consider the evidence resulting from system failure as malicious behavior in order to separate out unreliable nodes along with untrustworthy nodes. Also, trust management systems [24–34, 37–40, 42–46, 48–53, 55–58] allow nodes to exchange opinions (as reputation ratings) among themselves in order to enhance their decisions. However, they neither rely on higher authorities nor they are required to reach consensus with other nodes to make a decision. The reputation ratings that are disseminated to communicate opinions with other nodes are referred as *recommendations*, and they facilitate nodes to establish indirect relationships.

In MANET, trust management systems employ various mathematical models, such as graph theory–based models [37], entropy–based models [56], and Bayesian logic–based models [31], to quantify evidence into opinion and to represent the resulting opinion as a relationship between nodes. Finally, these systems execute simple decision policies such as whether to: (a) accept or reject a newly discovered route, (b) send or forward a packet on behalf of other nodes, (c) accept or ignore a recommendation from another node, (d) delete or retain paths containing misbehaving nodes, (e) abstain or warn others about the misbehaving nodes, and (f) which path to choose for communication. Few systems [73, 74] consider to model trust management systems based on human behavior; however, it is not certain whether such systems will meet the requirements as humans do not seem to always make fully rational trust decisions [75].

## 19.4 Trust Management Systems in MANET

Several trust management systems [24–58] have been proposed for MANET in recent years, and we choose few recently proposed [45, 50, 53, 55–58] and some well-reviewed systems [26–28, 30, 31, 46–48, 76] for analysis. In our analysis, we use the terms: *opinion*, *reputation*, and *relationship* interchangeably based on the hypothesis established in Section 19.3.1.

## 19.4.1 Quantifying Trust in Mobile Ad Hoc Networks

Virendra et al. [55] have proposed a trust model in which trust relationships are utilized to enable nodes to establish keys with other nodes. In their proposal, they eliminate the following considerations: (a) need for a centralized trusted authority to establish keys, and (b) general assumption that nodes are non-malicious at the time of key establishment.

At every node, the decentralized trust model adopts a five-phased approach to establish and maintain trust relationship with other nodes, and the phases are: (a) *initiation and monitor*, (b) *query and evaluation*, (c) *update*, (d) *restructure*, and (e) *re-establishment*. The model implements a trust metric to represent opinions (or relationships) in the range of 0–1, and further subdivides the range into three regions: (a) *bad*, (b) *uncertain*, and (c) *good*, respectively. However, the model lacks sound mathematical proof for the chosen metric.

During the initiation and monitor phase, a node fixes its relationship with other nodes to the median of uncertain region. It promiscuously monitors the packet transmissions of neighbors to collect evidence of trustworthiness for their benign and malicious behaviors. Also, it abstains from sending sensitive data to neighbors until keys are established with them, for which its relationship with them has to be in good region. Query and evaluation phase facilitates the node to collect recommendations from neighbors. However, recommendations are collected only from neighbors with whom the node has good relationship. Further, recommendations are treated secondary to the evidence collected from promiscuous monitoring. During update phase, the node regularly evaluates the relationship depending on the evidence received through promiscuous monitoring and recommendations. As the relationship shifts toward good region, the periodicity of evaluation is reduced. Restructure phase addresses the uncertainty that may creep in the relationship when: (a) neighbors move out of the node's transmission range, and (b) previously moved out neighbors are now back into the node's transmission range. When neighbors are out of range, the node exponentially decays its relationship to the floor of region (bad, uncertain, or good) to which the trust relationship belongs. In sequence, the decayed relationship is resumed from the floor of region (bad, uncertain, or good) when neighbors come back into the node's vicinity. Finally, the re-establishment phase is an optional phase in which a malicious neighbor re-establishes the relationship with the node by performing a linear benign behavior.

The node then utilizes its trust relationship to establish pair-wise and group keys, where keys are established with neighbors for whom trust relationships are in good region. Pair-wise keys are established with them using the mechanism such as *duckling* [77]. In group key establishment, one of the nodes in a group invites all the other neighbors to form a cluster depending on its trust relationships with them and the cluster is known as *Physical Logical Trust Domains (PLTD)*. The neighbors may accept the invitation to join the group or in turn request the inviting node to join their PLTD. The model also permits a node to participate in more than one PLTD.

The model's novelty rests on its scheme to employ *behavioral and recommendation based trust relationships* to establish pair-wise and group keys among nodes. However, it is unclear how the model defends against nodes that perform malicious attacks by spoofing packets with the identities of other nodes prior to the key establishment phase. Other drawbacks include the model's inability to handle recommendation issues such as honest-elicitation,[1] free-riding,[2] and recommender's bias. It is unclear how nodes select a malicious node for forwarding packets, while trustworthy neighbors exist in the neighborhood during re-establishment phase.

## 19.4.2 Establishing Trust in Pure Ad Hoc Networks

Pirzada and McDonald [47] have proposed a trust model to improve reliability in pure ad hoc networks.[3] The model performs three tasks, namely: (a) *trust derivation*, (b) *trust quantification*, and (c) *trust computation*. Trust derivation is synonymous to evidence manager and collects evidence for each type of packet (e.g., route request and route reply packets) within a category such as forwarding data packets, forwarding routing packets without modification, providing gratuitous route replies, and salvaging route errors. Here the opinion is referred as trust and measured in continuous values between $-1$ and $+1$. Similar to [55], the model fails to provide a sound mathematical proof for the chosen trust metric. Trust quantification formulates opinion for each category by combining the trust metric held for each packet type within the category. For this, a weight is assigned for each packet type based on the concepts, *utility* and *importance*, discussed by Marsh in [78]. The value for weight ranges from 0 (unimportant) to $+1$ (important). Here, utility refers to the cost and benefits associated with the context, and importance refers to the context's significance with respect to time. Finally, trust computation is responsible for computing the trustworthiness of a node by combining the

---

[1] A node is subject to honest-elicitation, when it forwards a high recommendation for a malicious node in order to avoid itself from being labeled with a low recommendation by the same malicious node. A malicious node may also exhibit honest-elicitation by forwarding low recommendations for colluding malicious nodes.

[2] A node is subject to free-riding, when it accepts recommendation from other nodes, but fails to reciprocate with recommendations when requested by them.

[3] A network in which centralized authority is absent for managing network services and establishing secure associations.

trust quantified for each category. For this, it adopts the same strategy of assigning weights to each category before combining them together.

Unlike the previous model [55], the system design is composed of only evidence manager, and weak computational logic. It overlooks one of the trust management components: decision manager, since it fails to take advantage of the established trust relationships between nodes. It is not clear whether the evidence manager, i.e., the trust derivation, considers recommendations, and also how the gratuitous route replies and salvaged route errors are accurately verified especially in a pure MANET where there is no pre-established relationship among nodes. This is because nodes can maliciously report a genuine link as broken link and introduce tampered routes through gratuitous route replies and salvaged route errors. Similar to [55], Pirzada and McDonald also fail to demonstrate the strengths and limitations of their proposed model through simulation results.

### 19.4.3 Propagating Trust in Ad Hoc Networks for Reliable Routing

Pirzada et al. enhanced their previous trust model [47] in [46] to address the question of decision manager and recommendations. They incorporated the previous model as a trust agent to collect behavioral evidence and subsequently to compute *direct trust* for a node. Reputation agent is responsible for sharing direct trust values as recommendations with other nodes, and the opinion derived from recommendations is known as *derived trust*. To achieve this, a recommendation is scaled with the direct trust held for recommender. Authors suggest HashCash[4] in order to keep fallacious recommendation requests and recommendations at bay. Combiner then combines direct and derived trusts from trust and reputation agents, respectively, to arrive at the *aggregate trust*. Finally, trustworthy routes are chosen by assigning weights to each link in the route depending on the aggregate trust and then employing shortest path algorithm using assigned weights.

Pirzada et al. further extended the current model in [48] to elaborate the dissemination of recommendations and demonstrated the effectiveness of their extended model through simulation results. In the extended model, nodes disseminate recommendations by appending their direct trust for other nodes along with data packets. This successfully reduces the overhead that may otherwise occur by forwarding separate packets to disseminate recommendations. The simulation results show that DSR coupled with trust model perform 10% better than the normal DSR.

---

[4] A CPU cost function that computes a proof-of-effort token using a cryptographic hash function. The requester tries different combinations of trial string to find a token, and then sends the token to the recommender, which can easily verify the token by performing a single hash operation.

Since the model excludes malicious nodes by resorting to only trustworthy routes, it is interesting to know how the model will include repenting malicious nodes in the future. Also, it is speculative whether a node would append its complete direct trust table (which contains its direct trust for all nodes) or only a fragment of its direct trust table (which corresponds to its direct trust for few selected nodes) along with every data packet. Finally, the notion of disseminating recommendations opens door to honest-elicitation, free-riding, and bias of the recommender.

## 19.4.4 Performance Analyses of the CONFIDANT Protocol (Cooperation Of Nodes: Fairness In Dynamic Ad-Hoc NeTworks)

Buchegger and Boudec have adopted defense-in-depth strategy to enhance the strength of prevention mechanism by proposing a trust management system known as *CONFIDANT* [26, 27]. They utilize the trust model to defend against attacks such as traffic deviation, route salvaging, unusual frequent route updates, silent route tampering, and abstinence from forwarding error, control or data packets. They derive motivation from Richard Dawking's *The Selfish Gene* [79], in which *suckers* and *cheaters* are demonstrated as losers, while *grudgers* as winners.

CONFIDANT collects evidence from direct experiences and recommendations. Later trust relationships are established between nodes based on collected evidence and accordingly used for making trust decisions. These operations are achieved through four interdependent modules: (a) *monitor*, (b) *reputation system*, (c) *path manager*, and (d) *trust manager*. Monitor collects evidence by passively monitoring the transmission of a neighbor after forwarding a packet to the neighbor. It then reports to the reputation system (which consists of a rating list and blacklist) only if the collected evidence represents a malicious behavior. Reputation system changes the rating for a node if the evidence collected for a node's malicious behavior exceeds the pre-defined threshold value. It is important to note that the evidence collected for a malicious behavior through direct experience has more influence than the evidence collected from recommendation. In sequence, path manager makes a decision to delete the malicious node from the path. Also path manager assists the node in making decision such as whether to forward a received packet by cross-checking the upstream node's identity (previous-hop) in the blacklist. The reputation system calls for a time-out operation to handle false accusations, fault rating, and list blow-up. Finally, trust manager is responsible for forwarding and receiving recommendations to and from trustworthy nodes. Here recommendations are known as ALARM messages and trustworthy nodes are referred as *friends*. The ALARM messages received from friends are evaluated for trustworthiness before being sent to the reputation system. Trust manager assists in making trust decisions for the following, whether to: (a) provide and accept routing information, (b) accept a node as a part of route, and (c) take part in a route originated by some other node. Authors demonstrate the effectiveness of their model through extensive simulation results. Interestingly, CONFIDANT nodes drop around 100 packets in the presence of 90% malicious nodes, while DSR nodes drop 10,000 packets in the presence of 10% malicious nodes. Also, CONFIDANT nodes deliver 75% of data packets in the presence of 60% malicious nodes.

Similar to the models discussed earlier, a friend's ALARM message may not reflect the actual status since the friend may be biased. Authors have extended the model in [30, 31] in order to address honest-elicitation issue, where a node: (a) accommodates recommendations only if those recommendations are in agreement with its opinion, and (b) in addition separately measures the trustworthiness of friends for forwarding honest recommendations. However, such a design fails to consider recommendations that report unusual behavior because of the disagreement. Further, the model fails to handle free-riding behavior. The notion of refreshing rating list and blacklist using time-outs is revised in their later proposal [28] to avoid re-entry of malicious nodes.

### 19.4.5  A Reputation-Based Mechanism for Isolating Selfish Nodes in Ad Hoc Networks

Refaei et al. [53] have proposed an autonomous evaluation of a neighbor's reputation depending on the completion of requested service. They then employ the reputation-based trust relationships to isolate selfish nodes. Although evidence collection and decision-making approaches are synonymous to related trust models, they differ from related models by analyzing the impact of approach deployed for computing reputation index.

At every node, the trust model maintains a reputation table which contains the reputation of all neighbors. In the case of packet propagation from a source to destination, all nodes that are positioned along the path increment the reputation index of their next-hop once they have received evidence for the completed propagation from destination in the form of acknowledgement. Nodes along the path verify the reputation of their previous-hop to decide whether to forward or discard the received packet prior to forwarding the packet to their next-hop. In this way, the model isolates selfish nodes which have failed to forward packets for others nodes. Alternatively, if a duplicate packet is received from previous-hop for retransmission, then the reputation index for next-hop (through which the original packet has been forwarded) is accordingly decremented. This is carried out even if the next-hop has not dropped the packet, but one of the downstream nodes that are positioned after the next-hop has dropped the packet. Such a design principle is adopted to make the next-hop responsible for the correct delivery of packet.

Let us now see the three reputation functions that are used to increment or decrement the values of a neighbor's reputation index. The model compares reputation functions by taking the ratio of decrement value to increment value from each function. Note that higher the ratio faster is the isolation of selfish nodes and also higher the number of false positives. The first reputation function is known as *Double Decrement to the Single Increment Ratio (DDSIR)*, in which a node increments the reputation index of next-hop by a positive constant "$n$" for every successfully forwarded packet. For failed delivery, it decrements the reputation index by a constant, "$2n$". Although it is not an aggressive scheme, it mandates the next-hop to forward packets in order to gain reputation. Second function is known as *Hops Away From Source (HAFS)*, and it is similar to DDSIR. However, it is aggressive since it decrements the reputation index of next-hop based on the number of hops "$h$" between the next-hop and source node, i.e., "$2n + m \cdot h$", where "$m$" is a constant. This is to ensure that the reputation index of packet dropping node is decremented more in comparison to the reduction of reputation index of other nodes, which are positioned previous to the dropping node along the route. Such a design prevents the reputation exhaustion of nodes that are positioned along with the dropping node in the route. Finally, in *Random Early Probation (REP)*, a node declines to participate in a route when the reputation index of previous-hop or next-hop is just above a predefined threshold but lesser than the initial reputation index, otherwise it imitates DDSIR. Refusing to participate in such a route is based on the belief that there is a high probability for low reputed neighbors to exhibit selfish behavior.

Simulation results demonstrate that the choice of reputation function has an influence in the isolation of selfish nodes and reducing false positives. The results show that the reputation functions follow the ascending order: (a) HAFS, (b) DDSIR, and (c) REP. Finally, authors recommend DDSIR since it consistently generates low false positives. The results also give an insight into mobility where average increase in mobility decreases the average number of packets dropped by a selfish node. This is due to the decrease in average interaction time. Interesting finding from the results is that the size of a reputation table is inversely proportional to the isolation of selfish nodes.

The design principle to make next-hop responsible for the correct delivery of a packet provided a selfish node that is positioned after the next-hop drops the packet may not hold in the case of malicious nodes. This is because the model assumes selfish nodes to retain the integrity of routing messages and not to generate spoofed packets. Also, it is unclear how the model would adapt for bidirectional transmissions, since its operation is completely dependent on the acknowledgements received from destination. Note that nodes decrement the reputation index of their next-hop after receiving a duplicate packet from previous-hop. Such a deduction may be prone to false positives because one or more nodes following the next-hop may be buffering the packet as a result of heavy contention and congestion.

### 19.4.6 Dynamic Trust Model for Mobile Ad Hoc Networks

In [45], Liu et al. have developed a trust model to establish trust relationships depending on the evidence collected through passive monitoring and recommendations. Unlike related models, trust relationships are measured in discrete values and applied to evaluate the security of routes. In addition, the model explores various approaches for distributing and evaluating recommendations. In the first approach, a node reports its neighbor's malicious behavior to other neighbors. In addition, it also forwards an error message back to the source if the next-hop's trust level falls below the threshold value. In the second approach, a node disseminates reports using controlled flooding, say for three hops. In such case, the receiving nodes consider reports only if the latter originate from trustworthy nodes and forwarded by intermediate nodes, which are also trustworthy. In the final approach, a node maintains a tuple containing: (a) source, (b) partial route, and (c) next-hop for each transmitted packet. Whenever a next-hop behaves benign, the node forwards its direct trust value held for the next-hop back to source. This eliminates hop-by-hop evaluation because the source can now choose a route that is more trustworthy. The model also recommends various strategies for merging direct trust and received reports. In the first approach, the model performs an average of direct trust and reports received from others. In the second approach, it evaluates the report received from a node based on the trust held for that node. In the final approach, it considers the aging of reports and the distance traveled by the report.

In comparison with related models, the proposed approach fails to establish the basis for measuring trust relationships in discrete values, given that the trust evolves continuously in MANET. Since the model has been proposed with an aim to replace secure routing protocols, it is unclear how nodes are authenticated and hence how the collected evidence is accurately measured. Although the trust model presents design alternatives for distributing and evaluating recommendations, those design alternatives remain skeptical in the absence of detailed simulations.

### 19.4.7 Trust Model Based Routing Protocol for Secure Ad Hoc Networks

Xiaoqi Li et al. [76] have proposed a trust model to address the issue of ignorance in trust relationships. Similar to related models, they establish trust relationships based on the evidence collected through passive monitoring and recommendations. The trust relationships are then utilized to choose trustworthy routes, isolate malicious nodes, and forward or request recommendations. The model differs from other models by representing a node's ignorance with respect to the behavior of other nodes as uncertainty. Since traditional probabilistic model can express only either belief or disbelief and not ignorance, the model utilizes subjective logic [80–82], which has sound mathematical foundation in dealing with evidential beliefs to represent trust relationships between nodes. In subjective logic, opinion is three-dimensional and they are – (a) *belief,* (b) *disbelief,* and (c) *uncertainty.* The evidence collected for malicious and benign behaviors are mapped to the components belief and disbelief, while a node's ignorance is represented as uncertainty in an ρpinion. Further, subjective logic provides sound mathematical proof for combining and deriving opinions. During the initial stages of deployment, the model enables a node to set its opinion for other nodes to uncertainty. Unlike related models, the authors take advantage of the secure routing protocol, SAODV [8], to authenticate intermediate nodes and to verify the integrity of routing messages. In addition, they also deploy a separate protocol to communicate recommendations. The protocol consists of three types of messages and they are (a) *Trust Request (TREQ),* (b) *Trust Reply (TREP),* and (c) *Trust Warning (TWARN)* messages. TREQ enables a node to request opinions for other nodes from a recommender, and the recommender responds back to requesting node using TREP. Alternatively, a node proactively disseminates TWARN when it identifies a node to be malicious.

This model is different from the trust models seen so far since it addresses a node's ignorance in its trust relationships with other nodes. However, its conventional style of exchanging recommendations is prone to recommender's bias, honest-elicitation, and free-riding problems. Since authors fail to demonstrate the effectiveness of model using simulation results, the benefit and impact of measuring uncertainty in trust relationships are unknown.

### *19.4.8 Information Theoretic Framework of Trust Modeling and Evaluation for Ad Hoc Networks*

Yan Lindsay et al. have developed a trust model for improving the security of MANET routing protocols in [56]. Similar to related models, the proposed trust model establishes trust relationships using the evidence collected through monitoring and recommendations. The established trust relationships are then used to choose trustworthy routes and to isolate malicious nodes. The model differs from related model by laying axioms for measuring, propagating, and combining trust. This is because the authors believe that it is hard to compare trust metrics from various trust models and believe that the fundamental definition of trust is incomplete. Similar to [76], they define trust as uncertainty and measure uncertainty using entropy. The evidence collected for malicious and benign behaviors are probabilistically mapped by following a modified Bayesian approach. The modification allows the model to give low importance to past evidence and more importance to recent evidence. The probabilistic estimate of Bayesian approach is then mapped to entropy.

In the case of recommendations, authors ascertain that the trust established through a recommendation should not be more than the trust held for the recommender or recommendation. Also, they state that the trust built by using multiple recommendations received from the same node should not be greater than the trust built using the recommendations received from independent nodes. To prevent nodes from knowing a requesting node's favorite recommenders, the node disseminates a *Trust Recommendation Request (TRR)* that contains a chosen list of nodes for which its recommendation trust is greater than threshold, and also includes few more nodes for which it is interested to update its recommendation trust. The requesting node receives recommendations only from the list of nodes that are specified in TRR packet and restricted by Time-To-Live (TTL) value. Note that it is up to the recommender to respond back to TRR. Later, the recommendation trust of recommenders is updated depending on the interaction with the node for which recommendation has been received.

Authors have performed intensive simulations using three types of malicious nodes: (a) nodes that drop packets, (b) nodes that forward wrong recommendations, and (c) nodes that perform both actions. Simulation results show that malicious nodes do not form clusters because once nodes are identified for malicious behaviors then there will be no more routes through them. In the case of selectively malicious behaviors, few of the benign nodes are mistakenly considered as malicious nodes. Finally, authors observe mobility to produce false negatives, for example, benign nodes are labeled as packet droppers due to mobility. However, mobility is believed to decrease the need for recommendations since nodes can move around and meet other nodes frequently.

Interestingly, the paper lays definition for trust and specifies the properties for propagating trust. However, recommendations are prone to recommender's bias, honest-elicitation, and free-riding problems. In addition, it is unclear how TRR and TTL are protected against modification attacks.

### 19.4.9 *Trust Evaluation in Ad Hoc Networks*

In [50], Theodorakopoulos and Baras formulate trust relationships as a shortest path problem in a weighted directed graph. Nodes are represented as vertices and their opinions as weighted edges in the graph. The model considers promiscuously collected evidence to be uncertain and incomplete, and hence represents opinion as a pair of values: (a) *trust* and (b) *confidence*. Trust value represents the trustworthiness of a node and confidence value gives the accuracy of trustworthiness. The model relies on semiring property to combine opinions and to select opinions among multiple opinions. Instances where opinions disagree, confidence value is then taken into account to resolve the disagreement.

For simulations, malicious nodes are designed to provide best opinion for other malicious nodes and worst opinion for benign nodes. The opinions are intermixed for both malicious and benign nodes. As the simulation progresses, the distinction between malicious and benign nodes becomes clear. The other factor influencing the distinction is the proportion of malicious nodes in the network.

The main strength of the model is that it presents a mathematical proof for mapping evidence into opinion and establishing trust relationships. However, it fails to elaborate how recommendations are communicated and how efficient decisions are made using the established trust relationships.

### 19.4.10  Trust-Enhanced Secure Ad Hoc Network Routing

In [57, 58], we have proposed a trust model known as *Secure MANET Routing with Trust Intrigue (SMRTI)* to enhance security decisions in the network. The decisions include whether to – *accept or reject a route from a route discovery, record or ignore a route from a forwarded packet, to forward or discard a packet, to forward a packet for a previous-hop, to send a packet to a next-hop, refresh or revoke the key for a node,* and *which route to choose for the communication.* Note that these decisions are based on the trust relationships established between nodes. The model enables a node to establish trust relationships with other nodes using three types of reputations – (a) *direct,* (b) *observed,* and (c) *recommended.* The direct reputation is formulated from the evidence collected during one-to-one interactions with a node. Observed reputation is unique to this model because nodes observe the interactions between its neighbors to formulate such an opinion. Finally, recommended reputation establishes indirect trust relationship between two nodes through positive recommendations derived from a recommender.

Similar to most of the trust models, trust relationships are measured in continuous values between $[-1, +1]$. SMRTI sets precise rules to map specific values to the collected evidence. For example, the evidence is mapped either to a positive or to a negative value within $[-1, +1]$ depending on whether the evidence represents either a malicious or a benign behavior. The magnitude of the positive value is proportional to the type of event (route request, route reply, route error, or data flow). In contrast, the magnitude of negative value is a function of both the type of event and the malicious behavior (flooding, packet dropping, modification of route sequence number, addition or deletion of routes, and fabrication). Once a node has been identified for a malicious behavior, SMRTI excludes it from the corresponding communication flow until the completion of flow, regardless of the relationship maintained with the node.

The main feature of SMRTI is its ability to prevent a node's opinion from being corrupted by the recommender, and thereof enforces the node to rely on its opinion. Therefore, it better resolves the issues concerned with recommender's bias. Further, it adopts a novel approach to communicate recommendations between nodes that eliminate the need to disseminate explicit packets as recommendations. This prevents recommenders from exhibiting both honest-elicitation and free-riding behaviors. Let us now consider the approach of deriving recommendations. For ease of explanation, the node that provides a recommendation is referred as *recommender* and the recommended node is referred as *recommendee*. SMRTI derives recommendations from a recommender for a recommendee using the route contained in a data packet. It deduces a node's intention to forward or discard an upstream packet received from its previous-hop, as the node's opinion for its previous-hop. It then derives the deduced opinion as the node's recommendation for its previous-hop. Since recommendations are derived from the route of a received packet, the trustworthiness of previous-hop, next-hop, source, destination, and route are evaluated before deriving recommendations. For example, let us consider the scenario where node X unicasts a packet to node N, which contains the route S → O → X → N → C → D. Node N will forward the received packet, only if X, C, S, D, and the route are trustworthy. N derives X's willingness to forward the packet on behalf of O, as X's recommendation for O. Similarly, N derives O's willingness to forward the packet on behalf of S, as O's recommendation for S. The process of deriving recommendations terminates at S as there is no previous-hop for S.

In the absence of secure routing protocols, SMRTI demonstrates efficient performance up to 70% of malicious nodes; however, it is hard for SMRTI to produce similar performance against spoofed packets in the absence of secure routing protocols. Similar to few related models, SMRTI takes advantage of the secure routing protocols to authenticate intermediate nodes and to verify the integrity of routing messages. The main shortcoming of SMRTI is lack of sound mathematical proof for its evidence mapping into trust relationships. However, it stands out from other trust models in proposing a novel approach to eliminate honest-elicitation, free-riding, and recommender's bias.

## 19.5 Thoughts for Practitioners

In this section, we summarize our analysis of various trust models that have been discussed above. Although trust management systems in MANET have introduced defense-in-depth strategy to enhance the security of communications, they are still incomplete due to some certain limitations and shortcomings.

### 19.5.1 Behavioral Evidence

Most of the trust management systems in MANET consider only behavioral evidence since they are observable and classifiable. However, it is not possible to collect evidence for all observable behaviors. Further, collected evidence may not be accurate due to: (a) *ambiguous collisions*, (b) *receiver collisions*, and (c) *limited transmission power*. Ambiguous collisions occur when a node receives a packet while it passively monitors the transmission of another node. Alternatively, receiver collision occurs when a node transmits a packet to its next-hop at the same time the next-hop receives another packet from a different neighbor. In such situation, even if the node fails to retransmit the packet, it can increase the reputation at its previous-hop because the previous-hop would have captured the transmission (that has undergone receiver collision) as a valid transmission. Similarly, a selfish node can circumvent a monitoring previous-hop by varying its transmission power such that the signal only reaches the monitoring previous-hop but not to its next-hop. Note that the selfish node requires the positional knowledge of each of its neighbors in order to vary the transmission power.

The severity increases when trust management systems collect only behavioral evidence and fail to consider the credentials or secret associations to authenticate nodes. In the absence of authentication, evidence may be collected from a spoofed packet, and henceforth corresponding trust relationships and resulting decisions may be inaccurate. This clearly advocates the need to authenticate intermediate nodes and to verify routing messages using secure routing protocols, and only few models [57, 58, 76] acknowledge such requirement.

## *19.5.2 Recommendations*

In MANET, if nodes happen to rely only on the evidence collected from direct interactions, then they can defend against malicious nodes only after making interactions with them. For this reason, nodes exchange their opinions of other nodes with their neighbors in the form of recommendations. This way they partially avoid encountering malicious nodes even without interacting with them. However, recommendations are effective only if the trust model is capable of deploying some strategies to eliminate or at least cope with – (a) false accusations or false praise, which we collectively refer as *honest-elicitation*, (b) *free-riding* in which neighbors fail to reciprocate with recommendations, and (c) *recommender's bias* resulting from the recommender's subjective evaluation of evidence.

A straightforward defense against honest-elicitation is to consensus the received recommendations so that valid recommendations are differentiated from falsified recommendations. However, this approach is vulnerable to collusion attack, if the total number of recommendations generated by colluding malicious neighbors surpass the recommendations disseminated by benign neighbors. Hence, it becomes important not to penalize any recommender as honest recommendations may fail to emerge due to the consensus approach. Alternatively, a node may defend against honest-elicitation by only considering recommendations that are in agreement with its belief. Such an approach overlooks the recommendation that may otherwise report the infrequent malicious behavior of a neighbor. Recommender's bias is another issue pertaining to recommendations. Unlike honest-elicitation, here the recommender may disseminate an honest recommendation, which is derived from either an optimistic or a pessimistic evaluation of collected evidence. Such recommendations are critical in transitive trust relationships, where a node that receives a recommendation believes the recommender to evaluate the trustworthiness of recommended node in the same manner it evaluates the trustworthiness of recommender. Free-riding confronts the advantages of recommendations as nodes adhere to the specifications of routing protocol in forwarding route discovery and data packets but fail to follow the specifications of trust model. One of the solutions to thwart free-riding is to opt for recursive design so that nodes are enforced to communicate recommendations with other nodes in the same manner they are enforced to forward packets for other nodes. Finally, propagation of recommendations opens door to many questions such as whether to: (a) restrict recommendations to single or multiple hops and (b) attach opinions for all nodes or report only malicious nodes. Answers for such questions rests on the balance between the overhead incurred in disseminating recommendations and the importance of informing other nodes regarding malicious nodes.

These issues lead us to question whether recommendations need to be considered at all in trust management systems. However, the benefit of defending against malicious nodes even without making interactions with them prioritizes the need to consider recommendations. Perhaps, the direction to look for solution is to investigate the trade-off between the advantage of considering recommendations to establish trust relationships with unknown nodes and the benefit of ignoring recommendations as a result of their vulnerabilities. Among related trust models, our model [57, 58] establishes the balance by addressing the issues pertaining to recommendations and at the same time taking advantage of the benefits of communicating recommendations.

### 19.5.3 Modeling Ignorance in Trust Relationships

Most if not all trust models, often fail to represent the aspect of ignorance and the associated uncertainty, which are intrinsic to the establishment of trust relationships in MANET. Hence, a trust relationship between two nodes may not always reflect the actual relationship and consequently the executed decision may not always be accurate. For example, consider a new node joining the network. In such instance, existing nodes in the network may not have a record of past evidence to trust or distrust a newly joining node. Assigning an arbitrary level of trust for the new node poses several issues. The trust models address this issue either by pessimistically assigning a low level of trust or by optimistically assigning a neutral or high level of trust to the new node. The purpose of pessimistic approach is to compel the new node to exhibit a consistent benign behavior from the point it enters the network. However, in some of these models, it is not always clear as to how the less trusted new node is selected for a communication when nodes with high trust values exist in network. If a new node is not preferred for a communication due to its low level of trust, then it lacks the opportunity to gain trust with existing nodes. Alternatively, with the optimistic approach, the aim is to promptly identify whether the new node exhibits malicious behavior from the point it enters into the network. Prompt identification is feasible because the neutral or high level of trust assigned for a new node decreases rapidly as the malicious behaviour increases. However, the optimistic approach favors existing malicious nodes to re-enter the network with a new identity. These issues arise as trust models explicitly fail to represent an existing node's ignorance about a newly joining node's behaviour. The issue also extends to nodes that have already established trust relationship with one another depending on their interactions and recommendations. For example, when a node moves away from a neighbor (due to mobility), it is unclear whether to consider the neighbor with the same level of trust or distrust during the next interaction (when it returns). It may be that the neighbour could retain its current behaviour, which may be either benign or malicious. Considering the neighbor to be benign, there is a chance for the neighbor to be compromised prior to the next interaction. Alternatively, if the neighbor is considered to be malicious, then it may be repenting and expecting for an interaction to improve its relationship. Another possibility is that the neighbor may be malicious as a result of compromise, and it may also be redeemed prior to the next interaction. These issues are often addressed by either increasing or decreasing the trust or distrust of the nodes in proportion to the duration for which they are out of communication. The seriousness of this approach is that a node's ignorance of other nodes is represented by either increasing or decreasing its trust or distrust for them, which should denote to their benign or malicious behaviour, respectively. Hence the failing to explicitly represent the notion of ignorance and the associated uncertainty has a fundamental impact on the trust model. Recently, few trust models [50, 56, 76] have considered this issue by modeling ignorance in the established trust relationships.

### 19.5.4 Selective Malicious Behaviors

Majority of trust models consider nodes to exhibit persistent malicious or benign behaviors. This is not true in MANET where nodes are either prone to capture and compromise or redeemed after being compromised. Modeling trust management systems to handle such selective malicious behaviors enable prompt detection and isolation of compromised benign nodes. Alternatively, it enables the detection of malicious nodes that may perform benign behaviors to remain unidentified or to gain trustworthiness in its neighborhood for a future objective. Similar approach can be employed to detect selfish nodes that aim for maximum utility by selectively dropping the packets received from neighbors.

## 19.5.5  Evidence to Opinion Mapping

Only few models [26–28, 30, 31, 45, 50, 56, 76] are built on strong mathematical proofs to map evidence to opinion. Remaining models provide weak basis for evidence to opinion mapping and also fail to provide valid reasoning for continuous or discrete values, which are chosen as the trust metric. This consequently leads to inconsistencies in the evaluations and introduces difficulty in comparing the results with other models. The other issue in managing opinions is whether to give more weightage to recent or past behaviors. Past behaviors are given more weightage in order to avoid recent wrong observation to overshadow past benign behaviors; however, it is also important to prevent nodes from taking leverage of past benign behaviors. Since trust evolves continuously in a dynamically changing MANET, it is advantageous to give more importance to recent behaviors. In such case, even if a benign node's behavior has been mistakenly evaluated for malicious behavior, the model should be designed to allow benign nodes to regain their reputation through their future benign behaviors. In other words, the logic should not reverse the existing opinion upside down. Another important factor that needs consideration while choosing the mathematical logic is the ability of the logic to prioritize different types of opinion. For example, opinion formulated from one-to-one interactions should take more weightage than the opinion formulated from recommendations.

Finally, not many models [59] enable nodes to call for a strict decision to exclude malicious nodes in their communication and to deny requests that have been received from malicious nodes. Such models have to be designed to take advantage of the established trust relationships so that they could efficiently make decisions to: (a) accept or reject a discovered route, (b) record or ignore a route from a forwarded packet, (c) forward or discard a packet, (d) forward a packet for a previous-hop, (e) send a packet to a next-hop, (f) refresh or revoke a secret association with a node, and (g) which route to choose for the communication.

## 19.6  Directions for Future Research

Although few trust models [57, 58, 76] address the issue related to spoofed behavioral evidence by taking advantage of secure routing protocols, they are incomplete and require additional research along the following directions. Trust models need to collect evidence from secure routing protocols, apart from using the latter for authentication and verification of messages. Further research is also required to improve the design of trust models so that they can enhance the security decisions of secure routing protocols and key management models.

As mentioned earlier, few models [26–28, 30, 31] measure the relative rate of selective malicious behaviors. Nevertheless, they fail to take advantage of their measurement by not defining efficient policies and categorizing such nodes into granular categories. Further research is required to explore the contexts for which selectively misbehaving nodes are suitable. In sequence, it is necessary to investigate appropriate mathematical models for classifying selectively misbehaving nodes into granular categories. Since IDS evolves with emerging attacks, thorough research is necessary to select suitable IDS for MANET and accordingly to evolve the decision policies of evidence manager. Finally, adequate simulation study is needed to understand the impact of ignorance in trust relationships as pointed out by the recently proposed trust models [50, 56, 76].

## 19.7 Conclusion

In this chapter, we have presented the state of the art in trust management systems in MANET. Initially, we have demonstrated the shortcomings of secure routing protocols and incentive-based systems to set up the rationale for trust management systems. We have then established the context for trust management systems in MANET by exploring the concepts and methodology of traditional trust management systems. Various trust models including recently proposed and few well reviewed have been analyzed and compared to study the state of trust management system in MANET. Finally, we have presented the analysis of these trust models and discussed their limitations and shortcomings. In addition, we have also highlighted the recent research directions toward the development of efficient trust management systems.

## Terminologies

*Trust management system*: A system that decides whether an entity can perform an action depending on the policies defined and the opinion derived using the credentials presented by entity.

*Trust*: It is a subjective belief that an entity will act as expected and depends on the opinion held for the entity.

*Reputation*: It is the opinion held for an entity based on the credentials collected or presented by the entity, and those credentials include behavioral evidence, certificates, etc.

*Security*: The process of denying access to unauthorized users and preventing unauthorized access to data.

*Trust-enhanced security system*: A security system that enhances its actions based on the decisions provided by a trust management system.

*Self-policing system*: A system in which evidence is reactively collected and decisions are proactively executed without the intervention of any higher authority.

*Detection reaction system*: It collects evidence by persistently monitoring the actions and then reacts depending on the policies defined for handling an anomaly.

*Mobile ad hoc networks (MANET)*: Network in which nodes are mobile and capable of establishing wireless communication with other mobile nodes (which are beyond their transmission range) in the absence of any infrastructure.

*Intrusion detection system*: A non-intrusive system that detects malicious or abnormal actions.

*Secure routing*: A system that deploys security mechanism to authenticate and verify the integrity of communication.

## Questions

1. What are the inherent issues in MANET?
2. What are the various security trends in MANET?
3. Why secure routing protocols do not serve as a complete security solution for MANET?
4. Why incentive-based system fails to complement secure routing protocols?
5. What is the difference between malicious and selfish nodes?
6. Why trust management system is necessary for MANET?
7. What are the shortcomings and limitations of trust management systems in MANET?
8. Why behavioral evidence is not sufficient to establish trust relationships?
9. What is the necessity to represent ignorance in trust relationships?
10. Why past evidence is given more weightage than recent evidence, and why it is important to give more weightage to recent evidence?

## References

1. S. Capkun and J.-P. Hubaux. "BISS: Building Secure Routing out of an Incomplete Set of Security Associations", *Proceedings of the 2003 ACM Workshop on Wireless Security*, San Diego, CA, USA, pp. 21–29, 2003
2. Y.-C. Hu and A. Perrig. "A Survey of Secure Wireless Ad Hoc Routing". *IEEE Security and Privacy*, 2(3), 28–39, 2004
3. Y.-C. Hu, A. Perrig and D. B. Johnson. "Packet Leashes: A Defense against Wormhole Attacks in Wireless Networks", *Proceedings of the 22nd Annual Joint Conference of the IEEE Computer and Communications Societies*, pp. 1976–1986, 2003

4. Y.-C. Hu, A. Perrig and D. B. Johnson. "Ariadne: A Secure On-demand Routing Protocol for Ad Hoc Networks", *Proceedings of the International Conference on Mobile Computing and Networking*, Atlanta, Georgia, USA, pp. 12–23, 2002

5. Y.-C. Hu, A. Perrig and D. B. Johnson. "Efficient Security Mechanisms for Routing Protocols", *Proceedings of the Network and Distributed System Security Symposium (NDSS'03)*, San Diego, USA, pp. 57–73, 2003

6. Y.-C. Hu, A. Perrig and D. B. Johnson. "Rushing Attacks and Defense in Wireless Ad hoc Network Routing Protocols", *Proceedings of the 2003 ACM workshop on Wireless security*, San Diego, CA, USA, pp. 30–40, 2003

7. P. Papadimitratos and Z. J. Haas. "Secure Routing for Mobile Ad hoc Networks", *Proceedings of the SCS Communication Networks and Distributed Systems Modeling and Simulation Conference*, San Antonio, TX, 2002

8. K. Sanzgiri, B. Dahill, B. N. Levine, C. Shields and E. M. Belding-Royer. "A Secure Routing Protocol for Ad Hoc Networks", *Proceedings of the 10th IEEE International Conference on Network Protocols (ICNP'02)*, Paris, France, pp. 78–89, 2002

9. V. Varadharajan, R. Shankaran and M. Hitchens. "Security for cluster based ad hoc networks". *Elsevier-Science Direct, Computer Communications*, 27(5), 488–501, 2004

10. H. Yang, X. Meng and S. Lu. "Self-Organized Network-Layer Security in Mobile Ad hoc Networks", *Proceedings of the ACM workshop on Wireless Security. International Conference on Mobile Computing and Networking*, Atlanta, GA, USA, pp. 11–20, 2002

11. M. G. Zapata. "Secure Ad hoc On-Demand Distance Vector (SAODV) Routing". *IETF Internet Draft (Work in Progress), draft-guerrero-manet-saodv-00.txt*, pp. 2001

12. M. G. Zapata. "Secure Ad hoc On-Demand Distance Vector Routing". *ACM SIGMOBILE Mobile Computing and Communications Review*, 6(3), 106–107, 2002

13. M. G. Zapata and N. Asokan. "Securing Ad hoc Routing Protocols", *Proceedings of the ACM workshop on Wireless Security. International Conference on Mobile Computing and Networking*, Atlanta, GA, USA, pp. 1–10, 2002

14. L. Zhou and Z. J. Haas. "Securing Ad Hoc Networks". *IEEE Network*, 13(6), pp. 24–30, 1999

15. H. Luo and S. Lu. "Ubiquitous and Robust Authentication Services for Ad Hoc Wireless Networks". Technical Report (200030), UCLA Computer Science Department, 2000

16. H. Luo, P. Zerfos, J. Kong, S. Lu and L. Zhang. "Self-securing Ad Hoc Wireless Networks", *Proceedings of the IEEE ISCC*, 2002

17. S. Capkun, L. Buttyan and J.-P. Hubaux. "Self-Organized Public-Key Management for Mobile Ad Hoc Networks", *Proceedings of the ACM International Workshop on Wireless Security (WiSe 2002)*, 2002

18. L. Buttyan and J. Hubaux. "Nuglets: A virtual currency to stimulate cooperation in self-organized ad hoc networks". Technical Report (DSC/2001/001), Swiss Federal Institute of Technology, 2001

19. J. Crowcroft, R. Gibbens, F. Kelly and S. Östring. "Modelling Incentives for Collaboration in Mobile Ad Hoc Networks". *Elsevier – Performance Evaluation*, 57(4), 427–439, 2004

20. Q. He, D. Wu and P. Koshla. "A Secure Incentive Architecture for Ad-hoc Networks". *Wiley's Wireless Communications and Mobile Computing (Special Issue: Wireless Networks Security)*, 2006

21. S. Zhong, J. Chen and Y. R. Yang. "Sprite: A Simple, Cheat-proof, Credit-based System for Mobile Ad-hoc Networks", *Proceedings of the IEEE 22nd Annual Joint Conference of the IEEE Computer and Communications Societies (INFOCOM 2003)*, pp. 1987–1997, 2003

22. N. B. Salem, L. Butty'an, J.-P. Hubaux and M. Jakobsson. "A Charging and Rewarding Scheme for Packet Forwarding in Multi-hop Cellular Networks", *Proceedings of the 4th ACM international symposium on Mobile Ad hoc Networking & Computing*, Annapolis, Maryland, USA, pp. 13–24, 2003

23. E. Huang, J. Crowcroft and I. Wassell. "Rethinking Incentives for Mobile Ad hoc Networks", *Proceedings of the ACM SIGCOMM Workshop on Practice and Theory of Incentives in Networked Systems (SIGCOMM'04 Workshops)*, Portland, Oregon, USA, pp. 191–196, 2004

24. S. Buchegger and J.-Y. L. Boudec. "Nodes Bearing Grudges: Towards Routing Security, Fairness, and Robustness in Mobile Ad Hoc Networks", *Proceedings of 10th Euromicro PDP (Paralel, Distributed and Network-based Processing)*, Gran Canaria, Spain, pp. 403–410, 2002

25. S. Buchegger and J.-Y. L. Boudec. "Cooperative Routing in Mobile Ad-hoc Networks: Current Efforts Against Malice and Selfishness", *Proceedings of Mobile Internet Workshop (Informatik 2002)*, Dortmund, Germany, 2002

26. S. Buchegger and J.-Y. L. Boudec. "Performance analysis of the CONFIDANT protocol", *Proceedings of 3rd ACM International Symposium on Mobile Ad hoc Networking & Computing*, Lausanne, Switzerland, pp. 226–236, 2002

27. S. Buchegger and J.-Y. L. Boudec. "Performance Analysis of the CONFIDANT Protocol. Cooperation Of Nodes – Fairness In Dynamic Ad-hoc NeTworks." Technical Report (IC/2002/01), EPFL I&C, 2002

28. S. Buchegger and J.-Y. L. Boudec. "Coping with False Accusations in Misbehavior Reputation Systems for Mobile Ad-hoc Networks". Technical Report (IC/2003/31), EFPL, 2003

29. S. Buchegger and J.-Y. L. Boudec. "The Effect of Rumor Spreading in Reputation Systems for Mobile Ad-hoc Networks", *Proceedings of Modeling and Optimization in Mobile, Ad Hoc and Wireless Networks (WiOpt 2003)*, Sophia-Antipolis, France, 2003

30. S. Buchegger and J.-Y. L. Boudec. "A Robust Reputation System for Mobile Ad-hoc Networks". Technical Report (IC/2003/50), EPFL-IC-LCA, Lausanne, Switzerland, 2003

31. S. Buchegger and J.-Y. L. Boudec. "A Robust Reputation System for P2P and Mobile Ad-hoc Networks", *Proceedings of the Second Workshop on the Economics of Peer-to-Peer Systems (P2PEcon 2004)*, Harvard University, Cambridge MA, USA, pp. 2004

32. S. Buchegger and J.-Y. L. Boudec. "Self-Policing Mobile Ad-hoc Networks" in *Mobile Computing Handbook*, M. Ilyas and I. Mahgoub, Ed.: CRC Press, USA, 2004

33. P. Lamsal. "Requirements for Modeling Trust in Ubiquitous Computing and Ad Hoc Networks". Technical Report (HUT TML – Course T-110.557 – ISBN 951-22-6309-2 ISSN 1456-7628 TML-C8), Helsinki University of Technology, 2002

34. P. Michiardi and R. Molva. "Core: A COllaborative Reputation mechanism to enforce node cooperation in Mobile Ad Hoc Networks", *Proceedings of the IFIP – Communication and Multimedia Security Conference*, 2002

35. S. Bansal and M. Baker. "Observation-based Cooperation Enforcement in Ad Hoc Networks". Technical Report (CoRR cs.NI/0307012), Stanford University, 2003

36. T. Hughes, A. Muckelbauer, J. Denny and J. Etzl. "Dynamic Trust Applied to Ad Hoc Network Resources", *6th International Workshop on Trust, Privacy, Deception, and Fraud in Agent Societies. Proceedings of Autonomous Agents & Multi-Agent Systems Conference (AAMAS 2003)*, Melbourne, Australia, pp. 1–7, 2003

37. Y. Liu and Y. R. Yang. "Reputation Propagation and Agreement in Mobile Ad-hoc Networks", *Proceedings of IEEE Wireless Communications and Networking (WCNC 2003)*, New Orleans, USA, pp. 1510–1515, 2003

38. D. Senn. "Reputation and Trust Management in Ad Hoc Networks with Misbehaving Nodes", *Diploma Thesis*. Information Security, Swiss Federal Institute of Technology Zurich 2003

39. P. Dewan, P. Dasgupta and A. Bhattacharya. "On Using Reputations in Ad hoc Networks to Counter Malicious Nodes", *Proceedings of 10th IEEE International Conference on Paralel and Distributed Systems (ICPADS 2004)*, California, USA, pp. 665–672, 2004

40. L. Eschenauer, V. D. Gligor and J. Baras. "On Trust Establishment in Mobile Ad-Hoc Networks", *Proceedings of 10$^{th}$ International Security Protocols Workshop*, Cambridge, UK, pp. 47–66, 2004

41. A. Fernandes, E. Kotsovinos, S. Ostring and B. Dragovic. "Pinocchio: Incentives for Honest Participation in Distributed Trust Management", *Proceedings of 2nd International Conference on Trust Management (iTrust 2004)*, Oxford, UK, pp. 63–77, 2004

42. Q. He, D. Wu and P. Khosla. "SORI: A Secure and Objective Reputation-based Incentive Scheme for Ad-hoc Networks", *Proceedings of IEEE Wireless Communications and Networking Conference (WCNC 2004)*, Atlanta, USA, pp. 825–830, 2004

43. T. Jiang and J. S. Baras. "Ant-based Adaptive Trust Evidence Distribution in MANET", *Proceedings of 24th International Conference on Distributed Computing Systems Workshops (ICDCSW 2004)*, Tokyo, Japan, pp. 588–593, 2004

44. J. Liu and V. Issarny. "Enhanced Reputation Mechanism for Mobile Ad Hoc Networks", *Proceedings of the 2nd International Conference on Trust Management (iTrust 2004)*, Oxford, UK, pp. 48–62, 2004

45. Z. Liu, A. W. Joy and R. A. Thompson. "A Dynamic Trust Model for Mobile Ad Hoc Networks", *Proceedings of 10th IEEE International Workshop on Future Trends of Distributed Computing Systems (FTDCS 2004)*, Suzhou, China, pp. 80–85, 2004

46. A. A. Pirzada, A. Datta and C. McDonald. "Propagating trust in Ad-hoc Networks for Reliable Routing", *Proceedings of IEEE International Workshop on Wireless Ad-Hoc Networks*, Oulu, Finland, pp. 58–62, 2004

47. A. A. Pirzada and C. McDonald. "Establishing Trust In Pure Ad-hoc Networks", *Proceedings of 27th conference on Australasian Computer Science*, Dunedin, New Zealand, pp. 47–54, 2004

48. A. A. Pirzada, C. McDonald and A. Datta. "Dependable Dynamic Source Routing without a Trusted Third Party", *Proceedings of 28th Australasian conference on Computer Science*, Newcastle, Australia, pp. 79–85, 2005

49. N. Pissinou, T. Ghosh and K. Makki. "Collaborative Trust-Based Secure Routing in Multihop Ad Hoc Networks", *Proceedings of the 3rd International IFIP-TC6 Networking Conference*, Athens, Greece, pp. 1446–1451, 2004

50. G. Theodorakopoulos and J. S. Baras. "Trust Evaluation in Ad-hoc Networks", *Proceedings of ACM Workshop on Wireless security (WiSe'04)*, Philadelphia, USA, pp. 1–10, 2004

51. S. Buchegger and J.-Y. Le Boudee. "Self-Policing Mobile Ad Hoc Networks by Reputation Systems". *IEEE Communications Magazine*, 43(7), 101–107, 2005

52. J. Mundinger and J.-Y. Le Boudec. "Analysis of a Reputation System for Mobile Ad-Hoc Networks with Liars", *Proceedings of the 3rd International Symposium on Modeling and Optimization in Mobile Ad Hoc and Wireless Networks (WIOPT 2005)*, Trentino, Italy, pp. 41–46, 2005

53. Y. Rebahi, V. E. Mujica-V and D. Sisalem. "A Reputation-Based Trust Mechanism for Ad hoc Networks", *Proceedings of 10th IEEE Symposium on Computers and Communications (ISCC 2005)*, Cartagena, Spain, pp. 37–42, 2005

54. M. T. Refaei, S. Vivek, L. DaSilva and M. Eltoweissy. "A Reputation-based Mechanism for Isolating Selfish Nodes in Ad Hoc Networks", *Proceedings of 2nd Annual international Conference on Mobile and Ubiquitous Systems: Networking and Services (MOBIQUITOUS 2005)*, Washington, DC, USA, pp. 3–11, 2005

55. M. Virendra, M. Jadliwala, M. Chandrasekaran and S. Upadhyaya. "Quantifying Trust in Mobile Ad-hoc Networks", *Proceedings of International Conference on Integration of Knowledge Intensive Multi-Agent Systems (KIMAS)*, Waltham, Massachusetts, USA, pp. 65–70, 2005

56. S. Yan Lindsay, Y. Wei, H. Zhu and K. J. R. Liu. "Information Theoretic Framework of Trust Modeling and Evaluation for Ad Hoc Networks". *IEEE Journal on Selected Areas in Communications*, 24(2), 305–317, 2006

57. V. Balakrishnan, V. Varadharajan, P. Lucs and U. Tupakula. "Trust Enhanced Secure Mobile Ad hoc Network Routing", *2nd IEEE International Symposium on Pervasive Computing and Ad Hoc Communications (PCAC 2007), Proceedings of the 21st IEEE International Conference on Advanced Information Networking and Applications Workshops(AINAW 2007)*, Niagara Falls, Canada, pp. 27–33, 2007

58. V. Balakrishnan, V. Varadharajan, U. Tupakula and P. Lucs. "Trust and Recommendations in Mobile Ad hoc Networks", *Proceedings of the 3rd International Conference on Networking and Services (ICNS 2007)*, Athens, Greece, pp. 64–69, 2007

59. S. Marti, T. J. Giuli, K. Lai and M. Baker. "Mitigating Routing Misbehavior in Mobile Ad Hoc Networks", *Proceedings of the 6th annual international conference on Mobile Computing and Networking (MOBICOM)*, Boston, Massachusetts, United States, pp. 255–265, 2000

60. Y.-a. Huang and W. Lee. "A Cooperative Intrusion Detection System for Ad Hoc Networks Security", *Conference on Computer and Communications, Proceedings of the 1st ACM workshop on Security of Ad hoc and Sensor Networks*, Fairfax, Virginia, pp. 135–147, 2003

61. D. B. Johnson, D. A. Maltz and J. Broch. "DSR: The Dynamic Source Routing Protocol for Multihop Wireless Ad hoc Networks" in *Ad hoc Networking*, C. E. Perkins, Ed.: Addison-Wesley Longman Publishing Co., Inc., Boston, MA, USA, 2001

62. C. E. Perkins, E. M. Royer, S. R. Das and M. K. Marina. "Performance Comparison of two On-demand Routing Protocols for Ad hoc Networks". *IEEE Personal Communications*, 8(1), 16–28, 2001

63. S. Capkun, J.-P. Hubaux and L. Buttyán. "Mobility Helps Security in Ad Hoc Networks", *Proceedings of the 4th ACM International Symposium on Mobile Ad hoc Networking & Computing.*, Annapolis, Maryland, USA, pp. 46–56, 2003

64. M. Grossglauser and D. N. C. Tse. "Mobility Increases the Capacity of Ad Hoc Wireless Networks". *IEEE/ACM Transactions On Networking*, 10(4), 477–486, 2002

65. P. Michiardi and R. Molva. "Simulation-based Analysis of Security Exposures in Mobile Ad Hoc Networks". *Proceedings of the European Wireless Conference,* Florence, Italy, 2002

66. P. Ning and K. Sun. "How to Misuse AODV: A Case Study of Insider Attacks against Mobile Ad-hoc Routing Protocols", *Proceedings of the IEEE Workshop on Information Assurance United States Military Academy*, West Point, NY, USA, pp. 60–67, 2003

67. S. Ruohomaa and L. Kutvonen. "Trust Management Survey", *Proceedings of the 3rd International Conference on Trust Management (iTrust 2005)*, Rocquencourt, France, pp. 77–92, 2005

68. L. Mui, M. Mohtashemi and A. Halberstadt. "A Computational Model of Trust and Reputation", *Proceedings of the 35th Annual Hawaii International Conference on System Sciences (HICSS 2002)*. Hawaii, USA, pp. 2431–2439, 2002

69. F. Azzedin and M. Maheswaran. "Evolving and Managing Trust in Grid Computing Systems", *Proceedings of the IEEE Canadian Conference on Electrical & Computer Engineering (CCECE '02)*, pp. 1424–1429, 2002

70. M. Blaze, J. Feigenbaum and J. Lacy. "Decentralized Trust Management", *Proceedings of the IEEE Symposium on Security and Privacy*, Oakland, CA, USA, pp. 164–173, 1996

71. M. Blaze, J. Feigenbaum and A. D. Keromytis. "KeyNote: Trust Management for Public-Key Infrastructures (Position Paper)", *Proceedings of the 6th International Workshop on Security Protocols*, Cambridge, UK, pp. 59–63, 1998

72. C. Lin and V. Varadharajan. "Modelling and Evaluating Trust Relationships in Mobile Agents Based Systems", *Proceedings of the International Conference on Applied Cryptography and Network Security*, Kunming, China, pp. 176–190, 2003

73. L. Capra. "Towards a Human Trust Model for Mobile Ad-hoc Networks". *Proceedings of the 2nd UK-UbiNet Workshop,* London, UK, 2004

74. L. Capra. "Engineering Human Trust in Mobile System Collaborations", *Proceedings of the 12th International Symposium on the Foundations of Software Engineering (SIGSOFT)*, Newport Beach, CA, USA, pp. 107–116, 2004

75. C. M. Jonker, J. J. P. Schalken, J. Theeuwes and J. Treur. "Human Experiments in Trust Dynamics", *Proceedings of the 2nd International Conference (iTrust 2004)*, Oxford, UK, pp. 206–220, 2004

76. L. Xiaoqi, M. R. Lyu and L. Jiangchuan. "A Trust Model Based Routing Protocol for Secure Ad Hoc Networks", *Proceedings of IEEE Aerospace Conference*, Big Sky, Montana, USA, pp. 1286–1295, 2004

77. F. Stajano and R. J. Anderson. "The Resurrecting Duckling: Security Issues in Ad-Hoc Wireless Networks", *Proceedings of the 7th Security Protocols Workshop*, Berlin, Germany, pp. 172–182, 2000

78. S. Marsh. "Formalizing Trust as a Computational Concept", *Ph.D. Thesis.* Department of Computer Science, University of Stirling 1994

79. R. Dawkins. "The Selfish Gene". Oxford University Press, 1989 Edition, 1976

80. A. Josang. "A Logic for Uncertain Probabilities". *International Journal of Uncertainty, Fuzziness and Knowledge-Based Systems*, 9(3), 279–311, 2001

81. A. Jøsang, L. Gray and M. Kinateder. "Simplification and Analysis of Transitive Trust Networks ". *Journal on Web Intelligence and Agent Systems*, 4(2), 139–161, 2006

82. A. Jøsang and S. Pope. "Semantic Constraints for Trust Tansitivity", *Proceedings of the Asia-Pacific Conference of Conceptual Modelling (APCCM)*, Australia, 2005

# Chapter 20
# Vehicular Ad Hoc Networks

**Yu Wang and Fan Li**

**Abstract** A *vehicular ad hoc network* consists of smart vehicles on the road and provides communication services among nearby vehicles or with roadside infrastructure. It is envisioned to provide numerous interesting services in the near future. This chapter first introduces the basic application scenarios of vehicular ad hoc network and its unique characteristics, and then provides a brief survey of several research issues in vehicular ad hoc networks, such as routing, data sharing, mobility models, and security.

## 20.1 Introduction

A *Vehicular Ad hoc Network* (VANET) is a form of wireless ad hoc network to provide communications among vehicles and nearby roadside equipments. It is emerging as a new technology to integrate the capabilities of new generation wireless networking to vehicles. The major purpose of VANET is to provide (1) ubiquitous connectivity while on the road to mobile users, who are otherwise connected to the outside world through other networks at home or at the work place, and (2) efficient vehicle-to-vehicle communications that enable the *Intelligent Transportation Systems* (ITS). ITS includes a variety of applications such as cooperative traffic monitoring, control of traffic flows, blind crossing (a crossing without light control), prevention of collisions, nearby information services, and real-time detour routes computation.

The idea of *Inter-vehicle Communications* (IVC) or *Vehicle-to-Vehicle* (V2V) communications has been proposed and studied for several decades. One of the

Y. Wang (✉)
Department of Computer Science, The University of North Carolina at Charlotte, 9201 University City Blvd. Charlotte, NC 28223, USA
e-mail: yu.wang@uncc.edu

F. Li (✉)
Department of Computer Science, Beijing Institute of Technology, Beijing, China
e-mail: fli@bit.edu.cn

S. Misra et al. (eds.), *Guide to Wireless Ad Hoc Networks*,
Computer Communications and Networks, DOI 10.1007/978-1-84800-328-6_20,
© Springer-Verlag London Limited 2009

earliest studies on IVC was started by JSK (Association of Electronic Technology for Automobile Traffic and Driving) of Japan in the early 1980s. Later, California PATH [1] and Chauffeur of EU [2] have also demonstrated the technique of coupling two or more vehicles together electronically to form a train. Europe has several large-scale programs in progress under the umbrella of Road Transport Informatics (RTI), which is the equivalent of the US ITS. Their main programs are Dedicated Road Infrastructures for Vehicle safety in Europe (DRIVE) and the Program for European Traffic with Highest Efficiency and Unprecedented Safety (PROMETHEUS). DRIVE was adopted by Council Decision 88/416/ EEC in 1988. The intention of DRIVE is to improve traffic efficiency and safety and reducing the adverse environmental effects of the motor vehicle. It focuses on the infrastructure requirements, traffic operations, and technologies of interest to public agencies responsible for the European road transport systems. PROMETHEUS [3] was started in 1986. The project is led by 18 European automobile companies, state authorities, and over 40 research institutions. Recently, the CarTALK 2000 [4] tries to investigate the problems related to safe and comfortable driving based on inter-vehicle communications. Since 2002, with the rapid development of wireless ad hoc networking technologies, VANETs have drawn a significant research interests from both academia and industry. The number of papers on VANETs has been dramatically increased. On the other hand, several major automobile manufacturers have already begun to invest real inter-vehicle networks. Audi, BMW, DaimlerChrysler, Fiat, Renault, and Volkswagen have united to create a non-profit organization called *Car2Car Communication Consortium* (C2CCC) [5] which is dedicated to the objective of further increasing road traffic safety and efficiency by means of inter-vehicle communications. IEEE also formed the new IEEE 802.11p task group, which focuses on providing wireless access for the vehicular environment. According to the official IEEE 802.11 work plan predictions, the formal 802.11p standard is scheduled to be published in April 2009.

Even though VANET is a form of wireless ad hoc networks, it has its unique characteristics due to the high nodes mobility and unreliable channel conditions. These characteristics pose many challenging research issues, including routing, data dissemination, data sharing, and security issues. For example, finding and maintaining routes becomes a very challenging task in VANETs. Many routing protocols have been developed for *Mobile Ad hoc Networks* (MANETs), and some of them can be applied directly to VANETs. However, simulation results showed that they suffer from poor performances because of the characteristics of fast vehicles movement and dynamic information exchange.

This chapter focuses on the survey of new research issues in VANETs. The remainder of this chapter is divided as follows. Section 20.2 discusses the application scenarios of VANETs. Section 20.3 briefly summarizes the network architectures and the unique characteristics of VANETs. Section 20.4 surveys the recent routing protocols for VANETs. Section 20.5 discusses data sharing and data dissemination in VANETs. Section 20.6 reviews existing mobility models for VANETs, and Section 20.7 introduces the security issue.

Section 20.8 and Section 20.9 provide some thoughts for practitioners and point out some future research directions of VANETs. Finally, Section 20.10 concludes this chapter.

## 20.2 Applications

With quick development of new techniques in vehicular ad hoc communications, various new applications are emerging in this area. Most VANET applications can be categorized into two major groups: *intelligent transportation applications* and *comfort applications*.

*Intelligent transportation applications* are the major applications of VANETs, which are used as components of the Intelligent Transportation Systems (ITS). Examples include on-board navigation, cooperative traffic monitoring, control of traffic flows, analysis of traffic congestion on the fly, and detour routes computation based on traffic conditions and the destination. In ITS, existing road-side sensors monitor traffic density and vehicular speeds and send them to a central authority that uses them to compute traffic flow controls and optimal traffic light schedules. This kind of an extended "feedback" loop can be greatly reduced by VANETs, where vehicular nodes share road conditions among themselves. In case of a road accident, the mobile nodes can even relay this information to road-side sensors, which then warn oncoming traffic about congestion or contact emergency response teams. VANETs can also be used for the implementation of blind crossing and highway entries to prevent collisions, and provide information query services of nearby points of interest on a given route by interacting with fixed road-side gateways, for example, upcoming gas stations or motels. These kinds of applications usually use broadcast or geocast routing schemes to exchange and distribute messages.

*Comfort applications* are applications to allow the passengers to communicate either with other vehicles or with Internet hosts, which improve passengers' comfort. For example, VANETs provide Internet connectivity to vehicular nodes while on the move so the passenger can download music, send e-mails, watch online movies, or play Internet games. Usually, some fixed or dynamic assigned network-to-Internet gateways are added to the networks, so they can deliver the messages between the VANET and the Internet. These applications use unicast routing as the primary communication method.

## 20.3 Network Architectures and Characteristics

Wireless ad hoc networks generally do not rely on fixed infrastructure for communication and dissemination of information. VANETs follow the same principle and apply it to the highly dynamic environment of surface transportation. As shown in Fig. 20.1, the architecture of VANETs mainly falls within three categories: *pure cellular/WLAN*, *pure ad hoc*, and *hybrid*.

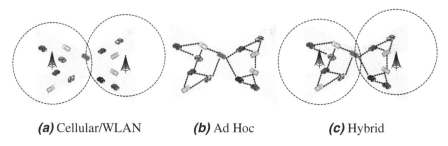

**(a)** Cellular/WLAN          **(b)** Ad Hoc          **(c)** Hybrid

**Fig. 20.1** Network architectures for VANETs

VANETs may use fixed cellular gateways and WLAN/WiMax access points at traffic intersections to connect to the Internet, gather traffic information, or for routing purposes. The network architecture under this scenario is a pure cellular or WLAN structure as shown in Fig. 20.1(a). VANETs can combine both cellular network and WLAN to form the networks so that a WLAN is used where an access point is available and a 3G connection otherwise.

Stationary or fixed gateways around the sides of roads could provide connectivity to mobile nodes (vehicles[1]), but are eventually unfeasible considering the infrastructure costs involved. In such a scenario, all vehicles and road-side wireless devices can form a pure mobile ad hoc network (Fig. 20.1(b)) to perform vehicle-to-vehicle communications and achieve certain goals, such as blind crossing.

A hybrid architecture (Fig. 20.1(c)) of combining infrastructure networks and ad hoc networks together has also been a possible solution for VANETs. Namboodiri et al. [7] proposed such a hybrid architecture, which uses some vehicles with both WLAN and cellular capabilities as the gateways and mobile network routers so that vehicles with only WLAN capability can communicate with them through multihop links to remain connected to the world. The hybrid architecture can provide better coverage, but also arises new problems, such as the seamless transition of the communication among different wireless systems.

This chapter will focus only on VANETs with pure ad hoc architectures. Such VANETs comprise radio-enabled vehicles, which act as mobile nodes as well as routers for other nodes. In addition to the similarities to ad hoc networks, such as short radio transmission range, self-organization and self-management, and low bandwidth, VANETs can be distinguished from other kinds of ad hoc networks as follows:

*Highly dynamic topology*: Due to high speed of movement between vehicles, the topology of VANETs is always changing. For example, assume that the wireless transmission range of each vehicle is 250 m, so that there is a link between two cars if the distance between them is less than 250 m. In the worst case, if two cars with the speed of 60 mph (25 m/sec) are driving in opposite directions, the link will last only for at most 10 sec.

---

[1] Hereafter, the terms "vehicle" and "node" are used interchangeably.

*Frequently disconnected network*: Due to the same reason, the connectivity of the VANETs could also be changed frequently. Especially when the vehicle density is low, it has higher probability that the network is disconnected. In some applications, such as ubiquitous Internet access, the problem needs to be solved. However, one possible solution is to pre-deploy several relay nodes or access points along the road to keep the connectivity.

*Mobility modeling and predication*: Due to highly mobile node movement and dynamic topology, mobility model and predication play an important role in network protocol design for VANETs. Moreover, vehicular nodes are usually constrained by pre-built highways, roads, and streets, so given the speed and the street map, the future position of the vehicle can be predicated.

*Geographical type of communication*: Compared to other networks that use unicast or multicast where the communication end points are defined by ID or group ID, the VANETs often have a new type of communication that addresses geographical areas where packets need to be forwarded (e.g., in safety driving applications).

*Various communications environments*: VANETs are usually operated in two typical communications environments. In highway traffic scenarios, the environment is relatively simple and straightforward (e.g., constrained one-dimensional movement), while in city conditions it becomes much more complex. The streets in a city are often separated by buildings, trees, and other obstacles. Therefore, there is not always a direct line of communications in the direction of intended data communication.

*Sufficient energy and storage*: A common characteristic of nodes in VANETs is that nodes have ample energy and computing power (including both storage and processing), since nodes are cars instead of small handheld devices.

*Hard delay constraints*: In some VANETs applications, the network does not require high data rates but has hard delay constraints. For example, in an automatic highway system, when brake event happens, the message should be transferred and arrived in a certain time to avoid car crash. In this kind of applications, instead of average delay, the maximum delay will be crucial.

*Interaction with on-board sensors*: It is assumed that the nodes are equipped with on-board sensors to provide information that can be used to form communication links and for routing purposes. For example, GPS receivers are increasingly becoming common in cars, which help to provide location information for routing purposes.

## 20.4  Routing Protocols

Because of the dynamic nature of the mobile nodes in the network, finding and maintaining routes is very challenging in VANETs. Routing in VANETs (with pure ad hoc architectures) has been studied recently and many different protocols were proposed. These protocols can be classified into five categories as follows: ad hoc, position-based, cluster-based, broadcast, and geocast routing.

## 20.4.1 Ad Hoc Routing

As mentioned earlier, VANET and MANET share the same principle: not relying on fixed infrastructure for communication, and have many similarities, e.g., self-organization, self-management, low bandwidth, and short radio transmission range. Thus, most ad hoc routing protocols are still applicable, such as AODV [8] and DSR [9]. AODV and DSR are designed for general-purpose mobile ad hoc networks and do not maintain routes unless they are needed.

However, VANET differs from MANET by its highly dynamic topology. A number of studies have been done to simulate and compare the performance of routing protocols in various traffic conditions in VANETs [7, 10–13]. The simulation results showed that most ad hoc routing protocols (e.g., AODV and DSR) suffer from highly dynamic nature of node mobility because they tend to have poor route convergence and low communication throughput. In [13], AODV is evaluated with six sedan vehicles. It showed that AODV is unable to quickly find, maintain, and update long routes in a VANET. Also in their real-world experiment, because packets are excessively lost due to route failures under AODV, it is almost impossible for a TCP connection to finish its three-way handshake to establish a connection. Thus, certain modification of the existing ad hoc routing protocols to deal with highly dynamic mobility or new routing protocols need to be developed.

Namboodiri et al. [7] considered routing from a vehicle to a gateway vehicle, which is expected to be only a few hops away. The highway scenario is highly partitioned and the probability of forming long paths is small. Thus the issue of scalability is not a problem and traditional reactive routing protocols (e.g., AODV) are still considered in small-scale networks with path lengths of only a few hops. However, the routes created by AODV can break very frequently due to the dynamic nature of mobility involved. To reduce the ill effects of frequent route breakages, thus increasing routing performance, two prediction-based AODV protocols, PRAODV and PRAODVM, are introduced. PRAODV and PRAODVM use the speed and location information of nodes to predict the link lifetimes. Their simulations showed some slight improvements regarding packet delivery ratio. With overhead not being as a major concern in vehicular networks, their protocols could have great utility. However, their methods depend heavily on the accuracy of the prediction method.

In [14], AODV is modified to only forward the route requests within the *Zone of Relevance* (ZOR). The basic idea is the same as the *location-aided routing* (LAR) [15]. ZOR is usually specified as a rectangular or circular range, it is determined by the particular application [16]. For example, for the road model of the divided highway, the ZOR covers the region behind the accident on the side of the highway where the accident happens.

## 20.4.2 Position-Based Routing

Node movement in VANETs is usually restricted in just bidirectional movements constrained along roads and streets. So routing strategies that use geographical location information obtained from street maps, traffic models, or even more prevalent navigational systems on-board the vehicles make sense. This fact receives support from a number of studies that compare the performance of topology-based routing (such as AODV and DSR) against position-based routing strategies in urban as well highway traffic scenarios [10, 11]. Therefore, geographic routing (position-based routing) has been identified as a more promising routing paradigm for VANETs.

Even though vehicular nodes in a network can make use of position information in routing decisions, such algorithms still have some challenges to be overcomed. Most position-based routing algorithms base forwarding decisions on location information. For example, greedy routing always forwards the packet to the node that is geographically closest to the destination. GPSR (*Greedy Perimeter Stateless Routing*) [17] is one of the best-known position-based protocols in literature. It combined the greedy routing with face routing by using face routing to get out of the local minimum where greedy fails. It works best in a free open space scenario with evenly distributed nodes. GPSR is used to perform simulations in [11], and its results were compared to DSR in a highway scenario. It is argued that geographic routing achieves better results because there are fewer obstacles compared to city conditions and is fairly suited to network requirements. However, when applied it to city scenarios for VANETs [10, 11, 18], GPSR suffers from several problems. First, in city scenarios, greedy forwarding is often restricted because direct communications between nodes may not exist due to obstacles such as buildings and trees. Second, if apply first the planarized graph to build the routing topology and then run greedy or face routing on it, the routing performance will degrade, i.e., packets need to travel a longer path with higher delays. See Example 1 for an example of a disconnected VANET due to the first phase of planarization in GPSR. Third, mobility can also induce routing loops for face routing; and, last, sometimes packets may get forwarded to the wrong direction, leading higher delays or even network partitions.

*Example 1:* Figure 20.2 is an example of a disconnected VANET due to the first phase of planarization in GPSR. The Relative Neighborhood Graph (RNG) shown in Fig. 20.2(a) is a planar topology used by GPSR, which consists a link $uv$ if the intersection of two circles centered at $u$ and $v$ with radius $||uv||$ (shaded area) does not contain any other nodes. In GPSR (Fig. 20.2(b)), link $uv$ is removed by RNG, since nodes $a$ and $b$ are inside the intersection of two circles

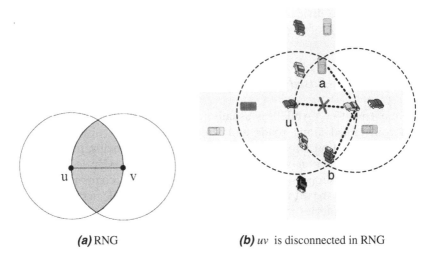

(a) RNG                                        (b) *uv* is disconnected in RNG

**Fig. 20.2** An example of GPSR's failure

centered at *u* and *v*. However, due to obstacles (such as buildings), there is no direct link *ua* or *ub*. Thus the network is disconnected between *u* and *v*, which causes GPSR's failure.

Various techniques have been proposed to deal with these challenges. Some use the digital map in the navigation system to calculate a preferred route from source to the destination [18–20]. Lochert et al. [18] proposed *Geographic Source Routing* (GSR) that assumes the aid of a street map in city environments. GSR essentially uses a *Reactive Location Service* (RLS) to get the destination position. The algorithm needs global knowledge of the city topology as it is provided by a static street map. Given this information, the sender determines the junctions that have to be traversed by the packet using the Dijkstra's shortest path algorithm. Forwarding between junctions is then done in a position-based fashion. By combining the geographic routing and topological knowledge from street maps, GSR proposes a promising routing strategy for VANETs in city environments.

Lochert et al. [21] also proposed another solution GPCR (*Greedy Perimeter Coordinator Routing*) later without the use of either source routing or the availability of street maps. It utilizes the fact that the nodes at a junction in the street follow a natural planar graph. Thus a restricted greedy algorithm can be followed as long as the nodes are in a street. Junctions are the only places where actual routing decisions are taken. Therefore, packets should always be forwarded to a node on a junction (called *Coordinator*) rather than being forwarded across the junction. See Example 2 for an example. Despite of the improved greedy routing strategy, GPCR uses a *repair strategy* to get out of the local minimum, i.e., no neighbor exists which is closer to the destination than the intermediate node itself. The repair strategy (1) decides, on each junction, which street the packet should follow next (by right-hand rule) and (2) applies greedy routing, in between junctions, to reach the next junction. Example 3 shows an example of the repair strategy. The simulation in [21] is done in the NS-2 simulator with a real city topology, which is a part of Berlin, Germany. The authors show GPCR has higher delivery rate than GPSR, with larger average number of hops and slight increase in latency.

*Example 2:* Source $S$ wants to forward the packet to the destination $D$ (Fig. 20.3(a)). If a regular greedy forwarding is used, the packet will be forwarded beyond the junction (Coordinator $C1$) to $N1$, then it will be lead to a local minimum at $N3$. But by forwarding the packet to coordinator $C1$, an alternative path to the destination can be found without getting stuck in a local minimum.

*Example 3:* Figure 20.3(b) is an example of using the right-hand rule to decide which street the packet should follow in the repair strategy of GPCR. Node $S$ is the local minimum since no other nodes is closer to the destination $D$ than itself. The packet is routed to the first coordinator $C1$. Node $C1$ receives the packet and decides which street the packet should follow by the right-hand rule. It chooses the street that is the next one counter-clockwise from the street the packet has arrived on. The packet is forwarded to the next coordinator $C2$ through the intermediate node $N1$ along the street. Then the coordinator $C2$ decides to forward the packet to node $N2$. At this moment, the distance from $N2$ to $D$ is closer than at the beginning of the repair strategy at node $S$. Hence GPCR is switched back to modified greedy routing. The packet reaches $D$.

Position-based routing for VANETs also faces other challenges in a built-up city environment. Generally, vehicles are more unevenly distributed due to the fact that they tend to concentrate more on some roads than others. And their

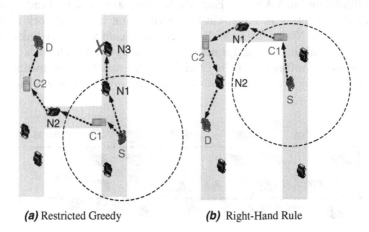

*(a)* Restricted Greedy              *(b)* Right-Hand Rule

**Fig. 20.3** (a) Greedy routing vs. restricted greedy routing in the area of a junction. (b) *Right-hand* rule is used to decide which street the packet should follow in the repair strategy of GPCR

constrained mobility by the road patterns, along with more difficult signal reception due to radio obstacles such as high-rise buildings, may lead VANETs unconnected. A new position-based routing technique called A-STAR (Anchor-*based Street and Traffic Aware Routing*) [10] has been proposed for such city environments. A-STAR uses the street map to compute the sequence of junctions (anchors) through which a packet must pass to reach its destination. But unlike GSR, A-STAR computes the anchor paths with traffic awareness. A-STAR differs from GSR and GPSR in two main aspects. Firstly, it incorporates traffic awareness by using statistically rated maps (counting the number of city bus routes on each street to identify anchor paths of maximum connectivity) or dynamically rated maps (dynamically monitoring the latest traffic condition to identify the best anchor paths) to identify an anchor path with high connectivity for packet delivery. Secondly, A-STAR employs a new local recovery strategy for packets routed to a local minimum, which is more suitable for a city environment than the greedy approach of GSR and the perimeter-mode of GPSR. In the local recovery state, the packet is salvaged by traversing the new anchor path. To prevent other packets from traversing through the same void area, the street at which local minimum occurred is marked as "out of service" temporarily. With traffic awareness, A-STAR shows the best performance compared to GSR and GPSR, because it can select paths with higher connectivity for packet delivery.

### 20.4.3 Cluster-Based Routing

In cluster-based routing, a virtual network infrastructure must be created through the clustering of nodes in order to provide scalability. See Fig. 20.4 for an illustration in VANETs. Each cluster can have a cluster-head, which is responsible for intra- and inter-cluster coordination in the network management functions. Nodes inside a cluster communicate via direct links. Inter-cluster communication is performed via the cluster-heads. The creation of a virtual network infrastructure is crucial for the scalability of media access protocols, routing protocols, and the security infrastructure. The stable

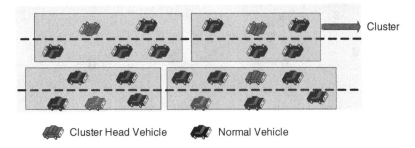

**Fig. 20.4** Vehicles form multiple clusters in cluster-based routing

clustering of nodes is the key to create this infrastructure. Many cluster-based routing protocols [22–24] have been studied in MANETs. However, VANETs behave in different ways than the models that predominate in MANETs research, due to driver behavior, constraints on mobility, and high speeds. Consequently, current MANETs clustering techniques are unstable in vehicular networks. The clusters created by these techniques are too short-lived to provide scalability with low communications overhead.

Blum et al. [25] proposed a *Clustering for Open IVC Networks* (COIN) algorithm. Cluster-head election is based on vehicular dynamics and driver intentions, instead of ID or relative mobility as in classical clustering methods. This algorithm also accommodates the oscillatory nature of inter-vehicle distances. They show that COIN produces much more stable structures in VANETs while introducing little additional overhead. Santos et al. [12] presented a reactive location-based routing algorithm that uses cluster-based flooding for VANETs called LORA_CBF. Each node can be the cluster-head, gateway, or cluster member. Each cluster has exactly one cluster-head. If a node is connected to more than one cluster, it is called a gateway. The cluster-head maintains information about its members and gateways. Packets are forwarded from a source to the destination by protocol similar to greedy routing. If the location of the destination is not available, the source will send out the *Location Request* (LREQ) packets. This phase is similar to the route discovery phase of AODV, but only the cluster-heads and gateways will disseminate the LREQ and LREP (*Location Reply*) messages. The performances of LORA_CBF, AODV, and DSR are evaluated in typical urban and highway traffic scenarios. Simulation results demonstrate that network mobility and network size affect the performance of AODV and DSR more significantly than LORA_CBF. Cluster-based method has also been used in data dissemination and information propagation for VANETs, such as in [26] the authors described a cluster-based message dissemination method using opportunistic forwarding.

In summary, cluster-based routing protocols can achieve good scalability for large networks, but a significant hurdle for them in fast-changing VANET systems is the delay and overhead involved in forming and maintaining these clusters.

### 20.4.4 Broadcast Routing

Broadcast is a frequently used routing method in VANETs, such as sharing traffic, weather, emergency, road condition among vehicles, and delivering advertisements and announcements. Broadcast is also used in unicast routing protocols (routing discovery phase) to find an efficient route to the destination. When the message needs to be disseminated to the vehicles beyond the transmission range, multihop is used.

The simplest way to implement a broadcast service is flooding in which each node re-broadcasts messages to all of its neighbors except the one it got this message from. Flooding guarantees the message will eventually reach all nodes in the network. Flooding performs relatively well for a limited small number of nodes and is easy to be implemented. But when the number of nodes in the network increases, the performance drops quickly. The bandwidth requested for one broadcast message transmission can increase exponentially. As each node receives and broadcasts the message almost at the same time, this causes contentions and collisions, broadcast storms, and high bandwidth consumption. Flooding may have a very significant overhead and selective forwarding can be used to avoid network congestion.

Durresi et al. [27] presented an emergency broadcast protocol, named BROADCOMM, which is based on a hierarchical structure for a highway network. In BROADCOMM, the highway is divided into virtual cells, which moves as the vehicles move. The nodes in the highway are organized into two level of hierarchy: the first level includes all the nodes in a cell; the second level is represented by the *cell reflectors*, which are a few nodes usually located closed to the geographical center of the cell. Cell reflector behaves for a certain time interval as a cluster-head that will handle the emergency messages coming from members of the same cell, or close members from neighbor cells. Besides that, the cell reflector serves as an intermediate node in the routing of emergency messages coming from its neighbor cell reflectors and decides which will be the first to be forwarded. This protocol outperforms similar flooding-based routing protocols in the message broadcasting delay and routing overhead. However, it is very simple and works only in simple highway networks.

*Urban MultiHop Broadcast* protocol (UMB) [28] is designed to overcome interference, packet collisions, and hidden nodes problems during message dissemination in multihop broadcast. In UMB, the sender nodes try to select the furthest node in the broadcast direction to assign the duty of forwarding and acknowledging the packet without any a priori topology information. At the intersection, repeaters are installed to forward the packets to all road segments. UMB protocol has much higher success percentage at high packet loads and vehicle traffic densities than 802.11-distance and 802.11-random protocols, which are flooding-based, modified IEEE 802.11 standards to avoid collisions among rebroadcast packets by forcing vehicles to wait before forwarding the packets.

*Vector-based TRAcking DEtection* (V-TRADE) and *History-enhanced V-TRADE* (HV-TRADE) [29] are GPS-based message broadcasting protocols. The basic idea is similar to the unicast routing protocol *Zone Routing Protocol* (ZRP) [30]. Based on position and movement information, their methods classify the neighbors into different forwarding groups. For each group, only a small subset of vehicles (called border vehicles) is selected to rebroadcast the message. They show significant improvement of bandwidth utilization with slight loss of reachability, because the new protocols pick fewer vehicles to rebroadcast the messages. But they still have routing overhead as long as the forwarding nodes are selected in every hop.

### 20.4.5  Geocast Routing

Geocast routing [31] is basically a location-based multicast routing. The objective of a geocast routing is to deliver the packet from a source node to all other nodes within a specified geographical region [*Zone of Relevance* (ZOR)]. Many VANET applications will benefit from geocast routing. For example, a vehicle identifies itself as crashed by vehicular sensors that detect events like airbag ignition, then it can report the accident instantly to nearby vehicles. Vehicles outside the ZOR are not alerted to avoid unnecessary and hasty reactions. In this kind of scenarios, the source node is usually inside the ZOR. See Fig. 20.5 for an illustration of the difference among unicast, broadcast, and geocast in VANETs.

Geocast can be implemented with a multicast service by simply defining the multicast group to be the certain geographic region. Most geocast routing methods are based on directed flooding, which tries to limit the message overhead and network congestion of simple flooding by defining a forwarding zone and restricting the flooding inside it. Non-flooding approaches (based on unicast routing) are also proposed, but inside the destination region, regional flooding may still be used even for protocols characterized as non-flooding.

In [16], a simple geocast scheme is proposed to avoid packet collisions and reduce the number of rebroadcasts. When a node receives a packet, it does not rebroadcast it immediately, but has to wait for some time to make a decision about rebroadcast. The waiting time depends on the distance of this node to the sender. The waiting time is shorter for more distant receiver. Thus mainly nodes at the border of the reception area take part in forwarding the packet quickly. When this waiting time expires, if it does not receive the same message from another node, it will rebroadcast this message. By this way, a broadcast storm is avoided and the forwarding is optimized around the initiating vehicle. Bachir and Benslimane [32] proposed an *Inter-Vehicles Geocast* protocol, called IVG, to broadcast an alarm message to all the vehicles being in risk area based on defer time algorithm in a highway. The main idea is very similar to [16].

Maihöfer and Eberhardt [33] concerned with cache scheme and distance-aware neighborhood selection scheme to deal with the situation of high velocities in VANET compared to the regular geocast protocols. The main idea of their cached greedy geocast inside the ZOR is to add a small cache to the routing layer that holds those packets which a node cannot forward instantly due to a local minimum. When a new neighbor comes into reach or known neighbors change their positions, the cached message can be possibly

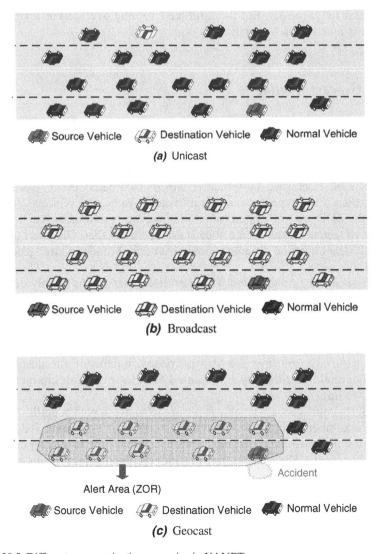

**Fig. 20.5** Different communication scenarios in VANETs

forwarded to the newly discovered node. Simulation results show that a cache for presently unforwardable messages caused by network partitioning or unfavorable neighbors can significantly improve the geocast delivery success ratio. The improved neighborhood selection taking frequent neighborhood changes into account significantly decreases network load and decreased end-to-end delivery delay.

Besides classical geocast routing, recently, Maihöfer et al. [34] also studied a special geocast, called abiding geocast, where the packets need to be delivered to all nodes that are sometime, during the geocast lifetime (a certain period of time), inside the geocast destination region. Services and applications like position-based advertising, publish-and-subscribe, and many others profit from abiding geocast. In [34], the authors provided three solutions: (1) a server is used to store the geocast messages; (2) an elected node inside the geocast region stores the messages; (3) each node stores all geocast packets destined for its location and keeps the neighbor information.

In this section, the challenges of designing routing protocols in VANETs and several routing protocols have been discussed. In general, position-based routing and geocasting are more promising than other routing protocols for VANETs because of the geographical constrains. However, the performance of a routing protocol in VANETs depends heavily on the mobility model, the driving environment, the vehicular density, and many other facts. Therefore, having a universal routing solution for all VANETs application scenarios or a standard evaluation criterion for routing protocols in VANETs is extremely hard. In other words, for certain VANETs application, a customizing routing protocol and mobility model need to be designed to fulfill its requirements.

## 20.5  Data Sharing and Data Dissemination

With advances in the field of VANETs, where vehicles continually exchange information, there is a need to address major database-related issues such as data freshness, consistency, replication, and query optimization. A consistent global view of the network, though desirable, is not feasible because no single node has all the node information in such dynamic networks. Constructing a distributed view of the whole network is simpler, but the local copies may have stale data then.

Bejan and Lawrence [35] described how to deploy a peer-to-peer communication architecture to provide information for cooperative driving. At any point of time, each vehicle has global information of two relations, which hold current and all previous location information about all the nodes in the network. Whenever a node changes location, both these relations get updated accordingly and distributively. The authors propose a query answer mechanism that takes care of query duplication and propagation issues. The queries can either be location-based (e.g., "where am I?") or alert queries to warn other nodes about a road event ahead.

To enable cooperating and sharing of content while on the move, a Swarming Protocol for VANETs (SPAWN) [36] has been proposed. SPAWN is similar to swarming file-sharing protocols in Internet, which is peers downloading a file form a mesh and exchanging pieces of the file amongst themselves. However, wireless networks are limited by their bandwidth and intermittent connectivity. SPAWN tries to achieve scalability and improve perceived performance for individual clients in such highly mobile scenario, by using a gossip mechanism that leverages the inherent broadcast nature of the wireless medium, and a proximity-driven content selection strategy.

Recently, Nandan et al. [37] proposed a new integrated system called AdTorrent for search, ranking, and content delivery in car networks. AdTorrent builds on a scalable push model architecture for content delivery. They presented a detailed analysis of the performance impact of key design parameters such as scope of the query flooding on the query hit ratio.

Data dissemination is to transport the information to intended receivers with certain design objectives, such as high delivery ratio and low delay. The best approach to data dissemination using vehicle forwarding remains an open issue. MDDV [38], a mobility-centric data dissemination algorithm, has been proposed for a partitioned and highly vehicle network. The algorithm combines the ideas of opportunistic forwarding, trajectory-based forwarding, and geographic forwarding. Messages are stored and forwarded opportunistically along a pre-defined geographical path. To improve the delivery reliability, (1) vehicles are supplied with dissemination status information so that they can make appropriate decisions based on it, and (2) active propagation is limited to the area near the message holder who is closest to the destination region. An analytical model has been described in [39] to study the performance of information propagation in the highly mobile VANETs. Their model reveals that vehicle traffic characteristics, such as the vehicle density, average vehicle speed, and relative vehicle speed, impact the information propagation significantly.

## 20.6 Mobility Model

This section will briefly review the *mobility model*[2] used by VANET routing protocols. A realistic mobility model is not only very important for getting accurate results in routing performance evaluation but also a necessary component to predict the next positions of vehicles and make smarter route decisions in many VANET routing protocols. Choffnes et al. [40] showed protocol performance varies with the mobility models and traffic scenarios. Realistic mobility models for VANETs need to take into account street conditions, urban conditions, traffic speed, vehicle density, and obstacles such as buildings.

One of the simplest and the earliest mobility models is *Random WayPoint* (RWP) Mobility model [41], where nodes randomly choose a destination and continue to move toward that destination at a uniform speed. When the destination is reached, another destination is chosen at random and so forth. RWP is widely used in ad hoc network simulations (such as NS-2), but it does not attempt to model any real mobility situation since street-bound vehicles follow a completely different movement pattern. Nadeem et al. [6] modified

---

[2] Some papers refer it as *traffic model*, however, to distinguish it from the network traffic model, *mobility model* is used in this chapter.

RWP model by accepting road length, average speed, number of lanes, and average gap between vehicles as parameters.

Saha and Johnson [42] first attempted to propose a realistic street mobility model where they used the road information from the TIGER (*Topologically Integrated Geographic Encoding and Referencing*) [43] US road map by US Census Bureau. In their model, they convert the map into a graph. Then they assume that each node starts at some random point on a road segment and moves toward a random destination following shortest path algorithm. The speed on each road is uniformly distributed within 5 mph above and below the speed limit. A more realistic mobility model, *STreet RAndom Waypoint* (STRAW) [40], is also based on the road information from the US road map by US Census Bureau. STRAW uses a simple car-following model to simulate realistic traffic congestion in an urban environment. Compared with the model in [42], STRAW considers the interaction among cars, traffic congestion, and traffic controls. Both AODV and DSR are used to compare performances of STRAW against RWP under varying traffic conditions in Chicago and Boston. It is concluded that significantly different results are obtained by the use of a realistic mobility model.

A new trend of building mobility model is using the realistic vehicular trace data. Fübler et al. [44] used a set of movement traces derived from typical situations on German Autobahns to simulate the traffic movement on a highway. The movement of cars is defined as tuples of a one-dimensional position and a lane on the highway for discrete time steps of 0.5 seconds. They cut those movement trace data into valid portions and combine them into certain movement scenarios. Jetcheva et al. [45] recorded the movement traces of the buses of the public transportation system in Seattle, Washington. However, these traces describe only the movement of the buses; they represent a tiny fraction of the total number of road traffic participants. Recently, Naumov et al. [46] introduced a new source of realistic mobility traces for simulation of inter-vehicle networks. Their traces are obtained from a *Multi-agent Microscopic Traffic Simulator* (MMTS) [47], which is capable to simulate public and private traffic over real regional road maps of Switzerland with a high level of realism.

## 20.7 Security

The wide applications of VANETs discussed in Section 20.2 make VANETs likely target for all forms of attacks. Since many applications will affect life-or-death decisions, illicit tampering can have devastating consequences. The characteristics of VANETs (e.g., the speed of the vehicles, the size of the network, and the relevance of their geographic position) make the security problem more challenging and novel than it is in other communication networks.

Golle et al. [48] classified the attacks in VANETs based on a number of characteristics, such as attack nature, attack target (local or projected), attack scope (limited or extended), and attack impact (undetected, detected, or corrected). VANETs require frequent node-to-node cooperative data exchange to facilitate traffic monitoring, route planning, and information querying. They proposed a general approach to evaluate the validity of VANETs data by analyzing sensor data. A key component of their approach is that each node maintains a model of the VANETs containing all the knowledge that the node has of the VANETs. Individual node uses this model to test the validity of sensor data received from other nodes. If all the data agrees with the model (or with high probability), the node accepts and trusts the data. When inconsistence arises, an adversarial parsimony model is used to search the explanations of the errors. Explanations are scored and the node accepts the data as valid by the highest scoring explanations.

Hubaux et al. [49] described some of the attacks vehicular networks may face and propose the mechanisms for providing secure positioning and electronic license plate. Ray and Habaux [50] considered the issues involved with key management for vehicular networks, as well as the use of anonymous public keys. They also analyze the feasibility of using a public key infrastructure to support the security requirements of VANETs. Recently, Parno and Perrig [51] also analyzed the security challenges specific to vehicular networks and proposed a set of security primitives that can be used as the building blocks of secure applications.

## 20.8 Thoughts for Practitioners

Above, we briefly review several networking research issues and some current solutions in VANETs. The actual realization of VANETs poses more challenging issues at different layers (beyond the network layer we discussed in this chapter, such as physical layer and MAC layer) and different levels (system architecture level or algorithm deployment level or physical device level). Many of these challenging problems are not yet solved or even addressed. In addition, a considerable amount of work is still needed in piecing together all solutions for these problems and making the entire system work well. However, the general feeling of all practitioners (networking researchers and automobile manufacturers) is that vehicles could benefit from wireless communications, making VANETs a reality in a near future.

## 20.9 Directions for Future Research

In the following we present some possible directions for future research in VANETs:

1. Even though many routing protocols for VANETs have been proposed, how to efficient routing in partitioned networks and how to handle dynamic topology in VANETs still need more research.
2. Many VANET applications have either delay constraints or other QoS requirements. How to provide QoS and reliability in VANET routing is still a challenging problem.

3. Efficient broadcasting of safety messages for getting full coverage and low latency should also be addressed.
4. Traffic flow (both in time and space) need to be studied and integrated in the design of reliable and high-performance protocols and applications for VANETs.
5. Since mobility of VANETs cannot be captured by general mobility models, special mobility models by making use of traffic flow information and map information should be proposed.
6. Since experimental evaluation of VANETs is expensive, most current protocols are evaluated in existing network simulators. However, these simulators are not designed for VANETs, they only provide simple mobility model and poor physical radio models. Thus, simulation techniques for VANETs should be improved.
7. New MAC protocol may be needed for VANETs to provide high channel utilization and enable channel access for different priority levels of traffic.
8. Cooperation among inter-vehicular networks and sensor networks placed within the vehicles or along the road need to be further investigated.
9. Security and privacy issues are critical in many VANET applications, simple but efficient protocols to secure the communication and preserve user-privacy are needed.
10. New heterogeneous VANET systems and applications need to be studied.

## 20.10  Conclusion

As a new emerging wireless networking area, VANET has received increasing attention in the wireless network community recently. Many network protocols and systems have been proposed and studied as shown in this chapter. However, due to highly dynamic environment inherent to the VANETs, there are still quite a few challenges and research issues that have not been fully solved yet (in all the research issues discussed here). More research needs to be conducted before the VANET can be widely deployed. In addition, other kinds of communication technologies can also be integrated with VANETs to support vehicle-to-vehicle communications. The vision for future intelligent transportation systems is the support of the co-existence and interoperability of heterogeneous wireless technologies with varying requirements.

## Terminologies

*Vehicular ad hoc network (VANET)* is a form of mobile ad hoc network to provide communications among nearby vehicles and between vehicles and nearby fixed equipment (such as roadside equipment).

*Intelligent transportation system (ITS)* is a transportation system that adds information and communications technology to transport infrastructure and vehicles. It aims to manage factors that are typically at odds with each other such as vehicles, loads, and routes to improve safety and reduce vehicle wear, transportation times, and fuel consumption.

*Intelligent transportation applications* are the applications of VANETs, which are used as components of the Intelligent Transportation Systems.

*Comfort applications* are the applications of VANETs, to allow the passengers to communicate either with other vehicles or with Internet hosts, which improve passengers' comfort.

*Position-based routing* is a kind of routing protocol that uses the node position information to help making routing decision. It is also called georouting or geographic routing.

*Cluster-based routing* is a kind of routing protocol that uses a hierarchical routing structure. The hierarchical structure is constructed via clustering.

*Geocast routing* is a special case of multicast routing that delivers information to a group of destinations in a network identified by their geographical locations.

*Greedy perimeter stateless routing (GPSR)* is an efficient position-based routing protocol for wireless networks. It uses *greedy forwarding* to forward packets to nodes that are always progressively closer to the destination. In regions of the network where such a greedy path does not exist, GPSR recovers by forwarding in *perimeter mode*, in which a packet traverses successively closer *faces* of a planar subgraph.

*Greedy perimeter coordinator routing (GPCR)* is a position-based routing protocol for VANETs, which utilizes the fact that the nodes at a junction in the street follow a natural planar graph. It uses a restricted greedy algorithm with a repair strategy and prefers a node on a junction (called Coordinator).

*Random WayPoint model (RWP)* is a commonly used synthetic model for mobility. It is an elementary mobility model where nodes randomly choose a destination and continue to move toward that destination at a uniform speed. When the destination is reached, another destination is chosen at random and so forth.

## Questions

1. What is Vehicular Ad Hoc Network?
2. What are the unique characteristics of VANETs?
3. What are the two types of applications of VANETs? Give examples for them.
4. What are the possible network architectures of VANETs? What are the advantages and disadvantages of them?
5. What is position-based routing? Why position-based routing is suitable for VANETS?
6. Does GPSR (Greedy Perimeter Stateless Routing) work well in VANETs? If not, what is the reason?

7. Explain how GPCR (Greedy Perimeter Coordinator Routing) works differently from GPSR. You can use examples.
8. What are the three communication scenarios in VANETs? Can you explain each scenario respectively?
9. What are the differences between geocast and multicast?
10. What is Random WayPoint Mobility model?

# References

1. J.K. Hedrick, M. Tomizuka, and P. Varaiya, "Control issues in automated highway systems," *IEEE Control Systems Magazine*, vol. 14, no. 6, pp. 21–32, December 1994.
2. O. Gehring and H. Fritz, "Practical results of a longitudinal control concept for truck platooning with vehicle to vehicle communication," in *Proceedings of the 1st IEEE Conference on Intelligent Transportation System (ITSC'97)*, pp. 117–122, October 1997.
3. M.Williams, "PROMETHEUS-The European research programme for optimizing the roadtransport system in Europe," in *Proceedings of the IEEE Colloquium on Driver Information*, pp. 1/1–1/9, December 1988.
4. D. Reichardt, M. Miglietta, L. Moretti, P. Morsink, and W. Schulz, "Cartalk 2000 – safe and comfortable driving based upon inter-vehicle communication," in *Proceedings of the IEEE Intelligent Vehicle Symposium (IV02)*, 2002.
5. "Car-to-car communication consortium," http://www.car-to-car.org.
6. T. Nadeem, S. Dashtinezhad, C. Liao, and L. Iftode, "Trafficview: traffic data dissemination using car-to-car communication." *Mobile Computing and Communications Review*, vol. 8, no. 3, pp. 6–19, 2004.
7. V. Namboodiri, M. Agarwal, and L. Gao, "A study on the feasibility of mobile gateways for vehicular ad-hoc networks.," in *Proceedings of the First International Workshop on Vehicular Ad Hoc Networks*, pp. 66–75, 2004.
8. C.E. Perkins and E.M. Royer, "Ad-hoc on demand distance vector routing," in *Proceedings of the 2nd IEEE Workshop on Mobile Computing Systems and Applications*, pp. 90–100, February 1999.
9. D.B. Johnson and D.A. Maltz, "Dynamic source routing in ad hoc wireless networks," in *Mobile computing* edited by T. Imielinski and H.F. Korth, Springer, 1996.
10. G. Liu, B.-S. Lee, B.-C. Seet, C.H. Foh, K.J. Wong, and K.-K. Lee, "A routing strategy for metropolis vehicular communications," in *International Conference on Information Networking (ICOIN)*, pp. 134–143, 2004.
11. Holger Füßler, Martin Mauve, Hannes Hartenstein, Michael Kasemann, and Dieter Vollmer, "Location-based routing for vehicular ad-hoc networks," *ACM SIGMOBILE Mobile Computing and Communications Review (MC2R)*, vol. 7, no. 1, pp. 47–49, January 2003.
12. Raúl Aquino Santos, Arthur Edwards, Robert Edwards, and Luke Seed, "Performance evaluation of routing protocols in vehicular adhoc networks," *The International Journal of Ad Hoc and Ubiquitous Computing*, vol. 1, no. 1/2, pp. 80–91, 2005.
13. S.Y. Wang, C.C. Lin, Y.W. Hwang, K.C. Tao, and C.L. Chou, "A practical routing protocol for vehicle-formed mobile ad hoc networks on the roads," *in Proceedings of the 8th IEEE International Conference on Intelligent Transportation Systems*, 2005, pp. 161–165.
14. C.-C. Ooi and N. Fisal, "Implementation of geocast-enhanced AODV-bis routing protocol in MANET," in *Proceedings of the IEEE Region 10 Conference*, vol. 2, pp. 660–663, 2004.
15. Y.-B. Ko and N.H. Vaidya, "Location-aided routing (LAR) in mobile ad hoc networks," *Wireless Networks*, vol. 6, no. 4, pp. 307–321, 2000.

16. L. Briesemeister, L. Schäfers, and G. Hommel, "Disseminating messages among highly mobile hosts based on inter-vehicle communication," in *proceedings of the IEEE Intelligent Vehicles Symposium*, pp. 522–527, 2000.

17. B. Karp and H.T. Kung, "GPSR: Greedy perimeter stateless routing for wireless networks," in *Proceedings of the ACM/IEEE International Conference on Mobile Computing and Networking (MobiCom)*, 2000.

18. C. Lochert, H. Hartenstein, J. Tian, D. Herrmann, H. Füßler, and M. Mauve, "A routing strategy for vehicular ad hoc networks in city environments," in *Proceedings of IEEE Intelligent Vehicles Symposium (IV2003)*, pp. 156–161, June 2003.

19. V. Dumitrescu and J. Guo, "Context assisted routing protocols for inter-vehicle wireless communication," in *Proceedings of the IEEE Intelligent Vehicles Symposium*, pp. 594–600, 2005.

20. J. LeBrun, C.-N. Chuah, D. Ghosal, and M. Zhang, "Knowledge-based opportunistic forwarding in vehicular wireless ad hoc networks," in *Proceedings of the IEEE Vehicular Technology Conference*, vol. 4, pp. 2289–2293, 2005.

21. C. Lochert, M. Mauve, H. Füßler, and H. Hartenstein, "Geographic routing in city scenarios," *ACM SIGMOBILE Mobile Computing and Communications Review (MC2R)*, vol. 9, no. 1, pp. 69–72, January 2005.

22. C.R. Lin and M. Gerla, "Adaptive clustering for mobile wireless networks," *IEEE Journal of Selected Areas in Communications*, vol. 15, no. 7, pp. 1265–1275, 1997.

23. B. Das and V. Bharghavan, "Routing in ad-hoc networks using minimum connected dominating sets," in *1997 IEEE International Conference on on Communications (ICC'97)*, vol. 1, pp. 376–380, 1997.

24. J. Wu and H. Li, "A dominating-set-based routing scheme in ad hoc wireless networks," *the special issue on Wireless Networks in the Telecommunication Systems Journal*, vol. 3, pp. 63–84, 2001.

25. J. Blum, A. Eskandarian, and L. Hoffman, "Mobility management in IVC networks," in *IEEE Intelligent Vehicles Symposium*, 2003.

26. T.D.C. Little and A. Agarwal, "An information propagation scheme for VANETs," in *Proceedings 8th International IEEE Conference on Intelligent Transportation Systems (ITSC 2005)*, 2005.

27. M. Durresi, A. Durresi, and L. Barolli, "Emergency broadcast protocol for inter-vehicle communications," in *ICPADS '05: Proceedings of the 11th International Conference on Parallel and Distributed Systems - Workshops (ICPADS'05)*, 2005.

28. G. Korkmaz, E. Ekici, F. Özgüner, and Ü. Özgüner, "Urban multi-hop broadcast protocol for inter-vehicle communication systems," in *ACM International Workshop on Vehicular Ad Hoc Networks*, pp. 76–85, 2004.

29. M. Sun, W. Feng, T.-H. Lai, K. Yamada, H. Okada, and K. Fujimura, "GPS-based message broadcasting for inter-vehicle communication," in *ICPP '00: Proceedings of the 2000 International Conference on Parallel Processing*, 2000.

30. Z.J. Haas and M.R. Pearlman, "The zone routing protocol(ZRP) for ad hoc networks," in *Internet draft - Mobile Ad hoc NETworking (MANET), Working Group of the Internet Engineering Task Force (IETF)*, Novermber 1997.

31. C. Maihöfer, "A survey of geocast routing protocols," *IEEE Communications Surveys & Tutorials*, vol. 6, no. 2, pp. 32–42, 2004.

32. A. Bachir and Abderrahim Benslimane, "A multicast protocol in ad hoc networks inter-vehicle geocast," in *Proceedings of the 57th IEEE Semiannual Vehicular Technology Conference*, vol. 4, pp. 2456–2460, 2003.

33. C. Maihöfer and R. Eberhardt, "Geocast in vehicular environments: Caching and transmission range control for improved efficiency," in *Proceedings of IEEE Intelligent Vehicles Symposium (IV)*, pp. 951–956, 2004.

34. C. Maihöfer, T. Leinmüller, and E. Schoch, "Abiding geocast: time–stable geocast for ad hoc networks," in *Proceedings of the 2nd ACM international workshop on Vehicular ad hoc networks (VANET '05)*, pp. 20–29, 2005.

35. A. Bejan and R. Lawrence, "Peer-to-peer cooperative driving," in *ISCIS2002 International Symposium on Computer and Information Sciences*, 2002.
36. A. Nandan, S. Das, G. Pau, M. Gerla, and M.Y.Sanadidi, "Co-operative downloading in vehicular ad-hoc wireless networks," in *Second Annual Conference on Wireless On-demand Network Systems and Services (WONS)*, pp. 32–41, 2005.
37. A. Nandan, S. Das, S. Tewari, M. Gerla, and L. Klienrock, "Adtorrent: Delivering location cognizant advertisements to car networks," in *The Third International Conference on Wireless On Demand Network Systems and Services(WONS 2006)*, 2006.
38. H. Wu, R.M. Fujimoto, R. Guensler, and M. Hunter, "MDDV: a mobility-centric data dissemination algorithm for vehicular networks," in *Vehicular Ad Hoc Networks*, 2004, pp. 47–56.
39. H. Wu, R. Fujimoto, and G. Riley, "Analytical models for data dissemination in vehicle-to-vehicle networks," in *IEEE 2004-fall Vehicle Technology Conference (VTC)*, 2004.
40. D Choffnes and F Bustamante, "An integrated mobility and traffic model for vehicular wireless networks," in *the 2nd ACM International Workshop on Vehicular Ad Hoc Networks*. ACM, 2005.
41. J. Broch, D.A. Maltz, D.B. Johnson, Y.-C. Hu, and J. Jetcheva, "A performance comparison of multi-hop wireless ad hoc network routing protocols," in *Mobile Computing and Networking*, pp. 85–97, 1998.
42. A.K. Saha and D.B. Johnson, "Modeling mobility for vehicular ad-hoc networks," in *ACM International Workshop on Vehicular Ad Hoc Networks*, pp. 91–92, 2004.
43. "U.S. Census Bureau. TIGER, TIGER/Line and TIGER-Related Products," http://www.census.gov/geo/www/tiger/.
44. H. Füßler, M. Torrent-Moreno, M. Transier, R. Krüger, H. Hartenstein, and W. Effelsberg, "Studying vehicle movements on highways and their impact on ad-hoc connectivity," in *ACM Mobicom 2005*, August 2005.
45. J. Jetcheva, Y. Hu, S. PalChaudhuri, A. Saha, and D. Johnson, "Design and evaluation of a metropolitan area multitier wireless ad hoc network architecture," in *Proceedings of 5th IEEE Workshop on Mobile Computing Systems and Applications (WMCSA 03)*, 2003.
46. V. Naumov, R. Baumann, and T. Gross, "An evaluation of inter-vehicle ad hoc networks based on realistic vehicular traces," in *MobiHoc '06: Proceedings of the 7th ACM International Symposium on Mobile Ad Hoc Networking and Computing*, pp. 108–119, 2006.
47. B. Raney, A. Voellmy, N. Cetin, M. Vrtic, and K. Nagel, "Towards a microscopic traffic simulation of all of Switzerland," in *International Conference on Computational Science (Part 1)*, pp. 371–380, 2002.
48. P. Golle, D.H. Greene, and J. Staddon, "Detecting and correcting malicious data in vanets.," in *Vehicular Ad Hoc Networks*, 2004, pp. 29–37.
49. J.-P. Hubaux, S. Capkun, and J. Luo, "The security and privacy of smart vehicles," *IEEE Security & Privacy Magazine*, vol. 2, no. 3, pp. 49–55, 2004.
50. M. Raya and J.-P. Hubaux, "The security of vehicular ad hoc networks," in *SASN '05: Proceedings of the 3rd ACM workshop on Security of ad hoc and sensor networks*, New York, NY, USA, ACM Press, 2005, pp. 11–21.
51. B. Parno and A. Perrig, "Challenges in securing vehicular networks," in *Fourth Workshop on Hot Topics in Networks (HotNets-IV)*, 2005.

# Chapter 21
# Integration of Mobile Ad Hoc Networks into IP-Based Access Networks

A. Triviño-Cabrera

**Abstract** The study of mobile ad hoc networks (MANET) is commonly focused on how nodes belonging to the same network communicate by means of ad hoc routing protocols. However, when integrated into the Internet, the ad hoc routing protocols become insufficient, so additional technologies are required. In the IPv6 context, technologies such as NDP (Neighbor Discovery Protocol), which plays a fundamental role in automatically configuring the mobile nodes, were initially conceived to operate in on-link communications. This assumption is a clear restriction in multihop networks so the technologies should be adapted to these new scenarios. In order to provide a framework where these adaptations could be accomplished, several mechanisms have been proposed. All of them introduce an additional element that plays the role of the Internet Gateway. The characteristics as well as the functionalities of the Internet Gateway differ in the integration supports. In this chapter, the problems that arise in the integration of MANET into IP-based access networks, such as the Internet, will be described as well as the solutions that the research community IS analyzing.

## 21.1 Introduction

The ubiquitous computing paradigm requires that mobile devices to be connected to the Internet anywhere and anytime. As Internet connection may be demanded in hostile environments where the existing infrastructure does not guarantee access to the Internet, the ad hoc networks provide a feasible solution to extend the coverage area of telecommunication networks due to their multihop capability [32, 61]. In turn, the use of ad hoc networks could lead to economical advantages as a result of the reduced infrastructure required. In spite of the advantages associated to their use, the integration of mobile ad hoc

A. Triviño-Cabrera (✉)
University of Málaga, Spain
e-mail: atrica@gmail.com

S. Misra et al. (eds.), *Guide to Wireless Ad Hoc Networks*,
Computer Communications and Networks, DOI 10.1007/978-1-84800-328-6_21,
© Springer-Verlag London Limited 2009

networks into IP-based access networks involves challenging aspects that must be solved. With this purpose, several works (most of them IETF Internet Drafts) have been published. These works affirm that the Access Router to the telecommunication network should be complemented by an Internet Gateway to make relevant Internet technologies operative in multihop scenarios. Among other functionalities, the Internet Gateway is responsible for delivering several address configuration parameters through multiple hops and also provides the ad hoc routing functionalities commonly absent in conventional Access Routers. Possible implementations of the Internet Gateway consider the inclusion of extra functionalities in the final telecommunication network Access Router, which becomes an Access Gateway. In contrast, some other proposals are supported by the introduction of an extra element connected to the Access Router (through a wireless or wired connection) and which enables the Gateway functionalities. Furthermore, some approaches suggest that the seamless integration of a mobile ad hoc network into an IP-based access network should be based on the opportunistic configuration of some MANET devices that act as the Internet Gateway on behalf of the rest of the ad hoc devices.

In addition to the presence of a Gateway, the technologies associated to the Internet access need to be adapted to operate properly in multihop wireless networks. In particular, the configuration mechanisms are receiving significant attention as these networks have some specific characteristics that must be taken into account. As summarized in the draft [6], conventional auto-configuration mechanisms should overcome the following drawbacks commonly present in the current networking technologies:

- Lack of MultiHop Support. Common Technologies in the Internet such as NDP (Neighbor Discovery Protocol), Stateless Address Auto-Configuration, or DHCP (Dynamic Host Configuration Protocol) were specifically designed for one-hop networks and, therefore, cannot work fully in conventional MANET contexts where mobile devices may be several hops away from the Gateway.
- Lack of Dynamic Topology Support. Some configuration supports establish a hierarchical scheme in order to distribute the configuration parameters (for instance, the Prefix Delegation procedures follow this approach). However, the mobile devices change their positions unpredictably, so static and fixed neighboring structures are not common in MANET. Therefore, the ad hoc configuration supports should establish the mechanisms necessary to cope with this dynamic topology. Additionally, these technologies should also operate even in the case of abrupt disconnections of mobile nodes, which are expected to be more frequent in wireless networks.
- Lack of Merging Support. IP communications are mainly supported by the fact that ad hoc nodes are univocally identified by an IP address. In conventional networks where a static topology is assumed, the uniqueness is achieved in a straightforward way. However, ensuring this condition in

mobile ad hoc networks is not a trivial issue as the mobility of nodes may cause some nodes with duplicate addresses to coexist in the same scope. Furthermore, the node movements may lead to the invalidation of some previous suitable IP addresses. These events commonly occur in the merging of independent ad hoc networks. The configuration mechanisms in ad hoc networks should incorporate the procedures to detect and solve these problems which arise as a consequence of the merging.

- Lack of Partitioning Support. The arbitrary movements of nodes in ad hoc networks may lead to new and different situations which are not common in conventional networks, specially the partitioning or the split of ad hoc networks. These events may lead to scenarios where some devices cannot reach the previously employed configuration servers. The configuration mechanisms in MANET should adapt themselves to these new conditions.

Section 21.2 will explain in detail why conventional technologies are not appropriate for multihop wireless networks. Therefore, in order to cope with the aforementioned problems, specific networking technologies have been proposed. These technologies are mainly related to the architecture demanded to enable the connection of ad hoc networks to the Internet as well as the procedures that permit the configuration of IP addresses in this multihop and highly dynamic scenario. Throughout the next sections, the main proposals concerning these aspects are described.

## 21.2 Background

In a context such as mobile computing where big amount of devices are expected to move frequently and unpredictably, the manual configuration of their IP addresses is unapproachable. NDP was proposed as the specifications to allow mobile nodes to configure their own global IPv6 addresses autonomously. The main significant NDP tasks are:

- Determine the link layer address associated to an IP address. In this sense, NDP implements the ARP (Address Resolution Protocol) functionalities common in IPv4. Nodes exchange Neighbor Solicitation and Neighbor Advertisement messages with this purpose.
- Discover Routers. Hosts can dynamically determine the Routers that are reachable in their neighborhood. With this purpose, they can generate Router Solicitation (RS) messages or receive unsolicited Router Advertisement (RA) messages generated by the Routers in their neighborhood. Then, NDP replaces the ICMP Router Discovery.
- From the information announced by the Routers, mobile nodes can autonomously configure their own global IPv6 addresses. For this task, Router Advertisement messages are necessary. These messages allow routers to inform hosts how to perform Address Autoconfiguration. In particular,

routers can specify whether hosts should use stateful (e.g., based on a DHCP server) or autonomous (stateless) address configuration [51]. With this purpose, routers can generate unsolicited RA messages or they can generate them on response to a Router Solicitation (RS) message.

As we can see, RA messages are necessary for the autonomous configuration of mobile nodes. However, NDP specifies that RA messages must contain link local addresses in the source address. The consequence of this condition is that these messages cannot be forwarded [23]. Under these circumstances, new techniques are demanded to allow nodes in a multihop wireless network to configure their own IPv6 addresses.

## 21.3  Architecture for Connecting MANET to IP-Based Access Networks

Access to the Internet or, in general terms, to IP-based access networks, in MANETs is based on integrating them into already connected networks. In this sense, cellular telecommunication networks or the WLAN controller offer an Access Router, the so-called GGSN (Gateway GPRS Support Node) in UMTS (Universal Mobile Telecommunications System) or Access Router in WLAN (Wireless Local Area Network), for those devices which are located in their coverage area. The mobile device directly connected to the Access Router can easily exchange its data with external hosts. However, those devices outside this area require ad hoc technology in order to become globally reachable.

In most cases, the Access Router cannot be customized to include the ad hoc technology, so an Internet Gateway is required. The Internet Gateway provides the ad hoc routing capabilities, which are absent in conventional Access Routers, so that the Internet Connection data can be forwarded through multiple hops. In order to disseminate these data, the Internet Gateway generates specific messages by which the mobile devices gain knowledge of some configuration parameters in order to communicate with external hosts. This information is useful for several tasks. Firstly, mobile nodes need to know the Internet Gateways IP address and the route to this element. On the other hand, when a mobile node decides to connect to the Internet, it should acquire a global IPv6 address appropriate to the domain of the selected Internet Gateway. With this purpose, the mobile nodes should be informed about the prefix information that the Internet Gateway manages as well as the mechanism by which an IPv6 address is obtained (stateless or stateful address configuration). Finally, the mobile node should also be notified of the time that the information announced by the Internet Gateway is considered to be valid [36].

In some proposals, the Internet Gateway also has several additional functionalities. For instance, the Internet Gateway may act as a DHCP Relay Agent to forward DHCP messages from MANET nodes to external DHCP servers or may even act as a DHCP server by itself [50].

The implementation and the characteristics of the Internet Gateways differ among the diverse mechanisms that support the integration of MANET into the Internet [57, 43, 26, 48, 11]. In the following sections, these two issues are studied separately.

### 21.3.1 Characteristics of the Internet Gateway

A great variety of integration supports have been proposed [57, 43, 26, 48, 11]. The Gateway characteristics such as mobility, implementation details or the availability to work with some other Gateways differentiate the proposals. Regarding the Gateway implementation, two very different approaches can be basically identified. Firstly, the Internet Gateway could be considered as a **dedicated** device that is specifically pre-installed in those environments where a MANET is expected to be connected to the Internet [57, 26, 11]. In this sense, the dedicated Internet Gateway has two interfaces: one to connect to the ad hoc network and another to communicate with the Access Router. Furthermore, the Internet Gateway could also be integrated into the Access Router. Figure 21.1 illustrates the connection scheme when dedicated gateways are employed. In this UML 2.0 diagram, only the main methods and attributes are included.

As mobile ad hoc networks could be spontaneously formed in scenarios where no specific ad hoc equipment has been previously set up, some works [3, 48, 44] propose the use of MANET nodes that configure themselves to operate as the Internet Gateway on behalf of the rest of the ad hoc devices. Under this scheme, the Internet Gateway is called **opportunistic**, non-dedicated, or occasional. The use of opportunistic gateways is appropriate for coverage area extensions in cellular networks where mobile nodes belonging to the same ad hoc network communicate with each other without the use of any previously deployed network, while the communication with an external host demands the use of a MANET node acting as the Gateway, which is neither provided nor controlled by the telecommunication operator. Moreover, these gateways could also be suitable for automotive scenarios where a vehicle temporarily connects

**Fig. 21.1** UML 2.0 class diagram for Internet connection in MANET using dedicated gateways

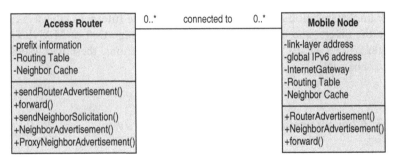

| Access Router | 0..* | connected to | 0..* | Mobile Node |
|---|---|---|---|---|

**Fig. 21.2** UML 2.0 class diagram for Internet connection in MANET by means of opportunistic gateways

to a petrol station or to a UMTS access while performing the Gateway functionalities on behalf of other vehicles [44]. Figure 21.2 represents the UML class diagram of the connection of ad hoc networks by means of opportunistic gateways.

The use of opportunistic gateways provides more flexibility and robustness for the MANET Internet connection. However, mobile nodes must implement an extra technique by which they may configure themselves as opportunistic gateways when they can act as opportunistic gateway, i.e., when they reside in the coverage area of the Access Router. Following the infrastructure-less nature of ad hoc networks, an opportunistic policy is preferred in the connection of MANET to external networks. Moreover, the configuration of the opportunistic gateway should be based on a distributed algorithm so the presence of centralized equipment to manage this procedure is recommended to be avoided. The algorithm, referenced in this text as the gateway configuration mechanism, may also take into account several parameters such as the mobility of the nodes or their energy level.

The gateway configuration mechanism offers the service for configuring opportunistic gateways, but it must also detect when a new gateway is necessary. With this purpose, the mechanism pays attention to the events that oblige the node acting as the Internet Gateway to stop providing these additional functionalities. Among these events, the movements to positions outside the coverage area of the Access Router or the disconnection of nodes are the most relevant. However, some supplementary factors may also be considered in the gateway configuration mechanism. For instance, a node playing the role of gateway may cease its Gateway functionalities when it detects that its energy-level has been considerably reduced or when the time that it has executed the Gateway role or the number of packets that it has forwarded on behalf of other MANET nodes has gone beyond a predefined threshold (for example, when fair energy consumption is desirable).

Figure 21.3 shows the UML 2.0 state diagram for the configuration of opportunistic gateways. The reception of Router Advertisement (RA)

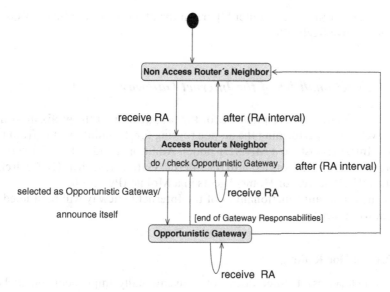

**Fig. 21.3** UML 2.0 state diagram for an opportunistic gateway in a MANET

messages, one-hop packets periodically emitted by the Access Router, provides the necessary information to know whether a mobile node is an access router's neighbor or not [36]. The distributed gateway configuration mechanism is responsible for selecting opportunistic gateways as well as shutting down the opportunistic gateways under some specific circumstances. The selected opportunistic gateway must announce its recently acquired functionalities to the rest of the MANET nodes. In order to detect the absence or disconnection of opportunistic gateways, this announcement should be periodically performed. By this technique, the neighbors of the Access Router continuously check if an Internet Gateway exists in the network and configure a new one when necessary. On the other hand, these neighbors know their reachability to the access router by the periodic reception of RA messages. Once an expected RA message is not received, the nodes assume to have escaped from the coverage area of the Access Router. This condition prevents them from operating as the Internet Gateway.

It is worth noting that, in some specifications [6, 12], the nature of the Internet Gateway (dedicated or opportunistic) is considered out of scope. Under these circumstances, the element that provides the Internet connectivity is commonly referred to as the MANET Border Router.

Another important characteristic of the Internet Gateways is related to its mobility. Some proposals consider that the Internet Gateway is a fixed device that is connected to the Access Router (through a wireless or wired connection) while some other works agree to the mobility of the Internet Gateway. In this sense, the mobility could be restricted to predetermined zones such as the coverage area of the access router [3] or, on the other hand, the Internet Gateway could move freely [48].

## 21.3.2 Functionalities of the Internet Gateway

The Internet Gateway is responsible for the propagation of the prefix information as well as for performing the ad hoc routing for downlink traffic (from the external Internet hosts to a MANET node). Additional and optional functionalities can also be incorporated into the Internet Gateway, such as DHCP Relay Agents, DHCP Server, or Home Agents in a Mobile IPv6 context.

The most relevant functionalities of the Internet Gateway will be studied in the following sections.

### 21.3.2.1 Ad Hoc Routing

An Internet Access Router does not conventionally implement an ad hoc routing protocol. Therefore, those packets that the Access Router receives and whose destination corresponds to a MANET node, which is placed outside its coverage area, cannot be forwarded. Key to this, the incorporation of an Internet Gateway is demanded. The Internet Gateway includes the implementation of an ad hoc routing protocol in order to discover, maintain, and manage the routes to all the mobile nodes that are not accessible by the Access Router.

Concerning this subject, the ad hoc routing protocol that the Internet Gateway executes should correspond to the ad hoc routing protocol used by the MANET.

Optionally, some integration supports employ the Internet Gateway's privileged view of the MANET in order to optimize the routing procedures [57]. In this sense, the Internet Gateway may participate in indicating the ad hoc nodes that optimal routes different to the ones that they are employing exist. This event usually occurs when a node that needs to be accessed through the Internet Gateway joins the MANET. Redirect ICMPv6 messages are used with this purpose [57]. In order to operate in multihop scenarios, a modified version of these ICMPv6 messages is proposed.

### 21.3.2.2 Propagation of the Prefix Information Through Multiple Hops

The Stateless Address Auto-configuration (SAA) technique is based on obtaining the prefix information that is appropriate to the domain where the configured IPv6 address is to be employed. As specified in RFC2461 [36], these data are carried by Router Advertisement (RA) messages that are generated by the routers following the NDP specifications. NDP operates with link local addresses so the RA messages cannot be forwarded through multiple hops. By these restrictions, those devices outside the coverage area of the router cannot get access to this information and, in turn, they cannot construct their own IP addresses.

In order to overcome this limitation, two possible strategies could be executed. Firstly, the ad hoc routing protocol could be extended in order to include the messages that are necessary for the exchange of the prefix information. On the other hand, NDP could be completed to incorporate additional messages that allow the multihop transmission. In both strategies, the incorporated messages are additionally employed to update the routes to external hosts or to update the routing information associated to the MANET nodes that have forwarded the messages.

Independent of the strategy followed to propagate the prefix information, in general terms, three additional messages are defined for this procedure. These messages are usually referenced differently in the proposed integration supports, and they may also have some properties that are private of each support. However, in order to describe a generic framework for the integration of ad hoc networks into the Internet, three generic messages are proposed in this chapter:

• Multicast Prefix Advertisement (MPA) Message

Internet Gateways generate this message with the information extracted from the Router Advertisements that they receive from the Access Router to which they are attached. Specifically, the data contained in the Prefix Information option in the RA message are employed for the MPA messages. However, in stateful configuration mechanisms, where the assignments of IP addresses are managed by centralized equipment such as in a DHCP-based scheme, the MPA messages may exclusively inform about the need of demanding an IP address to the entity responsible for managing the configuration of nodes.

The main difference between MPA and RA messages resides on the fact that these MPA messages can be forwarded through multiple hops to reach all nodes in the MANET (the multicast group).

In the related research works, the MPA message has received several names as Modified Router Advertisements (MRA) [48], Internet Gateway Advertisement (IGWADV) [57], Gateway Information (GW_INFO) [26], or Prefix Advertisement (PA) [43]. Depending on the integration support, this message could be emitted to reach all the nodes in the MANET (proactively) or to be received in a restricted area (a hybrid scheme), or even restricted to a one-hop area, but with the possibility of transferring copies of the message [26].

• Prefix Advertisement Solicitation (PAS) Message

It is emitted by those nodes that require the prefix information or need to discover the route to the Internet. When the ad hoc routing protocol is extended to include this message, it is commonly implemented as a Route Request packet (RREQ) with a determined bit set to 1 in order to be differentiated from a conventional RREQ. In this case, the packet is commonly cited as RREQ_I [57]. Additionally, the response to this message is based on a Route Reply packet (RREP), with the Internet bit set to 1. The response, which could be generated by the Internet Gateway as well as by those intermediate nodes that have the updated information on route to the Internet, is usually named as RREP_I. Additional names as Modified Router Solicitation Messages (MRS) [48] or the Gateway Solicitation (GWSOL) [26] messages are also employed to identify these messages.

- Unicast Prefix Advertisement (UPA) Message

It receives similar names as the multicast prefix advertisement message, but it is generated by the gateway on demand when it receives a Prefix Advertisement Solicitation. Under these circumstances, the gateway generates a unicast message that includes the prefix information that it is processing. Similar to the MPA messages, under stateful mechanisms the UPA messages just inform about the need of asking for an IP address to centralized equipment (e.g., a DHCP server). The technique by which the centralized entity is discovered depends on the stateful mechanism.

Although a generic definition of the gateway messages have been presented along this section, the terminology associated to the extension of NDP messages and the proposed one will be employed indifferently.

### 21.3.2.3 Gateway Discovery

The procedure by which the gateway announces its presence is referenced as the Gateway Discovery. As it has been already suggested, three different schemes for the gateway discovery have been proposed:

- *Proactive*: By this strategy, the gateway periodically emits multicast prefix advertisement messages. In order to reach all the nodes in the MANET, the TTL (time-to-live) value of these messages is set accordingly to the network diameter estimation. This scheme requires the configuration of the interval of emission of the MPA messages ($T_{MPA}$).
- *Reactive*: Mobile nodes trigger this procedure to obtain the required Internet connection data (prefix information or route to the Internet Gateway). Under these circumstances, the mobile devices emit a Prefix Advertisement Solicitation (PAS) message, which is forwarded to the Internet Gateway. On receipt, the Internet Gateway responds with a Unicast Prefix Advertisement (UPA) message, which is conveniently routed to the originator node.

Pure reactive schemes where the Internet Gateway just interacts with the MANET nodes on demand exist. However, this procedure could be also triggered in a proactive scheme when the employed $T_{MPA}$ does not guarantee the existence of a valid route to the Internet Gateway in the nodes. Under these circumstances, the mobile node directly demands the Gateway information (through a PAS message) to update its route to the Internet.

- *Hybrid*: This scheme combines the two previous strategies. As in the case of the proactive procedure, the Internet Gateway is responsible for emitting a

MPA message every configured $T_{MPA}$ interval. However, a finite value is set in the TTL field of these messages in order to deliver these packets just to the nodes that are closer to the MANET. Specifically, the nodes that receive the Multicast Prefix Advertisement messages belong to the $n$-neighborhood of the Gateway, that is, to say, the MPA message is just received by those nodes which are $n$ or less hops away from the Internet Gateway, being $n$ the value set in the TTL field. The rest of the nodes that do not receive this message proactively prompt the reactive procedure when no route to the Internet Gateway is known or when the available Internet connection data is stale.

#### 21.3.2.4 DHCP Functionalities

Under some stateful configuration schemes, the gateway also plays some DHCP functionalities that may include the roles of DHCP relay agent or DHCP server [50].

## 21.4 Technological Adaptations

The establishment of IP-based communications demands mobile devices to be configured with certain information such as IP addresses and the identification of key servers (e.g., DNS servers). Existing configuration protocols were initially conceived to be utilized in wired networks or infrastructure-based wireless networks. Due to this assumption, some protocols are supported by on-link communications, which clearly restricts their utilization in multihop networks. For instance, the NDP technology, which plays a fundamental role in address configuration, route discovery, and even address resolution [36], becomes insufficient in conventional MANET where routes usually consist of several hops [9].

In order to overcome these limitations, Internet Connectivity in ad hoc networks has prompted some adaptations in the already existing technologies as well as the proposal of new mechanisms.

The following sections explain the main modifications concerning this area.

### 21.4.1 Address Auto-Configuration: Problem Statement

Integration of MANET into external networks requires that ad hoc nodes should be provided an IP address appropriate to the domain where they are placed. In a pervasive environment, where mobile devices may change their points of attachments arbitrarily and where terminals are expected to participate in Internet communications, the manual configuration of nodes is clearly advised against. In contrast, an auto-configuration mechanism by which nodes configure themselves without human intervention is strongly recommended. In this sense, when a node moves to another network, it should configure its new IP address conveniently. To construct the IP address, nodes employ some information that they receive about the valid prefix and, optionally, it combines this information with local data (e.g., its MAC address). The process to construct the IP address is described in Section 21.4.2.

According to IPv6, address validity depends on the scope or area where this address is expected to be employed. Therefore, three different types of addresses could be configured in a MANET node:

- Link-local addresses for on-link communications
- MANET local addresses or Unique Local Addresses, which are valid in the entire ad hoc network
- Global addresses, which are employed for external communications.

For each of the considered scope, unicast communications demand the ad hoc nodes to be univocally identified. Due to the node freedom of movements and even to the absence of a centralized entity in charge of managing the IPv6 addresses in some scenarios, guaranteeing this condition is not a trivial task in conventional MANET applications where nodes configure their own IPv6 addresses. In order to obtain nodes configured with unique IP addresses, the authors in [62] highlight three events that the configuration mechanisms should deal with. These events are:

- A new node joins a configured MANET or a configured node leaves the MANET. The node incorporating could be configured with an IP address, which collides with the IP address already present in the MANET. On the other hand, when a node leaves the MANET, the IP addresses that it is managing (for its interfaces or to delegate to other neighbors) should be considered free.
- A partition of the MANET detaches from it and, after a while, it joins the previous MANET again. While the partitions are separated, new nodes may join them and, when reconnected, the collision of IP addresses may occur.
- Merging of independent MANET. Nodes in independent MANET could configure their same IP address and work with it correctly. However, once that they are merged, the collision may cause irregularities related to the network performance.

The previously explained events are associated to stand-alone MANETs. However, Internet-connected MANETs should also pay attention to these occurrences. Additionally, it is also possible to identify a new event that may provoke to invalid IP address assignments. Specifically, the problematic scenario is related to the presence of multiple gateways announcing different prefix information. Under these circumstances, a node configured with the prefix announced by one of the gateways may decide to change the gateway to employ and, therefore, a new IP address should be utilized. This "switching" of the gateway to employ and in their corresponding assigned IP addresses may lead to undesirable effects such as duplicity of addresses. Then, two decisions should be taken when a switching occurs. Firstly, it is necessary to decide whether the previous address is going to be assigned to the node in the future or not. Secondly, it is also necessary to confirm that the new address to employ is still unique, especially if the time elapsed since the moment of its construction is high.

### 21.4.2 Address Auto-configuration

Nowadays, alternative proposals to configure global IPv6 addresses in the MANET in a stateless way exist. In this sense, it is necessary to know that the range of valid IPv6 addresses that can be employed in a MANET constitutes an abstract entity called pool. The method by which this pool is managed differentiates two configuration strategies:

- *Conflict-detection allocation scheme*: By this technique, a node picks an IP address from the pool of addresses, configuring it as the tentative address. No node is in charge of managing the pool, so the node may become to know the pool of addresses that the network supports by receiving valid prefix information. From these data, it constructs the tentative IPv6 address by concatenating some information that is considered to be unique to the prefix. In order to confirm the uniqueness of the self-generated address, Duplicate Address Detection (DAD) techniques may be applied. In case of conflict, the node should pick a new address and repeat the procedure [40].
- *Conflict-free allocation scheme*: The range of valid IPv6 addresses or pool is successively fractionated and distributed among the nodes in the network. A new device intended to configure one of its interfaces accesses to the distributed pool of addresses and picks one. Further, the mobile nodes may obtain part of the already distributed pool in order to manage and distribute it to other nodes. This mechanism is supported by the concept that the addresses which are delegated are not being used by any node in the network. This can be achieved, for example, by ensuring that the nodes which participate in the delegation have disjoint address pools. In this way, there is no need of performing Duplicate Address Detection (DAD). However, ensuring the correct management of the partitioned pool of address becomes the main drawback of this strategy. This problem is especially conflictive when abrupt disconnection of the mobile nodes, which are controlling part of the available addresses, may take place [35, 49, 52].

The document *draft-bernardos-manet-autoconf-survey-01* [7] summarizes the main properties of the related work concerning this subject. In the following sections, taking as starting point the referenced document, the proposals will be further explained and complemented by some other technologies involved in the address configuration mechanism, which were not considered by the aforementioned text.

### 21.4.2.1 IP Address Conflict-Detection Mechanisms

This mechanism is based on a two-phase operation. Firstly, the mobile node constructs a valid IPv6 address (the tentative IP address), and then it confirms the uniqueness of the self-generated address by applying a DAD (Duplicate Address Detection) procedure. These two procedures are analyzed separately.

Construction of a Valid IPv6 Address

To autonomously construct a global IPv6 address, mobile nodes require a global prefix appropriate to the domain where they are placed. In this sense, Internet Gateways are responsible for propagating the prefix and its corresponding information in a multihop scenario. Several strategies have been proposed to disseminate this information and they mainly diverge in the mechanisms for integrating mobile ad hoc networks into the Internet. Once the device has the prefix information, it must concatenate some data to the prefix in order to generate the IPv6 address. In RFC4862 [51], the EUI-64 [20] identifier is suggested for this operation. However, the utilization of a random number is also recommended to prevent the user's tracking. With this target, [38] explains a methodology to derivate a random identifier from the IEEE identifier. Additionally, for secure communications as in the Secure Neighbor Discovery (SEND) [4], the IPv6 address may bind to a cryptographic key to univocally identify the sender. In order to obtain an appropriate address, the generation of the interface identifier could also be accomplished by computing a cryptographic hash of a public key as specified in RFC3972 [5].

However, not all the devices that are expected to operate in a MANET have a NIC (Network Interface Card) with a 48-bit IEEE-assigned unique MAC address, since mobile ad hoc networks may also be integrated by GSM or UMTS devices. In contrast, an IMEI (International Mobile Station Equipment Identities), i.e., a 14 decimal digit number, is attached to the hardware of any of these terminals [19, 1]. The IMEI plays a similar role as the IEEE MAC-48-bit address by informing about the producer and constituting a unique value, which univocally identifies the device. The document *draft-dupont-ipv6-imei-10.txt* [18] explains how to extract an IEEE MAC-48 address from an IMEI identifier in order to employ it in address configuration procedures. However, the resulting IEEE MAC-48 address may collide with the MAC addresses associated to operative NIC-based devices. In addition, some devices may not support the proposed procedure as no standardization has been defined yet.

In the previous schemes, a node concatenates local information to a global prefix to obtain an IP address. Nevertheless, some other proposals show alternative methods to construct the IPv6 address. For instance, in [39], the already configured nodes facilitate the configuration of recent arrived nodes. With this purpose, a node entering the network transfers the responsibility of selecting a new address to a configured device. By this delegation, the task to acquire an IP address is performed by a device that has more information, as it has been keeping track of the addresses that other nodes are employed in the network during its connection. A similar approach is followed in [16], where the authors also suggest that the already configured nodes should support the configuration of the recent devices. With this objective, the already configured nodes periodically announce themselves by means of ADDR_BEACON messages, which contain a range of temporary addresses that are assumed not to be employed by other nodes (a correct splitting of the address pool satisfies this restriction). On the receipt of these messages, the arriving node selects one of the already configured nodes and then it asks the elected one to assign it an IP address. When this process is completed, the node is supposed to have acquired a local address.

As we can see, there is a great variety of methods to construct an IP address. This diversity could lead to the generation of duplicated addressees. DAD procedures help detecting these duplications.

### Duplicate Address Detection (DAD)

As it has been previously explained, in a stateless address auto-configuration mechanism, mobile nodes receive the prefix information related to the network where they are placed. With this information, they can construct a MANET or global IPv6 address by concatenating certain data that are considered to be globally unique, for example, the IEEE MAC identifier. However, the self-generated IPv6 address is not guaranteed to be unique due to several reasons. Firstly, terminals with duplicate MAC address exist on the market because of non-registered or erroneously manufactured devices. On the other hand, users may intentionally alter the configured IPv6 or MAC address. In addition, the use of globally unique ID as part of the IPv6 address is not generally accepted as this utilization eases the users tracking, which may imply negative consequences on their privacy [37]. Furthermore, the MAC may not be available in certain device (e.g., cellular phones) [18].

Due to the multiple sources and methods that are available to generate the IPv6 addresses, duplicated addresses may become a reality. Therefore, if intrusions and misbehavings are required to be suppressed in the communications, a new mechanism should be applied after the address auto-configuration mechanism. This new technique is called *pre-service Duplicate Address Detection* (DAD) as the nodes check the validity of the self-configured IPv6 address before using it [6]. Additionally, DAD could also be applied continuously in order to ensure the uniqueness of the IPv6 address in those scenarios where mergings of MANETs that share the same prefix are expected to occur. This strategy is called the *in-service DAD*, as nodes should apply it while they used the self-generated IPv6 address [6].

Taking into account the fact that in-service DAD is only useful when mergings take place, a technique alternative to in-service DAD is based on the detection of MANET mergings. A remarkable procedure for detecting mergings in the network is proposed in [39]

As previously explained, DAD procedures can be classified according to the moment when the mechanisms are activated leading to pre-service and in-service DAD. However, it is also possible to establish a different classification by considering the messages that are employed by the DAD procedures. In this sense, under *active DAD*, specific messages are introduced in the network in order to detect duplicated addresses. On the other hand, *passive DAD* relies the detection of duplicated addresses on the analysis of the routing protocol messages, which are naturally introduced by the nodes [56].

Concerning active DAD, we should take into account that this procedure is not an exclusive MANET technique. Indeed, the stateless configuration mechanism recommends its utilization when configuring link local address [51]. In these specifications, a two-message protocol is proposed as follows. Initially, the node that is expected to confirm its tentative link-local address emits a Neighbor Solicitation message, indicating that the target destination is its tentative address. Then, the source attends to a potential Neighbor Advertisement message in case that an on-link device has already assigned the same IPv6 address. If this event takes place, the tentative address is not unique and, therefore, it cannot be employed. Otherwise, the process can be repeated a predetermined number of times before assuming that the tentative address is unique and then assign it to the interface. A similar approach is presented in RFC3926 [14], where ARP (Address Resolution Protocol) is employed to check the uniqueness of the self-configured address in an IPv4 context.

However, these previous methods cannot be fully applied into multi-hop networks as the employed messages are restricted to on-link communications. In order to overcome this limitation, several proposals exist in the literature. For example, the study in [40] suggests to extend the NDP messages and to embed them into the routing protocols. Following this technique (the so-called trial-and-error DAD), a node wishing to acquire a site-local IPv6 address requests a route to a hypothetic device that has the self-generated address. If no response is received during an interval of time ($T_{DAD}$), the node will assume that the address is unique within the MANET. An important design aspect in this DAD procedure is the selection of the value of $T_{DAD}$. The configuration of this parameter could greatly impact on the network performance. In this sense, too small values of $T_{DAD}$ may prevent from detecting duplicated addresses, as the response of the node with the duplicated address may take a longer time to traverse the network. On the other hand, it is not recommended to select high values for $T_{DAD}$ since nodes must interrupt external communications during the $T_{DAD}$ interval. In scenarios with multiple Internet Gateways where mobile nodes connect through different points of attachment, this time, usually 1 second [40], causes noticeable latencies and may even provoke the break of the on-going connections. In [39], authors devise that the suitable value of $T_{DAD}$ depends on the network diameter.

An alternative approach is presented in [39]. Under this scheme, the node that is selecting an IP address for a recently incorporated device needs that the rest of the MANET nodes confirm the availability of the IP address (either negatively or positively). Although it introduces more overhead than in the previous mechanisms, partitions and mergings may be detected easily by comparing the number of received messages to the number of the expected ones.

The aforementioned DAD techniques present clear limitations to work in ad hoc networks composed of hundreds of nodes, as they are supported by flooding the network with protocol messages. One solution is the utilization of the benefits related to MPR (Multi-Point Relay) in OLSR routing protocol [15], as suggested in [30]. On the other hand, in order to ease the network scalability, [58] recommends the establishment of a hierarchy in the MANET. Thus, certain devices, called the leader nodes, will be responsible for managing and propagating a random and already subscribed subnet ID (Identifier) by means of Router Advertisements, which will be modified to reach a region limited by a predefined number of hops (rs), in a similar way to the Modified Router Advertisements generated by the Internet Gateways. Nodes receiving these messages construct their IPv6 addresses from the known subnet ID and check the uniqueness of the self-generated address in an area restricted by rs-hops by means of extended NDP messages, which can traverse a number of rs-hop as this value is set in the TTL fields. In conclusion, the proposed cluster-based scheme restricts the flooding provoked by DAD operations and, due to this behavior, this strategy allows nodes to proactively check the validity of the constructed IPv6 addresses. On the other hand, one of the main issues of this proposal resides on the method to dynamically select the leader nodes. About this concern, the authors propose the employ of the number of neighbors that a node maintains as a promising heuristic to select the leaders. Another scalable solution is proposed in [13] where a ring-based topology is formed in the MANET.

Up to now, active DAD mechanisms have been studied. These previously explained strategies can be considered active techniques as specific control packets are introduced in the network in order to detect a conflict in the IPv6 addresses [56]. However, it is possible to discover this type of inconsistencies by monitoring the ad hoc routing packets. This strategy, referenced as passive DAD, is especially suitable for in-service DAD where the validity of IPv6 addresses could be continuously checked without any exchange of extra packets, which may lead to the consumption of unnecessary network resources.

One of the main proposals about passive DAD is auto-defined as *Weak DAD* [56]. Following this strategy, the demand on IPv6 addresses to be unique is relaxed compared to the previous strategies that are referenced as *Strong DAD*. Thus, the author defends that duplication is permissible whenever the packets are received by the intended destination nodes. As a significant drawback, we should remark that although this situation is not a conflict by itself, it is very sensitive to the movements of nodes that may invalidate the current routes. In that case, the required route discovery mechanism may be prone to establish routes to the undesired destination. Furthermore, we must remark that these situations, where duplicated IP address coexist meanwhile the routing is still adequate, is not possible in Internet connected MANET where the traffic flows through the Internet Gateway. Under these circumstances, only one destination is considered in the Gateway and, therefore, the address conflict may lead to the incorrect delivery of packets.

Aiming at detecting address conflicts in stand-alone MANETs, the author in [56] proposes the use of a specific key which is introduced in the routing protocol for the passive DAD operation and which is associated to each node in link-state routed MANETs. The key is assumed to be generated randomly or from the Interface Identifier at each node. This additional information is stored in the routing tables so, when intermediate nodes obtain the periodic link states, they can detect conflicts by analyzing the key's stored value and the received one. In [56], the author also analyzes how Weak DAD could be incorporated into DSR [28]. In DSR, as a source routing protocol, route discovery packets contain the list of nodes that the packet has already traversed. Typically, this list is composed of IP addresses that identify the nodes. The proposal suggests that this list should also include the key of each node from which the packet has been forwarded. In this sense, bigger routing headers are included and the routing protocol must be adapted to support passive DAD.

In order to overcome this limitation in link-state routing protocols, the original Weak DAD is optimized by linking the key to the sequence number commonly employed in this type of protocols to differentiate recent routing information [58]. By analyzing the sequence numbers in routing messages, MANET nodes receiving packets with its same IP address can easily detect whether the packet comes from another node with replicated IP address or not. On receipt of packets with a source address different to the own IP address, a device is also able to detect address conflicts. With this purpose, the sequence number of the received packet is compared to the corresponding value stored in its routing tables. When an abrupt disparity is observed, the node assumes the existence of a conflict so it informs one of the involved nodes with duplicated IP address about the need of configuring a new IP address (a new routing message is introduced with this objective). In contrast, when similar sequence numbers are compared, additional information is required to detect conflicts. In link-state algorithms, the most adequate and suggested data to employ for this purpose are the link states, so information about nodes with equal IP address but a different set of neighbor nodes will make the detection of conflicts possible. As stated in [8], this algorithm may lead to undesirable storm effects when multiple intermediate nodes detect conflicts and alert the conflicting node.

Under some circumstances, the detection of conflict in the IP addresses is also possible when determined anomalies are observed in the ad hoc routing protocol behavior. As specified in [60] devices must be alert to multiple events, which are symptoms of address collision in the network. Specifically, eight events are highlighted in an OLSR ad hoc network. Translating this strategy to on-demand protocols, in [59] the authors identify some erroneous events as indications of conflicts. For instance, when a node receives a Route Request, which contains its own IP address as source address but it has not generated any Route Request.

Despite of the simplicity of the proposal, the presence of multiple traffic sources in practical applications demand additional information to detect conflicts. Taking this fact into account, the authors in [29] suggest the inclusion of extra data in the route discovery messages so some conflictive events may also be tracked. In this sense, the localization information, obtained for example by GPS (Global Positioning System) devices, can help to detect duplicity easily. However, ad hoc nodes may not always be equipped with this technology so alternative methods are also proposed. In order to detect duplicate IP addresses in traffic sources, the neighbor list is included in the request messages so on receipt, intermediate nodes could alert about any existing conflict. Similarly, the route reply messages may also contain a subset of the neighbors that a node has, so a device could easily conclude the presence of duplicated addresses by receiving route replies from a theoretically unique address but with significant differences in the announced neighborhood. An alternative method consists in associating a sequence number to the route request messages so multiple copies of a route request are responded by means of route reply messages whose sequence number is incremented in each reply. By receiving a route reply message with a DAD sequence number, which is equivalent to a previously received route reply message for the same route request, the node would detect the existence of two responding nodes. In this sense, the multipath discovery is also supported by this passive DAD mechanism.

### 21.4.2.2 Conflict-Free Allocation Technique for Address Auto-Configuration

This scheme is supported by the distribution of a pool which contains all the IPv6 addresses which are valid in the MANET. The main concerns of this strategy reside on the methodology to distribute the address pool as well as on the mechanism that is triggered when a node that is responsible for part of the pool disconnects (abrupt or graceful disconnection should be considered). Several proposals about this scheme have been published. Next, the most remarkable approaches are described.

DCDP (Dynamic Configuration and Distribution Protocol) was originally conceived for address auto-configuration in hardwired networks [34]. Taking as basic support the distribution of addresses proposed in DAAP (Dynamic Address Allocation Protocol) [33], the available pool of addresses is recursively splitted along the DCDP servers. In this sense, DCDP behaves as a transactional model where nodes are either requestors of or responders to individual configuration requests. A requester asks for configuration information to a DCDP entity, which responds by sub-leasing part of its available address pool to the requester node. By recursively splitting the address pool, DCDP can automatically distribute address pools to each node. A configured node uses the first IP address of its IP addresses block as its own IP address. An important lack of this proposal resides on its unspecified treatment about network partitioning, merging, and disconnection of terminals. These drawbacks are overcome in the DAAP, as described in [52]. In this approach, a node entering the network and requesting an IP address must demand it to the leader of the network, which is considered to be the device with the highest IP address. In order to handle the partitioning and merging problems, the networks are identified by the MAC address of the first node that becomes part of it. However, as stated in [52], the movement of the initiator node could arise in multiple networks with the same identifier. Furthermore, as explained in the previous sections, MAC addresses may be duplicated or not available.

In [35], the address pool is also recursively splitted but a buddy algorithm is recommended for this operation. This technique, commonly employed in memory management, links the node offering half of its pool to the node that accepts and manages the supplied set of addresses. Due to the well-known characteristics of the buddy system, a graceful disconnection of nodes is easily managed by their buddy nodes, which conveniently recover the leased subset of addresses. These bindings are also necessary to properly verify that the assigned addresses are being employed, that is, to check if partitions or abrupt disconnection have occurred. With this objective, the device periodically initiates a synchronization procedure by which mobile devices announce the managed pool as well as the subsets of pools that have been delivered to other nodes. With this information, any node in the network can construct a tree of subset of addresses and detect a discontinuity in the assigned addresses. This fact indicates that certain range of IP addresses is wrongly managed and, therefore, the procedures for recovery are triggered.

The previous proposals assume that a configured node is going to be responsible for managing a part of the address pool. However, this assignation may depend on the capabilities, resources or even willingness of the mobile device. Taking into account this fact, the address pool is just splitted on demand in the proposal described in [49]. As neighboring nodes may not have any pool, a new node selects one of its neighbors as its leader and it starts a multihop handshaking protocol in order to obtain an IP address or a subset of the pool for the arrived node if the new node opts to manage a subset of IP addresses.

On the other hand, a different solution is presented in [62] where the construction of IP addresses is based on a sequence generation function. The function must be carefully chosen so that in the sequences of addresses that it generates, the interval between two occurrences of the same number is very large and the probability of more than one occurrence of the same number in a limited number of different sequences initiated by different seeds during some interval is extremely low. All the devices in the network manage this function, but the seed and state to employ in this mechanism must be communicated by a previous node in the network. As the device is not yet configured, a one-hop exchange protocol is proposed.

Along these proposals, the leader nodes are responsible for announcing certain information that they manage for address configuration and which is supposed to have been previously assigned by employing a distributed algorithm. However, it is important to notice that malicious nodes could misbehave in purpose by announcing false configuration parameters. For instance, these nodes could assign IP addresses already in use. Under these circumstances, the uniqueness of the assigned IP addresses cannot be guaranteed. Therefore, some additional mechanisms are recommended in order to prevent these malicious nodes from altering the network performance. In [10], a technique to avoid this drawback is described. By this method, only trusted nodes are able to participate in the auto-configuration service. With that purpose, a "$K$ out of $N$" trust model is employed. Under this scheme, a new node could be considered as trustworthy only if $K$ nodes in the MANET composed of $N$ devices trust in it. This technique may also be extended to conflict-detection configuration mechanisms, which usually require the acknowledgement to employ the selected IP address from the rest of the MANET nodes. In this sense, malicious nodes could also prevent the configuration in the network as they may reject all the proposed IP addresses.

## 21.5 Integration Supports

The integration supports define a framework where the gateway functionalities are defined. Additionally, they also specify how the nodes in the MANET operate with the Internet Gateway.

Nowadays, some integration supports exist. In the following sections, the most significant ones will be described.

### 21.5.1 Global Connectivity

This Integration Support, which is by far the most popular scheme, is based on the utilization of fixed gateways that are attached to the Access Router or even implemented on it [57].

In this scheme, the Gateway emits Internet Gateway Advertisements (IGWADV) proactively or on demand when receiving Internet Gateway Solicitation (IGWSOL) (the MPA and the MPS messages described in Section 21.3.2). A hybrid scheme was also proposed in [41]. Following a NDP-based terminology, the IGWADV will be also referenced as MRA (Modified Router Advertisement) messages.

The gateway discovery messages can be implemented by extending NDP (Neighbor Discovery Protocol) or the employed ad hoc routing protocol. Under both strategies, the gateway information must be similar. Additionally, the messages can inform whether the Gateway compels the self-configured nodes to inform the Internet Gateway about their constructed IPv6 address or not. This acknowledgment is carried out by means of a new message introduced in this gateway discovery mechanism, the Internet Gateway Confirmation messages (IGWCON). On their receipt, the IGWCON messages allow the Internet Gateway to maintain a list that associates the MANET local addresses to their global IPv6 addresses so the process under downlink routing may be improved.

The gateway advertisement processing leads to the update of the route to the Internet. In this context, the mechanism proposes the inclusion of a default route to forward packets to external host by means of the Internet Gateways, which perform as intermediate nodes. The main drawback of this policy is that a double lookup in the routing table is necessary: one to know the address of the Internet Gateway and another one to discover the address of the next hop to the selected Internet Gateway. In the forwarding process, the Internet Gateway can be explicitly mentioned in the optional proxy routing header when employed [25].

Due to its popularity, several research works have been developed assuming this mechanism as the Integration Support. Nevertheless, some of the conclusions extracted from these research works may be extended to some other mechanisms.

In this sense, one of the research lines is focused on the analysis of the adaptive gateway discovery techniques [31, 21]. The goal is to optimize the network performance by tuning the interval of emission of Multicast Prefix Advertisement messages (e.g., the IGWADV) properly. In the same context, [45, 46, 42] propose that the radius (scope) of the proactive zone under the hybrid gateway discovery can be set as a function of the position of nodes which are communicating with external hosts.

On the other hand, in scenarios where multiple gateways are reachable, nodes must select the gateway to employ. Consequently, the analysis of gateway selection criteria has become another important research topic. Thus, there are works that propose to select the gateway according to the expected lifetime of the route to the gateway [26] or to obtain an adequate balancing of traffic load [24, 47].

### 21.5.2 Prefix Continuity

This integration support, specified in [26], focuses on scenarios where multiple gateways announcing even different prefix information exist. In these environments, the mobile node must select which gateway to employ. The main contribution of this mechanism relays on the fact that it relates the gateway selection to the propagation of the gateway announcements.

Following a proactive scheme, the gateways are responsible for emitting periodic Multicast Prefix Advertisement messages, which are referenced as GW_INFO. One main feature of this message, which differentiates this mechanism from others, is that the GW_INFO messages are one-hop limited. Local decision on which gateway to employ is followed by the update of one GW_INFO message (the one related to the selected gateway) among all the received messages. Therefore, just this selected GW_INFO message is again propagated to the node neighbors. Making so, the overhead in the network is expected to be reduced.

The mechanism introduces the prefix continuity concept in MANET scenarios. This characteristic ensures that each connected ad hoc node has a neighbor which employs its same prefix information. Extending this behavior to the global ad hoc network, each device in the MANET holds a complete Internet communication path, which is formed by mobile nodes sharing the same prefix information.

The use of the prefix continuity leads to a topological division into subnetworks where all nodes share the same prefix.

### 21.5.3  Extended Support for Global Connectivity

This mechanism offers a solution for a stateful address configuration in MANET [11]. The Internet Gateways are responsible for managing the assigned IPv6 addresses. Thus, the Internet Gateway stores the binding of MANET local address and assigned global IPv6 address in a specific structure referenced as *MANET-node-cache* [11].

As a centralized architecture controls the assignment of IPv6 addresses, DAD can be avoided. In addition, unnecessary route discoveries are avoided when the gateway receives a packet whose destination does not correspond to an already assigned IPv6 address. In this case, the packet is silently dropped.

The gateway discovery is supported by extending ad hoc routing protocols, as suggested in the Global Connectivity mechanism. Specifically, the MANET nodes emits GW_SOL (similar to the Prefix Advertisement solicitation explained in Section 21.2.2) messages to obtain a valid route to the Internet or to acquire a valid IPv6 address while the Internet Gateway replies with a GW_ADV message (this messages will correspond to the Unicast Prefix Advertisement message explained in Section 21.2.2). To respond, the gateway must look up in its MANET-node cache. If no address was assigned to the node, a new IPv6 address is provided in the GW_ADV message. In other case, the validity of the already assigned IPv6 address is refreshed.

An important characteristic of this Integration Support is its capability to easily include some features of the Mobile IPv6 services. In this sense, a mobile node can register its home address by using LOC_UPDATE messages, which are responded by the gateway with LOC_UPDATE_REPLY messages. By this, the previous messages will play a similar role to the Binding Update message and the Binding Acknowledgment message defined in Mobile IPv6 [27], respectively.

### 21.5.4  Automatic Configuration with Multiple Gateways

This Integration Support [43] provides a solution to operate in scenarios where multiple Internet Gateways announcing different prefixes exist. Opportunistic or dedicated gateways are considered in this mechanism where gateways are supposed to announce themselves proactively. For this task, the Internet Gateways periodically generate Multicast Prefix Advertisement Messages, which are referenced as Prefix Advertisement Messages.

As a novelty, each mobile node configures one global IPv6 address for each prefix information received following a stateless scheme. Then, it chooses the IPv6 address to employ depending on the employed criterion for the gateway selection.

Since additional IPv6 addresses are available, a seamless gateway switching is supported when the currently employed IPv6 becomes invalid due to any reason such as a disconnection of the employed gateway or movements of the nodes. Under these circumstances, the mobile node can employed one of its global IPv6 addresses, which were previously configured. This operation by itself allows mobile nodes to change the IPv6 address quickly. However, routing may become inefficient as the learnt routes are not associated to these new addresses. In order to overcome this limitation, the mobile nodes are responsible for broadcasting all the self-configured IPv6 addresses, so the rest of the MANET can bind the node MANET local address to its global IPv6 addresses. With this purpose, the device can multicast the MID (Multiple Interface Declaration) messages considered in OLSR [15].

Among the configured IPv6 addresses, the device must select one of them for its external communications. For this selection, this support proposes two criteria:

- Default Gateway Method. The gateway currently in use gateway is preferred to the rest of the reachable gateways.
- Threshold Method. A new gateway is preferred when its announced metric compared to that of the present gateway exceeds a predetermined threshold.

The work does not detail any specific metric for the gateway selection, but it points out that several factors such as the traffic which the gateways are supporting or the distance could be employed for this purpose [43].

### 21.5.5 Mobile Multi-Gateway Support

As the previous mechanisms, this Integration Support enables ad hoc networks deployed at the edge of the Internet to get connectivity with the IPv6 hosts located on the Internet. As a particularity, the mechanism proposes the use of mobile gateways. The main concern of this support is the configuration of one or more ad hoc nodes as mobile Internet gateways.

This integration support was specified in [48]. The mechanism is based on opportunistic Internet Gateways, i.e., some MANET nodes configure themselves as the Internet Gateway in order to supply Internet access to the rest of the MANET nodes. The configuration of opportunistic gateways does not restrict the movements of the nodes acting as the Internet Gateways. So, any node can freely move, even to some areas where they could not act as the

Internet Gateway any longer. Then, this support employs temporal opportunistic gateways, which dynamically configure themselves to proceed as the Internet Gateways.

In this mechanism, all the nodes that are in the coverage area of the Access Router are potential Internet Gateways. Among them, one is selected to act as the Internet Gateway. This node is called the Default Gateway (DG). The rest of the nodes that are placed in this area are referenced as Candidate Gateways (CG) and they must be alert to detect if the DG shuts down its role (for example, because it escapes from the transmission range of the Access Router). So, at a given instant of time, one Default Gateway and none or multiple Candidate Gateways could coexist. Figure 21.4 describes the configuration proposed by this support.

The method for temporally and dynamically selecting the node to operate as the opportunistic gateway as well as the treatment associated to its mobility are the main concerns of this proposal. These subjects are explained in the following sections.

### 21.5.5.1 Distributed Algorithm for Opportunistic Gateway Configuration

Among all the devices that are in the coverage area of the Access Router (those that could operate as opportunistic Gateways), one of them is temporally selected to behave as the Internet Gateway.

All the Gateways (Default or Candidate) receive RA messages generated by the Access Router every RA interval ($T_{RA}$). However, only the DG is responsible for propagating the received information from the Access Router. With this purpose, the DG periodically broadcasts MRA (Modified Router Advertisement) messages that can be forwarded through multiple hops. The period

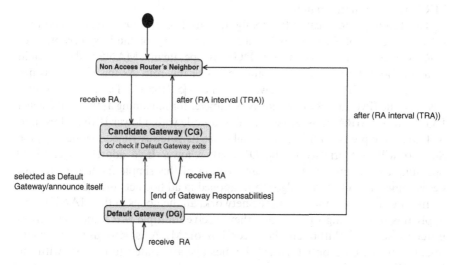

**Fig. 21.4** UML state diagram for MANET nodes under the mobile multi-gateway support

for this emission is fixed to a constant value called the MRA interval ($T_{MRA}$). By the reception of the MRA messages, the mobile devices can auto-configure their global IPv6 address (when necessary) and update their corresponding route to the DG. Then, when the terminal generates traffic for external hosts, it will forward the packets to the DG using the already learnt route. On the other hand, when the Access Router receives packets whose destination is a MANET component, it will forward the packets to the DG, which will start the ad hoc routing procedures to the destination ad hoc node.

The information contained in the MRA messages is considered to be valid during a period of time defined by its lifetime field. In this sense, we assume that the preferred lifetime employed in the NDP messages [36] is set to $T_{MRA}$. Before this time expires, the mobile nodes expect to receive a new MRA message. While the DG stays in the coverage area of the Access Router, the nodes in the MANET periodically receive the MRA message every $T_{MRA}$. However, as no mobility constraint is set on the DG, it could escape from the coverage area of the Access Router. In that case, the DG will stop receiving the RA messages and, consequently, it will disable its gateway functionalities immediately. When no MRA messages are received, the rest of the MANET nodes will assume that there is no DG configured. At that moment, the CGs should start the gateway selection procedure. This procedure is a distributed algorithm that respects the MANET philosophy as it does not require any centralized entity to select the future DG. Under this mechanism, the CGs trigger a random internal timer. When this random time is over, the CGs check if they have received any MRA message since the timer was started. The reception of a MRA would imply the presence of a new DG and, consequently, the CG will not configure itself as the DG. In other case, the CG will automatically configure itself as the new DG. To inform the MANET about its new role, it will start the periodic emission of MRA messages immediately.

This technique may cause the configuration of several DGs at the same time when the values of their random timers are very close or the loss/corruption of MRA messages occurs. As only one DG may operate in a MANET, there should be a new procedure to ensure the uniqueness of this role. As this problem is not considered in the IETF Draft [48], we recommend [53] the use of a simple algorithm based on the IPv6 address as a priority parameter for remaining as the DG: when a DG receives a MRA message generated at a node with a lower IPv6 address and with the same prefix information, it will cease its gateway functionalities. So, the Support will give priority to the DG with a lower IPv6 address to continue operating as the unique DG. The application of this simple method to resolve ties is common in clustering algorithms applied in ad hoc networks [2].

In order to reduce the load in the network, all the nodes in the MANET also apply this priority algorithm when they receive multiple MRA messages from different sources. Although the reception of MRA is also used for route updates, they would only forward the message associated to the DG with the lowest IPv6 address.

In this mechanism, it is possible that an expected MRA may not be received by the nodes (e.g., when the MRA is lost or corrupted). Under these specific circumstances, a MANET node demands the information contained in the MRA messages by the emission of a multicast Modified Router Solicitation (MRS) message. Once that the message is received by the DG, it replies by means of a unicast MRA message. If configured, the candidate gateways could also respond to the MRS message. This configuration is necessary for load balancing, as it will be shown in the next sections.

Additionally, a node generates an MRS message when the route to the DG is broken meanwhile the device is employing it.

### 21.5.5.2 Gateway Switching

The dynamic assignation of the default gateway functionalities to a mobile node adds flexibility in the procedures that integrate MANETs to external networks. However, in this Support the transfer of the gateway functionalities may introduce the existence of intervals of time where no default gateway is configured. During those intervals, the mobile nodes are unable to forward packets to the Internet, keeping them in an internal finite queue. As a consequence of this buffering, those queued packets will experiment a higher delay. Additionally, in those situations where the interval is too high or no space is available, packets will be dropped. Therefore, the switching times are responsible for the degradation of the network performance.

The conventional procedure exposed in the IETF Draft [48] describes how gateway switching proceeds. The default gateway notices that its role (and corresponding functionalities) has expired when no RA message is received after the corresponding period for the RA emission ($T_{RA}$), which is explicitly defined in each RA message. However, the candidate gateways will not initiate the distributed process for selecting the new default gateway until they detect that no MRA message has been received during the MRA interval. By following these specifications, the default gateway does not interact with the rest of the nodes when it detects to have escaped from the coverage area of the access router. Therefore, this is a passive strategy for the gateway switching.

An alternative scheme to improve the gateway switching is presented in [54]. Basically, the technique is based on forcing the DG that is escaping from the coverage area of the Access Router to announce its new state to the rest of the nodes in the MANET. By this announcement, the CG can anticipate the procedures to select the new DG.

## 21.6 Thoughts for Practitioners

MANETs are expected to operate without any manual configuration and even without any previous installation. Under these circumstances, the utilization of opportunistic gateways is necessary. However, some scenarios may be expecting to work with MANET technologies. In this sense, the practitioner may decide several issues. Firstly, the designer should determine where the Internet Gateways should be placed. Conventional MANET applications operate with routes of two or three hops [9]. In order to obtain a similar performance for communications in the MANET and sessions established with external host, the number of hops to the Internet Gateway should be limited to three hops. Additionally, routes composed of more links endure for shorter [55] so the traffic related to the solicitations to the Internet Gateway may become excessive if long routes are employed.

Additionally, the designer could decide which integration support to employ. So far, we cannot define the best integration support. Depending on the specific scenario, one mechanism could outperform the others or not. For instance, the prefix continuity mechanism outperforms the Global Connectivity in its proactive configuration when multiple gateways are present as the retransmission of MPA messages is optimized.

Then, the configuration parameters of the selected integration support should be carefully set. The work in [22] shows the impact of the interval of emission of MPA messages on the network performance when the Global Connectivity mechanism is employed, but there are more parameters to set.

Some integration supports also define the methodology to obtain a global IPv6 address. However, some others allow several schemes of address configuration.

## 21.7  Directions for Future Research

MANET technologies still need to be improved. Among others, third-party forwarding incentives are required. By this concept, we mean the incentives that mobile nodes receive as a consequence of forwarding the packets generated by other nodes. The success of MANETs in commercial applications is strongly linked to these incentives as no all the users in the MANET are willing to consume their batteries in order to guarantee the communication between two distant nodes.

Concerning the integration of MANET into IP-based access networks, the following research guidelines are identified:

- Automatic Configuration of the Integration Supports. The correct operation of the integration supports requires the configuration of some parameters. For instance, in a proactive gateway discovery, it is necessary to set the appropriate value of the interval of emission of the MPA messages.
- Secure Configuration. For the autonomous configuration of mobile nodes, especially in conflict-free allocation schemes, the mechanisms establish a hierarchy where the pool is distributed among the nodes. New incomings nodes rely on some neighbors to acquire a valid IPv6 address. However, how does the new incoming node know that the selected neighbor is trustworthy?

The techniques to identify the trustworthy nodes are necessary in order to guarantee a correct operation.

- Mesh Networks are composed of wireless Access Routers that form a backbone. A mobile node accesses an Access Router (accordingly to a determined criterion) and transmit their packets to it. The selected Access Router is, then, in charge of routing these packets to the corresponding Access Router (when the destination node is reachable in the Mesh Network) or to a Gateway when the destination is outside the Mesh Network. A MANET could be spontaneously formed around an Access Router so the technologies to integrate MANET into IP-based network are also necessary. However, these networks have specific characteristics that must be taken into account. For instance, the static structure could improve the network performance when the nodes change their points of attachment.
- NEMO (Network Mobility). So far, we have talked about the mobility of the nodes in the MANET and static point of attachment. However, a group of nodes (a network) can move together, establishing a mobile point of attachment [17]. In these scenarios, it is also necessary to guarantee the autonomous configuration of IPv6 addresses.

## 21.8 Conclusions

Mobile Ad hoc Networks technologies provide the capability of communicating distant nodes by means of some intermediate devices, which collaborate on behalf of other terminals. Key to the success of MANETs is that the relay nodes are dynamically selected without the need of any centralized equipment. This multihop and autonomous capability reports remarkable benefits such as reliability or the suppression of the investment necessary to deploy a telecommunication infrastructure. Although stand-alone MANET are by themselves economically attractive, the success of Web and, in general, Internet services in an all-connected world claims for the integration of MANET into external networks. In this sense, the utilization of multihop wireless networks could also introduce significant economical advantages in infrastructured networks by reducing the installation of the installed equipment. This advantage is reinforced by the fact that some scenarios where the access to the Internet is demanded are temporal or even hostile for the deployment of a new telecommunication infrastructure.

The promising benefits linked to the integration of multihop wireless networks have prompted the need of solving several concerns related to the present technologies. Basically, some conventional IP technologies such as stateless address configuration or DHCP were initially designed for one-hop networks so they need to be adapted to the multihop context. To proceed, architectural and technological modifications are then required. Along the last years, a considerable number of proposals (mainly IETF Internet drafts) have been published in order to offer a feasible solution to achieve the integration of MANET. Along this chapter, a comprehensible overview of these proposals has been formally presented. In this sense, the main developed technologies concerning the address configuration procedures in MANET are explained. Basically, two approaches are distinguished: stateful or centralized IP address assignment and stateless address configuration, which permits the nodes to autonomously configure their own IP address. Concerning the stateless mechanisms, two different techniques have been elaborated in the research literature. Firstly, the conflict-free allocation schemes distributed the available addresses along the network making some nodes be responsible for part of the address pool. On the other hand, in the conflict-detection procedures, the nodes construct their IP address by means of local information and the prefix announced by the Gateway. In order to verify the uniqueness of the self-generated IP address, the node may start a Duplicate Address Detection procedure. Commonly, this DAD procedure is associated to a flooding and the interruption of external communications.

Related to the architectural modifications that the connection of Mobile Ad hoc Networks to the Internet demands, all the integration proposals are based on the introduction of an Internet Gateway which provides the mechanisms necessary to make IP technologies operative in MANET. In this chapter, the most significant functionalities that the Internet Gateway is expected to perform are summarized.

Most of the integration supports assume that a dedicated and previously installed Internet Gateway is present in those environments where a MANET is expected to be connected. However, this condition may not always hold. Under these circumstances, it is necessary that a node in the MANET plays the role of the Internet Gateway on behalf of the rest of the devices in order to guarantee the access to the Internet. This elected node is referenced as occasional, non-dedicated, or opportunistic gateway.

To sum up, the integration of ad hoc networks into the Internet still offers some aspects that need to be solved.


## Terminologies

*Address auto-configuration*: The process of configuring an interface with a given address, using an automatic mechanism (contrary to manual configuration) [6].

*Stateless address auto-configuration*: Stateless Address Auto-configuration enables automatic configuration of an IP address to a host without contacting any kind of server [6].

*Internet gateway*: A router which provides Internet connectivity for nodes in the MANET. This router is located somewhere in a MANET and has a connection to both the Internet and the MANET [57].

*Network merging*: The process by which two or more previously disconnected MANETs get connected [6].

*Network partitioning*: The process by which a MANET splits into two or more disconnected MANETs [6].

*RA (router advertisement)*: One-Hop Message Generated by the Access Router, with the information of the prefix it works with and the mechanisms to configure the IP addresses. This message is defined in [36]

*RS (router solicitation)*: One-Hop Message Generated by the Mobile Node in order to solicit information about the prefix a reachable Router is working with. This message is defined in [36]

*NDP (neighbor discovery protocol)*: As defined in [36], NDP is a protocol employed by "nodes (hosts and routers) to determine the link-layer addresses for neighbors known to reside on attached links and to quickly purge cached values that become invalid. Hosts also use Neighbor Discovery to find neighboring routers that are willing to forward packets on their behalf. Finally, nodes use the protocol to actively keep track of which neighbors are reachable and which are not, and to detect changed link-layer addresses".

*MPA (multicast prefix advertisement)*: A Multicast message generated by the Internet Gateway to disseminate the gateway information.

*MPS (multicast prefix solicitation)*: A Multicast message generated by the mobile nodes to obtain the gateway information (the route, the prefix, the mechanism to configure the IP address, etc.)

## Questions

1. Identify the events that cause that a unique IPv6 address becomes duplicated.
2. Why RA messages do not guarantee the address auto-configuration of MANET nodes?
3. For the integration of MANET into IP-based networks could be the solution to extend the RA messages so they can be forwarded through multiple hops?
4. What kind of gateways does the mobile multi-gateway support employ?
5. What type of gateway discovery does the Global Connectivity mechanism support?
6. Could the mobile multi-gateway support employ reactive gateway discovery?
7. If the Global Connectivity support is employed with a hybrid gateway discovery and a TTL value of 2, does a node that is 3 hops away from the Access Router receive the periodic IGWADV?
8. What are the most significant schemes to configure an IP address without centralized equipment?

9. In conflict-detection allocation schemes, duplications may occur. In order to detect them, DAD procedures are executed. According to the time that these procedures are executed, what type of DAD procedures exist?
10. And, according to the packets that DAD employs, what type of DAD procedures exist?

## References

1. 3GPP, "3rd Generation Partnership Project: Technical Specification Group Core Network; Numbering, addressing and identification (Release 1999)", 3GPP TS 23.003, June 2001.
2. B. S. Abdallah Goncalves, N. Mitton, I. Guerin-Lassous, "Comparison of two Self-Organization and Hierarchical Routing Protocols for Ad Hoc Networks", in the Proceedings of The 2nd International Conference on Mobile Ad-hoc and Sensor Networks (MSN 2006), December 2006.
3. H. Ammari, H. El-Rewini, "Integration of Mobile Ad Hoc Networks and the Internet using Mobile Gateways", in Proceedings of the 18th International Parallel and Distributed Processing Symposium (IPDPS'04), Santa Fe (Mexico), April 2004.
4. J. Arkko, J. Kempf, B. Zill, P. Nikander, "SEcure Neighbor Discovery (SEND)", IETF RFC 3971, March 2005.
5. T. Aura, "Cryptographically Generated Addresses (CGA)", IETF RFC 3972, March 2005.
6. E. Baccelli, "Address Autoconfiguration for MANET: Terminology and Problem Statement", IETF Internet-Draft, February 2008 (work in progress).
7. C. Bernardos, M. Calderón, H. Moustafa, "Survey of IP address autoconfiguration mechanisms for MANETs", IETF Internet Draft, July 2007 (work in progress).
8. S. Boudjit, C. Adjih, P. Mühlethaler, A. Laouiti, "Duplicate Address Detection and Autoconfiguration in OLSR", in Journal of Universal Computer Science, vol. 13, no. 1, pp. 4–31, January 2007.
9. J. Broch, D. A. Maltz, D. B. Johnson, Y. C. Hu, J. Jetcheva, "A performance comparison of multi-hop wireless ad hoc network routing protocols", in the Proceedings of the 4th Annual ACM/IEEE International Conference on Mobile Computing and Networking, pp. 85–97, 1998.
10. F. Buiati, R. Staciarini-Puttini, R. T. de Sousa, C. J. Barenco-Abbas, L. J. García-Villalba, "Authentication and Autoconfiguration for MANET Nodes", in the Proceedings of the International Conference on Embedded and Ubiquitous Computing (EUC 2004), Japan, August 2004.
11. H. W. Cha, J. S. Park, H. J. Kim, "Extended Support for Global Connectivity for IPv6 Mobile Ad Hoc Networks", IETF Internet Draft, October 2003 (work in progress).
12. I. Chakeres, J. Macker, T. Clausen, "Mobile Ad hoc Network Architecture", IETF Internet Draft, August 2007 (work in progress).
13. Y.S. Chen, T.H. Lin, S. M. Lin, "RAA: a ring-based address autoconfiguration protocol in mobile ad hoc networks", in Wireless Personal Communications, April 2007.
14. S. Cheshire, B. Aboba, E. Guttman, "Dynamic Configuration of IPv4 Link-Local Addresses", IETF RFC 3927, March 2005.
15. T. Clausen, P. Jacquet, "Optimized Link State Routing Protocol (OLSR)", IETF RFC 3626, October 2003.
16. T. Clausen, E. Baccelli, "Simple MANET Address Autoconfiguration", IETF Internet Draft, January 2005 (work in progress).

17. V. Devarapalli, R. Wakikawa, A. Petrescu, P. Thubert, "Network Mobility (NEMO) Basic Support Protocol", IETF RFC 3963, January 2005.
18. F. Dupont, L. Nuaymi, "IMEI-based universal IPv6 interface IDs", IETF Internet Draft, October 2005 (work in progress).
19. ETSI, "Digital cellular telecommunications system: International Mobile station Equipment Identities (IMEI)", GSM 02.16, February 2000.
20. IEEE, "Guidelines for 64-bit Global Identifier (EUI-64) Registration Authority", http://standards.ieee.org/db/oui/tutorials/EUI64.html, March 1997.
21. M. Ghassemian, V. Friderikos, A. Aghvami, "A Generic Algorithm to Improve the Performance of Proactive Ad hoc Mechanisms" in Proceedings of the IEEE International Symposium on a World of Wireless, Mobile and Multimedia Networks, Italy, June 2005.
22. Ali Hamidian, Ulf Körner, Anders Nilsson, "Performance of Internet Access Solutions in Mobile Ad Hoc Networks", in the Proceedings of Workshop Mobility and Wireless in Euro-NGI and also published in G. Kotsis and O. Spaniol (Eds.): Mobile and Wireless Systems, LNCS 3427, pp. 189–201, 2005.
23. R. Hinden, S. Deering, "IP Version 6 Addressing Architecture", IETF RFC 4291, February 2006.
24. C.F. Huang, H.W. Lee, Y.S. Tseng, "A Two-Tier Heterogeneous Mobile Ad Hoc Network Architecture and Its Load-Balance Routing Problem", in Mobile Networks and Applications, vol. 9, Issue 4, pp. 379–391, 2004.
25. C. Huitema, "IPv6 : The New Internet Protocol", Second edition, Prentice Hall, November 1997, ISBN: 0138505055.
26. C. Jelger, T. Noël, "Proactive Address Autoconfiguration and Prefix Continuity in IPv6 hybrid ad hoc networks", in the Proceedings of the Sensor and Ad Hoc Communications and Networks (SECON), pp. 107–117, September 2005.
27. D. Johnson, C. Perkins, J. Arkko, "Mobility Support in IPv6", IETF RFC 3775, June 2004.
28. D. Johnson, Y. Hu, D. Maltz, "The Dynamic Source Routing Protocol (DSR) for Mobile Ad Hoc Networks for IPv4", IETF RFC 4728, February 2007.
29. D. Kim, H. J. Jeong, J. C. Cano, "Improving the Accuracy of Passive Duplicate Address Detection Algorithms over MANET On-demand Routing Protocols", in the Proceedings of the 8th International Symposium on Autonomous Decentralized Systems (ISADS'07), pp. 534–542, 2007.
30. A. Laouiti, "Address autoconfiguration in Optimized Link State Routing Protocol", February 2005 (work in progress).
31. J. Lee, D. Kim, J. J.Garcia-Luna-Acebes, Y. Choi, J. Choi, S. Nam, "Hybrid Gateway Advertisement Scheme for Connecting Mobile Ad Hoc Networks to the Internet", in Proceedings of the 57th IEEE Semiannual Vehicual Technology Conferences (VTC), vol. 1, pp. 191–195, April 2003.
32. Y. D. Lin, Y. C. Hsu, "Multihop cellular: A new architecture for wireless communications", in Proceedings of the 19th Annual Joint Conference of the IEEE Computer and Communications Societies (INFOCOM 2000) pp. 1273–1282, 2000.
33. A. McAuley, K. Manousakis, "Self Configuring Networks," in Proceedings of the 21st Century Military Communications Conference (MILCOM), vol. 1, pp. 315–319, 2000.
34. A. Misra, S. Das, A. McAuley, "Autoconfiguration, Registration, and Mobility Management for Pervasive Computing", in IEEE Personal Communications, August 2001.
35. M. Mohsin, R. Prakash, "IP Address Assignment in a Mobile Ad Hoc Network", in Proceedings of MILCOM 2002, 2002.
36. T. Narten, E. Nordmark, W. Simpson, "Neighbor Discovery for IP Version 6", IETF RFC 2461, December 1998.
37. T. Narten, R. Draves, "Privacy Extensions for Stateless Address Autoconfiguration in IPv6", IETF RFC 3041, January 2001.

38. T. Narten, R. Draves, S. Krishnan, "Privacy Extensions for Stateless Address Autoconfiguration in IPv6", IETF Internet Draft, August 2006 (work in progress).
39. S. Nesargi, R. Prakash, "MANETconf: Configuration of Hosts in a Mobile Ad Hoc Network", in Proceedings of INFOCOM 2002.
40. C. Perkins, R. Wakikawa, J. Malinen, E. Belding-Royer, Y. Suan, "IP Address Autoconfiguration for Ad Hoc Networks", IETF Internet Draft, November 2001 (work in progress).
41. P. Ratanchandani, R. Kravets, "A Hybrid Approach to Internet Connectivity for Mobile Ad Hoc Networks", in Proceedings of the IEEE WCNC, vol. 3, pp. 1525–1527, New Orleans (USA), March 2003.
42. F. J. Ros, P. Ruiz, "Low Overhead and Scalable Proxied Adaptive Gateway Discovery for Mobile Ad Hoc Networks", in Proceedings of the 3rd IEEE International Conference on Mobile Ad-hoc and Sensor Systems (MASS 2006), pp. 226–235, Vancouver (Canada), October 2006.
43. S. Ruffino, P. Stupar, "Automatic configuration of IPv6 addresses for MANET with multiple gateways (AMG)", IETF Internet-Draft, June 2006 (work in progress).
44. S. Ruffino, P. Stupar, T. Clausen, S. Singh, "Connectivity Scenarios for MANET", IETF Internet Draft, January 2006.
45. P. M. Ruiz, A. F. Gómez-Skarmeta, "Maximal Source Coverage Adaptive Gateway Discovery for Hybrid Ad Hoc Networks", Lecture Notes in Computer Science, vol. 3158, pp.28–41, July 2004.
46. P. M. Ruiz, A. F. Gómez Skarmeta, "Enhanced Internet Connectivity for Hybrid Ad hoc Networks Through Adaptive Gateway Discovery", in Proceedings of the 29th Annual IEEE Conference on Local Computer Networks (LCN-2004), Tampa (USA), November 2004.
47. J. Shin, H. Lee, J. Na, A. Park, S. Kim, "Load Balancing among Internet Gateways in Ad Hoc Networks", in the Proceedings of the IEEE Vehicular Technology Conference, vol. 3, pp. 1677–1680, September 2005.
48. S. Singh, J. H. Kim, Y. G. Choi, K. L. Kang, Y. S. Roh, "Mobile multi-gateway support for IPv6 mobile ad hoc networks", IETF Internet Draft, June 2004 (work in progress).
49. A. Tayal, L. Patnaik, "An address assignment for the automatic configuration of mobile ad hoc networks", in the Proceedings of Personal Ubiquitous Computing, 2004.
50. F. Templin, S. Russert, S. Yi, " MANET Autoconfiguration", IETF Internet Draft, February 2008 (work in progress).
51. S. Thompson, T. Narten, T. Jinmei "IPv6 Stateless Address Autoconfiguration", IETF RFC 4862, September 2007.
52. M. Thoppian, R. Prakash, "A Distributed Protocol for Dynamic Address Assignment in Mobile Ad Hoc Networks", IEEE Transactions on Mobile Computing, vol. 5, Issue 1, pp. 4–19, January 2006.
53. A. Triviño-Cabrera, S. Singh, E. Casilari, F. J. González-Cañete, "Integration of Mobile Ad Hoc Networks into the Internet without Dedicated Gateways", in the Proceedings of International Conference on Wireless and Mobile Communications (ICWMC 2006) , Bucharest (Romania), July 2006.
54. A. Triviño-Cabrera, E. Casilari, F. J. González-Cañete, "Active gateway switching in hybrid ad hoc networks", in Electronic Letters, vol. 42, No. 21, pp. 1252–1254, October 2006.
55. A. Triviño-Cabrera, J. García-de-la-Nava, E. Casilari, F. J. González-Cañete,"An analytical model to estimate path duration in MANETs", in the Proceedings of the 9th ACM international symposium on Modeling analysis and simulation of wireless and mobile systems, Torremolinos (Spain), October 2006.
56. N. H. Vaidya, "Weak duplicate address detection in mobile ad hoc networks", in Proceedings of the 3rd ACM international symposium on Mobile ad hoc networking & computing, pp. 206–216, 2002.

57. R. Wakikawa, J. T. Malinen, A. Nilsson, A. J. Tuominen, "Global connectivity for IPv6 Mobile Ad Hoc Networks", IETF Internet-Draft, March 2006 (work in progress).
58. K. Weniger, M. Zitterbart, "IPv6 Autoconfiguration in Large Scale Mobile Ad-Hoc Networks", in the Proceedings of European Wireless 2002, 2002.
59. K. Weniger, "PACMAN: Passive Autoconfiguration for Mobile Ad hoc Networks", in IEEE Journal on Selected Areas in Communications, vol. 23, Issue 3, pp. 507–519, March 2005.
60. K. Weniger, K. Mase, "PDAD-OLSR: Passive Duplicate Address Detection for OLSR", IETF Internet Draft, June 2006, (work in progress).
61. H. Wu, C. Qiao, S. De, O. Tonguz, "Integrated cellular and ad hoc relaying systems: iCAR", in IEEE Journal on Selected Areas in Communications, pp. 2105–2115, October 2001.
62. H. Zhou, L. Ni, M. Mutka, "Prophet Address Allocation for Large Scale MANETs", in Proceeding of IEEE Infocom, 2003, vol. 2, pp. 1304–1311, April 2003.

# Index

# Author Biographies

**Dr. Sudip Misra** is an Assistant Professor in the School of Information Technology at the Indian Institute of Technology Kharagpur, India. Prior to this he worked in Cornell University (USA), Yale University (USA), Nortel Networks (Canada), Ryerson University (Canada) and the Government of Ontario (Canada). He received his Ph.D. degree in Computer Science from Carleton University, in Ottawa, Canada, and the masters and bachelors degrees, respectively, from the University of New Brunswick, Fredericton, Canada and the Indian Institute of Technology, Kharagpur, India. He has several years of experience work-

ing in the academia, government, and the private sectors in research, teaching, consulting, project management, architecture, software design, and product engineering roles.

His current research interests include algorithm design and engineering for telecommunication networks, software engineering for telecommunication applications, and computational intelligence and soft computing applications in telecommunications.

Dr. Misra is the author/editor of over *100 scholarly research papers and books*. He has won *five research paper awards* in different conferences. He was also the recipient of several academic awards and fellowships such as the *(Canadian) Governor General's Academic Gold Medal* at Carleton University, the *University Outstanding Graduate Student Award* in the Doctoral level at Carleton University. In 2008, he was conferred *The National Academy of Sciences, India – Swarna Jayanti Puraskar* (Golden Jubilee Award).

He was also awarded the Canadian Government's prestigious *NSERC Post Doctoral Fellowship*. His biography was also selected for inclusion in the 2006–2007 edition of Marquis Who's Who in Science and Engineering, and the 25th Edition of the Marquis Who's Who in the World, California, USA. A mention about him and his work has also appeared in the Ottawa Citizen newspaper.

Dr. Misra is the *Editor-in-Chief* of 2 journals – the *International Journal of Communication Networks and Distributed Systems* (IJCNDS) and the *International Journal of Information and Coding Theory* (IJICoT), UK. He is an Associate Editor of the *Telecommunication Systems Journal* (Springer SBM), *Security and Communication Networks Journal* (Wiley), *International Journal of Communication Systems* (Wiley), and the *EURASIP Journal of Wireless Communications and Networking*. He is also an Editor/Editorial Board Member/ Editorial Review Board Member of the *IET Communications Journal, Computers and Electrical Engineering Journal* (Elsevier), *International Journal of Internet Protocol Technology,* the *International Journal of Theoretical and Applied Computer Science,* the *International Journal of Ad Hoc and Ubiquitous Computing, Journal of Internet Technology*, and the *Applied Intelligence Journal* (Springer).

Dr. Misra is an editor of 6 books in the areas of wireless ad hoc networks, wireless sensor networks, wireless mesh networks, communication networks and distributed systems, network reliability and fault tolerance, and information and coding theory, published by reputed publishers such as Springer and World Scientific.

He was invited to chair several international conference/workshop programs and sessions. He has been serving in the program committees of over a dozen international conferences. He was also invited to deliver *keynote lectures* in around a dozen international conferences in USA, Canada, Europe, Asia and Africa.

**Dr. Isaac Woungang** received his M.A.Sc and Ph.D degrees, all in Applied Mathematics, from the Université du Sud, Toulon-Var, France, in 1990 and 1994, respectively. In 1999, he received an M.A.Sc degree from INRS-Materials and Telecommunications, University of Quebec, Canada. From 1999 to 2002, he worked as a software engineer at  Nortel Networks, Ottawa, Canada. Since 2002, he has been with Ryerson University, where he is now an Assistant Professor of computer science. In 2004, he founded the DABNEL (the Distributed Applications and Broadband NEtworks Laboratory) R&D group at Ryerson University, Canada. His research interests are telecommunications network design, network security, and computational intelligence applications in telecommunications.

**Dr. Subhas C. Misra** is a Visiting Assistant Professor at Indian Institute of Technology, Kanpur, India. Earlier he was at Harvard University, USA and the State University of New York, Buffalo, NY, USA. He received his Ph.D. degree from Carleton University, in Ottawa, Canada, and M.S. and M.Tech. degrees, respectively, from the University of New Brunswick, in Fredericton, Canada, and the Indian Institute of Technology (IIT), at Kharagpur, India. He is the author of over 50 scholarly research papers and is an editor of 3 books. He has won Best Research Paper Award in an international conference held in the United States. He was also the recipient of more than 15 reputed academic awards and fellowships. A mention about him and his work has also appeared in one of the Canadian newspapers.

Dr. Misra is the Managing Editor of the *International Journal of Information and Coding Theory (IJICoT)*, and the *International Journal of Communication Networks and Distributed Systems*. He is also an Associate Editor of the *ICIC Express Letters* (An international journal motivated to Innovative Computing, Information, and Control; published from Japan). He is an Editor of the *International Journal of Systemics, Cybernatics, and Informatics,* and an Editorial Board Member of the *International Journal of Theoretical and Applied Computer Science.* He has been invited internationally to deliver *keynote/invited lectures* in several conferences.